AF323153

Radar Technology

Radar Technology

Edited by
Guy Kouemou

Radar Technology,
Edited by Dr. Guy Kouemou

CBS, Edition **2016**

Published by InTech
Janeza Trdine 9, 51000 Rijeka, Croatia

Notice

Statements and opinions expressed in the chapters are these of the individual contributors and not necessarily those of the editors or publisher. No responsibility is accepted for the accuracy of information contained in the published chapters. The publisher assumes no responsibility for any damage or injury to persons or property arising out of the use of any materials, instructions, methods or ideas contained in the book.

Additional hard copies can be obtained from orders@intechopen.com

Radar Technology, Edited by Dr. Guy Kouemou
p. cm.
ISBN 978-953-307-029-2

Preface

One of the most important inventions for the development of radars was made by Christian Huelsmeyer in 1904. The German scientist demonstrated the possibility of detecting metallic objects at a distance of a few kilometres. While many basic principles of radar, namely using electromagnetic waves to measure the range, altitude, direction or speed of objects are remained up to now practically unchanged, the user requirements and technologies to realise modern radar systems are highly elaborated. Nowadays many modern types of radar (originally an acronym for Radio Detection and Ranging) are not designed only to be able to detect objects. They are also designed or required to track, classify and even identify different objects with very high resolution, accuracy and update rate.

A modern radar system usually consists of an antenna, a transmitter, a receiver and a signal processing unit. A simple signal processing unit can be divided into parameter extractor and plot extractor. An extended signal processing unit consists additionally of a tracking module. The new category of signal processing units has also the possibility of automatically target classification, recognition, identification or typing.

There are numerous classifications of radars. On the one hand they can be classified by their platform as ground based, air borne, space borne, or ship based radar. On the other hand they can be classified based on specific radar characteristics, such as the frequency band, antenna type, and waveforms utilized. Considering the mission or functionality one may find another categorization, such as weather, tracking, fire control, early warning, over the horizon and more.

Other types are phased array radars, also called in some literatures as multifunction or multimode radars (not necessary the same). They use a phased array antenna, which is a composite antenna with at least two basic radiators, and emit narrow directive beams that are steered mechanically or electronically, for example by controlling the phase of the electric current.

Mostly radars are classified by the type of waveforms they use or by their operating frequency.

Considering the waveforms, radars can be Continuous Wave (CW) or Pulsed Radars (PR). CW radars continuously emit electromagnetic energy, and use separate transmit and receive antennas. Unmodulated CW radars determine target radial velocity and angular position accurately whereas target range information only can be gathered by using some form of modulation. Primarily unmodulated CW radars are used for target velocity search and track and in missile guidance. Pulsed radars use a train of pulsed waveforms, mainly with modulation. These systems can be divided based on the Pulse Repetition Frequency

(PRF) into low PRF (mainly for ranging; velocity is not of interest), medium PRF, and high PRF (primarily for velocity measurement) radars. By using different modulation schemes, both CW and PR radars are able to determine target range as well as radial velocity.

These radar bands from 3 MHz to 300 MHz have a long historically tradition since the World War II. These frequencies are well known as the passage from radar technology to the radio technology. Using electromagnetic waves reflection of the ionosphere, High Frequency (HF, 3MHz – 30MHz) radars such as the United States Over the Horizon Backscatter (U.S. OTH/B, 5 – 28 MHz), the U.S. Navy Relocatable Over the Horizon (ROTHR) and the Russian Woodpecker radar, can detect targets beyond the horizon. Today these frequencies are used for early warning radars and so called Over The Horizon (OTH) Radars. By using these very low frequencies, it is quite simple to obtain transmitters with sufficient needed power. The attenuation of the electromagnetic waves is therefore lower than using higher frequencies. On the other hand the accuracy is limited, because a lower frequency requires antennas with very large physical size which determines the needed angle resolution and accuracy. But since the most frequency-bands have been previously attributed to many operating communications and broadcasting systems, the bandwidth for new radar systems in these frequencies area is very limited. A comeback of new radar systems operating in these frequency bands could be observed the last year. Many radar experts explain this return with the fact that such HF radars are particularly robust against target with Stealth based technologies.

Many Very High Frequency (VHF, 30MHz – 300MHz) and Ultra High Frequency (UHF, 300MHz – 1GHz) bands are used for very long range Early Warning Radars. A well known example in these categories of radar is for example, The Ballistic Missile Early Warning System (BMEWS). It is a search and track monopulse radar that operates at a frequency of about 245 MHz. The also well known Perimeter and Acquisition Radar (PAR), is a very long range multifunction phased array radar. The PAVE PAWS is also a multifunction phased array radar that operates at the UHF frequencies.

The UHF operating radar frequency band (300 MHz to1 GHz), is a good frequency for detection, tracking and classification of satellites, re-entry vehicles or ballistic missiles over a long range. These radars operate for early warning and target acquisition like the surveillance radar for the Medium Extended Air Defence System (MEADS). Some weather radar-applications like the wind profilers also work with these frequencies because the electromagnetic waves are very robust against the volume clutter (rain, snow, graupel, clouds).

Ultra wideband Radars (UWB-Radars) use all HF-, VHF-, and UHF- frequencies. They transmit very low pulses in all frequencies simultaneously. Nowadays, they are often used for technically material examination and as Ground Penetrating Radar (GPR) for different kind of geosciences applications.

Radars in the L-band (1GHz – 2GHz) are primarily ground and ship based systems, used in long range military and air traffic control search operations. Their typical ranges are as high as about 350-500 Kilometres. They often transmit pulses with high power, broad bandwidth and an intrapulse modulation. Due to the curvature of the earth the achievable maximum range is limited for targets flying with low altitude. These objects disappear very fast behind the radar horizon. In Air Traffic Management (ATM) long-range surveillance radars like the Air Surveillance Radar the ASR-L or the ASR-23SS works in this frequency band. L-band radar can also be coupled with a Monopulse Secondary Surveillance Radar

(MSSR).They so use a relatively large, but slower rotating antenna. One well known maritime example in this category is the SMART-L radar systems.

S-band (2GHz – 4GHz), are often used as airport surveillance radar for civil but also for military applications. The terminology S-Band was originally introduced as counterpart to L-Band and means "smaller antenna" or "shorter range". The atmospheric and rain attenuation of S-band radar is higher than in L-Band. The radar sets need a considerably higher transmitting power than in lower frequency ranges to achieve a good maximum range. In this frequency range the influence of weather conditions is higher than in the L-band above. For this raison, many weather forecast and precipitation radars in the subtropics and tropic climatic zone (Africa, Asia, Latin America) operate in S-band. One advantage here is that these radars (usually Doppler-radar) are often able to see beyond typical severe storm or storm system (hurricane, typhoon, tropical storm, cyclonic storm). Special Airport Surveillance Radars (ASR) are used at airports to detect and display the position of aircraft in the terminal area with a medium range up to about 100 kilometres. An ASR detects aircraft position and weather conditions in the vicinity of civilian and military airfields.

The most medium range radar systems operate in the S-band (2GHz – 4GHz), e.g. the ASR-E (Air Surveillance Radar) or the Airborne Warning And Control System (AWACS).

The C-band radar systems (4GHz – 8GHz) are often understand in the radar community as a kind of compromising frequency-band that is often used for medium range search. The majority of precipitation radars used in the temperate climate zones (e.g. Europe, Nord America) operates in this frequency band. But C-band radars are also often used for fire control military applications. There exist many mobile battlefield radars with short and medium range. For these defence application for example, C-band antennas are used for weapon guidance. One of the reasons is that, additionally to there high precision, they are also small and light enough for usually transport systems (Truck, Small boat). The influence of weather phenomena is also very large and for this reason, the C-band antennas air surveillance purposes mostly operate with circular polarization. In the C-band radar series, the TRML-3D (Ground) and the TRS-3D (Naval) Surveillance Radar are well known operating systems.

The X-band (8GHz – 12.5GHz) radar systems are often used for applications where the size of the antenna constitutes a physical. In this frequency band the ratio of the signal wavelength to the antenna size provide a comfortable value. It can be achieved with very small antennas, sufficient angle measurement accuracy, which favours military use for example as airborne radar (airborne radar).This band is often used for civilian and military maritime navigation radar equipment. Several small and fast-rotating X-band radar are also used for short-range ground surveillance with very good coverage precision. The antennas can be constructed as a simple slit lamp or patch antennas. For space borne activities, X-band Systems are often used as Synthetic Aperture Radars (SAR). This covers many activities like weather forecast, military reconnaissance, or geosciences related activities (climate change, global warming, and ozone layer). Special applications of the Inverse Synthetic Aperture Radar (ISAR) are in the maritime surveillance also to prevent environmental pollution.

Some well known examples of X-band Radar are: the APAR Radar System (Active Phased Array, Ship borne multi-function Radar), The TRGS Radar Systems (Tactical Radar Ground Surveillance, active phased array system), The SOSTAR-X (Stand-Off Surveillance and Target Acquisition Radar), TerraSAR-X (Earth observation satellite that uses an X-band SAR to provide high-quality topographic information for commercial and scientific applications).

In the higher frequency bands (Ku (12.5GHz – 18GHz), K (18GHz – 26.5GHz), and Ka (26.5GHz – 40GHz)) weather and atmospheric attenuation are very strong which leads to a limitation to short range applications, such as police traffic radars. With expectant higher frequency, the atmosphrerical attenuation increases, but the possible range accuracy and resolution also augment. Long range cannot be achieved. Some well known Radar applications examples in this frequency range are: the airfield surveillance radar, also known as the Surface Movement Radar (SMR) or the Airport Surface Detection Equipment (ASDE). With extremely short transmission pulses of few nanoseconds, excellent range resolutions are achieved. On this mater contours of targets like aircraft or vehicles can briefly be recognised on the radar operator screen.

In the Millimetre Wave frequency bands (MMW, above 34GHz), the most operating radars are limited to very short range Radio Frequency (RF) seekers and experimental radar systems.

Due to molecular scattering of the atmosphere at these frequencies, (through the water as the humidity here) the electromagnetic waves here are very strong attenuated. Therefore the most Radar applications here are limited to a range of some few meters. For frequencies bigger than 75 GHz two phenomena of atmospheric attenuation can be observed. A maximum of attenuation at about 75 GHz and a relative minimum at about 96 GHz. Both frequencies are effectively used. At about 75 to 76 GHz, short-range radar equipment in the automotive industry as a parking aid, brake assist and automatic avoidance of accidents are common. Due to the very high attenuation from the molecular scattering effects (the oxygen molecule), mutual disturbances of this radar devices would occur. On the other side, the most MMW Radars from 96 to 98 GHz exist as a technical laboratory and give an idea of operational radar with much greater frequency.

Nowadays, the nomenclature of the frequency bands used above originates from world war two and is very common in radar literature all over the world. They vary very often from country to country. The last year's efforts were made in the world radar community in order to unify the radar frequency nomenclature.

In this matter the following nomenclature convention is supposed to be adapted in Europe in the future: A-band (< 250MHz), B-band (250MHz – 500MHz), C-band (500MHz – 1GHz), D-band (1GHz – 2GHz), E-band (2GHz- 3GHz), F-band (3GHz – 4GHz), G-band (4GHz – 6GHz), H-band (6GHz – 8GHz), I-band (8GHz – 10GHz), J-band (10GHz – 20GHz), K-band (20GHz – 40GHz), L-band (40GHz – 60GHz), M-band (> 60GHz).

In this book "Radar Technology", the chapters are divided into four main topic areas:

Topic area 1: "Radar Systems" consists of chapters which treat whole radar systems, environment and target functional chain.

Topic area 2: "Radar Applications" shows various applications of radar systems, including meteorological radars, ground penetrating radars and glaciology.

Topic area 3: "Radar Functional Chain and Signal Processing" describes several aspects of the radar signal processing. From parameter extraction, target detection over tracking and classification technologies.

Topic area 4: "Radar Subsystems and Components" consists of design technology of radar subsystem components like antenna design or waveform design.

The editor would like to thank all authors for their contribution and all those people who directly or indirectly helped make this work possible, especially Vedran Kordic who was responsible for the coordination of this project.

Editor

Dr. Guy Kouemou
EADS Deutschland GmbH,
Germany

Contents

TOPIC AREA 3: Radar Functional Chain and Signal Processing

TOPIC AREA 4: Radar Subsystems and Components

TOPIC AREA 1: Radar Systems

1

Radar Performance
of Ultra Wideband Waveforms

Svein-Erik Hamran
FFI and University of Oslo
Norway

1. Introduction

In the early days of radar, range resolution was made by transmitting a short burst of electromagnetic energy and receiving the reflected signals. This evolved into modulating a sinusoidal carrier into transmitting pulses at a given repetition interval. To get higher resolution in the radars the transmitted pulses got shorter and thereby the transmitted spectrum larger. As will be shown later the Signal-to-Noise Ratio (SNR) is related to the transmitted energy in the radar signal. The energy is given by the transmitted peak power in the pulse and the pulse length. Transmitting shorter pulses to get higher range resolution also means that less energy is being transmitted and reduced SNR for a given transmitter power. The radar engineers came up with radar waveforms that was longer in time and thereby had high energy and at the same time gave high range resolution. This is done by spreading the frequency bandwidth as a function of time in the pulse. This can be done either by changing the frequency or by changing the phase.

If the bandwidth is getting large compared to the center frequency of the radar, the signal is said to have an Ultra Wide Bandwidth (UWB), see (Astanin & Kostylev, 1997) and (Taylor, 2001). The definition made by FFC for an UWB signal is that the transmitted spectrum occupies a bandwidth of more than 500 MHz or greater than 25% of the transmitted signal center frequency. UWBsignals have been used successfully in radar systems for many years. Ground Penetrating Radar (GPR) can penetrate the surface of the ground and image geological structures. Absorption of the radar waves in the ground is very frequency dependent and increases with increasing frequency. Lower frequencies penetrate the ground better than higher frequency. To transmit a low frequency signal and still get high enough range resolution calls for a UWB radar signal. The interest in using UWB signals in radar has increased considerably after FFC allocated part of the spectrum below 10 GHz for unlicensed use. Newer applications are through the wall radar for detecting people behind walls or buried in debris. Also use of UWB radar in medical sensing is seeing an increased interest the later years.

UWB radar signal may span a frequency bandwidth from several hundred of MHz to several GHz. This signal bandwidth must be captured by the radar receiver and digitized in some way. To capture and digitize a bandwidth that is several GHz wide and with sufficient resolution is possible but very costly energy and money wise. This has been solved in the impulse waveform only taking one receive sample for each transmitted pulse. In the Step-Frequency (SF) waveform the frequencies are transmitted one by one after each other. A general rule for UWB radars is that all of the different waveform techniques have different methods to reduce the sampling requirement. The optimal would be to collect the entire reflected signal in time and frequency at once and any technique that is only collecting part of the received signal is clearly not optimal.

This chapter will discuss how different UWB waveforms perform under a common constraint given that the transmitted signal has a maximum allowable Power Spectral Density (PSD). The spectral limitations for Ground Penetration Radars (GPR) is given in Section 2 together with a definition on System Dynamic Range (SDR). In Section 3 a short presentation on the mostly used UWB-radar waveforms are given together with an expression for the SDR. An example calculation for the different waveforms are done in Section 4 and a discussion on how radar performance can be measured in Section 5.

2. Radar performance

There are different radar performance measures for a given radar system. In this chapter only the SDR and related parameters will be discussed. Another important characteristic of a radar waveform is how the radar system behave if the radar target is moving relative to the radar. This can be studied by calculating the ambiguity function for the radar system. In a narrow band radar the velocity of the radar target gives a shift in frequency of the received waveform compared to the transmitted one. For a UWB-waveform the received waveform will be a scaled version of the transmitted signal. This is an important quality measure for a radar system but will not be discussed in this chapter.

2.1 Radiation limits

A comparison between an Impulse Radar (IR) and a Step Frequency (SF) radar was done by (Hamran et al., 1995). No constraint on the transmitted spectrum was done in that comparison. The new licensing rules for UWB signals put a maximum transmitted Power Spectral Density (PSD) on the Equivalent Isotropically Radiated Power (EIRP) peak emissions from the UWB device. The unit of the PSD is dBm/Hz and is measured as peak

Frequency Range (MHz)	Maximum Peak PSD
30 to 230	-44.5 dBm/120 kHz
230 to 1 000	-37.5 dBm/120 kHz
1 000 to 18 000	-30 dBm/MHz

Table 1. The maximum allowed measured radiated PSD for GPR/WPR imaging systems according to European rules

Frequency Range (MHz)	Maximum Mean PSD (dBm/MHz)
<230	-65
230 to 1000	-60
1000 to 1600	-65
1600-3400	-51.3
3400-5000	-41.3
5000-6000	-51.3
>6000	-65

Table 2. The maximum allowed mean PSD for GPR/WPR imaging systems according to European rules

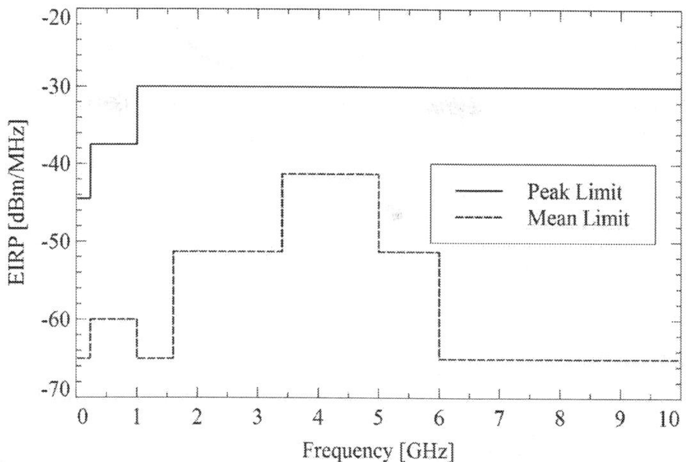

Fig. 1. The maximum allowed measured radiated PSD and mean PSD for GPR/WPR imaging systems

power in a 1 MHz measurement bandwidth (mB) above 1 GHz. There is no single set of emission limits for all UWB devices as both the level in dBm/Hz and the mB is changing with device and frequency. Table 1, Table 2 and Figure 1 give an example on how the PSD varies with frequency for GPR under European rules, (ETSI EN 302 066-1, 2007).

The limits in Table 1 are in principle the same for time domain systems like impulse radars and for frequency domain systems like SF-systems. The way the PSD is measured is however different see (ETSI EN 302 066-1, 2007) and (Chignell & Lightfoot, 2008).

For impulse systems the following formula should be used:

$$PSD_{mean} = P_{peak} + conversion\_factor \qquad (1)$$

with:

$$conversion\_factor = 10log(PRF \times \tau)$$

where τ is the pulse width of the GPR transmitter measured at 50 % amplitude points and PRF is the pulse repetition frequency of the pulse. The peak power and pulse length is measured using a wide bandwidth oscilloscope.

For systems using SF-waveforms the signal is formed by transmitting a sequence of discrete frequencies each having a Dwell Time (DT) and a total sequence length called Scan Time (ST). Calculating the mean PSD for a SF-system the following formula should be used:

$$PSD_{mean} = P_{peak} + conversion\_factor \qquad (2)$$

with:

$$conversion\_factor = 10log(DT/ST)$$

where DT is measured at 50 % amplitude points at a frequency near the maximum of the radiated spectrum, using a spectrum analyser in zero span mode and 1 MHz resolution bandwidth.

A simplified PSD spectrum will be used during the discussion in this chapter. The objectives are to see how the different type of UWB-waveforms performs under a given constraint on the radiated PSD. The simplified spectrum mask is just a constant maximum limit on the PSD over a bandwidth B. The PSD is measured for a given receiver measurement bandwidth, mB. For the mean PSD given in Table 2 the measurement bandwidth is $mB = 1MHz$.

2.2 System Dynamic Range (*SDR*)

The *SDR* is the ratio between the peak radiated power from the transmitting antenna and the minimum detectable peak signal power entering the receiver antenna. This number quantifies the maximum amount of loss the radar signal can have, and still be detectable in the receiver. Since the minimum detectable signal is strongly related to the integration time, the performance number should be given for a given integration time. In the literature the *SDR* measure is defined in many different ways and many names are used such as System Performance or System Q. Most of them excludes the antenna gain on the radar side and are different from what is used here. We will use the definition given in (Hamran et al., 1995) with the exception that we will call it the System Dynamic Range, *SDR*, of the radar system instead of only Dynamic Range. This number tells us that if a reflector has a maximum loss which is less than the radar system *SDR*, then the reflector can be detected by the radar system. This assumes that the reflected signal is within the Receiver Dynamic Range (*RDR*) of the radar system, thus some gain parameters in the radar system may have to be varied to fully make use of the *SDR*. To be able to calculate the *SDR* of a given radar system the matched filter radar equation must be used. Using the simple radar equation for a transmitting pulse having a time-bandwidth product equal unity makes one run into problems when trying to compare different system on a general basis. The following discussion is based on (Hamran et al., 1995) The Signal-to-Noise ratio (*SNR*) for a radar receiver that is matched to the transmitted signal is given by the following equation:

$$SNR = \frac{2E_R}{N_0} = \frac{2E_T G^2 \lambda^2 \sigma}{N_0 (4\pi)^3 R^4} \tag{3}$$

This is the matched receiver radar equation, where E_T = transmitted signal energy, E_R = received signal energy, N_0 = noise spectral density, G = transmit/receive antenna gain, λ = wavelength, σ = target radar cross section and R = range to target.

For the matched filter in the white noise case, the *SNR* is dependent only on the signal energy and the noise spectral density, and is otherwise independent of the signal waveform, provided that the filter is matched to the signal, (DiFranco & Rubin, 1980). The power spectral density of the noise can be expressed by:

$$\frac{N_0}{2} = k_B T_0 F \tag{4}$$

where k_B is Boltzmann's constant ($1.3806503 \cdot 10^{-23} m^2 kg s^{-2} K^{-1}$), T_0 is the room temperature in Kelvins (typically 290 K) and F is the dimensionless receiver noise figure. The signal-to-noise ratio is directly proportional to the transmitted energy. Thus, the longer the receiver integration time, the higher the signal-to-noise-ratio will be. The average transmitting power over a time period is given by:

$$\hat{P}_T = \frac{E_T}{\tau_i}.$$ (5)

Bringing all this into Equation 3 and splitting the radar equation in two separate parts where the first contains the radar system dependent parameters and the second the medium dependent parameters we have

$$\frac{\hat{P}_T \tau_i G^2}{k_B T_0 F(SNR)} = \left[\frac{\lambda^2 \sigma}{(4\pi)^3 R^4} \right]^{-1}.$$ (6)

The left side contains the radar system dependent parts and the right the propagation and reflection dependent parts. The SDR of the radar system is defined as:

$$SDR = \frac{\hat{P}_T \tau_i G^2}{k_B T_0 F(SNR)}.$$ (7)

When comparing different radar systems, the same integration or observation time should be used. Comparing two radar systems having the same antennas and integration time, the following expression is obtained assuming the same SNR for both systems:

$$\frac{SDR_1}{SDR_2} = \frac{\hat{P}_{T1}}{F_1} / \frac{\hat{P}_{T2}}{F_2}.$$ (8)

We see that only the transmitted average power and receiver noise figure need to be considered. This is however only correct if the radar receiver is matched to the transmitted waveform. If the receiver is not matched, a mismatch loss must be introduced. Several factors are involved in this loss and could be taken into account by introducing a matching coefficient (=1 if matched) in the equations above.

In the following discussion we assume that all radars have the same antenna and the antenna gain has been set to 1. It is further assumed that all radars have the same noise figure that has been set equal to 5. The SNR is set to 1. The transmitted spectrum is the same for all waveforms and is given in Figure 2. The spectrum is flat with a power spectral density of PSD and a bandwidth of B.

3. UWB waveforms

Figure 3 show a pulse train that is made up of pulses with a 1 GHz center frequency and a Gaussian amplitude shape with a bandwidth of 80 % of the carrier frequency. The pulse repitition interval (PRI) is 10 ns. The frequency spectrum of the pulse train is shown in the lower plot in the figure. We see that the spectrum of the pulse train is a line spectrum centered at 1 GHz and a -4 dB bandwidth of 800 MHz. The distance between the lines is given by $1/PRI$ and is 100 MHz. The pulse train and its spectrum represents the same signal. In stead of sending the pulse train one could send the line spectrum. In fact, if a line spectrum made by several fixed oscillator transmitting at the same time and amplitude weighted accordingly, the different sinusoidal signals would sum up and produce the pulse train. If the radar scattering scene does not move we could send out one line at a time and measure the reflected signal for each frequency component. A Fourier transform of the measured frequency response would produce the reflected time signal just like transmitting

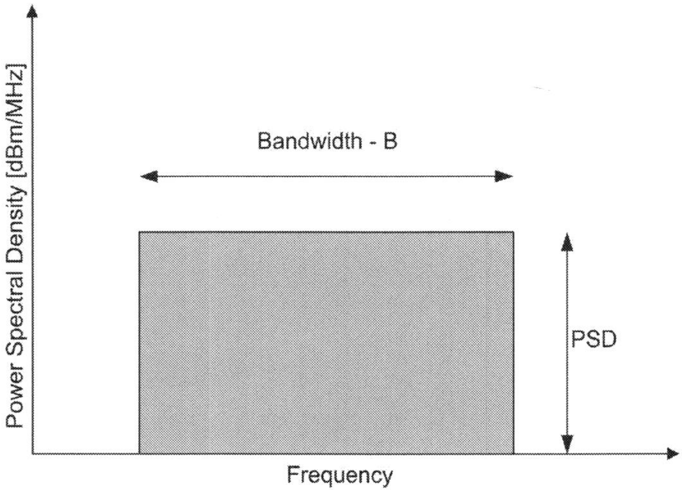

Fig. 2. For the comparison between the different radar waveforms the simple spectrum showed in the figure will be used. It has a flat spectrum with a power spectral density of *PSD* over the bandwidth *B*.

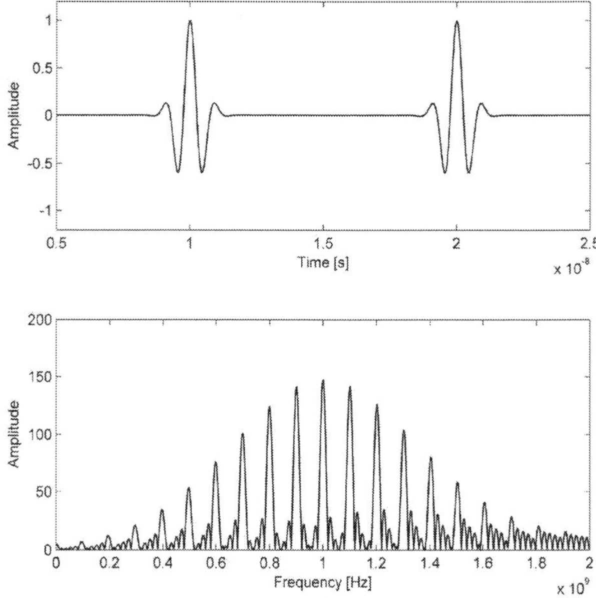

Fig. 3. The figure show a pulse train and its frequency spectrum. The pulse train has a Gaussian pulse with center frequency 1 GHz and a relative bandwidth of 80%.

a pulse and measuring the reflected signal. This method is used in radar and is called a Step Frequency (SF) radar.

All UWB radar waveforms have a way to reduce sampling speed in the receiver so that the full transmitted signal bandwidth does not need to be sampled directly. In impulse radars only one sample per transmitted pulse can generally be taken. The time between the transmitted pulse and the received sample is changed between each pulse and the range profile is then built up. Alternatively, several range cells are sampled with only one bit resolution and several transmitted pulses are needed to get more resolution. In Frequency Modulated Continous Wave (FMCW) and Step-Frequency radars the frequencies are changed linearly with time or stepped with time respectively. Only the beat frequency between the transmitted signal and the received signal is sampled. In continuation of this section a short description of commonly used UWB radar waveforms will be given.

3.1 Impulse radar

In a normal pulsed radar system a local oscillator is switched on and off to generate the transmitted pulse. In impulse radar a short DC pulse is applied directly to the transmitter antenna. The transmitter antenna works as a filter and a band pass signal is radiated from the antenna. The shape and bandwidth of the radiated signal is determined by the antenna transfer function, in other word the parts of the spectrum that the antenna effectively can transmit. A filter may also be put in front of the antenna to shape the spectrum. The received signal needs to be sampled and stored for further treatment. Figure 4 illustrates the sampling process in impulse radar.

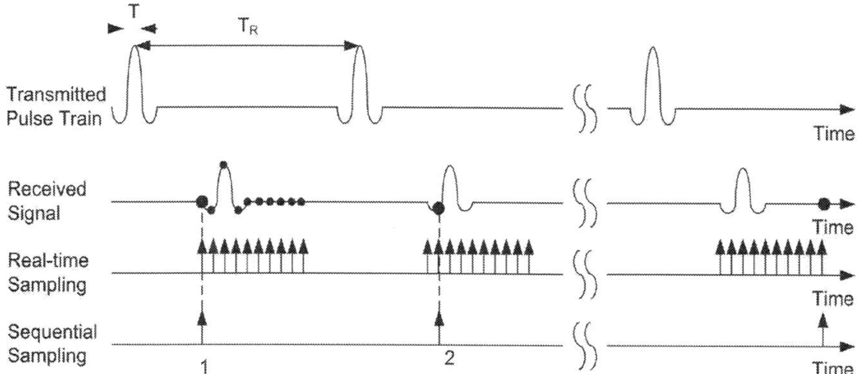

Fig. 4. Illustration of impulse radar with real time or sequential sampling, The transmitted pulse has a pulse width T and a pulse repetition interval T_R.

The matched filter for a pulse is a bandpass filter having the same bandwidth as the pulse and sampling the pulse at the maximum amplitude point. This can be done by sampling the received waveform with a sampling frequency of at least twice the highest frequency in the waveform. In impulse GPR they normally sample at least 10 times the center frequency of the transmitted waveform. In UWB impulse radars the highest frequency can be several GHz so that an Analog to Digital Converter (ADC) that can sample at this speed with several bit resolution is very difficult to manufacture and would be expensive and power consuming. This is called real time sampling and is illustrated in Figure 4. A solution to this problem is to take only one receive sample for each transmitted pulse. Then the ADC only needs to take one sample every pulse repetition interval (PRI) so the need for high speed

ADC is avoided. The lower part of Figure 4 illustrates the sequential sampling technique also called transient repetitive sampling or stroboscopic sampling. This technique is currently being used in oscilloscopes for capturing wideband repetitive signals.

Figure 5 show a block diagram of impulse radar with sequential sampling, see (Daniels, 2004). A noise generator is controlling the trigger of the transmitted pulse. This is to randomize the PRI of the pulse train and thereby reduce peaks and smooth the transmitted spectrum. The PRI-generator triggers the impulse transmitter. The trigger signal is delayed a certain amount before being used to trigger the receiver sampler to take one range sample. The delay is shifted before the next sample is taken. In this radar a multi bit high resolution ADC can be used.

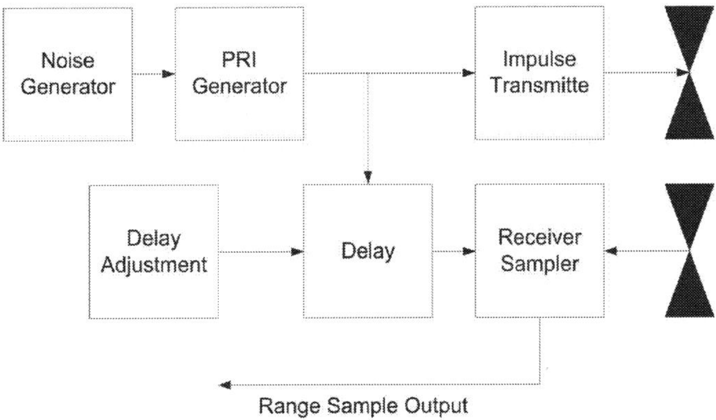

Fig. 5. Impulse radar with sequential sampling.

Another way to capture the reflected signal is to use a single bit ADC and collect several range cells for each transmitted impulse. Figure 6 show a block diagram of this type of impulse receiver. In a practical realization of this receiver one single bit ADC with dither is used and several delays are implemented with a counter on each tap, see (Hjortland et al., 2007). Since the delay is needed only on a single bit signal this is easily implemented using CMOS logic. However several transmitted impulse signals is needed to get sufficient resolution in the received radar profile. Each single bit signal from a range cell is fed to a counter effectively integrating up the single bit signal to yield a higher resolution and thereby higher dynamic range.

In both realizations of impulse radar described above several transmitted pulses are needed to build up one range profile. This means that the receiver is not optimally matched to the transmitted signal and that if the reflector is moving during the profile acquisition it may blur the resulting radar image.

The average transmitted power for an impulse radar with peak power P_T, pulse length T and pulse repetition interval T_R is:

$$\hat{P}_T = P_T \frac{T}{T_R} = PSD \times B. \tag{9}$$

The SDR, given in equation 7, for an impulse radar that has a real time sampling of all the receiver range gates and is therefore matched is given as:

$$SDR = \frac{PSB \times B\tau_i G^2}{k_B T_0 F(SNR)}.$$ (10)

If the impulse radar has a sequential sampling the effective sampling is reduced by the number of range samples in the range profile, n_p. The radar transmits n_p pulses before it returns to the same range sample. The SDR for a sequential sampled impulse radar is given by:

$$SDR = \frac{PSB \times B\tau_i G^2}{n_p k_B T_0 F(SNR)}.$$ (11)

If the radar has a single bit ADC, one transmitted pulse is needed for each signal level that needs to be discretized. If a 12-bits resolution is needed in the digitized range profile, $n_b = 2^{12}$ transmitted pulses need to be transmitted before all the levels are measured. This gives the following expression for the SDR:

$$SDR = \frac{PSB \times B\tau_i G^2}{n_b k_B T_0 F(SNR)}.$$ (12)

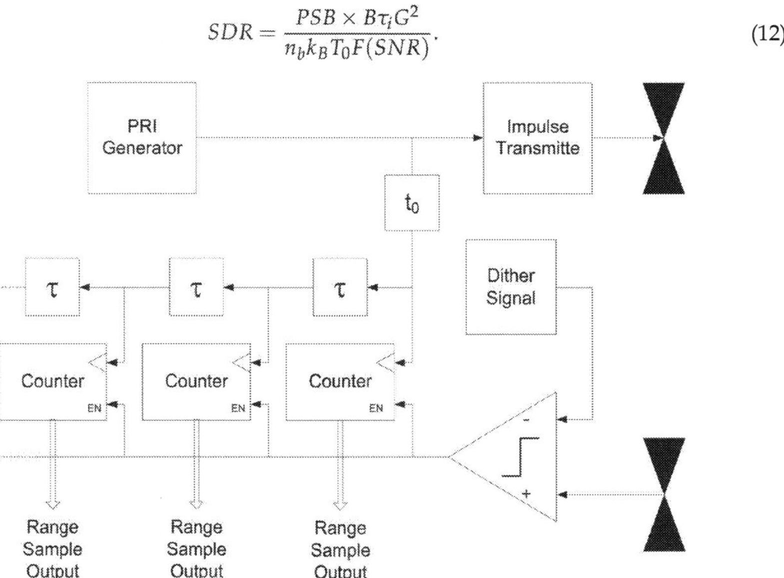

Fig. 6. Impulse radar with single bit ADC receiver

3.2 Step frequency radar (SFR)

A pulse train can be synthesized by sending single tone frequencies simultaneously as illustrated in Figure 3. The different sinusoidal frequencies will sum up, and the transmitted signal will be a pulse train. The width of each pulse is determined by the span between the highest and lowest frequency transmitted. The repetition interval is determined by the distance between the transmitted frequencies. A pulse train can also be synthesized by transmitting the different frequencies one at a time and summing them up afterwards. The step frequency radar does this and transmits single tone frequencies in sequence where only one frequency is transmitted at a time. The phase and amplitude for each frequency is measured of the reflected signal and the radar reflectivity profile is obtained by taking and inverse Fourier transform of the received frequencies.

Figure 7 show a block diagram of a step frequency radar. An oscillator generates the correct frequency that is transmitted by an antenna. The received signal is captured by the receiver antenna and amplified before it is multiplied with the same signal as transmitted. The signal after the mixer is low pass filtered and sampled in an I- and Q-detector. The I- and Q-detector can be implemented after the signal has been sampled. In a homo dyne receiver as shown in Figure 7 the signal oscillator signal need to have a low second and third harmonic content. If this is difficult to achieve a heterodyne receiver should be made.

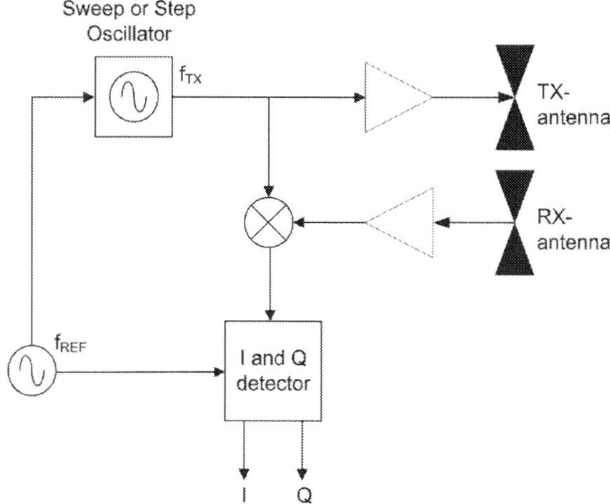

Fig. 7. Block diagram of a step frequency or FMCW radar with homo dyne receiver

When stepping to a new frequency by changing the frequency in the oscillator, it may take some time before the new frequency is stable and can be used. Figure 8 illustrates this where the oscillator needs T_{su} seconds before it is stable. If the effective integration time is T_i the reduction in integration time given by $T_{su}/(T_{su} + T_i)$. The T_{su} interval depends on the oscillator. If a Direct Digital Synthesizer - DDS is used the interval is very small and can be neglected. If however a PLL is used in the oscillator it will take some time before it is locked depending on the loop filter values.

The SF signal has a Dwell Time equal $DT = T_{su} + T_i$ for each frequency and a Scan Time equal $ST = DTn_f$ where n_f is the number of transmitted frequencies. If we assume that the dwell time is the inverse of the measurement bandwidth for the PSD, the average transmitted power for the step frequency signal is:

$$\hat{P}_T = P_T = PSD \times mB. \tag{13}$$

If this radar is used as a GPR the average power can be increased according to Equation 2 by a factor given by:

$$\frac{ST}{DT} = \frac{N(1/mB)}{1/mB} = \frac{B}{mB}. \tag{14}$$

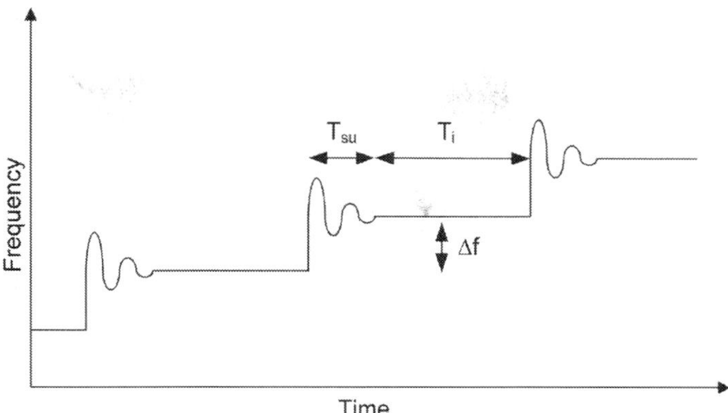

Fig. 8. Changing the frequency in a SF radar system may result in a need to wait until the frequency is stable

The new allowable average transmitted power is:

$$\hat{P}_T = PSD \times mB \frac{B}{mB} = PSD \times B \tag{15}$$

being the same as for the impulse radar in Equation 9. If the correction is not done the SFwaveform would have a reduced average transmitted power of mB/B compared to impulse radar.

The SDR for a SF-radar is:

$$SDR = \frac{PSB \times B\tau_i T_i G^2}{k_B T_0 F(T_{su} + T_i)(SNR)}. \tag{16}$$

3.3 Frequency modulated continuous wave (FMCW)

Another widely used radar technique in UWB systems is the Frequency-Modulated Continuous-Wave (FMCW) radar. This radar is also collecting the data in the frequency domain as with the SFR technique. In stead of changing the frequency in steps, as with SF-radars, the frequency is changed linearly as a function of time. The schematic block diagram is the same as shown in Figure 9 for the SFR except that the frequency oscillator is sweeping. The received signal, a delayed version of the transmitted signal is multiplied with the transmitted signal. The received signal will due to the propagation delay have a different frequency than the transmitted signal so the signal after the mixer will have a beat frequency proportional to the range of the reflector. The signal after digitization is the same as for the SF-radar and an inverse Fourier transform can be used to get the radar reflectivity profile.

Figure 9 show the frequency as a function of time. The parameters that characterize the sweep is the Sweep Time to sweep the whole bandwidth B, $ST = B/\alpha$ where α is the sweep frequency rate of change. The Dwell Time is defined at the time the sweep signal stays within the measurement bandwidth mB and is given as $DT = mB/\alpha$.

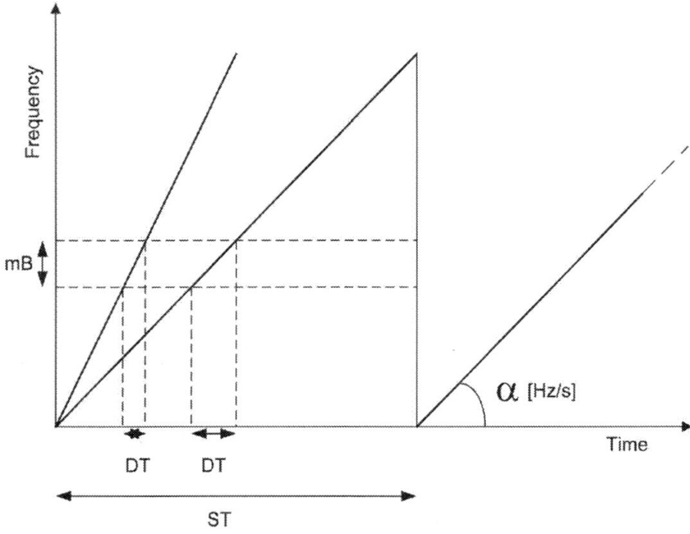

Fig. 9. The dwell time will depend on the sweep

The allowable average transmit power for the sweep is given as the PSD times the inverse of the time the sweep need to cross the mB:

$$\hat{P}_T = P_T = PSD\frac{1}{DT} = PSD\frac{\alpha}{mB}. \tag{17}$$

If the same correction that can be applied to the SF-radar also can be applied to the FMCW we get the following correction factor:

$$\frac{ST}{DT} = \frac{B/\alpha}{mB/\alpha} = \frac{B}{mB}. \tag{18}$$

The corrected average power then becomes:

$$\hat{P}_T = PSD\frac{\alpha}{mB}\frac{ST}{DT} = PSD\frac{\alpha}{mB}\frac{B}{mB} = PSD \times B\frac{\alpha}{mB^2}. \tag{19}$$

We see that the measurements bandwidth of the PSD and the sweep rate enters into the calculation. If $mB = 1/DT$ we get $\hat{P}_T = PSD \times B$.
The SDR for a FMCW-radar is:

$$SDR = \frac{PSB \times B\alpha\tau_i G^2}{mB^2 k_B T_0 F(SNR)}. \tag{20}$$

3.4 Noise radar

A block diagram of a noise radar is shown in Figure 10. A random noise signal is transmitted by the antenna. The transmitted signal is reflected back and enters the receiver after a time given by the two way travel time delay. The received signal is amplified and correlated with a reference signal that is a delayed version of the transmitted noise signal. When the reference signal is delayed the same amount as the received signal a strong correlation value would be expected.

The delay line in Figure 10 can be implemented in different ways. The most straight forward is a cable with a given length providing the correct delay. A mixer and a low pass filter can be used as a correlator. Another implementation is using a digital RF-memory where the transmitted signal is sampled and delayed digitally before converted to analog signal and correlated with the received signal. A fully digital correlation can be done where both the transmitted signal and the received signal is digitized and stored in a computer. Then the cross-correlation can be done by calculating the cross-spectrum by taking a Fourier transform of the transmitted and received signal. The correlation peak-to-noise ratio for a noise radar is close to the time bandwidth product. For a 40 dB peak-to-noise ratio 10 000 samples must be collected. For an analog correlator this number of new samples must be collected for each range cell. If a digital correlator is used, the same collected samples can be used for all the range cells. Thus the PRF is much higher for a digital correlator than for an analog correlator. The sampling in a noise radar can be done with a single bit ADC, see (Axelsson, 2003).

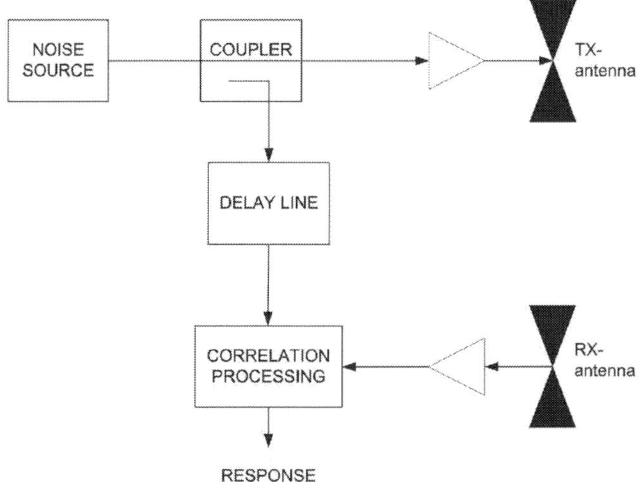

Fig. 10. Block diagram of a noise radar

If the transmitted noise signal is a pseudo random noise generated by a shift register a repetitive signal can be made. When the transmitted signal is repetitive a sequential sampling receiver like for the impulse radar can be made. This technique is described in (Sachs, 2004) and a schematic of the radar is given in Figure 11. The RF-clock generating the pseudo random noise signal is divided to make up the sampling clock. If the clock is divided by 4 it will take 4 pseudo random noise sequences to make up one measurements, see (Sachs, 2004) for more details.

The transmitted average power for a noise radar is:

$$\hat{P}_T = PSD \times B. \tag{21}$$

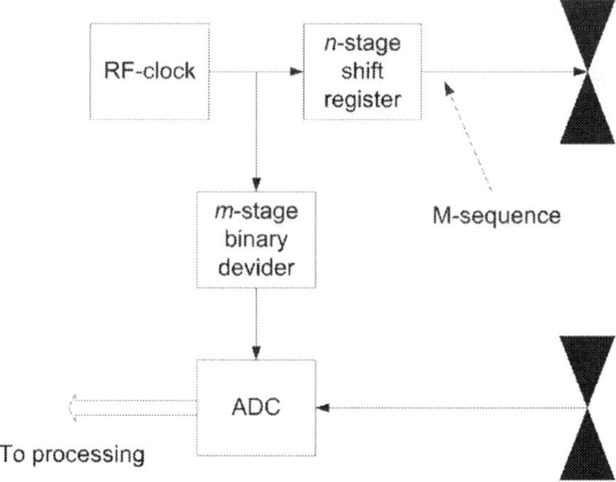

Fig. 11. Block diagram of a noise radar using pseudo random noise and sequential sampling in the receiver

If both the transmitted noise signal and the reflected signal is digitized and sampled the SDR is given by:

$$SDR = \frac{PSD \times B\tau_i G^2}{k_B T_0 F(SNR)}.$$
(22)

If pseudo random noise and sequential sampling is used the above equation must be multiplied with a factor $1/n_m$ where n_m is the number of pseudo random sequences needed to collect all the data. The SDR becomes:

$$SDR = \frac{PSD \times B\tau_i G^2}{n_m k_B T_0 F(SNR)}.$$
(23)

4. Examples on radar performance

A general comparison between the different waveforms is not possible as the SDR depends on the number of range samples needed and the bandwidth of the radar signal. To compare the waveforms a radar with the parameters given in Table 3 will be used. The radar has a bandwidth of 1 GHz and a center frequency of 1 GHz. The range window that should be imaged is 200 ns long. The integration time for collecting one trace is 1 ms. Table 4 gives the calculated SDR for the different waveforms. The impulse real time sampling waveforms has an ADC that can collect all samples in the radar trace for each transmitted impulse. The receiver is matched and the radar system gets the maximum possible SDR = 131.5 dB. The impulse sequential sampling is having an equivalent sampling at 10 times the radar center frequency. This means that it needs n_p = 2000 transmitted pulses to build one radar trace. This gives a reduction of 33 dB compared to the real time sampling. The single bit receiver changes the threshold levels to get 12-bit resolution in the receiver. The radar needs to transmit $n_b = 2^{12}$ pulses to get one radar trace at 12-bit resolution. This gives a reduction of 36.1 dB compared to the real time sampling, only 3.1 dB lower than the sequential sampling receiver.

Radar parameters	Value
f_c	1 GHz
B	1 GHz
τ_i	1 ms
G	1
F	1
SNR	1
T_0	300 K
Range window	200 ns

Table 3. General radar parameters used for all the radar systems under study

Radar waveform	SDR
Impulse real time sampling	131.5 dB
Impulse sequential sampling	98.5 dB
Impulse single bit sampling	95.4 dB
Step frequency	91.5 dB
Step frequency corrected	121.5 dB
FMCW	101.5 dB
FMCW corrected	131.5 dB
Noise	131.5 dB
Noise sequential sampling	115.5 dB

Table 4. SDR for the different waveforms for the radar parameters given in Table 3

In the step frequency waveform the oscillator set up time is $T_{su} = 100$ ns and the integration time per frequency is $T_i = 900$ ns. The dwell time is $DT = 1/mB$. The SDR is calculated to be 91.5 dB that is 40.4 dB lower than impulse real time performance. This is due to the much lower allowable transmitted average power given by Equation 13. If the correction factor in Equation 2 is applied the SDR is 121.5 dB, still 10 dB lower that impulse real time sampling. If a DDS is used as a oscillator with practically no set up time between the frequency steps the, SDR would increase and be the same as for the impulse real time sampling.

The sweep in the FMCW is adjusted so that $DT = 1/mB = 1\mu s$. This gives $\alpha = 1MHz/\mu s$ and one sweep takes 1 ms. The SDR for the FMCW is 101.5 dB and if corrected by Equation 13 becomes 131.5 dB.

A noise radar that collects all the bandwidth with a real sampling techniques gets a SDR of 131.5 dB If a sequential sampling where 40 sequences are needed to get one radar trace the performance is reduced by 16 dB.

The SDR numbers given in Table 4 are depending on several parameters and will change depending on radar center frequency, signal bandwidth and radar range window. Table 4 show that the SDR is varying more than 40 dB for the different waveforms.

5. Measuring radar performance

Figure 12 illustrates the different signal levels present in the radar receiver. The signal levels will be referenced to the transmitter signal level P_T either in dBm or in dB relative to the transmitter power level. The transmitted power level is indicated by a horizontal dashed line. The next level in the receiver is the level where the ADC has reached full scale, ADC-

FS. The ADC-FS should be given by what signal level in dBm at the receiver antenna input is driving the ADC full scale. This level is also indicated by a horizontal line in Figure 12. The next level is the thermal noise, N, in the receiver. This is the thermal noise at the receiver input given by Equation 4. There is also an internal coupling in the radar system that is illustrated on the left side as a sloping dashed line. The internal coupling level will decrease with time in the receiver time window. The internal coupling may be characterized by the maximum signal level in the receiver and the slope given in $[dB/ns]$. The coupling ring down depends among other things on the oscillator phase noise and the receiver IF-mixer. The different levels indicated in Figure 12 can be measured by some simple measurements.

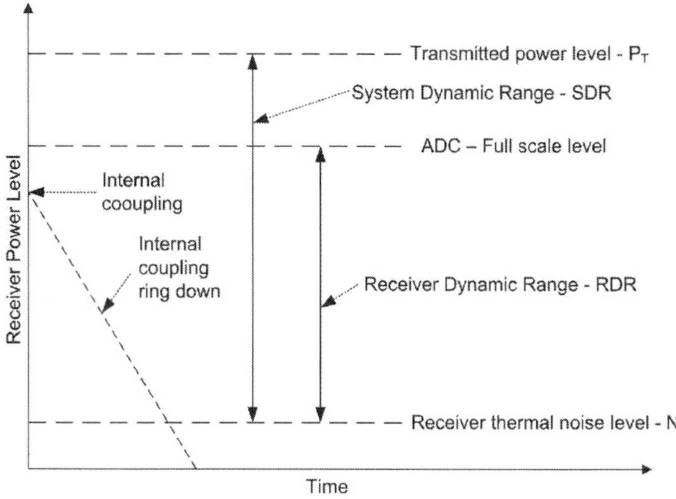

Fig. 12. Illustration of the different signal levels in the radar receiver

5.1 Transmitter power

UWB radars use wide band microwave amplifiers that do not necessarily have the same gain for all the frequencies. Normally the gain decrease as a function of frequency. If the gain change by 2 dB over the used bandwidth for each amplifier and the radar have 5 such gain stages in the transmitter chain the transmitted spectrum will change by 10 dB over the band. The transmitted spectrum should be measured using a calibrated spectrum analyzer.

5.2 Receiver gain

The receiver level should be calibrated relative to the transmitted power measured in 5.1. If the transmitter power is known in dBm the receiver can be calibrated in dBm. This can be done by connecting a cable with a known attenuator as shown in Figure 13. The cable will give a signal in the receiver with a level given as $P_T - Att$. The cable signal level should be much higher than the thermal noise level to give a high signal-to-noise ratio. If the cable and attenuator are frequency independent the receiver gain can be measured. Measuring the received signal from the cable and subtracting the input signal to the receiver given by $P_T - Att$ the receiver gain can be calculated as a function of frequency. The gain is the total receiver gain from the receiver antenna input to the ADC.

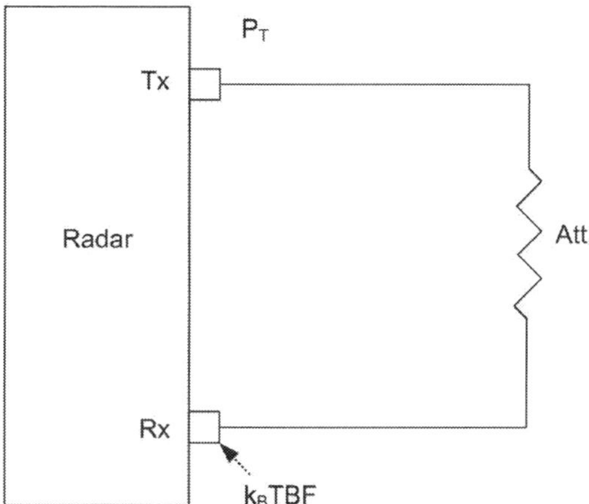

Fig. 13. Measuring radar system performance. The transmitter output is connected to the receiver input via a cable and an attenuator

5.3 Internal coupling
The internal coupling should be measured by terminating both the transmitter output and the receiver input to their characteristic impedances. Then a normal radar measurement should be done. The coupling should be characterized by its maximum signal level given in dBm and its slope given in [dB/ns]. The internal coupling should be as low as possible and the slope as steep as possible.

5.4 Receiver noise
The measurement above can also be used to measure the receiver noise level. The internal coupling will slope downward as a function of travel time in the radar receiver window. At a certain time the signal will level out and become horizontal. The level of this horizontal part in the receiver is the receiver noise level. Since the receiver is calibrated in dBm the noise level can be measured in dBm and should be compared with the theoretical value given by equation 4. The receiver thermal noise level will depend on the integration time of the radar system.

5.5 System Dynamic Range - SDR
The difference between the measured transmitted power and the measured receiver noise is the System Dynamic Range. The SDR gives a measure for how much loss the radar wave can have and still be detectable in the radar receiver. This is a key measure for the performance of the radar system. The SDR is dependent on the radar integration time. If the integration time is made 10 times longer then the SDR will increase by 10 dB.

5.6 Receiver Dynamic Range - RDR
The difference between the ADC full scale value and the noise level is called the Receiver Dynamic Range. This value tells us what signal difference there can be between two reflected signals and that both can be imaged in the receiver window. Increasing the integration increases the RDR.

5.7 Spurious free receiver dynamic range

Increasing the integration will increase the RDR and eventually lead to spurious signals to show up in the receiver window. The difference between the ADC-FS and the peak of the spurious signals will give the spurious free receiver dynamic range. This tells us how large the RDR can be made without having unwanted signal in the receiver window. If the spurious signals are stable and does not change from measurement to measurement they can in principle be removed by subtracting an average of the measurements.

6. Conclusion

UWB-waveforms are very often used in radar system that are looking though structures like the ground or walls. The propagation of the waves through these structures may attenuate the waves considerably. For the radar system to be successful in these applications it must have a high SDR. This chapter have developed equations for some popular UWB waveforms so that the SDR can be estimated based on parameters for the different waveforms. The SDR varies more than 40 dB between the different UWB-waveforms under study here. The achieved radar performance for a given system can be measured and a simple way to do this has been explained.

7. Acknowledgement

The author would like to thank his colleagues Trygve Sparr and Tor Berger for valuable discussions when preparing this chapter.

8. References

Astanin, L.Y. & Kostylev, A.A. (1997). Ultrawideband radar measurements analysis and processing, *IEE*, ISBN 0 85296 894 9

Axelsson, S.R.J. (2003). Noise radar for range/Doppler processing and digital beamforming using low-bit ADC, *IEEE Trans. on Geoscience and Remote Sensing*, VOL. 41, NO. 12

Chignell, R.J. & Lightfoot, N. (2008). GPR Radio Licences and Product Standards within Europe, *12th Int. Conf. on GPR*, June 16-19, Birmingham, UK

Daniels, D.J. (2004). Ground Penetration Radar 2nd Edition, *IEE*, ISBN 0 86341 360 9

DiFranco, J.V. & Rubin, W.L. (1980). Radar detection, *Artec House, Inc*, ISBN 0-89006-092 4

ETSI EN 302 066-1 (2007). Electromagnetic compatibility and Radio spectrum Matters (ERM); Short Range Devices (SRD); Ground- and Wall- probing Radar applications; Part 1: Technical characteristics and test methods.

Hjortland, H.A.; Wisland, D.T.; Lande, T.S.; Limbodal, C. & Meisal, K. (2007). Thresholded samplers for UWB impulse radar, *IEEE International Symposium on Circuits and Systems*, ISCAS 2007.

Hamran, S-E.; Gjessing, D.T.; Hjelmstad, J. & Aarholt, E. (1995). Ground penetrating synthetic pulse radar: dynamic range and modes of operation, *Journal of Applied Geosciences*, 33, pp. 7-14

Sachs, J. (2004). M-sequence radar, In Ground Penetration Radar 2nd Edition, *IEE*, ISBN 0 86341 360 9

Taylor, J.D. (2001). Ultra-Wideband Radar Technology, *CRC Press*, ISBN 0-8493-4267-8

Short Range Radar Based on UWB Technology

L. Sakkila[1,2,3], C. Tatkeu[1,2], Y. ElHillali[1,3], A. Rivenq[1,3],
F. ElBahhar[1,2] and J-M. Rouvaen[1,3]
[1]Univ. Lille Nord de France, F-59000 Lille,
[2]INRETS, LEOST, F-59666 Villeneuve d'Ascq,
[3]UVHC, IEMN-DOAE, F-59313 Valenciennes,
France

1. Introduction

In this Chapter, a short range radar system based on ultra-wideband (UWB) technology is presented. The radar and its applications were reserved during a long time to national defence, air security or weather services domains. Since a few years, with the emergence of new technologies, the radar applications were developed and become known in many sectors of daily life. The arrival of a new technology called Ultra-Wideband (UWB) allows in particular the development of compact and low-cost radar with multiple fields of applications. UWB uses very short non-sinusoidal pulses that have widths less than 1.5 ns, so that the spectrum of the transmitted signals may spread over several Gigahertz. This radar offers a resolution in distance of about a few centimetres, for example 15 cm for a pulse width of 1ns, making this system very interesting in several short range applications.

UWB radar has many applications in Medical, Building, Surveillance, Security and Monitoring applications [Dam2007] and will appear more and more in our daily life. For example, UWB radar systems for local monitoring allow creating dome radar surveillance around a sensitive object or subject. These compact systems contain a small UWB radar with a range of about 10 meters, a standard radio system for transmitting the alarm in case of intrusion and a GPS system for localisation function. These systems can be engaged in public safety functions in buildings, aircrafts or for artwork protection in a museum, but also as an alarm system around a house or near a swimming pool to avoid too frequent small children drowning. Thanks to their sensitivity the UWB radar can detect movement as slight as a heart beat or breathing rate. These systems is responsive enough to accurately depict a heart rate as compared to others existing systems. UWB radar could be used as ground penetrating radar (GPR). GPR systems can obtain very precise and detailed images sub-soil. UWB radar is moved along surface and sends electromagnetic pulses into the ground. The analysis of received echoes can produce a very specific profile of underground. The investigation depth varies, depending on the type of ground, from a few meters in asphalt or clay to more than a hundred meters in limestone or granite, or even several kilometres into the ice. Finally, UWB radar could be used for short range collision avoidance as mentioned in this paper.

This collision avoidance system, 24 GHz UWB Short Range Radar (SRR), was developed principally by European car manufacturers. It is a combination of an UWB radar and a conventional Doppler radar to measure vehicle speeds and to detect obstacles with a resolution in distance between 10 cm and 30 m. These systems are placed at the front and sides of the vehicle and warn the driver of potential impacts with other vehicles or pedestrians. They are also useful as parking assistance. Current systems warn the driver of a potential danger without intervening in the braking system. This system should allow a reduction in traffic accidents as the standard rear collisions often due to inattention, so these collisions, estimated 88%, could be avoided [ESaf2006].

In the following paragraphs, UWB technology, its advantages and disadvantages will be introduced, before presenting the UWB radar.

2. Ultra Wide Band technology

The radar studied exploits a new radio frequency technique called Ultra-Wide Band (UWB), to perform obstacles detection. The UWB technology is a radio modulation technique based on very short pulses transmission [Bar2000].

These pulses have typical widths of less than 1.5 ns and thus bandwidths over 1 GHz. This technique, as defined by the Federal Communication Commission (FCC) [Fcc2002], has a Fractional Bandwidth (FB) greater than 25%, this fractional bandwidth is defined as:

$$FB = \frac{signal\ bandwidth}{center\ frequency} = \frac{f_h - f_l}{f_h + f_l} * 100 \qquad (1)$$

where f_h and f_l represent respectively the highest and the lowest frequencies which are 10 dB below the maximum.

The key value of UWB is that its RF (Radio Frequency) bandwidth is significantly wider than the information bandwidth [Fcc2000].

The advantages of UWB technology are:

- Its exceptional multipath immunity.
- Its relative simplicity and likely lower cost to build than spread spectrum radios.
- Its substantially low consumed power, lower than existing conventional radios.
- Its implementation as a simple integrated circuit chipset with very few off-chip parts.
- Its high bandwidth capacity and multi-channel performance.
- Its high data rates for wireless communications.

However, UWB technology presents also some disadvantages among which are:

- A risk of being subjected to the dispersion and frequency distortion phenomena since they extend over a wide bandwidth [Tay1995]. Dispersion is the phenomenon whereby different frequency components of a signal do not propagate at the same speed in a channel. The distortion in frequency can be defined as non-uniform attenuation of different frequency components of a signal. UWB systems must therefore take account of these effects.
- Throughput performance will probably never exceed that of the optical systems at great rates

Moreover, thanks to advantages provided by this technology, the UWB radar presents good precision in distance calculation. Below, the principle of UWB radar and its benefits are introduced.

3. UWB radar

UWB radar sends very short electromagnetic pulses. This type of radar can employ traditional UWB waveforms such as Gaussian or monocycle pulses. To calculate the distance between radar and obstacle, the time delay Δt between emission and reception is measured. This distance is given by:

$$d = \frac{c.\Delta t}{2}$$

(2)

where c is the light speed.

This radar offers a resolution in distance of about 15 cm for a width pulse of 1ns, so that this system is very interesting for short range road safety applications.

This UWB radar presents good performances, so as firstly, the brevity of the pulses with strong spectral contents makes it possible to obtain information on the target with a rich transitory response content. This allows the dissociation of various echoes at the reception stage. Then the broad band spectrum authorizes to obtain results on the entire frequency band in a single measurement together with a strong capacity of detection. Finally the pulse spectrum has capabilities to penetrate through naturally screening materials (ground, vegetation...) [Dam2007].

Considering all these properties, UWB radar, using very short pulses, is of great interest for many applications of obstacle detection and target identification in the short range.

This UWB radar is operational for the short ranges where the conventional radars, using pulses of width from 1µs, are unable to detect obstacles below 15 meters. Moreover, these is not an intermediate electronic stage because of the non-existence of the carrier frequency. So, this system has a simpler implementation and lower cost compared to the traditional radar.

According to the application considered, the choice of the appropriate UWB waveform needed. In fact, each waveform gives a specific cross-correlation function and the obtained peaks of this function must be easily detectable at the receiver.

Different waveforms can be used for the UWB radar [Cham1992] for example: Gaussian pulse, monocycle pulse, Gegenbauer and Hermite functions.

4. UWB pulse radio

4.1 Gaussian pulse

The Gaussian pulse is a waveform described by the Gaussian distribution.

In the time domain, the expression of the Gaussian pulse waveform is given by [Bar2000]:

$$g(t) = A \exp[-(t/\sigma)^2]$$

(3)

where A stands for the maximum amplitude and σ for the width of the Gaussian pulse. The corresponding time and spectral representations are given in figure 1 assuming a sampling frequency of 20 GHz/Samples.

4.2 Monocycle pulse

The monocycle pulse is the first derivative of the Gaussian pulse. The expression for the monocycle pulse waveform is written as [Carl2000]

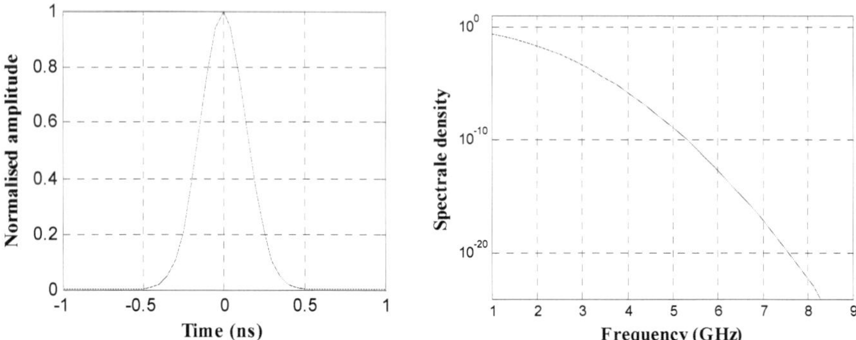

Fig. 1. Gaussian pulse time and spectral representations.

$$m(t) = \frac{t}{\tau}\exp[-(t/\tau)^2] \tag{4}$$

where τ is the pulse width (the centre frequency is then proportional to $1/\tau$).
The time and spectral representations are given in figure 2, again with a sampling frequency of 20GHz/Samples.

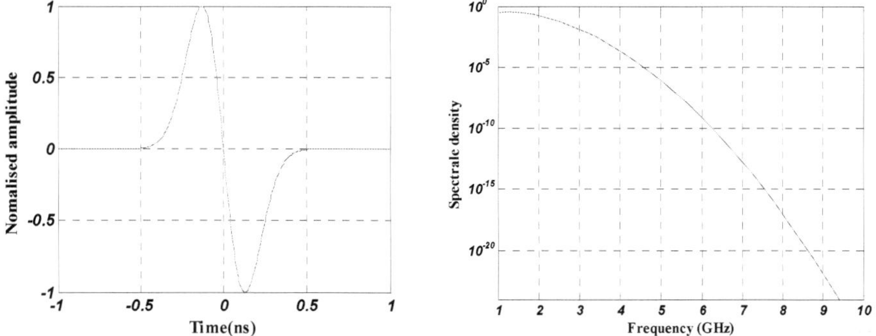

Fig. 2. Monocycle pulse time and frequency representations.

The nth derivate of Gaussian pulse can be obtained from the following expression:

$$p^n(t) = -\frac{n-1}{\sigma^2}g^{(n-2)}(t) - \frac{t}{\sigma^2}g^{(n-1)}(t)$$

In the simple case, single user or single radar system, the configuration of transmitter and receiver is given by the figure 3. So, due to the fact that Gaussian and its first derivative Monocycle pulses are easy to generate, they allow very simple and less expensive implementation of a radar in a single user case. The pulse leaving the generator is transmitted using the transmitter antenna. After reflection on the obstacle, the received signal echo is correlated with the reference pulse, in order to detect the peak, by using the threshold detection method.

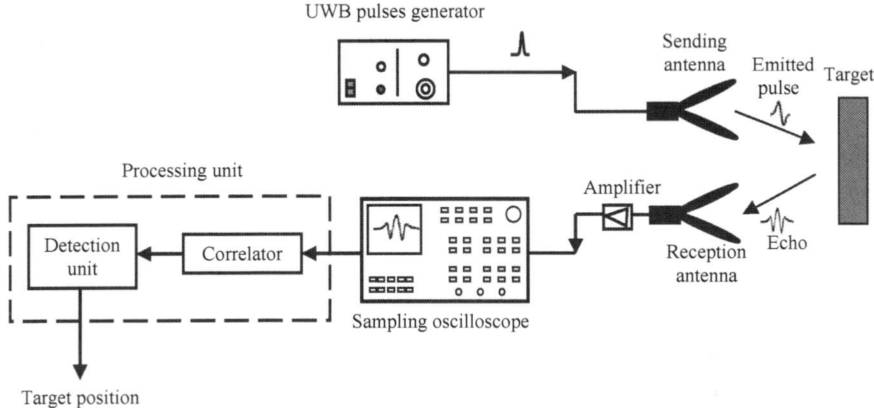

Fig. 3. The radar based on UWB pulses.

However, in a real scenario, several UWB radars may exist in the same propagation channel (environment). So, using these simple waveforms (Gaussian and Monocycle pulses), is not appropriate. In fact, it is necessary to encode the pulses emitted by different users in order to avoid interference between them. To do this, to each user or radar system a specific code is attributed. The coding is done by multiplying each bit of code by the UWB pulse as in CDMA (Code Division Multiple Access) technique (Figure 4).

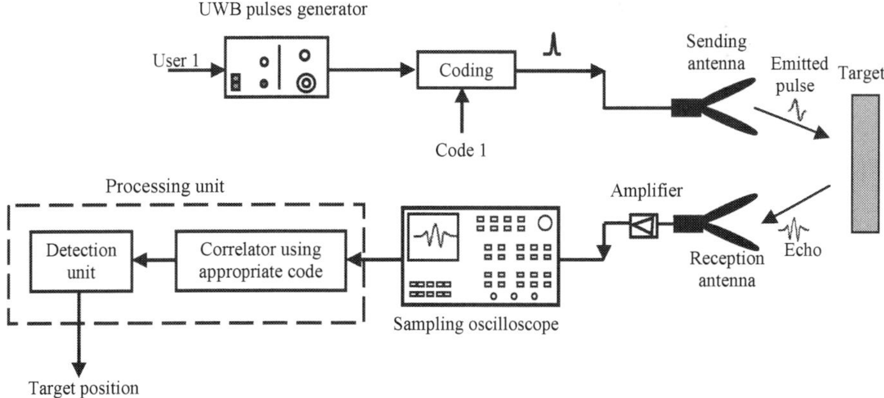

Fig. 4. The proposed radar based on UWB coded pulse.

Another way to perform multiple access exists; namely using waveforms based on orthogonal polynomials. Some UWB orthogonal waveforms, such as Gegenbauer and Hermite polynomials, could be used to ensure multiple access (each user has his own order of polynomial waveform) without resorting to the use of codes, which can reduce the complexity of the system compared with conventional systems. Multiple access based on the use of unique waveforms allows for up to 8 users.

4.3 Orthogonal waveforms for UWB radar system
4.3.1 Gegenbauer polynomials

The Gegenbauer polynomials are also called ultra-spherical polynomials. These polynomials are defined in the interval [- 1, 1] and they satisfy the following differential equation of second order [Elb2005]:

$$(1 - x^2)\ddot{G}(n,\beta,x) - (2\beta + 2)x\dot{G}_n(x) - n(n + 2\beta + 2)G_n(x) = 0 \tag{5}$$

with $\beta \succ -1/2$

n: order of Gegenbauer polynomial

The different orders of the Gegenbauer polynomials are bound by the following recurrence equation:

$$G(n,\beta,x) = 2\left(1 + \frac{n + \beta - 1}{n}\right)xG(n - 1,\beta,x) - \left(1 + \frac{n + 2\beta - 2}{n}\right)G(n - 2,\beta) \tag{6}$$

for $n \succ 1$

The weight function for these polynomials is given by:

$$w(x,\beta) = (1 - x^2)^{\beta - 1/2} \tag{7}$$

To be able to use these polynomials in an UWB communication system, the signals generated from polynomials must be very short. So, the orders of polynomials $G(n,\beta,x)$ are multiplied by a factor corresponding to the square root of the weight function of this polynomials family [Elb2005]. In general we have following:

$$\int w(x,\beta)G(m,\beta,x)G(n,\beta,x) = 0 \tag{8}$$

The modified Gegenbauer function is given by the following equation:

$$G_u(n,\beta,x) = \sqrt{w(x,\beta)} * G(n,\beta,x) \tag{9}$$

The first four orders of these functions for $\beta = 1$ are given by the following expressions:

$$\begin{aligned}
G_u(0,\beta,x) &= 1 * (1 - x^2)^{1/4} \\
G_u(1,\beta,x) &= 2x * (1 - x^2)^{1/4} \\
G_u(2,\beta,x) &= (-1 + 4x^2) * (1 - x^2)^{1/4} \\
G_u(3,\beta,x) &= (-4x + 8x^3) * (1 - x^2)^{1/4}
\end{aligned} \tag{10}$$

Figure 5 illustrates the time representation of the first four orders of modified Gegenbauer functions with $\beta = 1$.

The spectral representation is given in figure 6.

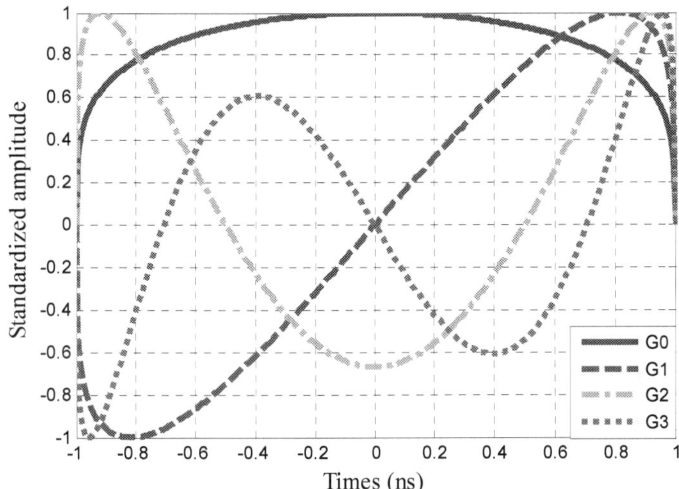

Fig. 5. Modified Gegenbauer functions of orders n = 0, 1, 2 and 3.

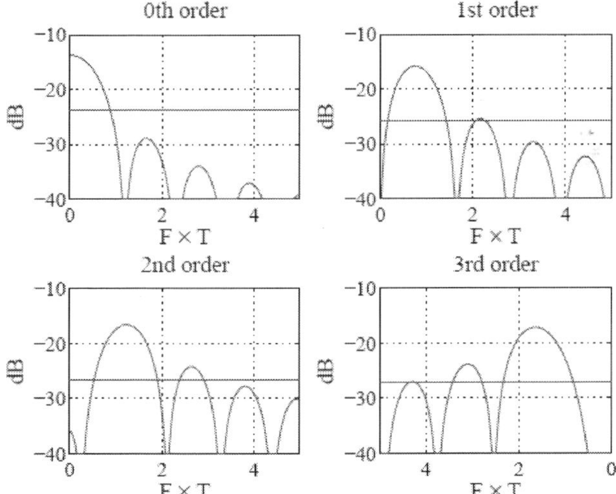

Fig. 6. Spectral representation for modified Gegenbauer functions of orders n = 0 to 3.

4.3.2 The modified Hermite functions

Ghavami, Michael and Kohno have proposed the Modified Hermite functions for a multi-user communication system [Lach2002]. These functions are defined in the interval]-∞, +∞ [by:

$$h_{e0}(t) = 1 \qquad (11)$$

$$h_{en}(t) = (-1)^n e^{t^2/4} \frac{d^n}{dt^n}(e^{-t^2/2}) \tag{12}$$

where n = 1, 2... , n represents the order.
Different orders of Hermite polynomials are associated by the following equations

$$h_{e_{n+1}}(t) = t\, h_{e_n}(t) - \dot{h}_{e_n}(t) \tag{13}$$

$$\dot{h}_{e_n}(t) = n\, h_{e_{n-1}}(t) \tag{14}$$

with $\dot{h}$ present the derivative of h
Using equations (13) and (14), the Hermite polynomials satisfy the differential equation of second order defined as:

$$\ddot{h}_{e_n} - t\,\dot{h}_{e_n} + n\, h_{e_n} = 0 \tag{15}$$

The Hermite functions are derived from these polynomials by multiplying them by the factor $e^{-\frac{t^2}{4}}$ as follows:

$$h_n(t) = e^{-\frac{t^2}{4}} h_{e_n}(t)$$
$$= (-1)^n e^{-\frac{t^2}{4}} \frac{d^n}{dt^n}(e^{-\frac{t^2}{2}}) \tag{16}$$

Therefore, the orthogonal Hermite functions must verify the following differential equations:

$$\ddot{h}_n + (n + \frac{1}{2} - \frac{1}{4}t^2)\, h_n = 0 \tag{17}$$

$$\dot{h}_n + \frac{t}{2}h_n = n\, h_{n-1} \tag{18}$$

$$h_{n+1} = \frac{t}{2}\, h_n - \dot{h}_n \tag{19}$$

If the Fourier transform of $h_n(t)$ is symbolized by $H_n(f)$, equations (17), (18) and (19) can be written respectively, as follows:

$$\ddot{H}_n + 16\pi^2\,(n + \frac{1}{2} - 4\pi^2 f^2)H_n = 0 \tag{20}$$

$$j8\pi^2 f\, H_n + j\,\dot{H}_n = 4\pi n\, H_{n-1} \tag{21}$$

$$H_{n+1} = j\,\frac{1}{4\pi}\,\dot{H}_n - j2\pi f H_n \tag{22}$$

For example, for n=0, $h_0(t) = e^{-\frac{t^2}{4}}$ and $H_0(f) = 2\sqrt{\pi}\ e^{-4\pi^2 f^2}$, all other orders can be found. The first three orders, after the order 0, are given by the following equation:

$$H_1(f) = (-4j\pi f)^2 \sqrt{\pi}\ e^{-4\pi^2 f^2}$$
$$H_2(f) = (1 - 16\pi^2 f^2)^2 \sqrt{\pi}\ e^{-4\pi^2 f^2} \qquad (23)$$
$$H_3(f) = (-12 j\pi f + 64 j\pi^3 f^3)^2 \sqrt{\pi}\ e^{-4\pi^2 f^2}$$

The first four Hermite time functions (n = 0 to 3), presented in Figure 7, are given by equations:

$$h_0(t) = e^{-t^2/4}$$

$$h_1(t) = te^{-t^2/4} \qquad (24)$$

$$h_2(t) = (t^2 - 1)e^{-t^2/4}$$

$$h_3(t) = (t^3 - 3t)e^{-t^2/4}$$

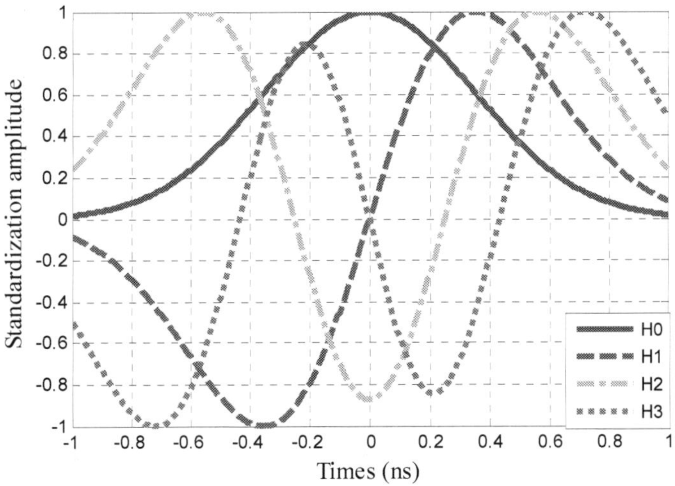

Fig. 7. Hermite functions, orders n = 0 to 3.

Their widths are normalized to 1 ns, the truncation being performed so that at least 99% of energy is kept for the fourth order function. The vertical units are chosen so that the energy of all functions is equal to unity.

The spectral representation is given in figure 8.

The realization of an UWB radar system requires the choice of the waveform. A good choice of this parameter allows optimisation of performances at the reception stage [Elb2001] and reduces the implementation complexity. In order to choose the most appropriate waveform

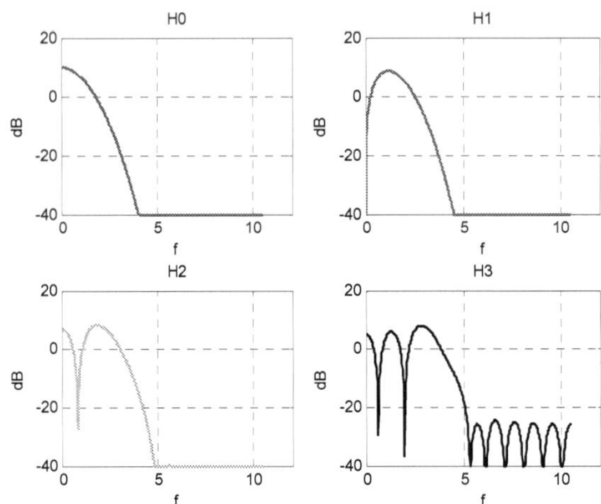

Fig. 8. Spectral representation for Hermite functions of orders n = 0 to 3.

for the radar function, in the following paragraphs, the autocorrelation functions of different waveforms are compared in simulations first and then using the real signals.

At each step, a comparison is made at first between the autocorrelation functions, in terms of dynamics (difference between the correlation peak amplitude and the maximum amplitude of noise), peak width measured at -3dB and a new criterion that represents the ratio between the dynamics and the peak width. Using this criterion we consider that the higher this ratio, the better will be the performances of the radar. The ratio between the dynamics and peak width of different orders of Gegenbauer and Hermite polynomials will be compared. The orders that give the best performance for the radar detection functionality is chosen, so the best ratio between the dynamics and peak width, and compared those selected with the Gaussian and monocycle pulses autocorrelation functions.

5. Comparison performances of UWB waveforms

The autocorrelation functions of different orders Gegenbauer functions are given in figure 9. The comparison between these functions is given in table 1. We can see that the highest ratio between dynamic and peak width belongs to the order 3 autocorrelation of Gegenbauer polynomials.
The same comparison is made for the Hermite polynomials. Results are shown in figure 10. The best autocorrelation is obtained with order 1 of Hermite function H1.
Then, a comparison between the orders of the Gegenbauer and Hermite functions chosen and the Gaussians and Monocycle pulses is made. The functions that offer the best autocorrelations is sought.
Considering the ratio between dynamics and the correlation peak width, the order 3 of Gegenbauer function followed by the monocycle pulse and the order 1 of Hermite function are better. In addition, taking the realisation complexity into account, monocycle offers advantages over other waveforms. So Monocycle pulse compared to order 1 of Hermite function, order 3 of Gegenbauer function and Gaussian pulse seems more adapted for our system in single-user case. However, all these waveforms are interesting in multi-users case.

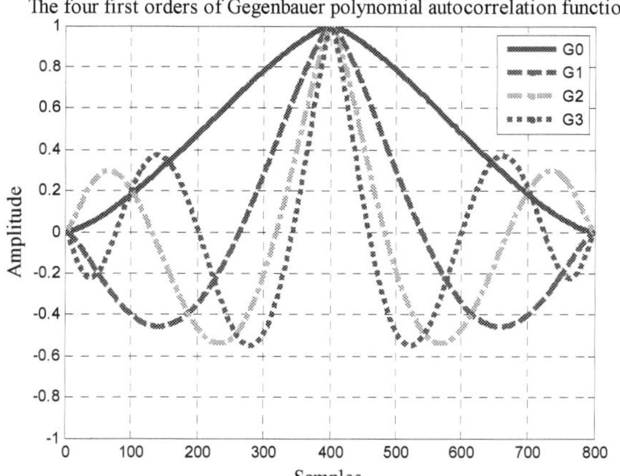

The four first orders of Gegenbauer polynomial autocorrelation function

Waveform	Dynamics (D)	Peak width (W)	Ratio (D/W)
G0	1	386	26.10^{-4}
G1	1	150	67.10^{-4}
G2	0,7	89	79.10^{-4}
G3	0,63	62	102.10^{-4}

Fig. 9. Comparison between autocorrelation functions of the first four orders of Gegenbauer functions.

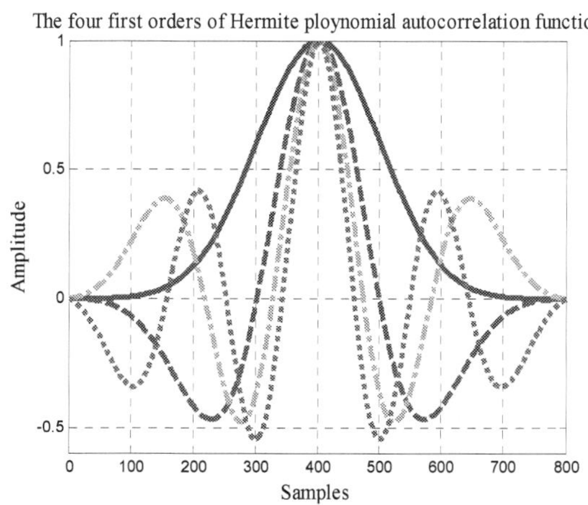

The four first orders of Hermite ploynomial autocorrelation function

Waveform	Dynamics (D)	Peak width (W)	Ratio (D/W)
H0	1	234	43.10^{-4}
H1	1	124	81.10^{-4}
H2	0,61	94	65.10^{-4}
H3	0,58	74	78.10^{-4}

Fig. 10. Comparison between autocorrelation functions of the first four orders of Hermite functions.

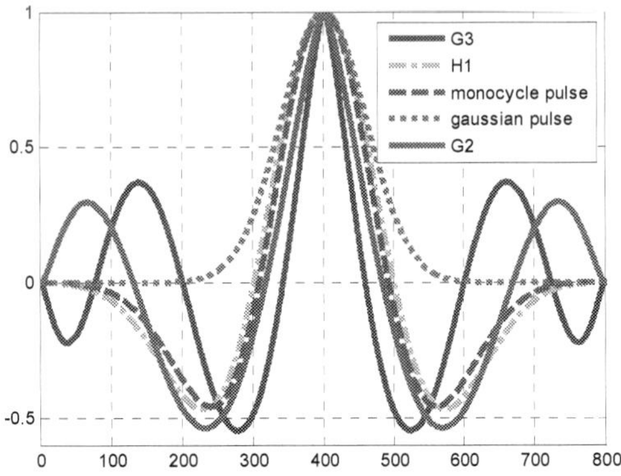

Waveform	Dynamics (D)	Peak width (W)	Ratio (D/W)
H1	1	124	81.10⁻⁴
G2	0,7	89	79.10⁻⁴
G3	0,63	62	102.10⁻⁴
Monocycle pulse	1	115	87.10⁻⁴
Gaussian pulse	1	142	70.10⁻⁴

Fig. 11. Comparison of G2, G3, H1, and Gaussian and Monocycle pulses autocorrelation functions.

In order to validate the simulation results and compare between different waveforms, some tests are performed in an anechoic chamber in order to isolate the effects of the equipment from that of noise. A UWB pulse generator lunches waveforms with a sampling frequency of 20 GS/sec. The receiver includes an amplifier and a direct sampling analyzer with 12 GHz bandwidth, 40GS/sec sampling rate and 8 bits precision. Two "Vivaldi" antennas are used for measurements, one for emission and the other for reception.

For experimental comparisons of waveforms, signals used correspond to the signals taken in the absence of target; they therefore correspond to the leakage between the antennas for transmission and reception.

The same comparisons, as in simulations, are made and experimental results confirm the simulations results.

6. UWB System for avoidance collision applications

To evaluate radar performances based on UWB technology in outdoor environment, a mock-up laboratory has been developed as described above and several tests, in real conditions, were realized in different configurations using different obstacles (Car, pedestrian, metal plate of dimension 1 squared meter, wooden plate, motorway barrier…). To determine the target position, the received signal is correlated with reference signal and position is then calculated by the threshold detection method.

UWB radar has been implemented using a monocycle pulse generator. It generates monocycle pulses of 3V peak to peak amplitude and 300 ps width. The pulse leaving the generator is transmitted using the transmitter antenna, this emitted signal of duration T is

noted s (t). After reflection on the obstacle, the received signal r(t) is correlated with the reference pulse c(t).
The received signal is given by the formula:

$$r(t) = A.s(t - \tau) + n(t) \tag{25}$$

where n(t) is the noise
τ : delay time
The most probable positions are those that make up the following expression:

$$R_{rc} = \int_0^T r(t)c(t)dt \tag{26}$$

This expression is nothing other than the correlation of the received signal with the reference. The value of this expression represents the possibility of the presence of the obstacle and maxima correspond to the most probable positions of the target. These maxima are detected using the method of threshold detection
The measurements have been performed using the monocycle pulse. In figure 12, the reference signal c(t) is presented. It is the monocycle pulse deformed successively first by the transmitter antenna and after by the receiver one, this signal is called here c(t). The pulse time sampling is 8.33 10⁻¹²s.

7. Test instrumentation

7.1 Single obstacle detection case

At a first time, a metal plate is placed at a distance of 8 metres. An example of the received signal is shown in figure 13.The first pulse in the received signal corresponds to the leakage between the transmitting/receiving antennas. The second pulse corresponds to the reflection on the obstacle. The correlation with the reference signal is presented in figure 14. The distance calculation gives a distance corresponding to 8.04 m instead of 8 meters, which is a very close fit.

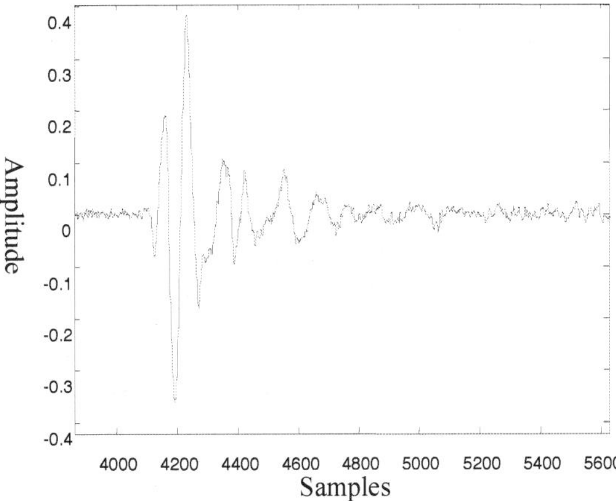

Fig. 12. Reference signal.

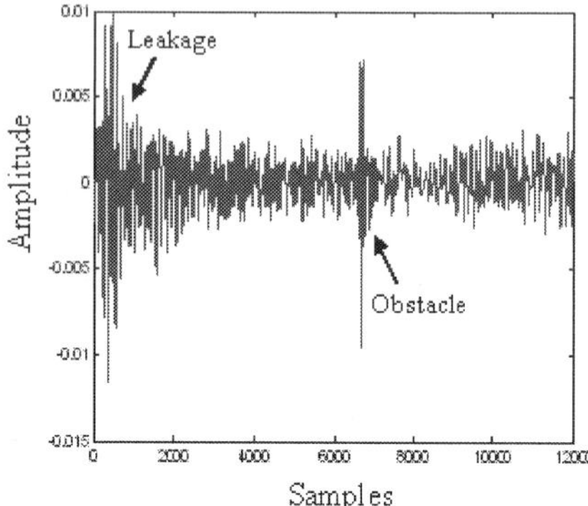

Fig. 13. The reflected echo by the metal plate placed at 8 m.

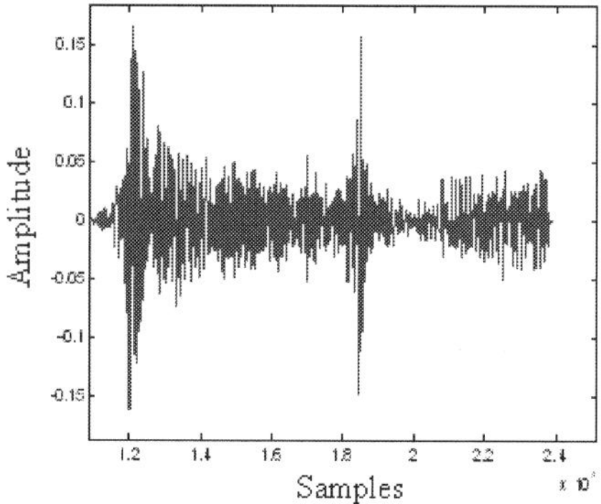

Fig. 14. Correlation signal result.
Then a car is placed at 10 metres in front of radar. The received signal and the calculation of correlation are presented respectively in figure 15 and 16.

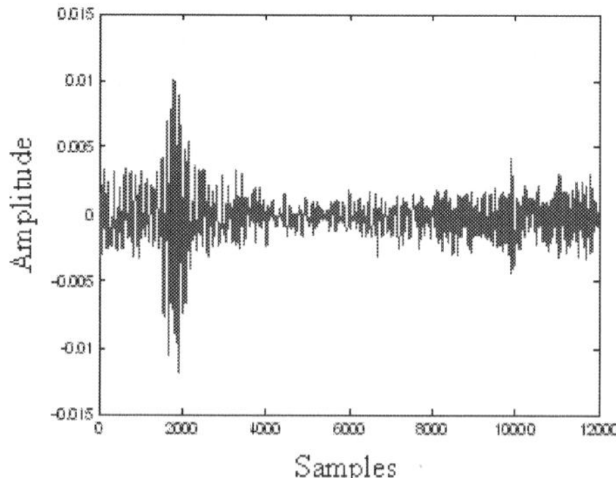

Fig. 15. Received echo reflected by car located at 10 m.

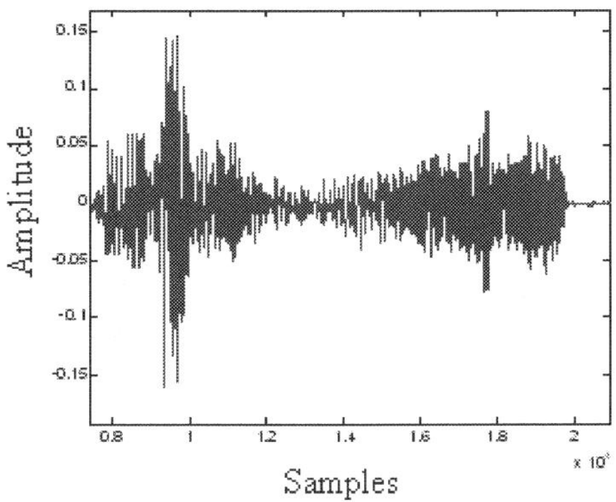

Fig. 16. Correlation signal.

The calculated distance obtained is 10.04 metres versus 10 metres theoretical; the calculated value is again very close to the real distance.

In order to place this radar in real road conditions, measurements involving a motorway barrier and a pedestrian are performed. First a motorway barrier is placed at 2.70 meters far from radar, the received signal is presented in figure 17 and the correlation is shown in figure 18.

The distance calculated using the developed radar is 2.85m, very near to the real distance of 2.70 m.

The reflected signal on pedestrian, placed at 2 meters, is presented in figure 19. The correlation signal is presented in figure 20.

The found distance is 1.87 metre for a real distance equal to 2 metres.

These previous measurements show that the developed UWB radar offers a great precision when using a single obstacle.

In addition, it is interesting to verify that the UWB radar is able to detect several obstacles at the same time with a good precision.

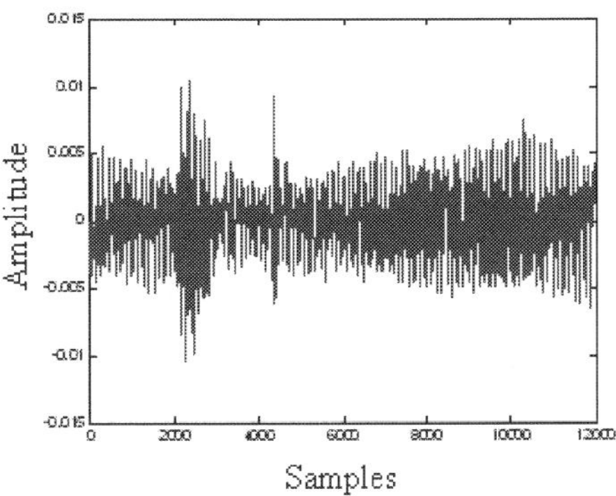

Fig. 17. Received echo in the case of a motorway barrier placed at 2.70 m.

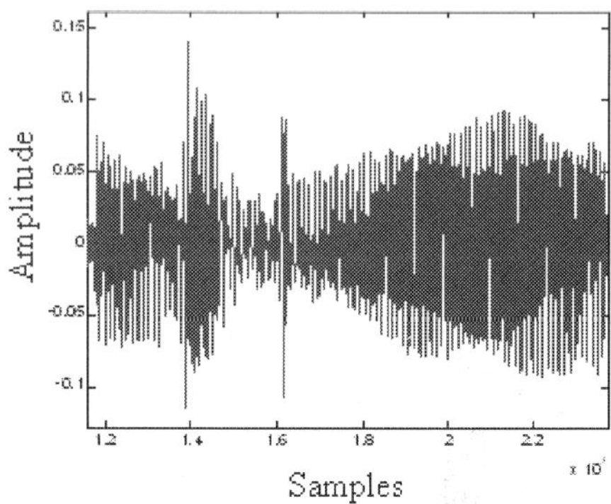

Fig. 18. Correlation signal.

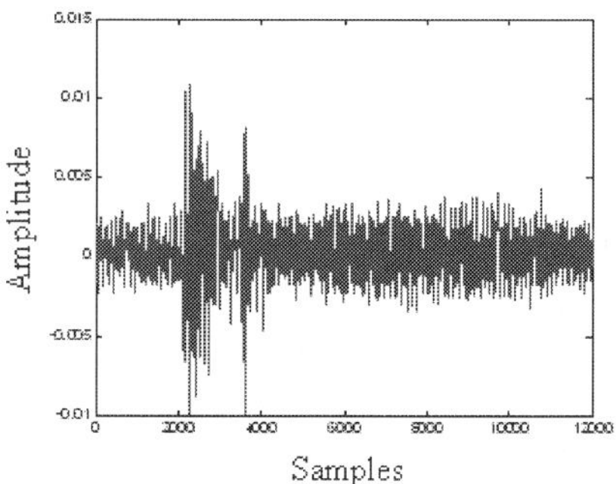

Fig. 19. Reflected signal by. pedestrian placed at 2 m.

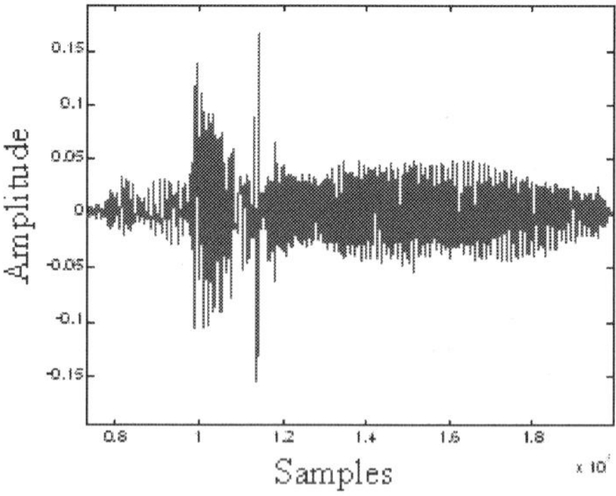

Fig. 20. Correlation signal using a pedestrian at 2 m far from radar.

7.2 The multiple obstacles detection case

A car is placed at 5 meters far from radar, a metal plate at 3 meters and a pedestrian at 1.70 meter.

On the received signal, presented in figure 21, the leakage signal between the antennas and the three echoes corresponding to the three obstacles can be distinguished easily. The corresponding correlation result is presented in figure 22.

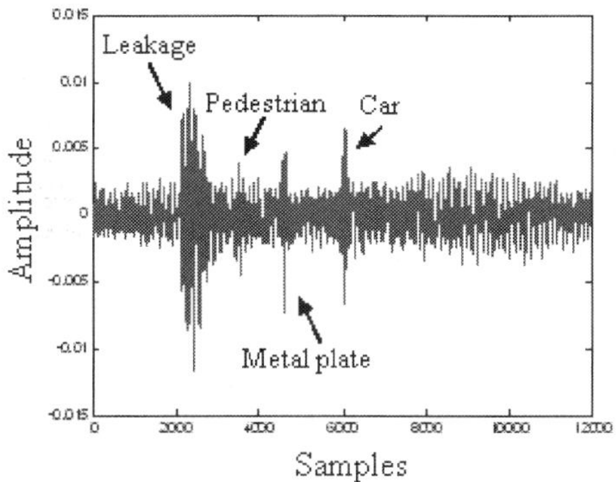

Fig. 21. Reflected echo by a metal plate, a car and a pedestrian.

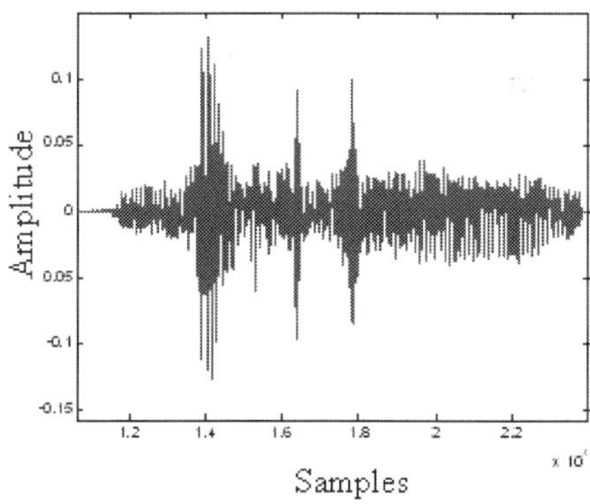

Fig. 22. Correlation signal.

A table comparing the real distances and those calculated is presented below (Table 1).

After processing the received signals, we notice that the UWB radar system offers a good precision in distance calculations.

However, we see that the correlation does not allow the automatic detection by the threshold method. In fact the correlation peak is drowned in the noise and the secondary lobes. It seems difficult to distinguish between the obstacle peaks and noise.

Obstacles	Real distance (m)	Calculated distance (m)	Precision (m)
Car	5	4.99	0.01
	10	10.04	0.04
Metal plate	3	2.88	0.12
	8	8.04	0.04
Pedestrian	1.70	1.57	0.13
Motorway barrier	2.70	2.85	0.15

Table 1. Distances comparisons

In order to solve this problem and to allow automatically detection, the High Order Statistics will be used. So to improve the detection performances of our system we look for new delay estimation algorithms. The concerned algorithms are based on the High Order Statistics. They give more performances than correlation (2nd statistics order) where the noises of the two distinct sensors are correlated. Detection is done by applying the algorithms to the two signals received by the two sensors.

In this study, to make detection we have only the received signal, our motivation to use the H.O.S (over 3) is in their capacities of the suppression of the Gaussian noise received attached to the useful signal. So, higher orders cumulants remove the Gaussian noise and keep the useful signal. These algorithms must be compatible with a use in real-time, which imposes operational limits of resource memory and in computing times, for this reason that we limit this study to the 4th order.

The expression of this algorithm is [Tugn1993]:

$$J_4(i_0) = \frac{cum_4(c(i-i_0), c(i-i_0), r(i), r(i))}{\sqrt{|cum_4((c(i), c(i))|.|cum_4(r(i), r(i))|}}$$

With:
c: is the reference signal
r: is received signal
i_0: the decision time index

$$cum_4(c(i-i_0), c(i-i_0), r(i), r(i)) = \frac{1}{N}\sum_{i=0}^{N-1} c^2(i-k).r^2(i) - \left[\frac{1}{N}\sum_{i=0}^{N-1} c(i-i_0).r(i)\right]^2 - \left[\frac{1}{N}\sum_{i=0}^{N-1} c^2(i-i_0)\right]\left[\frac{1}{N}\sum_{i=0}^{N-1} r^2(i)\right]$$

This algorithm is based on the assertion that the noise is Gaussian; its cumulants of order 4 is zero.

The result of these algorithms applied to the signal shown in figure 21 is shown in figure 23.

The result of the algorithms Tugnait 4, compared with those of the correlation for the same signal (figure 22), gives performances much better. In fact, these algorithms make well leave the obstacles peaks from the noise. For that these algorithms will be much useful for automating the device.

Ongoing works show the capacity of this radar to identify the types of obstacles detected, thanks to the radar signature. So, this radar, has the capacity to detect not only single obstacle with a great precision, but also is able to distinguish obstacles in case of several obstacles.

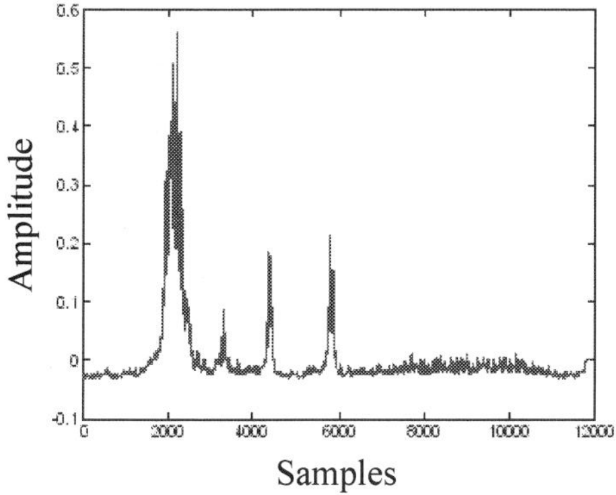

Fig. 23. The Tugnait 4 algorithms result

This study will be completed by establishing a database composed by recurrent obstacles signatures. This database could be obtained by elaborating correlation forms for different obstacles cases (metal, wood, pedestrian, wall...). Next step consists in developing signal processing algorithms able to perform automatic recognition by classification.

8. Conclusion

In this chapter, an original radar based on the new Ultra Wide Band technology is presented. This radar will be embedded on vehicles in order to increase road safety. We show the interest of this type of radar, thanks to the high degree of accuracy it offers and to its capacity to make the difference between various obstacles (cars, plates, pedestrians...). This study will be supplemented by a realization of the receiver and a method of adequate detection allowing the database classification of the obstacles, thanks to the UWB radar signature. This functionality allows better information to the driver or a follow-up of trajectory in the case of an automatic control application (Autonomous Cruise Control).

9. References

[Bar2000] Barret, T. W. "History of Ultra Wide Band (UWB) radar & communications: pioneers and inventors", Progress in Electromagnetics Symposium 2000, Cambridge, MA, July 2000

[Cham1992] Chambers,C.,S. R.Cloude and P. D. Smith (1992) "Wavelet processing of Ultra-WideBand radar signals", IEE Colloquium on Antenna and Propagation Problems of Ultra WideBand Radar, London, UK, 4p.

[Carl2000] Carlberg, T.: «Analysis of Ultra Wide Band (UWB) technology for indoor geolocation and physiological monitoring system», Master Thesis, Chalmers University of Technology (Sweden), 2000

[Dam2007] Damien Scherrer «OFCOM Infomailing N° 8», Federal Office of COMmunication, pp.11-15, September 2007

[Elb2001] Elbahhar, F. Rivenq-Menhaj, A. Rouvaen, J.M. Deloof, P. and Boukour, T. «An Ultra-Wide band system for vehicle-to-vehicle communication», Proc. ITS Conf. (on CD-ROM), Sydney, Australia, Sept. 2001.

[Elb2005] Elbahhar, F. Rivenq-Menhaj, A. Rouvaen, J.M, «Multi-user Ultra Wide Band communication system based on modified Gegenbauer and Hermite functions », Wireless Personal communications, Volume 34, Issue 3, August 2005

[ESaf2006] The 28 eSafety recommandations, *e*Safety Compendium, pp. 44-45, May 2006

[Fcc2002]http://www.fcc.gov/Bureaus/Engineering_Technology/Orders/2002/fcc02048.pdf

[Fcc2000] Federal Communication Commission. « Notice of proposed rule making », FCC 00-163, Et Docket 98-153, In the matter of revision of part 15 of the commission's rules regarding Ultra wideband transmission system, Washington, 11 may 2000.

[Lach2002] Lachlan, B.M. Ghavami, M. and Kohno, R. «Multiple Pulse Generator for Ultra-Wideband Communication using Hermite Polynomial Based Orthogonal Pulses », IEEE Conference on Ultra Wideband Systems and Technology, 2002.

[Tay1995] Taylor, James D. t al introduction to Ultra Wideband radar system, James D. Taylor Editor, CRC Press, Boca Raton, 1995, 670 p.

[Tugn1993] J. K. Tugnait. Timr delay estimation with unknow spatialy correlated gaussian noise. IEEE, Transaction on signal processing, vol. 42, n° 2, pages 549-558, Febrary 1993.

3

Wideband Noise Radar based in Phase Coded Sequences

Ana Vázquez Alejos[1], Manuel García Sánchez[1],
Iñigo Cuiñas[1] and Muhammad Dawood[2]
[1]University of Vigo
[2]New Mexico State University
[1]Spain
[2]USA

1. Introduction

At present, widely-used radar systems either use very short pulses or linear frequency modulated waveforms, and are based on either mono- or bi-static configurations. These systems suffer from having stronger sidelobes, thereby masking weaker returns from subtle changes and making it difficult to detect these changes. To overcome the problem of stronger sidelobes, various coding techniques have been proposed with varying degrees of success.

One such technique is based on the transmission of pseudorandom binary sequences (PRBS). Although, PRBS are considered a good option in terms of their autocorrelation function, these sequences are not optimal if sidelobes level is taken into account. This problem can be overcome if the transmit process is composed of complementary binary series of sequences known as Golay series [Golay, 1961; Sivaswamy, 1978; Alejos, 2005; Alejos, 2007; Alejos, 2008]. Golay codes are pairs of codes that have non-periodic autocorrelation functions with null sidelobes levels.

In the present chapter, we propose the improvement of PRBS-based noise radar sounders by using pairs of complementary binary series of Golay sequences series. We demonstrate, in qualitative and quantitative forms, the improvement reached by employing Golay sequences.

The larger dynamic range of the sounder and the better precision in wideband parameter estimation are the main benefits due to the use of Golay sequences. This is important when dealing with large attenuation, as at millimeter frequency bands.

Different schemes are explained in order to implement this kind of radar sounders, as well as the measurement procedure is detailed. The processing gain is also explained. Measurements, using both kinds of sequences, have been performed in actual scenarios. Channel functions and parameters have been calculated and compared. Finally the Fleury's limit [Fleury, 1996] is introduced as a formal way of checking the availability of the obtained results.

2. Fundamentals of wideband noise radar

A generic modulation technique habitually used in radar systems is the one known as noise modulated radar. This technique offers large number of advantages to the radar systems designers due to its robustness to the interferences. Nevertheless, until a few years ago it was very difficult to find practical implementations of these systems. One of the main problems with them is the ambiguity zone and the presence of sidelobes.

The development of the random radar signals generation techniques has impelled the development of systems based on this type of noise modulation. The ultrawideband random noise radar technique has received special importance. It is based on the transmission of a UWB signal, such as the Gaussian waveforms.

Other implementations of the ultrawideband random noise technique use waveforms based on pseudorandom binary sequences with maximum length, also named PRBS sequences (Pseudo Random Binary Sequences) or M-sequences, where M denotes the transmitted sequence length in bits. This technique, known as M-sequence radar, offers great advantages related to the hi-resolution of targets and its large immunity to detection in hostile surroundings but also against artificially caused or natural interferences.

Nevertheless, the radar technique by transmission of M-sequences presents serious limitation in the dynamic range offered, which goes bound to length M of the transmitted sequence. It makes difficult the detection of echoes, which in some scenes with large attenuation values will be confused with noise. These PRBS sequences present in addition a serious problem of large power sidelobes presence, which aggravates the problem of false echoes detection habitual in the applications radar.

The level amplitude of these sidelobes is directly proportional to length M of the sequence, so that if this length is increased with the purpose of improving the dynamic range, the sidelobes amplitude level will be also increased in addition. As well, an increase in the length of the transmitted sequence reduces the speed of target detection, limiting the answer speed of the radar device.

In this chapter it is considered the application to the noise modulation techniques of a type of pseudorandom binary sequences that contributes a solution to the problematic related to the M-length of the transmitted sequence. The used pseudorandom sequences are known as Golay series and consist of a pair of pseudorandom binary sequences with complementary phase properties. The autocorrelation properties of a pair of Golay sequences produces the sidelobes problem disappearance and in addition they cause that the dynamic range of the system is double to which would correspond to a PRBS sequence with the same length. This allows the use of smaller length sequences to increase the target detection speed.

3. Coded sequences

In this section, we introduce some of the more known sequences used for radio channel sounding and their main features. All sequences described present autocorrelation properties that try to fit the behaviour of white noise. So they are usually named as like-noise sequences.

3.1 Fundamentals

The radio channel impulse response is obtained by transmission of a signal which autocorrelation equals a delta function, like white noise autocorrelation function. As replication of a white noise signal for correlation at the receiver end is difficult, pseudo noise (PN) sequences are used instead. PN sequences are deterministic waveforms with a noise-like behaviour, easily generated by using linear feedback shift registers. They exhibit good autocorrelation properties and high spectral efficiency [Sarwate, 1980; Cruselles, 1996]. The best known examples of such waveforms are maximal length pseudorandom binary sequences (m-sequences or PRBS).

For a maximal length code, with chip period T_C and length M, its autocorrelation can be expressed by (1) and (2):

$$p(t) = \sum_{n=1}^{M} a_n c(t - nT_c) \qquad (1)$$

$$R_p(t) = \begin{cases} -(t+T_c)\dfrac{(M^2+1)}{MT_c} + M & 0 \le t \le T_c \\[2ex] (t+T_c)\dfrac{(M^2+1)}{MT_c} - \dfrac{1}{M} & -T_c \le t \le 0 \\[2ex] -\dfrac{1}{M} & \textit{otherwise} \end{cases} \qquad (2)$$

In (1) $p(t)$ denotes a train of periodic pulses with period T_C and amplitudes $a_n=\pm1$, $c(t)$ is the basic pulse with time duration T_C and amplitude unitary.

An adequate sequence for channel sounding, which will reassure efficient path delay recognition, should have an autocorrelation function with a narrow main lobe and low sidelobes. Traditionally, PRBS are considered a good option for channel sounding. However, if sidelobe level is taken into account, these sequences are not optimal. The sidelobe level for the correlation of PRBS is constant and equals -1/M, or -1 if normalization is not applied. The shape of the autocorrelation function can be seen in Fig. 1, for a PRBS with length $M = 7$ and chip period $T_C = 1$.

There are binary phase codes with non-periodic autocorrelation functions that have minimum sidelobe levels. **Barker** codes [Nathanson, 1999] are the binary phase codes which non-periodic autocorrelation exhibits minimum possible sidelobe levels. The peak of the sidelobes at their autocorrelation functions are all less than or equal to $1/M$, where M is the code length.

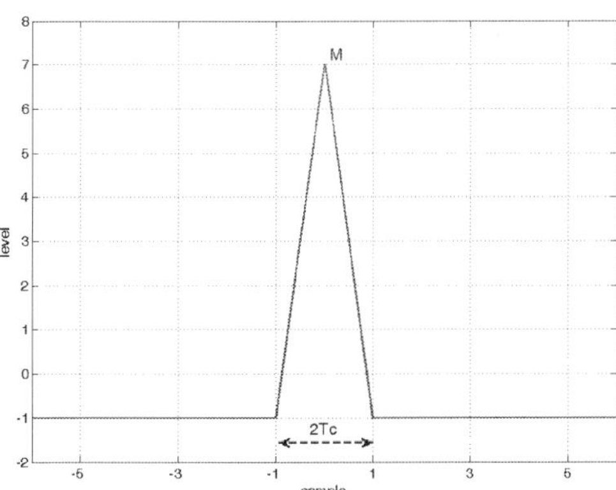

Fig. 1. Autocorrelation of a m-sequence. $M = 7$, $T_C = 1$

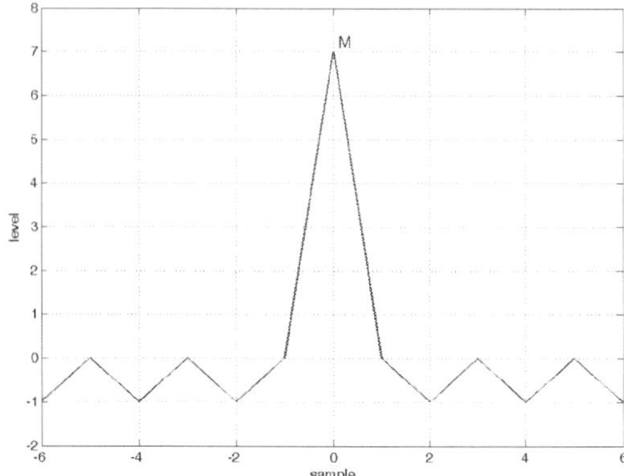

Fig. 2. Autocorrelation of a Barker sequence. $M = 7$, $T_C = 1$

The interesting properties of the Barker codes are that, firstly, the sidelobe structures of their autocorrelation function contain, theoretically, the minimum possible energy; and secondly, this energy is uniformly distributed among all the sidelobes. Barker codes are called perfect codes. An example of the autocorrelation function of a Barker code with $M = 7$ is shown in Fig. 2.

The disadvantage of Barker codes is that there exist no more than eleven known sequences with the longest one only having 13 elements. The relation of all known Barker codes, according to [Cohen, 1987], is given in Table I. In the second column, +/- represent two different code elements.

An important kind of binary phase codes are complementary codes, also known as **Golay** codes [Golay, 1961; Sivaswamy, 1978]. Complementary codes are a pair of equal length sequences that have the following property: when their autocorrelation functions are algebraically added, their sidelobes cancel. Moreover, the correlation peak is enlarged by the addition.

Length of the code	*Code elements*
1	+
2	+ - + +
3	+ + - + - +
4	+ + - + + + + -
5	+ + + - +
7	+ + + - - + -
11	+ + + - - - + - - + -
13	+ + + + + - - + + - + - +

Table 1. Barker codes known according to [Cohen, 1987].

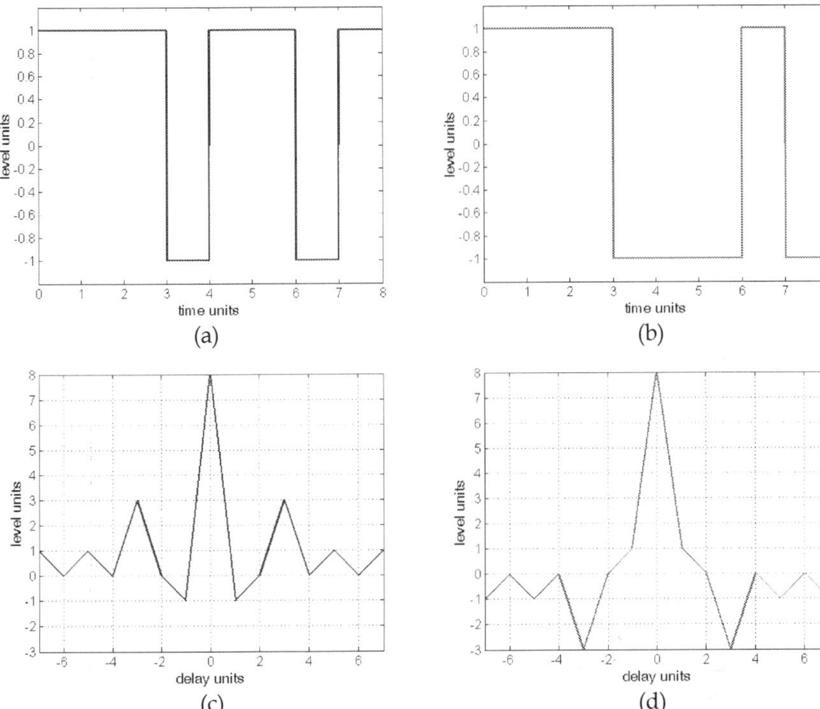

Fig. 3. (a) Complementary Code Golay A; (b) Complementary Code Golay B; (c) Autocorrelation of Code Golay A; (d) Autocorrelation of Code Golay B.

An example of these codes can be seen in Fig. 3(a-d), altogether with their autocorrelation functions. As shown in Fig. 4, the addition of both autocorrelation functions results in total cancellation of the sidelobes, because the correlation sidelobes complement each other. This correlation property of complementary sequences is used in communications systems, as spreading sequences, to allow several communication channels to use simultaneously the same frequency band [Golay, 1961; Díaz, 2002; Wong, 2003], for data encoding [Chase, 1976; Weng, 2000], and even for PAPR (Peak-to-Average-Power-Ratio) reduction [Popovic, 1999; Gil, 2002] in CDMA systems.

3.2 Comparison of complementary series and pseudonoise sequences

Complementary series present certain characteristics that do them more adequate than PRBS for some applications [Budisin, 1992]. These benefits are found only when Golay sequences are used in pairs, because the advantages arise from the complementary properties. Their individual properties are rarely considered. Some of the most important characteristics are:

i.   *Autocorrelation properties*: the sum of the autocorrelation of the sequences is exactly zero everywhere except at the origin. Therefore, they exhibit better performance than feedback shift register sequences, including PRBS.

SEQUENCE LENGTH	GOLAY SEQUENCE	PERIODIC PRBS
4/3	4	1
8/7	24	2
16/15	192	2
32/31	1920	6
64/63	23040	6
128/127	322560	18
256/255	5160960	16
512/511	92897280	48
1024/1023	1857945600	60
2048/2047	40874803200	176
4096/4095	980995276800	144

Table 2. Number of different PRBS and Golay sequences

ii.   *Sequence length*: the length of Golay sequences is 2^N, whereas the length of pseudorandom sequences is $2^N - 1$. Since power-of-two length is simpler to implement, complementary sequences are preferable to PN sequences.

iii. *Number of different sequences*: the number of different Golay sequences is larger than the number of different PRBS. Moreover, this number increases more quickly with sequence length, as can be seen in Table 2.

iv.   *Spectral properties*: spectral peaks of Golay sequence are not larger than 3 dB over the average of the spectrum. Although the spectrum of PN sequences is theoretically flat, the actual spectrum of the PN sequence is far from constant. The peak values of the spectrum of PN sequences often exceed the average value by more than 3 dB.

v.   *Merit Factor*: defined by Golay [Golay, 1983] as a qualitative measure to facilitate analytical treatment. It is closely related to the signal-to-noise ratio, and is defined as (3):

$$F = \frac{L^2}{2E} \tag{3}$$

where L is the sequence length, and E can be seen as an energy term because it is the quadratic sum of all autocorrelation functions (4):

$$E = \sum_{k=1}^{L-1} R_k \quad with \quad R_k = \sum_{i=1}^{L-k} s_i \cdot s_{i+k} \tag{4}$$

The merit factor for a Golay pair of codes doubles the merit factor of a PRBS sequence, and therefore the signal-to-noise ratio will be 3dB higher for the Golay case. This is a very interesting characteristic when dealing with impulse responses with low power echoes, as is the case for measurements in millimeter wave frequency band, where the contributions resulting from first or second-order reflections [García, 2003; Hammoudeh, 2002] may be difficult to detect. Due to the increased dynamic range, weak multipath can be detected.

3.3 Benefits of employing phase coded sequences

The autocorrelation function of a noise-like M-length sequence, with bit period Tc, presents a peak of amplitude M and duration $2 \cdot T_C$. Both PRBS and Golay sequences of length 2^{13} bits with chip period $T_C = 20\ ns$ have been computer generated, using a general purpose mathematical software, and their theoretical correlation functions have been calculated and compared.

The correlation functions associated to each sequence are shown in Fig. 5. In both cases, a main peak can be found. But for the Golay sequence correlation, the main peak has double amplitude and the noise level is lower. This indicates a theoretical improvement in dynamic range of 3dB.

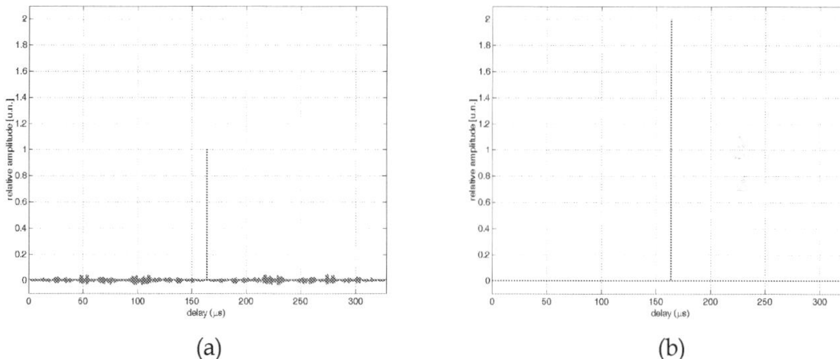

(a)   (b)

Fig. 5. Example of correlation functions of (a) PRBS with $M_{PRBS} = 2^{13}\text{-}1$, $T_C = 1$ and (b) Golay sequences with $M_{GOLAY} = 2^{13}$, $T_C = 1$.

These properties are kept when actual signals, instead of mathematical sequences, are generated using hardware generators. Other different phenomena may be observed if actual signals are used, main of them is the presence of additional peaks corresponding to the sidelobes problem.

Two signals corresponding to the previous theoretical sequences were generated by a general purpose signal generator and captured with a digital oscilloscope. Later, the correlation functions were calculated off-line. The dependency of the sidelobe level with the ratio E_b/N_0 has tried to be determined.

So, firstly, actual signals were obtained generating both codes with a ratio E_b/N_0 of 20dB and the related results are presented in Fig. 6. PRBS and Golay sequences were generated again, now with a ratio E_b/N_0 of 5dB. Results are shown in Fig. 7. The E_b/N_0 ratio is a configurable parameter of the generator. E_b is the energy in a single chip of the considered code, and N_0 is the power noise spectral density level.

A degradation of dynamic range when E_b/N_0 is reduced can be appreciated for both codes, but it is larger for the PRBS case. We can observe that autocorrelation peak of Golay codes always presents double amplitude independently of the ratio E_b/N_0.

It may be also seen spurious correlation peaks well above the noise level in the PRBS case. Then, we can conclude that theoretical dynamic range improvement is 3dB, but in actual cases, this improvement is larger due to the noise cancellation associated to Golay codes.

Definition given for dynamic range can be used to measure the improvement in the dynamic range. For a ratio E_b/N_0 of 20dB, the dynamic range is 13dB for PRBS and 23dB for Golay. The improvement achieved is 10dB. For a ratio E_b/N_0 of 5dB, the dynamic range values are 5dB and 19dB, respectively. The improvement is 14dB. We can conclude that the improvement achieved is the sum of the 3dB correlation peak increment and the correlation sidelobes reduction.

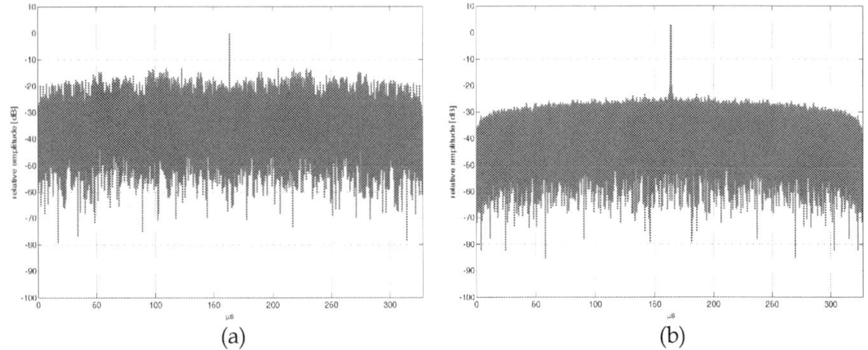

(a)   (b)

Fig. 6. Example of correlation functions of (a) captured PRBS signal with $M_{PRBS} = 2^{13}-1$, $T_C =$ 20 ns, E_b/N_0=20dB and (b) Golay captured signals with $M_{GOLAY} = 2^{13}$, $T_C = 20$ ns, E_b/N_0=20dB.

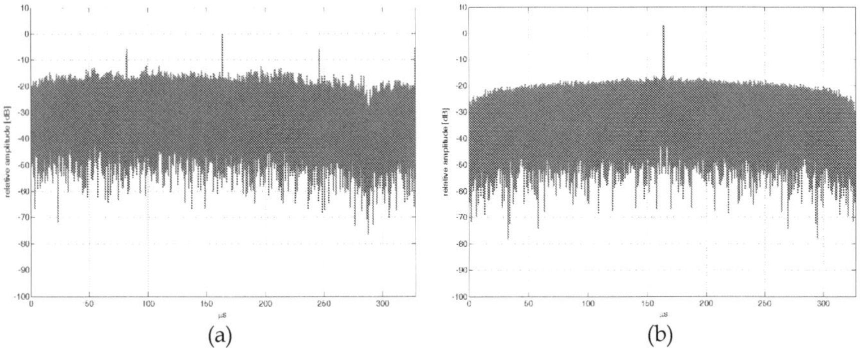

(a)   (b)

Fig. 7. Example of correlation functions of (a) captured PRBS signal with $M_{PRBS} = 2^{13}-1$, $T_C =$ 20 ns, E_b/N_0=5dB and (b) Golay captured signals with $M_{GOLAY} = 2^{13}$, $T_C = 20$ ns, E_b/N_0=5dB.

3.4 ISL, SSL and PSL comparison

Some software simulations were performed in Matlab to illustrate the robustness against noise interferences of PRBS and Golay sequences [Alejos, 2008]. Their capabilities will be measured in terms of peak-to-sidelobe, secondary-sidelobe and integrated-sidelobe levels, PSL, SSL and ISL respectively. This will show also better understanding of the advantage provided by reduced sidelobe levels.

Two length 2048 Golay sequences and one length 4096 PRBS sequence were software generated. White Random Gaussian noise was added to the sequences with E_b/N_0 ratio levels in the range of -50dB to 50dB. The correlation functions between noisy and original sequences were obtained. Figure 8 shows a comparison for PSL, SSL and ISL parameters obtained for PRBS and Golay sequences in presence of the mentioned E_b/N_0 ratio level range. None average has been performed.

From this figure, it can be noticed that, for a ratio E_b/N_0 larger than 3dB, the PSL levels of Golay and PRBS sequence are identical. It can be observed that SSL level of Golay sequences is up to 50dB smaller when compared to PRBS sequence. It is also clear that as the E_b/N_0 ratio increases, the SSL level difference between Golay and PRBS sequence decreases.

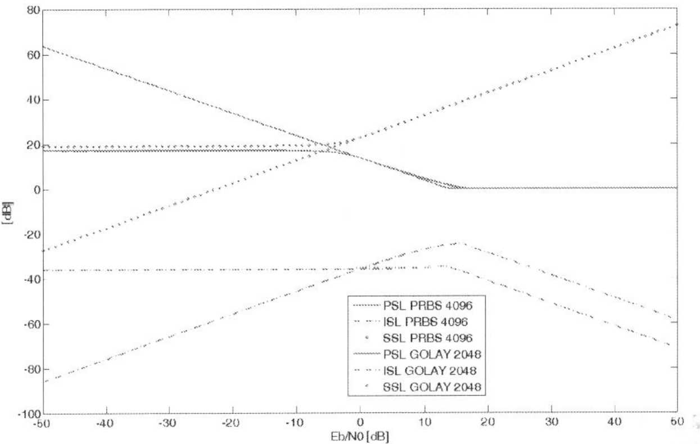

Fig. 8. PSL, SSL and ISL comparison for 2048-Golay and 4096-PRBS sequences.

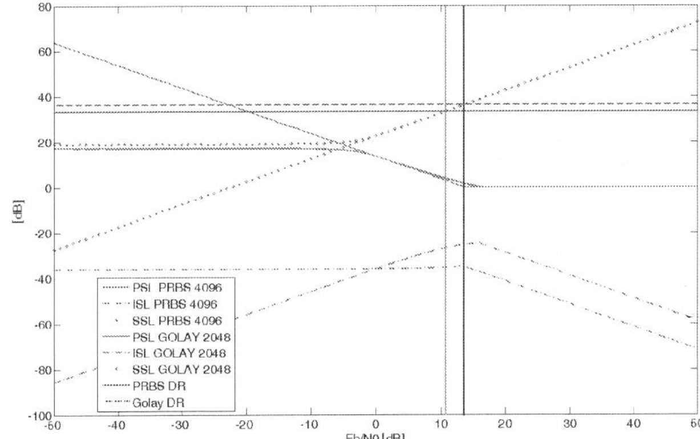

Fig. 9. PSL, SSL and ISL comparison including ideal dynamic ranges.

The figure also shows ISL level vs. E_b/N_0 ratio for PRBS and Golay sequences. It can be observed that the ISL level for PRBS sequence is almost 50dB larger than the ISL level for Golay case. Moreover, as the E_b/N_0 ratio increases the ISL level difference between Golay and PRBS sequences decreases. Whereas as the E_b/N_0 ratio level is increased to 16dB, the Golay case shows slightly larger ISL level than PRBS sequence. At this same point the PSL level is zero. It is produced when the AWGN power is larger than the sequence power, so the noise masks the signal.

In plots of Figure 9, we have included the ideal dynamic ranges corresponding to both codes. We can notice that the cross point between the DR and SSL lines for the Golay case is 3dB larger than for the PRBS case.

4. Estimation of improvement achieved by using Golay codes in noise radar

A controlled experiment of propagation was conducted in order to estimate the improvement achieved in the estimation of channel parameters. Two signals, one based in

PRBS and other resulting from Golay codes, were generated with a general purpose pattern generator, with chip period $T_C = 20$ *ns*. The signals were generated with two different values of E_b/N_0 ratio, which are 20dB and 5dB. Signals were captured with a digital oscilloscope. In order to reduce the effect of system noise, one hundred captures are taken for each signal and then averaged. These measured signals were correlated off-line with a replica of the original one.

The obtained impulse response was used to estimate the *mean delay*, τ_{mean}, and *rms delay*, τ_{rms}. Applying a Fourier transform to the averaged power-delay profile, an estimation of the frequency correlation function and the *coherence bandwidth*, CB, were calculated.

From each code, Golay and PRBS, six signals were created. The first one consisted in a simple sequence corresponding to single path propagation, whereas the other five include multipath components. In this way the second signal includes one echo; the third, two echoes; and successively until the sixth signal, which includes five multipath components.

The time delays and the relative levels and phases of the multipath components were established by setting the function generator properly. It was decided to assign to each new multipath component a level 3dB lower to the prior component. No phase distortion was added. The distribution of the echoes and their amplitudes can be seen in Fig. 10. This graphic represents the ideal impulse response for a signal with five multipath components.

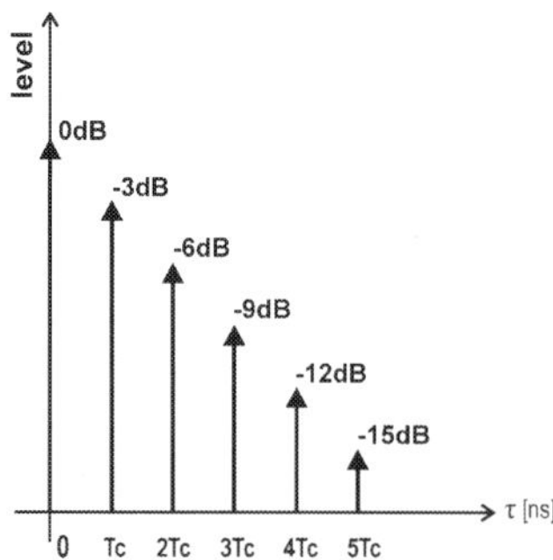

Fig. 10. Impulse response generated for improvement estimation with multipath components, $T_C = 20$ ns.

The error in the estimation of the parameters has been measured in terms of a mean quadratic error, *MSE*, and relative error $e \cdot 100\%$, according to the following expressions (5) to (8):

$$MSE\left(\tau_{mean/rms}\right)= \sqrt{\sum_{i=1}^{n=5 e\, cos}\left(\tau_{mean/rmsPRBS/Golayi} - \tau_{mean/rmsIDEALi}\right)^2 \Big/ n\cdot(n-1)} \qquad (5)$$

$$e \cdot 100\%\left(\tau_{mean/rms}\right)= MSE\,\tau_{mean/rmsPRBS/Golayi} \Big/ <\tau_{mean/rmsIDEALi}> \qquad (6)$$

$$MSE\left(CB_\alpha\right)= \sqrt{\sum_{i=1}^{n=5 e\, cos}\left(CB_{\alpha PRBS/Golayi} - CB_{\alpha IDEALi}\right)^2 \Big/ n\cdot(n-1)} \qquad (7)$$

$$e \cdot 100\%\left(CB_\alpha\right)= MSE\left(CB_\alpha\right)\Big/ <CB_{\alpha IDEALi}> \qquad (8)$$

Values obtained for errors MSE and $e \cdot 100\%$ are shown in Tables 3 and 4. The errors achieved are lower for the case of Golay codes, even when the ratio E_b/N_0 is as high as 20dB. From these values, we can predict a better performance of Golay codes for the measurements to be made in actual scenarios.

Sequence	Error	E_b/N_0 =20dB		E_b/N_0 =5dB	
		mean delay	rms delay	mean delay	rms delay
PRBS	MSE [ns]	1.2	1.6	1.4	2.4
	$e \cdot 100(\%)$	8.9%	11.5%	10.3%	15.4%
GOLAY	MSE [ns]	1.0	1.4	1.1	1.9
	$e \cdot 100(\%)$	7.5%	9.9%	8.2%	13.6%

Table 3. Mean Square Error, in *ns*, and Relative Error *(%)*, for time parameters estimations resulting from the experiment described in section 4, obtained with a ratio E_b/N_0 of 20dB and 5dB.

Sequence	Error	E_b/N_0 =20dB		E_b/N_0 =5dB	
		$CB_{0.9}$	$CB_{0.5}$	$CB_{0.9}$	$CB_{0.5}$
PRBS	MSE [MHz]	0.8	1.6	0.7	1.7
	$e \cdot 100(\%)$	17.6%	13.6%	18.6%	14%
GOLAY	MSE [MHz]	0.5	1.6	0.6	1.6
	$e \cdot 100(\%)$	13.7%	13.2%	16.4%	13.3%

Table 4. Mean Square Error, in *MHz*, and Relative Error *(%)*, for Coherence Bandwidth estimations resulting from the experiment described in section 4, obtained with a ratio E_b/N_0 of 20dB and 5dB.

5. Channel sounding procedure based in Golay series

Two important questions must be considered to employ Golay codes for channel sounding. The first of these questions is relative to the facility of generation of Golay sequences. Marcell Golay [Golay, 1961] developed a method to generate complementary pairs of codes. These codes have the following general properties:

i.　The number of pairs of similar elements with a given separation in one series is equal to the number of pairs of dissimilar elements with the same separation in the complementary series.

ii.　The length of two complementary sequences is the same.

iii.　Two complementary series are interchangeable.

iv.　The order of the elements of either or both of a pair of complementary series may be reversed.

In order to generate Golay sequences $\{a_i\}$ and $\{b_i\}$, the properties enunciated lead us to an iterative algorithm. It starts with the Golay pair $\{a_i\}_1 = \{+1, +1\}$, $\{b_i\}_1 = \{+1,-1\}$, and the following calculation is repeated recursively:

$$\{a_i\}_{m+1} = \{\{a_i\}_m | \{b_i\}_m\}$$
$$\{b_i\}_{m+1} = \{\{a_i\}_m | \{-b_i\}_m\}$$

$$(9)$$

where $|$ denotes sequence concatenation.

The second question to take into account is the procedure to be applied to obtain the channel impulse response. A method to measure impulse response of time invariant acoustic transducers and devices, but not for the time varying radio channel, has been proposed in [Foster86, Braun96] based on the use of Golay codes. In those occasions, the authors intended to employ the benefits of autocorrelation function of a pair of Golay codes to cancel a well known problem in magnetic systems. Along this chapter we present the benefits of the application of Golay codes in radio channel characterization to obtain more precise estimations of main parameters such as delay spread and coherence bandwidth. This method consists in three steps:

i.   Probe the channel with the first code, correlating the result with that code. This yields the desired response convolved with the first code.
ii.   Repeat the measurement with the second code, correlating the result whit that code and obtaining the response convolved with the second code.
iii. Add the correlations of the two codes to obtain the desired sidelobe-free channel impulse response.

For time variant radio channel sounding a variation of this method can be considered. A sequence containing in the first half, the first Golay code, and in a second part the complementary code is built. Between both parts a binary pattern is introduced to facilitate the sequence synchronization at reception, but this is not absolutely necessary. In any case the two sequences can be identified and separated with an adequate post processing, so each one can be correlated with its respective replica. Finally, the addition of the two correlation functions provides the channel impulse response.

The total sequence containing the two Golay codes, and the optional synchronization pattern, will present a longer duration that each one of the complementary codes. This increases the time required for the measurement and, consequently could reduce the maximum Doppler shift that can be measured or the maximum vehicle speed that can be used.

6. Processing gain

In the practical cases, we have to take care of one aspect which affects to the amplitude of the autocorrelation. This factor is the sample rate. The autocorrelation pertaining to the ideal sequences takes one sample per bit. But, in sampled sequences, there will be always more than one sample per bit, since we must have a sample rate, at least, equal to the Nyquist frequency to avoid the aliasing. This causes that we have more than two samples per bit and the amplitude of the autocorrelation of a sampled sequence will be double of the ideal. It takes place a processing gain due to the sampling process. It appears a factor that multiplies the autocorrelation peak value that is related to the sample rate. We have named this factor *'processing gain'*.

The value of the autocorrelation peak for a case with processing gain, M', can then be written as:

$$\left.\begin{array}{l} F_S \geq F_{Nyquist} = 2 \cdot \dfrac{1}{T_c} \\ M' = F_S \cdot T_C \cdot \left(2^n - 1\right) \end{array}\right\} \Rightarrow M' \geq 2 \cdot \left(2^n - 1\right) \Rightarrow M' \geq 2 \cdot M \tag{9}$$

If we employ a sample rate s times superior to the minimum necessary to avoid aliasing, $F_{Nyquist}$, we will increase the gain factor according to (10):

$$s = \frac{F_S}{F_{Nyquist}} \Rightarrow gain = 2 \cdot s \Rightarrow M' = gain \cdot M \qquad (10)$$

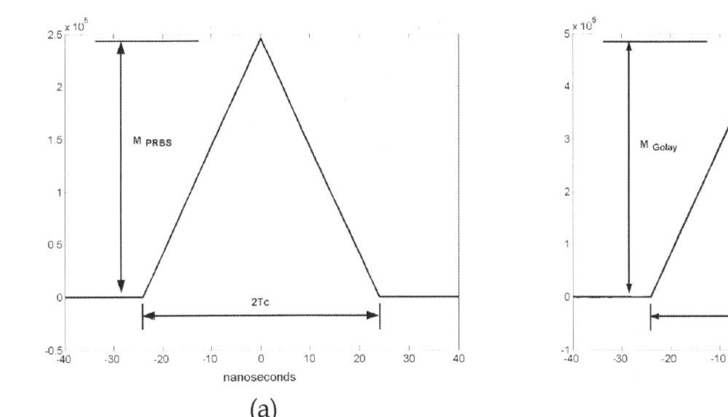

(a)   (b)

Fig. 11. Example of processing gain due to oversampling: (a) ideal PRBS signal with $M_{PRBS} = 2^{13}\text{-}1$, $T_C = 24$ ns, $Fs=1.25GS/s$, and (b) ideal Golay signals with $M_{GOLAY} = 2^{13}$, $T_C = 24$ ns, $Fs=1.25GS/s$.

In our particular case, the sample rate was 1.25GS/s, so we wait to obtain an autocorrelation peak value $M' = 2 \cdot 15 \cdot M$, where $M=2^{13}\text{-}1=8191$ for the PN sequence and $M=2^{13}=8192$ for the Golay codes. In below Fig. 11, we can see that, for the Golay case, this peak amplitude is double in relation to the expected value. We can give then the next expression:

$$M_{GOLAY} = 2 \cdot (2 \cdot s \cdot M) = 2 \cdot M_{PRBS} \qquad where: \quad M = 2^n \qquad (11)$$

Giving values to the variables of equation (11) we obtain

$$M_{GOLAY} = 2 \cdot (2 \cdot 15 \cdot 2^{13}) = 491520 \qquad M_{PRBS} = (2 \cdot 15 \cdot 2^{13}) = 245760$$

We can effectively test that the maximum of the peak autocorrelation is double in the Golay sequence case.

7. Noise radar set-up by using Golay codes

In the previous sections, we have analyzed the improvements introduced by Golay sequences. The most important among them is the double gain in the autocorrelation function. This gain is achieved with no need of changes in the hardware structure of classical PN radar sounders with respect to the hardware structure of a PRBS-based sounder.

Next section presents the adaptation of the radio channel sounder built described in [Alejos, 2005; Alejos, 2007] in order to obtain a radar sounder [Alejos, 2008]. The general measurement procedure has been detailed in previous section 5, but it will experience slight variations according to the selected type of hardware implementation. This question is analyzed in section 7.2.

7.1 Hardware sounder set-up

The wideband radar by transmission of waveforms based on series of complementary phase sequences consists in the transmission of a pair of pseudorandom complementary phase sequences or Golay sequences. These sequences are digitally generated and they are modulated transmitted. In their reception and later processing, the phase component is also considered and not only the envelope of the received signal.

For this purpose, different receiving schemes can be adopted, all focused to avoid the loss of received signal phase information. This receiver scheme, based on module and phase, is not used in UWB radars where the receiving scheme is centred in an envelope detector.

In the receiver end it is included, after the radio frequency stage, a received signal acquisition element based on an analogue-digital conversion. By means of this element, the received signal is sampled and the resulting values are stored and processed.

The radiating elements consist of antennas non-distorting the pulses that conforms the transmitted signal, arranging one in the transmitter and another one in the receiver. Three main types of antennas can be considered: butterfly, Vivaldi and spiral antennas.

The operation principle of the system is the following one. A pair of complementary sequences or Golay sequences is generated, of the wished length and binary rate. The first sequence of the pair is modulated, amplified and transmitted with the appropriate antenna. A generic transmitter scheme is shown in Figure 12.

This scheme is made up of a transmission carrier (b), a pseudorandom sequence generator(c), a mixer/modulator (d), a bandpass filter (e), an amplifier (f), and the radiation element (g).

In the receiver end, a heterodyne or superheterodyne detection is carried out by means of a baseband or zero downconversion. Any of the two receiving techniques can be combined with an I/Q demodulation.

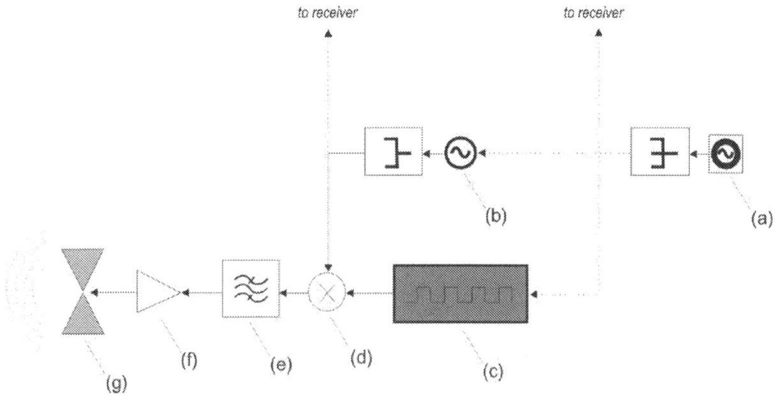

Fig. 12. Generic transmitter scheme for noise radar sounder

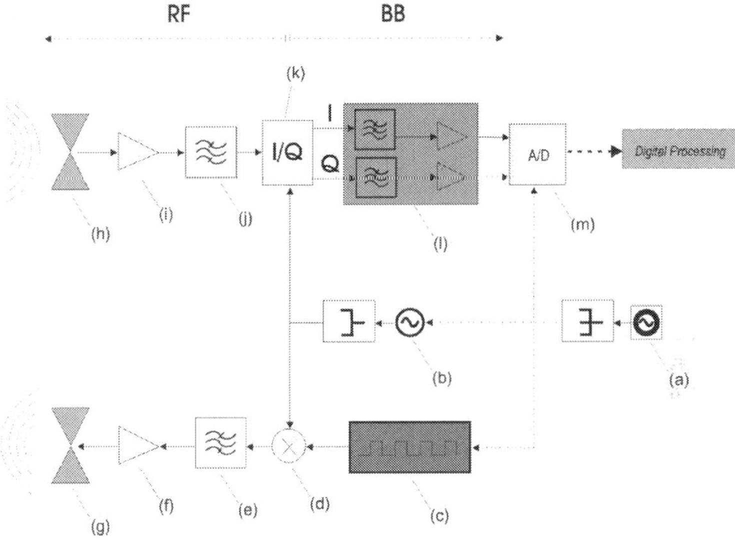

Fig. 13. Receiver scheme for noise radar sounder: I/Q superheterodyne demodulation to zero intermediate frequency

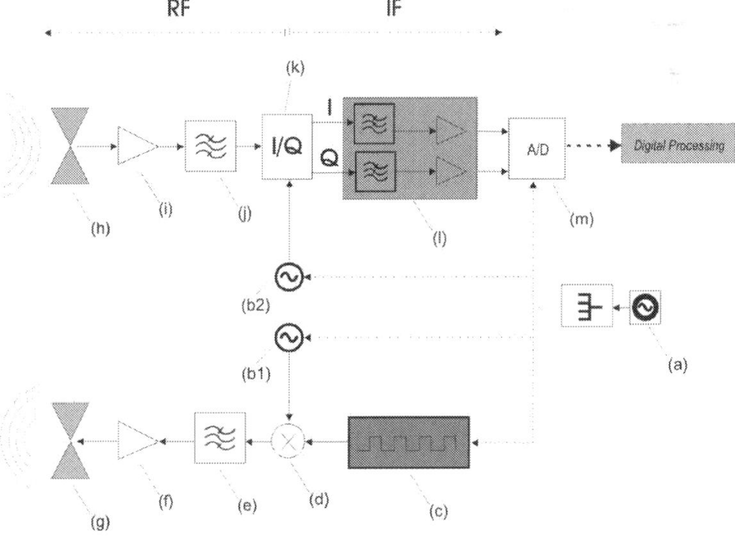

Fig. 14. Receiver scheme for noise radar sounder: I/Q superheterodyne demodulation to non-zero intermediate frequency

Four different schemes are proposed to implement the system, being different by the reception stage (Figures 13-15). All have in common the transmission stage and the used antennas. The diverse schemes can be implemented by hardware or programmable logic of type FPGA (Field Programmable Gate Array) or DSP (Digital Signal Processor).

In Figures 13-15, the transmission stage is included to emphasize the common elements shared by both ends of the radar. The main common element between transmitter and receiver is the phase reference (a), composed generally by a high stability clock, such as a 10 or 100MHz Rubidium oscillator.

The schemes of figures 13-15 have in common one first stage of amplification (f) and filtrate (h). Next in the scheme of Figure 13 is an I/Q heterodyne demodulation (h) to baseband by using the same transmitter carrier (b). The resulting signals are baseband and each one is introduced in a channel of the analogue-digital converter (m), in order to be sampled, stored and processed. Previously to the digital stage, it can be arranged an amplification and lowpass filtered stage (l).

In the scheme of Figure 13, an I/Q superheterodyne demodulation (k) with a downconversion to a non-zero intermediate frequency is performed. The outcoming signals are passband and each one of them is introduced in a channel of the analogue-digital converter (m) for its sampling, storing and processing. An amplification and passband filtered stage (l) can also be placed previously to the acquisition stage.

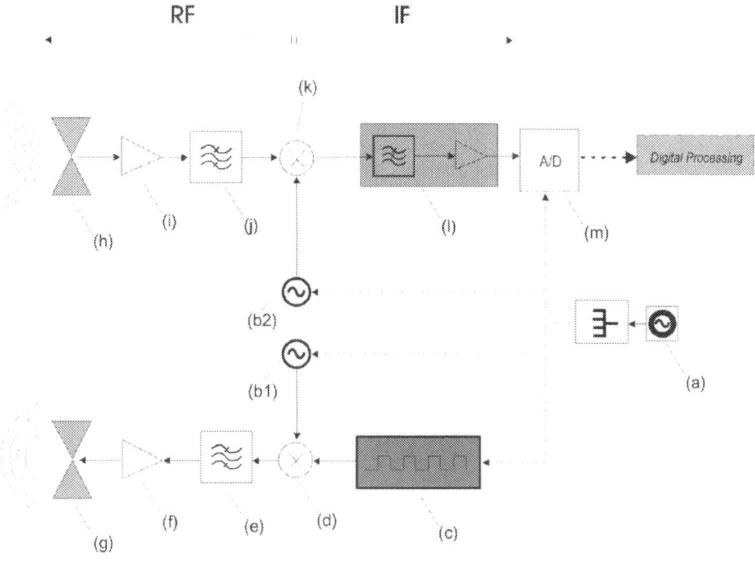

Fig. 15. Receiver scheme for noise radar sounder: superheterodyne demodulation to non-zero intermediate frequency

In scheme of Figure 15 a superheterodyne mixer (k) with downconversion to non-zero intermediate frequency is used. The resulting signal is passband and it is the input to a channel of the analogue-digital converter (m), to be sampled, stored and processed. The previous amplification and passband filtered stages (l) are also optional.

7.2 Measurement procedure

The processing algorithm is based on the sliding correlation principle, employee in the sector of radio channel sounding systems based on the transmission of pseudorandom binary sequences of PRBS type.

This processing can be implemented to work in real time or in off-line form. The processing requires the existence of a version previously sampled and stored of the transmitted signal. This version can also be generated in the moment of the processing, whenever the transmitted signal parameters are known. The first process to implement consists of carrying out a cross-correlation between the received signal and its stored version. From this first step, the time parameters and received echoes amplitudes are extracted.

The processing implemented in this description obtains an echo/multipath time resolution superior to the provided one by classic schemes. For that it is necessary to consider all the samples of the received and sampled signal. In the classic processing a sample by transmitted bit is considered. The fact to consider all the samples will allow obtaining a larger accurate parameter estimation of the channel under study.

From the sample processing, the corresponding radar section images are obtained. Algorithms widely described in literature will be used for it. When lacking sidelobes the received signal, many of the phenomena that interfere in the obtaining of a correct radar image, such as the false echoes, will be avoided.

The part corresponding to the processing and radar images obtaining closes the description of the hardware set-up implementation presented here. Following we will describe some experimental results corresponding to measurements performed in actual outdoor scenarios for a receiver scheme for the noise radar sounder similar to Figure 14.

8. Experimental results

The previously described wideband radar sounder was used to experimentally compare the performance of PRBS and Golay sequences in actual scenarios. Some results are here introduced from the measurement campaign performed in an outdoor environment in the 1GHz frequency band.

In the transmitted end, a BPSK modulation was chosen to modulate a digital waveform sequence with a chip rate of 250Mbps. The resulting transmitted signal presented a bandwidth of 500MHz. In the receiver end, an I/Q superheterodyne demodulation scheme to non-zero intermediate frequency (125MHz) was applied according to description given in section 7.1 for this hardware set-up.

Transmitter and receiver were placed according to the geometry shown in Fig. 16, with a round-trip distance to the target of 28.8m. The target consisted in a steel metallic slab with dimensions 1m². The experiment tried to compare the performance of Golay and PRBS sequences to determine the target range. Results with this single target range estimation are shown in Table 4.

From measurement it has been observed also the influence of the code length M in the detection of multipath component. As larger the code as larger number of echoes and stronger components are rescued in the cross-correlation based processing.

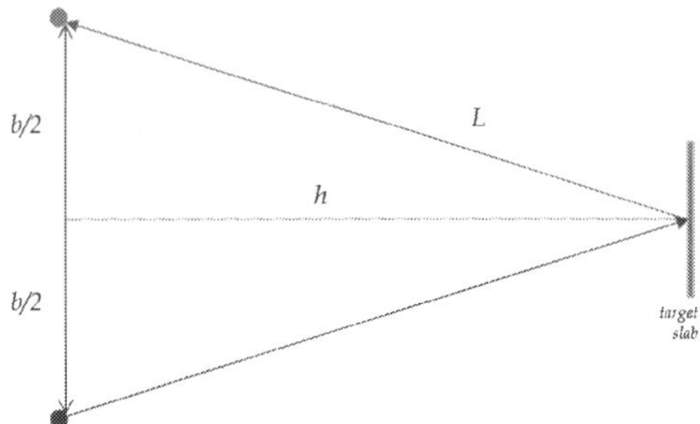

Fig. 16. Geometry of measurement scenario (b=2.25m, h=14.447m, L=14.4m)

Sequence transmitted		GOLAY	PRBS
M (sequence length)		4096	8192
Link range [m]		28.8	28.8
Link range [ns]		96	96
Measured	vertical	97	94
Delay [ns]	horizontal	97	94
Estimated	vertical	29.1	28.2
range [m]	horizontal	29.1	28.2
Relative	vertical	1.04	2.1
error [%]	horizontal	1.04	2.1

Table 4. Single target range estimation results.

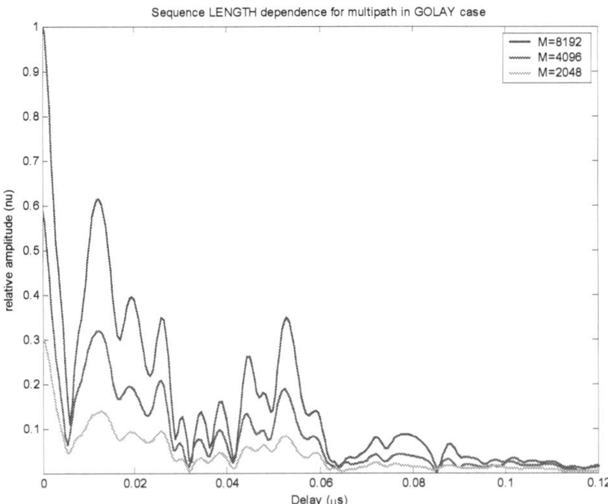

Fig. 17. Code length influence on multipath detection.

9. Relation between channel parameters: Fleury Limit

The only restriction to be satisfied by the values of *CB* is the known as *Fleury limit* [Fleury, 1996]. The expression for this limit is (12):

$$CB \;\geq\; \arccos(c)/(2 \cdot \pi \cdot \tau_{rms}) \tag{12}$$

where c is the level of correlation, and verifies $c \in [0,1]$. The theoretical values of the Fleury limit for the 0.5 and 0.9 correlation levels can be found by (12). These limit values can be later compared with those experimentally achieved. The results for values of coherence bandwidth measured should verify the theoretical limit given in [Fleury, 1996] to ensure the reliability of the experimental outcomes.

This limit for 50% and 90% CB is defined according to equation (12) and plotted in graphic of Fig. 18.

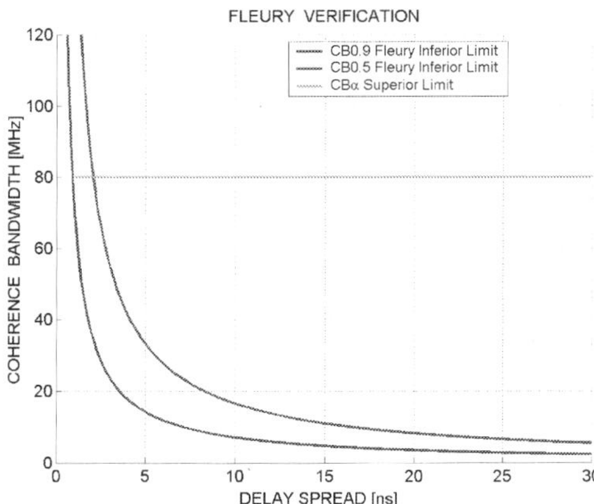

Fig 18. Graphical illustration of Fleury limit.

10. Conclusion

The improvement introduced by the hereby introduced noise radar in dynamic range thematic and sidelobes suppression is based on the autocorrelation properties of the used pseudorandom sequences: the Golay series.

The binary phase codes are characterized by a nonperiodic autocorrelation function with minimum sidelobes amplitude level. An important class of these binary codes is the denominated complementary codes, also known as Golay codes. The complementary codes consist of a pair of same length sequences that present the following property: when their respective autocorrelation functions are algebraically added, their sidelobes are cancelled. In addition, the amplitude level of the autocorrelation peak is increased by this sum doubling its value.

The complementary series have certain features that make them more suitable than PRBS sequences for some applications. Among them most important it is the double gain in the autocorrelation function. This increased gain is obtained with no need to introduce an additional hardware structure.

This behaviour of the Golay codes will provide a significant improvement at the detection level of the signal in the receiver. When increasing the dynamic range, is allowed to suitably separate the echoes level of background noise, avoiding the false echoes and allowing better target estimation. The improvement in the dynamic range is obtained thanks to a double effect: the autocorrelation peak is of double amplitude and the sidelobes are cancelled. This behaviour is very useful in extreme situations, like a scene where the transmitted and the reflected signals experience a large attenuation, as it is the case of ground penetrating radars application surroundings.

In [Alejos, 2007] it has been demonstrated, in quantitative and qualitative form that the improvement reached when using Golay sequences in the radio channel parameters estimation reaches values of up to 67.8% with respect to PRBS sequences. The only cost resulting of use Golay sequences is the double time required for the sequence transmission, since a pair of sequences instead of a single one is used.

The sounders or radar systems, for any application, that use binary sequences have been traditionally used because they are of easy implementation to obtain wide bandwidth. The sounders based in binary sequences implied the use of hardware shift registers of bipolar technology ECL (Emitter Coupled Logic), with reduced logical excursion, low switching speed and with non good noise threshold. All this makes difficult the attainment of quality sequences, greater level, low noise level and high binary rate.

Nowadays, nevertheless, it is possible to use hardware for arbitrary waveform generation of cheaper implementation and with a faster and versatile operation. In the present invention authors had boarded the generation of the used Golay sequences by means of the combination of algorithms programmed on a FPGA and a later analogue-digital conversion stage of great bandwidth and large amplitude resolution. As result, to use Golay sequences, previously stored or instantaneously generated, is easier and precise than it used to be.

Some advantages of the Golay codes have been practically demonstrated by means of the measurements described in section 8. The single target range estimation offers better outcomes for the Golay case even for the short round-trip here shown. It has been practically stated the influence of the code length in the multipath detection. This fact seems be due to that a longer code yields to a higher dynamic range cross-correlation.

11. References

A. Alejos, M. García Sánchez, I. Cuiñas (2005). "Benefits of Using Golay Sequences in Channel Swept Time Cross-Correlation Sounders", Proceedings of the *35th European Microwave Conference*, Paris, ISBN 2-9600551-2-8.

A. Alejos, M. García Sánchez, I. Cuiñas (2007). "Improvement of wideband radio channel swept time cross-correlation sounders by using Golay sequences". *IEEE Transactions on Vehicular Technology*, vol. 56, n° 1, ISSN 00189545.

A. Alejos, Dawood Muhammad, Manuel García Sánchez, Russell Jedlicka, Iñigo Cuiñas, Habeeb Ur Rahman Mohammed (2008). "Low Sidelobe Level Radar Techniques Using Golay Based Coded Sequences", Proceedings of the *2008 IEEE International Symposium on Antennas and Propagation*, San Diego (USA), ISBN 978-1-4244-2041-4.

Budisin, S. (1992). "Golay complementary sequences are superior to PN sequences", *IEEE International Conference on Systems Engineering 1992*, pp. 101-104, ISBN 0-7803-0720-8.

Cohen, M. N. (1987). *"Pulse compression in radar systems"*. Principles of modern radar, Van Nostrand Reinhold Company Inc., ISBN 0-9648312-0-1, New York.

Cruselles Forner Ernesto, Melús Moreno, José L. (1996). *"Secuencias pseudoaleatorias para telecomunicaciones"*, Ediciones UPC, ISBN 8483011646, Barcelona.

Chase D. (1976). "Digital signal design concepts for a time-varying Rician channel". *IEEE Transactions on Communication*, vol. COM-24, no. 2, pp. 164–172, ISSN 0090-6778.

Díaz, V. D., Hernández and J. Ureña (2002). "Using complementary sequences for direct transmission path identification," in *Proceedings of the 28th Annual Conference IEEE IECON*, vol. 4, pp. 2764–2767, ISBN 0-7803-7474-6.

Fleury, Bernard Henri (1996). "An uncertainty relation for WSS processes and its application to WSSUS systems". *IEEE Transactions on Communications*, vol. 44, issue 12, pp. 1632–1634, ISSN 00906778.

Gil, V.P., Jimenez, M.S. Fernandez, A.G. Armada (2002). "Study and implementation of complementary Golay sequences for PAR reduction in OFDM signals", *MELECON 02, 11th Mediterranean Electrotechnical Conference*, pp. 198-203, ISSN 0-7803-7527-0.

Golay, M. J. E. (1961). "Complementary series". *IEEE Transactions on Information Theory*, vol. 24, pp. 82-87, ISSN 0018-9448.

Golay, M.J.E. (1983). "The merit factor of Legendre sequences". *IEEE Transactions on Information Theory*, vol. IT-29, no. 6, pp. 934–936, ISSN 0018-9448.

Hammoudeh, Akram, David A. Scammell and Manuel García Sánchez (2003). "Measurements and analysis of the indoor wideband millimeter wave wireless radio channel and frequency diversity characterization". *IEEE Transactions on Antennas and Propagation*, vol. 51, issue 10, pp. 2974–2986, ISSN 0018-926X.

M.O. Al-Nuaimi and Andreas G. Siamarou (2002). "Coherence bandwidth characterisation and estimation for indoor Rician multipath wireless channels using measurements at 62.4GHz", *IEE Proceedings - Microwaves, Antennas and Propagation*, vol. 149, issue 3, pp. 181-187, ISSN 1350-2417.

Nathanson, F. E. ,M. N. Cohen and J. P. Reilly (1999). *"Phase-coding techniques. Signal processing and the environment (2nd edition)"*. SciTech Publishing, ISBN 978-1-891121-09-8, New York.

Popovic, B. M. (1999). "Spreading sequences for multicarrier CDMA systems". *IEEE Transactions on Communications*, vol. 47, no. 6, pp. 918–926, ISSN 0090-6778.

Sarwate, D. and M. Pursley (1980). "Crosscorrelation properties of pseudorandom and related sequences". *Proceedings of the IEEE*, vol. 68, no. 5, pp. 593--619, ISSN 0018-9219.

Sivaswamy, R. (1978). "Multiphase Complementary Codes". *IEEE Transactions on Information Theory*, vol. 24, no. 5, pp. 546-552, ISSN 0018-9448.

Weng, J. F. and Leung, S. H. (2000). "On the performance of DPSK in Rician fading channels with class A noise". *IEEE Transactions on Vehicular Technology*, vol. 49, no. 5, pp. 1934–1949, ISSN 00189545.

Wong, K. K. and T. O'Farrell (2003). "Spread spectrum techniques for indoor wireless IR communications". *IEEE Wireless Communications*, vol. 10, no. 2, pp. 54–63, ISSN 1536-1284.

4

Sensitivity of Safe Game Ship Control on Base Information from ARPA Radar

Józef Lisowski
Gdynia Maritime University
Poland

1. Introduction

The problem of non-collision strategies in the steering at sea appeared in the Isaacs works (Isaacs, 1965) called "the father of the differential games" and was developed by many authors both within the context of the game theory (Engwerda, 2005; Nowak & Szajowski, 2005), and also in the control under uncertainty conditions (Nisan et al., 2007). The definition of the problem of avoiding a collision seems to be quite obvious, however, apart from the issue of the uncertainty of information which may be a result of external factors (weather conditions, sea state), incomplete knowledge about other ships and imprecise nature of the recommendations concerning the right of way contained in International Regulations for Preventing Collision at Sea (COLREG) (Cockcroft & Lameijer, 2006). The problem of determining safe strategies is still an urgent issue as a result of an ever increasing traffic of ships on particular water areas. It is also important due to the increasing requirements as to the safety of shipping and environmental protection, from one side, and to the improving opportunities to use computer supporting the navigator duties (Bist, 2000; Gluver & Olsen, 1998). In order to ensure safe navigation the ships are obliged to observe legal requirements contained in the COLREG Rules. However, these Rules refer exclusively to two ships under good visibility conditions, in case of restricted visibility the Rules provide only recommendations of general nature and they are unable to consider all necessary conditions of the real process. Therefore the real process of the ships passing exercises occurs under the conditions of indefiniteness and conflict accompanied by an imprecise co-operation among the ships in the light of the legal regulations. A necessity to consider simultaneously the strategies of the encountered ships and the dynamic properties of the ships as control objects is a good reason for the application of the differential game model - often called the dynamic game (Osborne, 2004; Straffin, 2001).

2. Safe ship control

2.1 Integrated of navigation

The control of the ship's movement may be treated as a multilevel problem shown on Figure 1, which results from the division of entire ship control system, into clearly determined subsystems which are ascribed appropriate layers of control (Lisowski, 2007a), (Fig. 1).
This is connected both with a large number of dimensions of the steering vector and of the status of the process, its random, fuzzy and decision making characteristics - which are

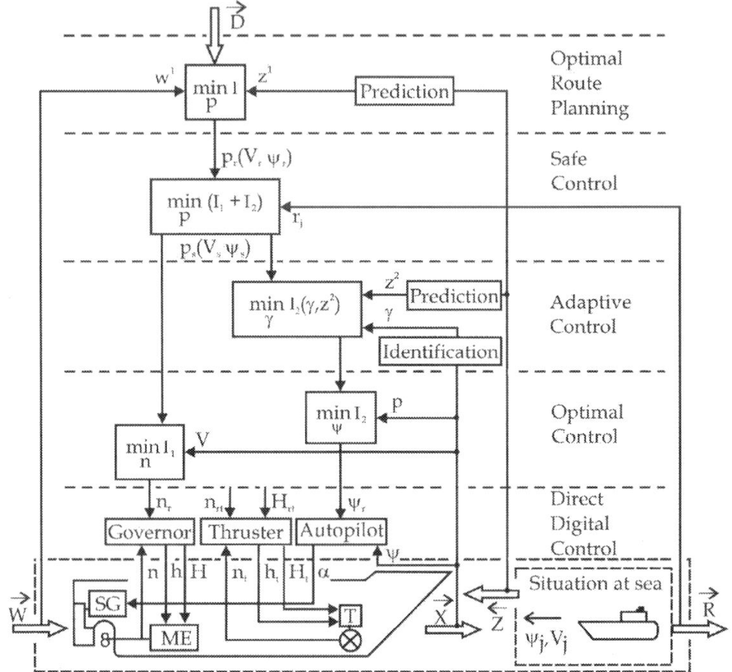

Fig. 1. Multilevel ship movement steering system

affected by strong interference generated by the current, wind and the sea wave motion on the one hand, and a complex nature of the equations describing the ship's dynamics with non-linear and non-stationary characteristics. The determination of the global control of the steering systems has in practice become too costly and ineffective (Lisowski 2002).

The integral part of the entire system is the process of the ship's movement control, which may be described with appropriate differential equations of the kinematics and dynamics of a ship being an object of the control under a variety of the ship's operational conditions such as:

- stabilisation of the course or trajectory,
- adjustment of the ship's speed,
- precise steering at small speeds in port with thrusters or adjustable-pitch propeller,
- stabilisation of the ship's rolling,
- commanding the towing group,
- dynamic stabilisation of the drilling ship's or the tanker's position.

The functional draft of the system corresponds to a certain actual arrangement of the equipment. The increasing demands with regard to the safety of navigation are forcing the ship's operators to install the systems of integrated navigation on board their ships. By improving the ship's control these systems increase the safety of navigation of a ship - which is a very expensive object of the value, including the cargo, and the effectiveness of the carriage goods by sea (Cymbal et al., 2007; Lisowski, 2005a, 2007b).

2.2 ARPA anti-collision radar system of acquisition and tracking

The challenge in research for effective methods to prevent ship collisions has become important with the increasing size, speed and number of ships participating in sea carriage. An obvious contribution in increasing safety of shipping has been firstly the application of radars and then the development of ARPA (Automatic Radar Plotting Aids) anti-collision system (Bole et al., 2006; Cahill, 2002), (Fig. 2).

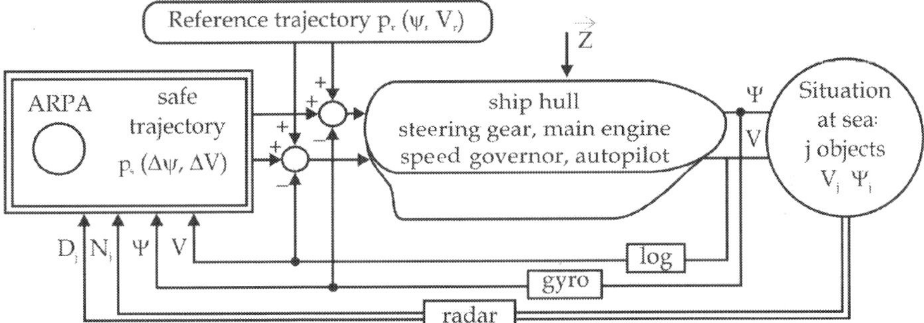

Fig. 2. The structure of safe ship control system

The ARPA system enables to track automatically at least 20 encountered j objects as is shown on Figure 3, determination of their movement parameters (speed V_j, course ψ_j) and elements of approach to the own ship ($D_{min}^j = DCPA_j$ - Distance of the Closest Point of Approach, $T_{min}^j = TCPA_j$ - Time to the Closest Point of Approach) and also the assessment of the collision risk r_j (Lisowski, 2001, 2008a).

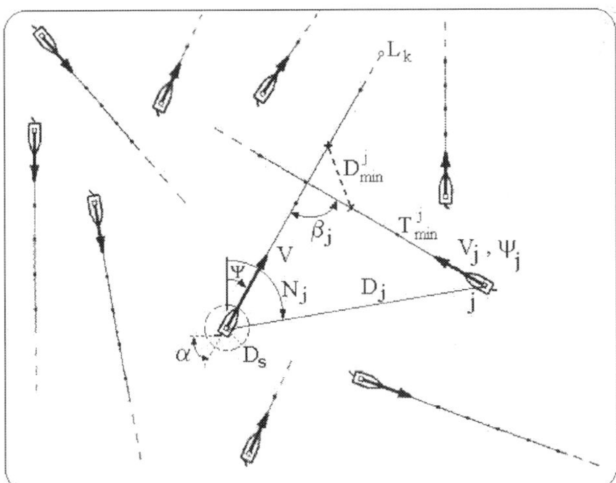

Fig. 3. Navigational situation representing the passing of the own ship with the j-th object

The risk value is possible to define by referring the current situation of approach, described by parameters D_{min}^j and T_{min}^j, to the assumed evaluation of the situation as safe, determined by a safe distance of approach D_s and a safe time T_s – which are necessary to execute a collision avoiding manoeuvre with consideration of distance D_j to j-th met object - shown on Figure 4 (Lisowski, 2005b, 2008c):

$$r_j = \left[k_1 \left(\frac{D_{min}^j}{D_s} \right)^2 + k_2 \left(\frac{T_{min}^j}{T_s} \right)^2 + \left(\frac{D_j}{D_s} \right)^2 \right]^{-\frac{1}{2}}$$ (1)

The weight coefficients k_1 and k_2 are depended on the state visibility at sea, dynamic length L_d and dynamic beam B_d of the ship, kind of water region and in practice are equal:

$$0 \leq [k_1(L_d, B_d), \ k_2(L_d, B_d)] \leq 1$$ (2)

$$L_d = 1.1 \ (1 + 0.345 \ V^{1.6})$$ (3)

$$B_d = 1.1 \ (B + 0.767 \ LV^{0.4})$$ (4)

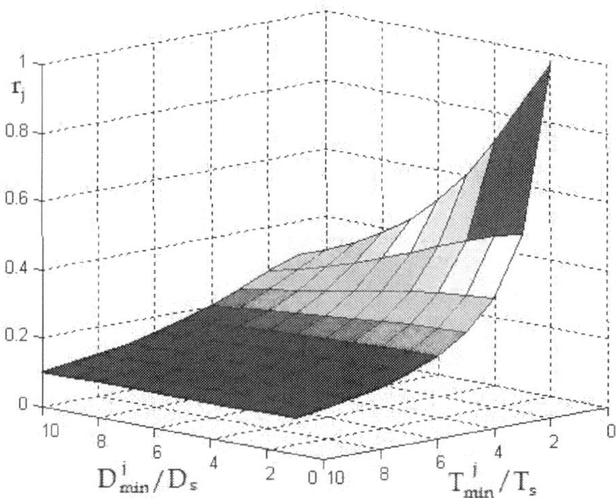

Fig. 4. The ship's collision risk space in a function of relative distance and time of approaching the j-th object

2.3 ARPA anti-collision radar system of manoeuvre simulation

The functional scope of a standard ARPA system ends with the simulation of the manoeuvre altering the course $\pm\Delta\psi$ or the ship's speed $\pm\Delta V$ selected by the navigator as is shown on Figure 5 (Pasmurow & Zimoviev, 2005).

2.4 Computer support of navigator manoeuvring decision

The problem of selecting such a manoeuvre is very difficult as the process of control is very complex since it is dynamic, non-linear, multi-dimensional, non-stationary and game making in its nature.

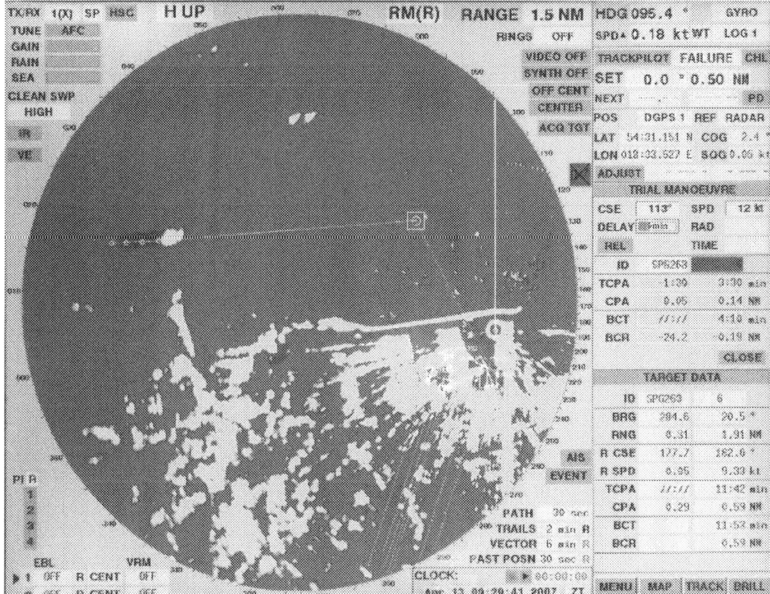

Fig. 5. The screen of SAM Electronics ARPA on the sailing vessel s/v DAR MLODZIEZY (ENAMOR Gdynia, Poland)

In practice, methods of selecting a manoeuvre assume a form of appropriate steering algorithms supporting navigator decision in a collision situation. Algorithms are programmed into the memory of a Programmable Logic Controller PLC (Fig. 6). This generates an option within the ARPA anti-collision system or a training simulator (Lisowski, 2008a).

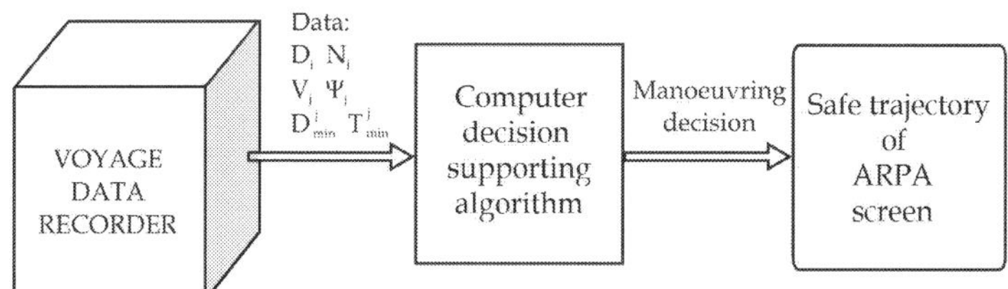

Fig. 6. The system structure of computer support of navigator decision in collision situation

3. Game control in marine navigation

3.1 Processes of game ship control

The classical issues of the theory of the decision process in marine navigation include the safe steering of a ship (Baba & Jain 2001; Levine, 1996).

Assuming that the dynamic movement of the ships in time occurs under the influence of the appropriate sets of control:

$$[U_0^{(\xi_0)}, U_j^{(\xi_j)}] \tag{5}$$

where:

$U_0^{(\xi_0)}$ - a set of the own ship strategies,

$U_j^{(\xi_j)}$ - a set of the *j-th* ships strategies,

$\xi(\xi_0, \xi_j) = 0$ - denotes course and trajectory stabilisation,

$\xi(\xi_0, \xi_j) = 1$ - denotes the execution of the anti-collision manoeuvre in order to minimize the risk of collision, which in practice is achieved by satisfying following inequality:

$$D_{min}^j = \min D_j(t) \geq D_s \tag{6}$$

where:

D_{min}^j - the smallest distance of approach of the own ship and the j-th encountered object,

D_s - safe approach distance in the prevailing conditions depends on the visibility conditions at sea, the COLREG Rules and ship dynamics,

D_j - current distance to the j-th ship taken from the ARPA anti-collision system (Fig. 2).

$\xi(\xi_0, \xi_j) = -1$ - refers to the manoeuvring of the ship in order to achieve the closest point of approach, for example during the approach of a rescue vessel, transfer of cargo from ship to ship, destruction the enemy's ship, etc.).

In the adopted describing symbols ξ we can discriminate the following type of steering ship in order to achieve a determined control goal:

-   basic type of control – stabilization of course or trajectory: $[U_0^{(0)} U_j^{(0)}]$

-   avoidance of a collision by executing:

a. own ship's manoeuvres: $[U_0^{(1)} U_j^{(0)}]$

b. manoeuvres of the j-th ship: $[U_0^{(0)} U_j^{(1)}]$

c. co-operative manoeuvres: $[U_0^{(1)} U_j^{(1)}]$

-   encounter of the ships: $[U_0^{(-1)} U_j^{(-1)}]$

-   situations of a unilateral dynamic game: $[U_0^{(-1)} U_j^{(0)}]$ and $[U_0^{(0)} U_j^{(-1)}]$

Dangerous situations resulting from a faulty assessment of the approaching process by one of the party with the other party's failure to conduct observation - one ship is equipped with a radar or an anti-collision system, the other with a damaged radar or without this device (Lisowski, 2001).

-   chasing situations which refer to a typical conflicting dynamic game: $[U_0^{(-1)} U_j^{(1)}]$ and $[U_0^{(1)} U_j^{(-1)}]$.

3.2 Basic model of dynamic game ship control

The most general description of the own ship passing the j number of other encountered ships is the model of a differential game of j number of moving control objects (Fig. 7).

The properties of control process are described by the state equation:

$$\dot{x}_i = f_i[(x_{0,\vartheta_0}, x_{1,\vartheta_1}, ..., x_{j,\vartheta_j}, ..., x_{m,\vartheta_m}), (u_{0,v_0}, u_{1,v_1}, ..., u_{j,v_j}, ..., u_{m,v_m}), t], \; i = 1, ..., (j \cdot \vartheta_j + \vartheta_0), \; j = 1, ..., m \tag{7}$$

where:

$\vec{x}_{0,\vartheta_0}(t)$ - ϑ_0 dimensional vector of process state of own ship determined in time $t \in [t_0, t_k]$,

$\vec{x}_{j,\vartheta_j}(t)$ - ϑ_j dimensional vector of the process state for the j-th ship,

$\vec{u}_{0,v_0}(t)$ - v_0 dimensional control vector of the own ship,

$\vec{u}_{j,v_j}(t)$ - v_j dimensional control vector of the j-th ship.

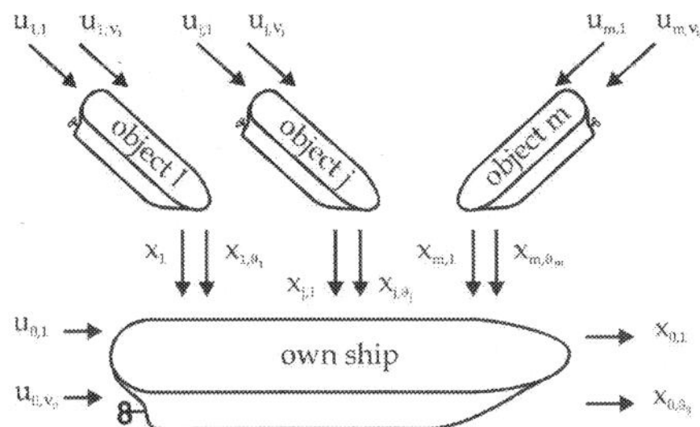

Fig. 7. Block diagram of a basic dynamic game model of ship control

The state equations of ship game control process take the following form:

$$\dot{x}_{0,1} = x_{0,2}$$

$$\dot{x}_{0,2} = a_1 x_{0,2} x_{0,3} + a_2 x_{0,3}|x_{0,3}| + b_1 x_{0,3}|x_{0,3}|u_{0,1}$$

$$\dot{x}_{0,3} = a_4 x_{0,3}|x_{0,3}|\|x_{0,4}|x_{0,4}(1+x_{0,4}) + a_5 x_{0,2} x_{0,3} x_{0,4}|x_{0,4}| + a_6 x_{0,2} x_{0,3} x_{0,4} + a_7 x_{0,3}|x_{0,3}| +$$
$$+ a_8 x_{0,5}|x_{0,5}|x_{0,6} + b_2 x_{0,3} x_{0,4}|x_{0,3} u_{0,1}|$$

$$\dot{x}_{0,4} = a_3 x_{0,3} x_{0,4} + a_4 x_{0,3} x_{0,4}|x_{0,4}| + a_5 x_{0,2} x_{0,2} + a_9 x_{0,2} + b_2 x_{0,3} u_{0,1}$$

$$\dot{x}_{0,5} = a_{10} x_{0,5} + b_3 u_{0,2}$$

$$\dot{x}_{0,6} = a_{11} x_{0,6} + b_4 u_{0,3}$$

$$\dot{x}_{j,1} = x_{0,3} + x_{j,2} x_{0,2} + x_{j,3}\cos x_{j,3}$$

$$\dot{x}_{j,2} = x_{0,2} x_{j,1} + x_{j,3}\sin x_{j,3}$$

$$\dot{x}_{j,3} = x_{j,4}$$

$$\dot{x}_{j,4} = a_{11+j} x_{j,4} + b_{4+j} u_{j,1}$$

$$\dot{x}_{j,5} = a_{12+i} x_{j,5} + b_{5+i} u_{j,2}$$

(8)

The state variables are represented by the following values:

$x_{0,1} = \psi$ - course of the own ship,

$x_{0,2} = \dot{\psi}$ - angular turning speed of the own ship,

$x_{0,3} = V$ - speed of the own ship,

$x_{0,4} = \beta$ - drift angle of the own ship,

$x_{0,5} = n$ - rotational speed of the screw propeller of the own ship,

$x_{0,6} = H$ - pitch of the adjustable propeller of the own ship,

$x_{j,1} = D_j$ - distance to j-th ship,

$x_{j,2} = N_j$ - bearing of the j-th ship,

$x_{j,3} = \psi_j$ - course of the j-th ship,

$x_{j,4} = \dot{\psi}_j$ - angular turning speed of the j-th ship,

$\dot{x}_{j,5} = V_j$ - speed of the j-th ship,

where: $\vartheta_0 = 6, \vartheta_j = 4$.

While the control values are represented by:

$u_{0,1} = \alpha_r$ - rudder angle of the own ship,

$u_{0,2} = n_r$ - rotational speed of the own ship screw propeller,

$u_{0,3} = H_r$ - pitch of the adjustable propeller of the own ship,

$u_{j,1} = \alpha_{rj}$ - rudder angle of the j-th ship,

$u_{j,2} = n_{r,j}$ - rotational speed of the j-th ship screw propeller,

where: $v_0 = 3, v_j = 2$.

Values of coefficients of the process state equations (8) for the 12 000 DWT container ship are given in Table 1.

Coefficient	Measure	Value
a_1	m^{-1}	$- 4.143 \cdot 10^{-2}$
a_2	m^{-2}	$1.858 \cdot 10^{-4}$
a_3	m^{-1}	$- 6.934 \cdot 10^{-3}$
a_4	m^{-1}	$- 3.177 \cdot 10^{-2}$
a_5	-	$- 4.435$
a_6	-	$- 0.895$
a_7	m^{-1}	$- 9.284 \cdot 10^{-4}$
a_8	-	$1.357 \cdot 10^{-3}$
a_9	-	0.624
a_{10}	s^{-1}	$- 0.200$
a_{11+j}	s^{-1}	$- 5 \cdot 10^{-2}$
a_{12+j}	s^{-1}	$- 4 \cdot 10^{-3}$
b_1	m^{-2}	$1.134 \cdot 10^{-2}$
b_2	m^{-1}	$- 1.554 \cdot 10^{-3}$
b_3	s^{-1}	0.200
b_4	s^{-1}	0.100
b_{4+j}	m^{-1}	$- 3.333 \cdot 10^{-3}$
b_{5+j}	$m \cdot s^{-1}$	$9.536 \cdot 10^{-2}$

Table 1. Coefficients of basic game model equations

In example for j=20 encountered ships the base game model is represented by i=86 state variables of process control.

The constraints of the control and the state of the process are connected with the basic condition for the safe passing of the ships at a safe distance D_s in compliance with COLREG Rules, generally in the following form:

$$g_j(x_{j,\vartheta_j}, u_{j,v_j}) \le 0 \tag{9}$$

The constraints referred to as the ships domains in the marine navigation, may assume a shape of a circle, ellipse, hexagon, or parabola and may be generated for example by an artificial neural network as is shown on Figure 8 (Lisowski et al., 2000).

The synthesis of the decision making pattern of the object control leads to the determination of the optimal strategies of the players who determine the most favourable, under given conditions, conduct of the process. For the class of non-coalition games, often used in the control techniques, the most beneficial conduct of the own ship as a player with j-th ship is the minimization of her goal function (10) in the form of the payments – the integral payment and the final one.

$$I_{0,j} = \int_{t_0}^{t_k} [x_{0,\vartheta_0}(t)]^2 dt + r_j(t_k) + d(t_k) \to \min \tag{10}$$

The integral payment represents loss of way by the ship while passing the encountered ships and the final payment determines the final risk of collision $r_j(t_k)$ relative to the j-th ship and the final deviation of the ship $d(t_k)$ from the reference trajectory (Fig. 9).

Generally two types of the steering goals are taken into consideration - programmed steering $u_0(t)$ and positional steering $u_0[x_0(t)]$. The basis for the decision making steering are the decision making patterns of the positional steering processes, the patterns with the feedback arrangement representing the dynamic games.

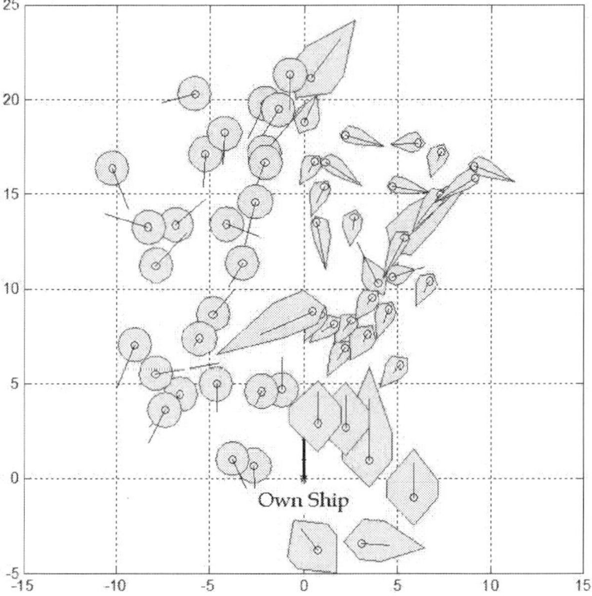

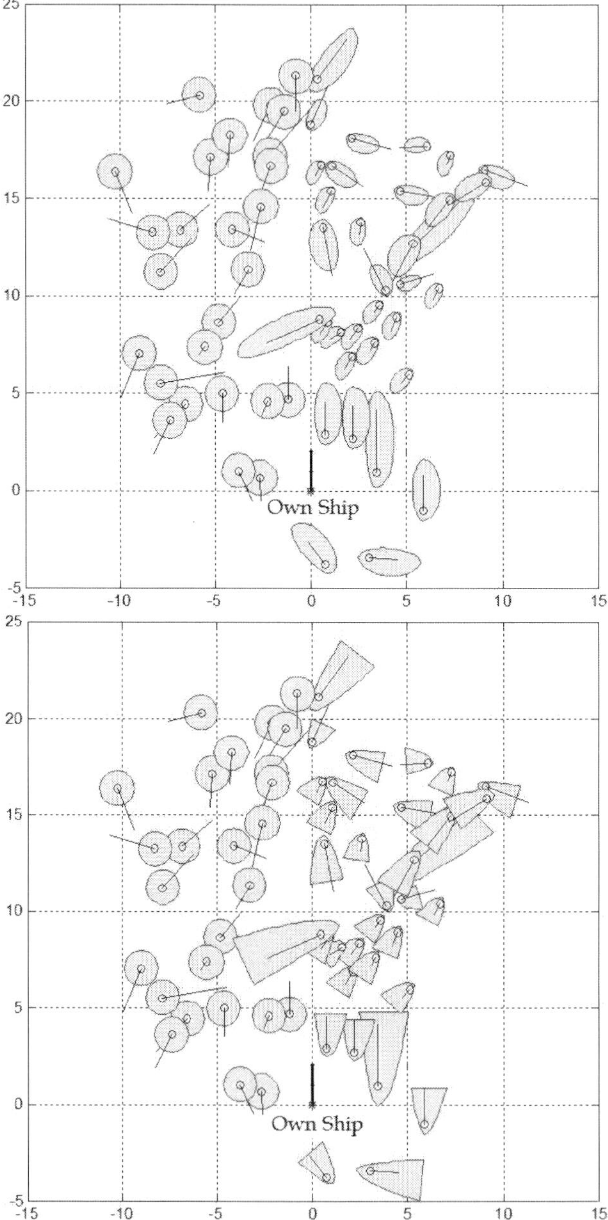

Fig. 8. The shapes of the neural domains in the situation of 60 encountered ships in English Channel

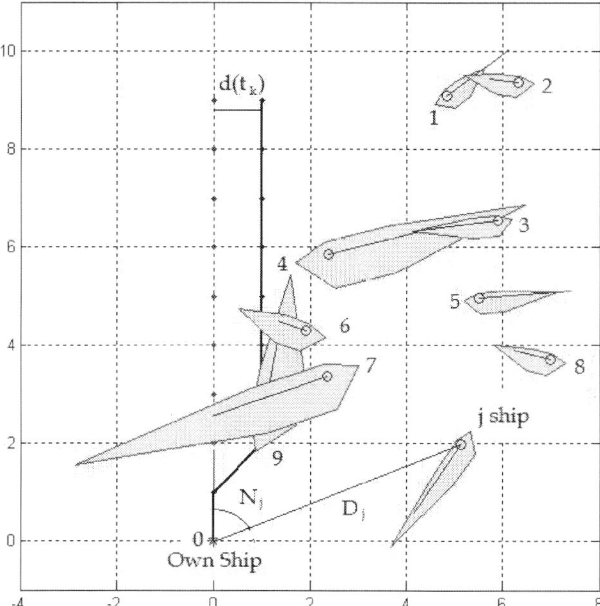

Fig. 9. Navigational situation representing the passing of the own ship with the j-th encountered ship

The application of reductions in the description of the own ship dynamics and the dynamic of the j-th encountered ship and their movement kinematics lead to approximated models: multi-stage positional game, multi-step matrix game, fuzzy matrix game, fuzzy dynamic programming, dynamic programming with neural state constraints, linear programming (LaValle, 2006; Lisowski, 2004).

4. Algorithms of safe game ship control

4.1 Multi-stage positional game trajectory POSTRAJ

The general model of dynamic game is simplified to the multi-stage positional game of j participants not co-operating among them, (Fig. 10).
State variables and control values are represented by:

$$\left.\begin{array}{c} x_{0,1} = X_0, \; x_{0,2} = Y_0, \; x_{j,1} = X_j, \; x_{j,2} = Y_j \\ u_{0,1} = \psi, \; u_{0,2} = V, \; u_{j,1} = \psi_j, \; u_{j,2} = V_j \\ j = 1, 2, \ldots, m \end{array}\right\} \quad (11)$$

The essence of the positional game is to subordinate the strategies of the own ship to the current positions $p(t_k)$ of the encountered objects at the current step k. In this way the process model takes into consideration any possible alterations of the course and speed of the encountered objects while steering is in progress. The current state of the process is determined by the co-ordinates of the own ship's position and the positions of the encountered objects:

$$\left. \begin{array}{l} x_0 = (X_0,\ Y_0),\ x_j = (X_j,\ Y_j) \\ j = 1,\ 2,\ ...,\ m \end{array} \right\} \tag{12}$$

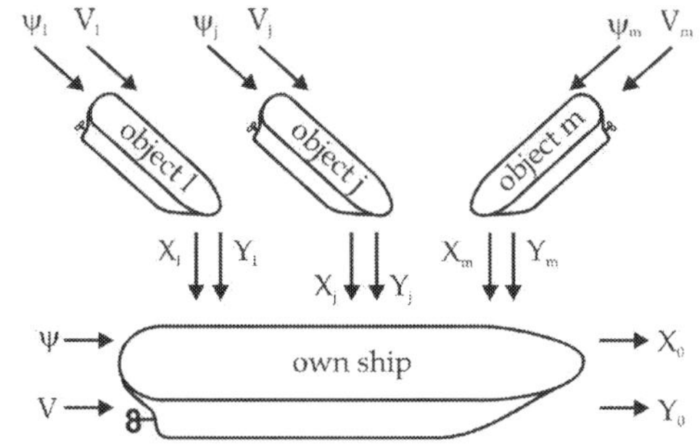

$\psi_1\ V_1$ $\quad$ $\psi_j\ V_j$ $\quad$ $\psi_m\ V_m$

object 1 $\quad$ object j $\quad$ object m

X_i Y_i $\quad$ X_j Y_j $\quad$ X_m Y_m

ψ $\quad$ own ship $\quad$ X_0

V $\quad$ Y_0

Fig. 10. Block diagram of the positional game model

The system generates its steering at the moment t_k on the basis of data received from the ARPA anti-collision system pertaining to the positions of the encountered objects:

$$p(t_k) = \begin{bmatrix} x_0(t_k) \\ x_j(t_k) \end{bmatrix} \quad j = 1,\ 2,...,\ m \quad k = 1,\ 2,...,\ K \tag{13}$$

It is assumed, according to the general concept of a multi-stage positional game, that at each discrete moment of time t_k the own ship knows the positions of the objects.
The constraints for the state co-ordinates:

$$\left\{ x_0(t),\ x_j(t) \right\} \in P \tag{14}$$

are navigational constraints, while steering constraints:

$$u_0 \in U_0,\ u_j \in U_j \quad j = 1,\ 2,...,\ m \tag{15}$$

take into consideration: the ships' movement kinematics, recommendations of the COLREG Rules and the condition to maintain a safe passing distance as per relationship (6).
The closed sets U_0^j and U_j^0, defined as the sets of acceptable strategies of the participants to the game towards one another:

$$\{U_0^j[p(t)],\ U_j^0[p(t)]\} \tag{16}$$

are dependent, which means that the choice of steering u_j by the j-th object changes the sets of acceptable strategies of other objects.
A set U_0^j of acceptable strategies of the own ship when passing the j-th encountered object at a distance D_s - while observing the condition of the course and speed stability of the own ship and that of the encountered object at step k is static and comprised within a half-circle of a radius V_r (Fig. 11).

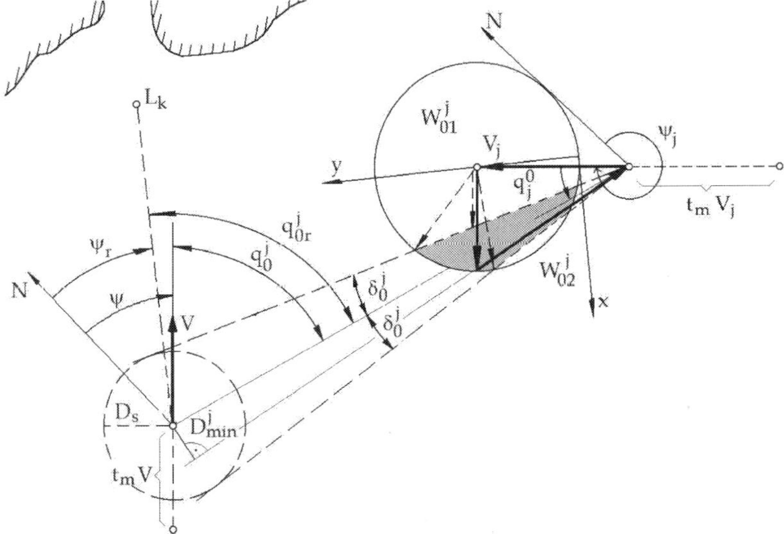

Fig. 11. Determination of the acceptable areas of the own ship strategies $U_0^j = W_{01}^j \cup W_{02}^j$

Area U_0^j is determined by an inequality (Fig. 12):

$$a_0^j \, u_0^x + b_0^j \, u_0^y \le c_0^j \tag{17}$$

$$\left(u_0^x\right)^2 + \left(u_0^y\right)^2 \le V_r^2 \tag{18}$$

where:

$$
\left.
\begin{aligned}
&\vec{V}_r = \vec{u}_0(u_0^x, \, u_0^y) \\
&a_0^j = -\chi_0^j \, \cos(q_{0r}^j + \chi_0^j \, \delta_0^j) \\
&b_0^j = \chi_0^j \, \sin(q_{0r}^j + \chi_0^j \, \delta_0^j) \\
&c_0^j = -\chi_0^j \begin{bmatrix} V_j \, \sin(q_j^0 + \chi_0^j \, \delta_0^j) + \\ V_r \, \cos(q_{0r}^j + \chi_0^j \, \delta_0^j) \end{bmatrix} \\
&\chi_0^j = \begin{cases} 1 & \text{dla } W_{01}^j \quad (\text{Starboard side}) \\ -1 & \text{dla } W_{02}^j \quad (\text{Port side}) \end{cases}
\end{aligned}
\right\}
\tag{19}
$$

The value χ_0^j is determined by using an appropriate logical function Z_j characterising any particular recommendation referring to the right of way contained in COLREG Rules.

The form of function Z_j depends of the interpretation of the above recommendations for the purpose to use them in the steering algorithm, when:

$$Z_j = \begin{cases} 1 & \text{then } \chi_0^j = 1 \\ 0 & \text{then } \chi_0^j = -1 \end{cases} \tag{20}$$

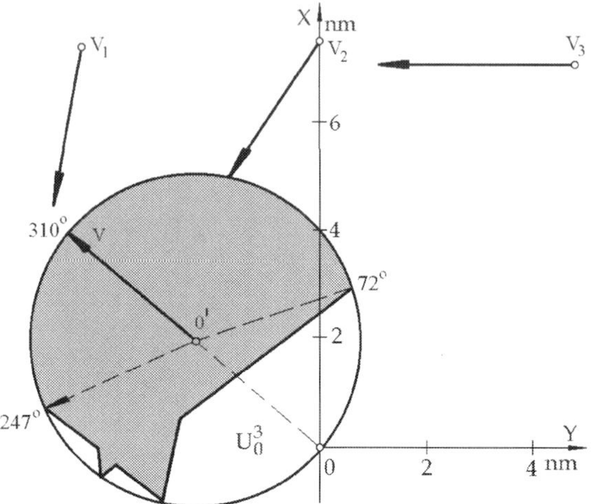

Fig. 12. Example of summary set U_0^3 of acceptable manoeuvres for three encountered ships

Interpretation of the COLREG Rules in the form of appropriate manoeuvring diagrams developed by A.G. Corbet, S.H. Hollingdale, E.S. Calvert and K.D. Jones enables to formulate a certain logical function Z_j as a semantic interpretation of legal regulations for manoeuvring.

Each particular type of the situation involving the approach of the ships is assigned the logical variable value equal to one or zero:

A – encounter of the ship from bow or from any other direction,

B – approaching or moving away of the ship,

C – passing the ship astern or ahead,

D – approaching of the ship from the bow or from the stern,

E – approaching of the ship from the starboard or port side.

By minimizing logical function Z_j by using a method of the Karnaugh's Tables the following is obtained:

$$Z_j = A \cup \overline{A}(\overline{B}\,\overline{C} \cup \overline{D}\,\overline{E}) \tag{21}$$

The resultant area of acceptable manoeuvres for m objects:

$$U_0 = \bigcap_{j=1}^{m} U_0^j \quad j = 1, 2, ..., m \tag{22}$$

is determined by an arrangement of inequalities (17) and (18).

A set for acceptable strategies U_j^0 of the encountered j-th object relative to the own ship is determined by analogy:

$$a_j^0 \, u_j^x + b_j^0 \, u_j^y \le c_j^0 \tag{23}$$

$$\left(u_j^x\right)^2 + \left(u_j^y\right)^2 \le V_j^2 \tag{24}$$

where:

$$\left.\begin{array}{l} \vec{V}_j = \vec{u}_j(u_j^x, u_j^y) \\[4pt] a_j^0 = -\chi_j^0 \cos(q_j^0 + \chi_j^0 \delta_j^0) \\[4pt] b_j^0 = \chi_j^0 \sin(q_j^0 + \chi_j^0 \delta_j^0) \\[4pt] c_j^0 = -\chi_j^0 V_0 \sin(q_0^j + \chi_j^0 \delta_j^0) \end{array}\right\} \tag{25}$$

The sing χ_j^0 is determined analogically to χ_0^j.

Taking into consideration of navigational constraints – shoal and shore line, presents additional constraints of the set of acceptable strategies:

$$a_0^{l,l-1} u_0^x + b_0^{l,l-1} u_0^y \le c_0^{l,l-1} \tag{26}$$

where: l – the closest point of intersection for the straight lines approximating the shore line (Cichuta & Dalecki, 2000).

The optimal steering of the own ship $u_0^*(t)$, equivalent for the current position $p(t)$ to the optimal positional steering $u_0^*(p)$, is determined in the following way:

- sets of acceptable strategies $U_j^0[p(t_k)]$ are determined for the encountered objects relative to the own ship and initial sets $U_0^{jw}[p(t_k)]$ of acceptable strategies of the own ship relative to each one of the encountered objects,
- a pair of vectors u_j^m and u_0^j relative to each j-th object is determined and then the optimal positional strategy for the own ship $u_0^*(p)$ from the condition:

$$I^* = \min_{u_0 \in U_0 = \bigcap_{j=1}^{m} U_0^j} \left\{ \max_{u_j^m \in U_j} \min_{u_0^j \in U_0^j(u_j)} S_0\left[x_0(t_k), L_k\right] \right\} = S_0^*(x_0, L_k) \quad U_0^j \subset U_0^{jw} \quad j = 1, 2, ..., m \tag{27}$$

where:

$$S_0[x_0(t), L_k] = \int_{t_0}^{t_{L_k}} u_0(t)\, dt \tag{28}$$

refers to the continuous function of the own ship's steering goal which characterises the ship's distance at the moment t_0 to the closest point of turn L_k on the assumed voyage route (Fig. 3).

In practice, the realization of the optimal trajectory of the own ship is achieved by determining the ship's course and speed, which would ensure the smallest loss of way for a safe passing of the encountered objects, at a distance which is not smaller than the assumed value D_s, always with respect to the ship's dynamics in the form of the advance time to the manoeuvre t_m, with element $t_m^{\Delta\psi}$ during course manoeuvre $\Delta\psi$ or element $t_m^{\Delta V}$ during speed manoeuvre ΔV (Fig. 13).

The dynamic features of the ship during the course alteration by an angle $\Delta\psi$ is described in a simplified manner with the use of transfer function:

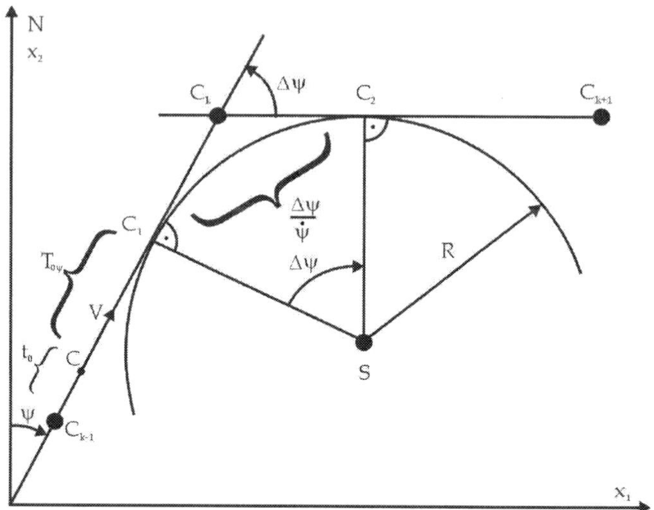

Fig. 13. Ship's motion during $\Delta\psi$ course changing

$$G_1(s) = \frac{\Delta\psi(s)}{\alpha(s)} = \frac{k_\psi(\alpha)}{s(1+T_\psi s)} \cong \frac{k_\psi(\alpha)\cdot e^{-T_{o\psi}s}}{s} \tag{29}$$

where:

T_ψ - manoeuvre delay time which is approximately equal to the time constant of the ship as a course control object,

$k_\psi(\alpha)$ - gain coefficient the value of which results from the non-linear static characteristics of the rudder steering.

The course manoeuvre delay time:

$$t_m^{\Delta\psi} \cong T_{o\psi} + \frac{\Delta\psi}{\dot\psi} \tag{30}$$

Differential equation of the second order describing the ship's behaviour during the change of the speed by ΔV is approximated with the use of the first order inertia with a delay:

$$G_2(s) = \frac{\Delta V(s)}{\Delta n(s)} = \frac{k_V e^{-T_{ov}s}}{1+T_V s} \tag{31}$$

where:

T_{oV} - time of delay equal approximately to the time constant for the propulsion system: main engine - propeller shaft – screw propeller,

T_V - the time constant of the ship's hull and the mass of the accompanying water.

The speed manoeuvre delay time is as follows:

$$t_m^{\Delta V} \cong T_{oV} + 3\,T_V \tag{32}$$

The smallest loss of way is achieved for the maximum projection of the speed vector maximum of the own ship on the direction of the assumed course ψ_r. The optimal steering of the own ship is calculated at each discrete stage of the ship's movement by applying Simplex method for solving the linear programming task.

At each one stage t_k of the measured position $p(t_k)$ optimal steering problem is solved according to the game control principle (27) (Fig. 14).

By using function *lp – linear programming* from Optimization Toolbox of the MATLAB software POSTRAJ algorithm was developed to determine a safe game trajectory of a ship in a collision situation (Łebkowski, 2001).

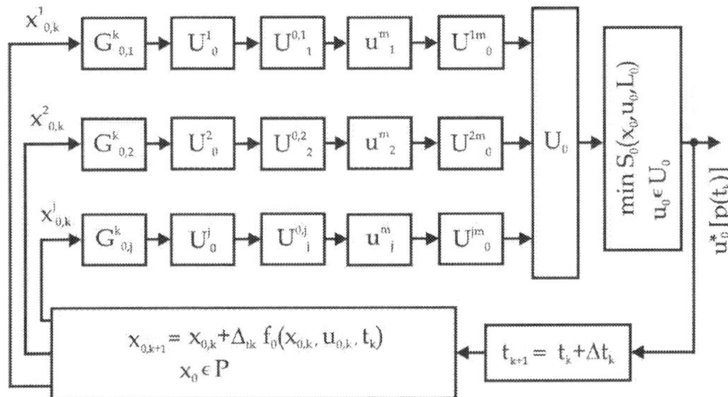

Fig. 14. Block diagram of the positional pattern for positional game steering: $G_{0,j}^k$ - a set of parameters of the own ship approach relative to j-th object taken from ARPA radar system

4.2 Multi-step matrix game trajectory RISKTRAJ

When leaving aside the ship's dynamics equations the general model of a dynamic game for the process of preventing collisions is reduced to the matrix game of j participants non-co-operating among them (Fig. 15).

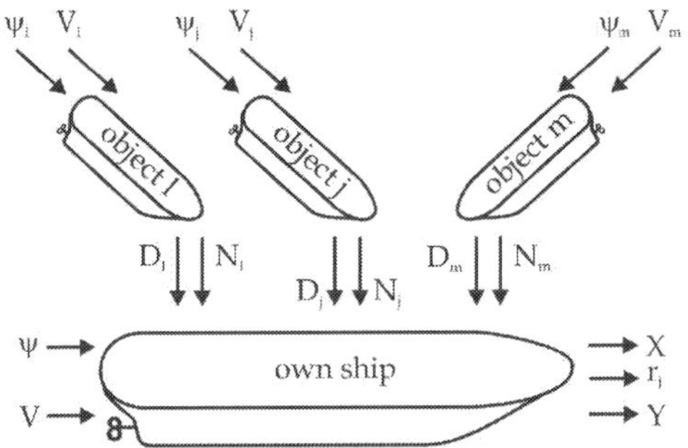

Fig. 15. Block diagram of a matrix game model

The state and steering variables are represented by the following values:

$$x_{j,1} = D_j, \ x_{j,2} = N_j, \ u_{0,1} = \psi, \ u_{0,2} = V, \ u_{j,1} = \psi_j, \ u_{j,2} = V_j \qquad j = 1, 2, ..., m \qquad (33)$$

The game matrix $\mathbf{R}\,[r_j(\upsilon_j,\upsilon_0)]$ includes the values of the collision risk r_j determined on the basis of data obtained from the ARPA anti-collision system for the acceptable strategies υ_0 of the own ship and acceptable strategies υ_j of any particular number of j encountered objects. The risk value is defined by equation (1). In a matrix game player I - own ship has a possibility to use υ_0 pure various strategies, and player II - encountered ships has υ_j various pure strategies:

$$R = [r_j(\upsilon_j,\upsilon_0)] = \begin{vmatrix} r_{11} & r_{12} & \cdots & r_{1,\upsilon_0-1} & r_{1\upsilon_n} \\ r_{21} & r_{22} & \cdots & r_{2,\upsilon_0-1} & r_{2\upsilon_n} \\ \cdots & \cdots & \cdots & \cdots & \cdots \\ r_{\upsilon_1 1} & r_{\upsilon_1 2} & \cdots & r_{\upsilon_1,\upsilon_0-1} & r_{\upsilon_1\upsilon_n} \\ \cdots & \cdots & \cdots & \cdots & \cdots \\ r_{\upsilon_j 1} & r_{\upsilon_j 2} & \cdots & r_{\upsilon_j,\upsilon_0-1} & r_{\upsilon_j\upsilon_n} \\ \cdots & \cdots & \cdots & \cdots & \cdots \\ r_{\upsilon_m 1} & r_{\upsilon_m 2} & \cdots & r_{\upsilon_m,\upsilon_0-1} & r_{\upsilon_m\upsilon_n} \end{vmatrix} \qquad (34)$$

The constraints for the choice of a strategy (υ_0,υ_j) result from the recommendations of the way priority at sea (Radzik, 2000). Constraints are limiting the selection of a strategy result from COLREG Rules. As most frequently the game does not have a saddle point, therefore the balance state is not guaranteed. In order to solve this problem we may use a dual linear programming.

In a dual problem player I aims to minimize the risk of collision, while player II aims to maximize the collision risk. The components of the mixed strategy express the distribution of the probability of using by the players their pure strategies. As a result for the goal control function in the form:

$$\left(I_0^{(j)}\right)^* = \min_{\upsilon_0}\ \max_{\upsilon_j}\ r_j \qquad (35)$$

probability matrix $\mathbf{P}$ of applying each one of the particular pure strategies is obtained:

$$P = [p_j(\upsilon_j,\upsilon_0)] = \begin{vmatrix} p_{11} & p_{12} & \cdots & p_{1,\upsilon_0\,1} & p_{1\upsilon_n} \\ p_{21} & p_{22} & \cdots & p_{2,\upsilon_0\,1} & p_{2\upsilon_n} \\ \cdots & \cdots & \cdots & \cdots & \cdots \\ p_{\upsilon_1 1} & p_{\upsilon_1 2} & \cdots & p_{\upsilon_1,\upsilon_0\,1} & p_{\upsilon_1\upsilon_n} \\ \cdots & \cdots & \cdots & \cdots & \cdots \\ p_{\upsilon_j 1} & p_{\upsilon_j 2} & \cdots & p_{\upsilon_j,\upsilon_0\,1} & p_{\upsilon_j\upsilon_n} \\ \cdots & \cdots & \cdots & \cdots & \cdots \\ p_{\upsilon_m 1} & p_{\upsilon_m 2} & \cdots & p_{\upsilon_m,\upsilon_0\,1} & p_{\upsilon_m\upsilon_n} \end{vmatrix} \qquad (36)$$

The solution for the control problem is the strategy representing the highest probability:

$$\left(u_0^{(\upsilon_0)}\right)^* = u_0^{(\upsilon_0)}\left\{ \left[p_j\left(\upsilon_j,\upsilon_o\right)\right]_{max} \right\} \qquad (37)$$

The safe trajectory of the own ship is treated as a sequence of successive changes in time of her course and speed. A safe passing distance is determined for the prevailing visibility conditions at sea D_s, advance time to the manoeuvre t_m described by equations (30) or (32) and the duration of one stage of the trajectory Δt_k as a calculation step. At each one step the most dangerous object relative to the value of the collision risk r_j is determined. Then, on the basis of semantic interpretation of the COLREG Rules, the direction of the own ship's turn relative to the most dangerous object is selected.

A collision risk matrix **R** is determined for the acceptable strategies of the own ship υ_0 and that for the j-th encountered object υ_j. By applying a principle of the dual linear programming for solving matrix games the optimal course of the own ship and that of the j-th object is obtained at a level of the smallest deviations from their initial values.

Figure 16 shows an example of possible strategies of the own ship and those of the encountered object while, Figure 17 presents the hyper surface of the collision risk for these values of the strategy.

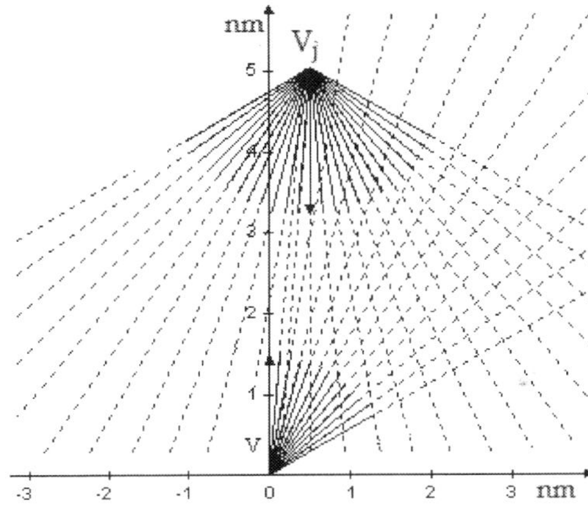

Fig. 16. Possible mutual strategies of the own ship and those of the encountered ship

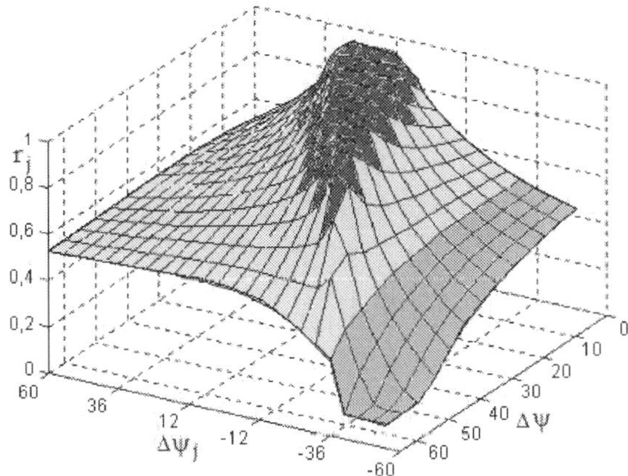

Fig. 17. Dependence of the collision risk on the course strategies of the own ship and those of the encountered ship

If, at a given step, there is no solution at the own ship's speed V, then the calculations are repeated for a speed decreased by 25%, until the game has been solved. The calculations are repeated step by step until the moment when all elements of the matrix **R** are equal to zero and the own ship, after having passed encountered objects, returns to her initial course and speed.

By using function *lp – linear programming* from Optimization Toolbox of the MATLAB software RISKTRAJ algorithm was developed to determine a safe game trajectory of a ship in a collision situation (Cichuta & Dalecki, 2000).

5. Sensitivity of game ship control

5.1 Definition of sensitivity

The investigation of sensitivity of game control fetch for sensitivity analysis of the game final payment (10) measured with the relative final deviation of $d(t_k)=d_k$ safe game trajectory from the reference trajectory, as sensitivity of the quality first-order (Wierzbicki, 1977). Taking into consideration the practical application of the game control algorithm for the own ship in a collision situation it is recommended to perform the analysis of sensitivity of a safe control with regard to the accuracy degree of the information received from the anti-collision ARPA radar system on the current approach situation, from one side and also with regard to the changes in kinematical and dynamic parameters of the control process.

Admissible average errors, that can be contributed by sensors of anti-collision system can have following values for:

- radar,
    - bearing: ±0,22°,
    - form of cluster: ±0,05°,
    - form of impulse: ±20 m,
    - margin of antenna drive: ±0,5°,
    - sampling of bearing: ±0,01°,
    - sampling of distance: ±0,01 nm,
- gyrocompas: ±0,5°,
- log: ±0,5 kn,
- GPS: ±15 m.

The sum of all errors, influent on picturing of the navigational situation, cannot exceed for absolute values ±5% or for angular values ±3°.

5.2 Sensitivity of control to inaccuracy of information from ARPA radar

Let $X_{0,j}$ represent such a set of state process control information on the navigational situation that:

$$X_{0,j} = \{V, \psi, V_j, \psi_j, D_j, N_j\} \tag{38}$$

Let then $X_{0,j}^{ARPA}$ represent a set of information from ARPA anti-collision system impaired by measurement and processing errors:

$$X_{0,j}^{ARPA} = \{V \pm \delta V, \psi \pm \delta \psi, V_j \pm \delta V_j, \psi_j \pm \delta \psi_j, D_j \pm \delta D_j, N_j \pm \delta N_j\} \tag{39}$$

Relative measure of sensitivity of the final payment in the game s_{inf} as a final deviation of the ship's safe trajectory d_k from the reference trajectory will be:

$$s_{inf} = (X_{0,j}^{ARPA}, X_{0,j}) = \frac{d_k^{ARPA}(X_{0,j}^{ARPA})}{d_k(X_{0,j})}$$

(40)

$$s_{inf} = \{s_{inf}^V, s_{inf}^\psi, s_{inf}^{V_j}, s_{inf}^{\psi_j}, s_{inf}^{D_j}, s_{inf}^{N_j}\}$$

5.3 Sensitivity of control to process control parameters alterations

Let X_{param} represents a set of parameters of the state process control:

$$X_{param} = \{t_m, D_s, \Delta t_k, \Delta V\}$$

(41)

Let then X'_{param} represents a set of information saddled errors of measurement and processing parameters:

$$X'_{param} = \{t_m \pm \delta t_m, D_s \pm \delta D_s, \Delta t_k \pm \delta \Delta t_k, \Delta V \pm \delta \Delta V\}$$

(42)

Relative measure of sensitivity of the final payment in the game as a final deflection of the ship's safe trajectory d_k from the assumed trajectory will be:

$$s_{dyn} = (X'_{param}, X_{param}) = \frac{d'_k(X'_{param})}{d_k(X_{param})}$$

(43)

$$s_{dyn} = \{s_{dyn}^{t_m}, s_{dyn}^{D_s}, s_{dyn}^{\Delta t_k}, s_{dyn}^{\Delta V}\}$$

where:
t_m - advance time of the manoeuvre with respect to the dynamic properties of the own ship,
Δt_k - duration of one stage of the ship's trajectory,
D_s - safe distance,
ΔV - reduction of the own ship's speed for a deflection from the course greater than 30°.

5.4 Determination of safe game trajectories

Computer simulation of POSTRAJ and RISKTRAJ algorithms, as a computer software supporting the navigator decision, were carried out on an example of a real navigational situation of passing j=16 encountered ships. The situation was registered in Kattegat Strait on board r/v HORYZONT II, a research and training vessel of the Gdynia Maritime University, on the radar screen of the ARPA anti-collision system Raytheon.
The POSGAME algorithm represents the ship game trajectories determined according to the control index in the form (27) (Fig. 18).
The RISKTRAJ algorithm was developed for strategies: $\upsilon_0 = 13$ and $\upsilon_j = 25$ (Fig. 19).

5.5 Characteristics of control sensitivity in real navigational situation at sea

Figure 20 represents sensitivity characteristics which were obtained through a computer simulation of the game control POSTRAJ and RISKTRAJ algorithms in the Matlab/Simulink software for the alterations of the values $X_{0,j}$ and X_{param} within ±5% or ±3°.

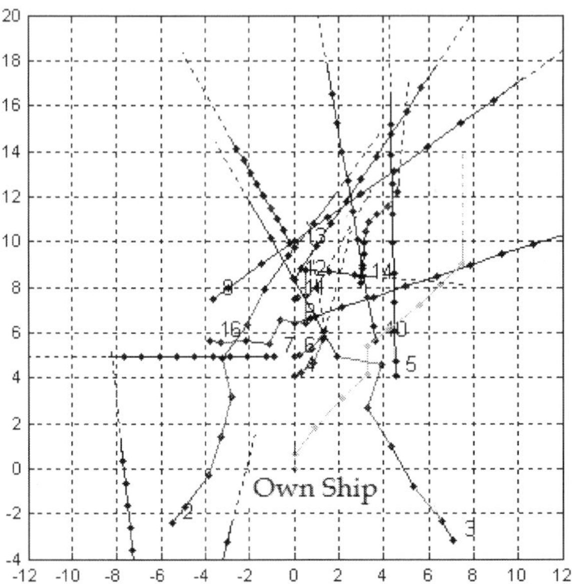

Fig. 18. The own ship game trajectory for the POSTRAJ, in good visibility, D_s=0,5 nm, $r(t_k)$=0, $d(t_k)$=7,72 nm, in situation of passing j=16 encountered ships

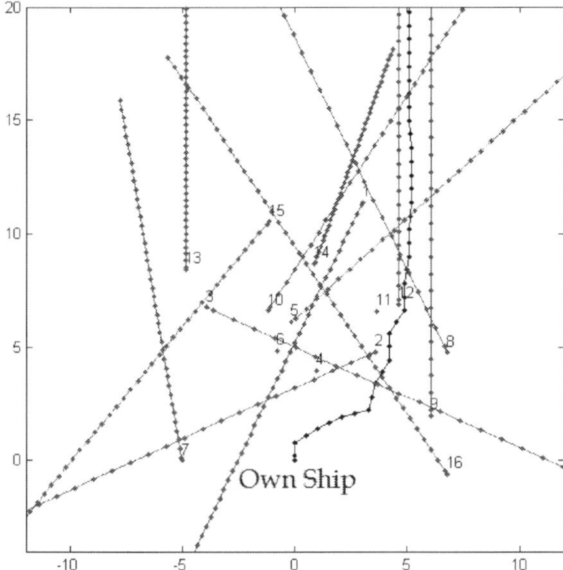

Fig. 19. The own ship game trajectory for the RISKTRAJ, in good visibility, D_s=0,5 nm, $r(t_k)$=0, $d(t_k)$=6,31 nm, in situation of passing j=16 encountered ships

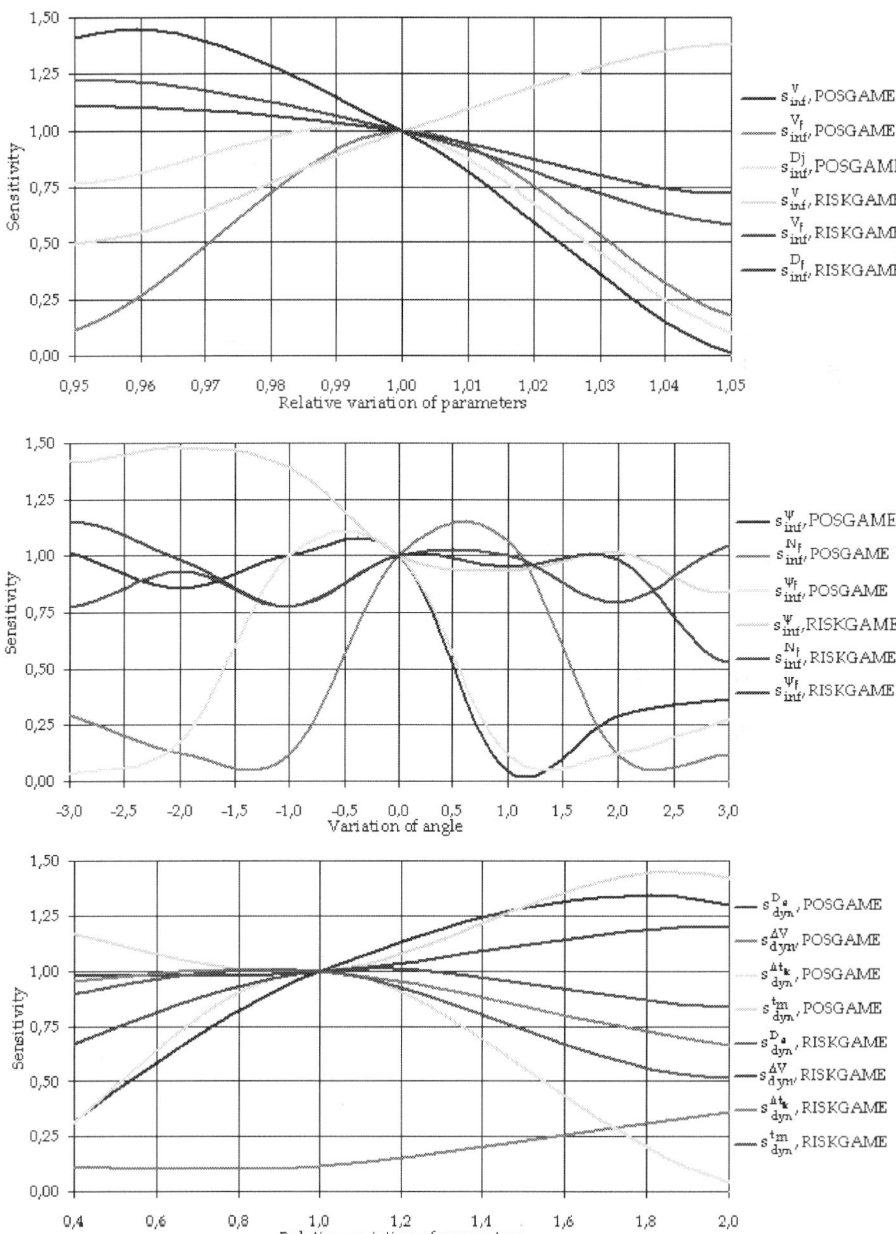

Fig. 20. Sensitivity characteristics of safe game ship control according to POSGAME and RISKTRAJ algorithms on an example of a navigational situation in the Kattegat Strait

6. Conclusion

The application of simplified model of the dynamic game of the process to the synthesis of the optimal control allows the determination of the own ship safe trajectory in situations of passing a greater number of the encountered ships as a certain sequence of the course and speed manoeuvres. The developed RISKTRAJ algorithm takes also into consideration the Rules of the COLREG Rules and the advance time of the manoeuvre approximating the ship's dynamic properties and evaluates the final deviation of the real trajectory from the reference value.

The sensitivity of the final game payment:

- is least relative to the sampling period of the trajectory and advance time manoeuvre,
- most is relative to changes of the own and met ships speed and course,
- it grows with the degree of playing character of the control process and with the quantity of admissible strategies.

The considered control algorithm is, in a certain sense, formal model of the thinking process of a navigator conducting a ship and making manoeuvring decisions. Therefore they may be applied in the construction of both appropriate training simulators at the maritime training centre and also for various options of the basic module of the ARPA anti-collision radar system.

7. References

Baba, N. & Jain, L. C. (2001). *Computational Intelligence in Games,* Physica-Verlag, ISBN 3-7908-1348-6, New York

Bist, D.S. (2000). *Safety and security at sea,* Butter Heinemann, ISBN 0-7506-4774-4, Oxford-New Delhi

Bole, A.; Dineley, B. & Wall, A. (2006). *Radar and ARPA manual,* Elsevier, ISBN 978-0-7506-6434-9, Amsterdam-Tokyo

Cahill, R. A. (2002). *Collisions and Thair Causes*, The Nautical Institute, ISBN 187-00-77-60-1, London

Cichuta, M. & Dalecki, M. (2000). *Study of computer program determining safe ship's game trajectory with consideration risk of collision",* M.Sc. thesis, Gdynia Maritime University, Poland, (in Polish)

Cockcroft, A.N. & Lameijer, J.N.F. (2006). *The collision avoidance rules,* Elsevier, ISBN 978-0-7506-6179-9, Amsterdam-Tokyo

Cymbal, N.N.; Burmaka, I.A. & Tupikow, I.I. (2007). *Elastic strategies of the passing ships,* KP OGT, ISBN 978-966-8128-96-7, Odessa (in Russian)

Engwerda, J. C. (2005). *LQ Dynamic Optimization and Differential Games,* John Wiley & Sons, ISBN 0-470-01524-1, West Sussex

Gluver, H. & Olsen, D. (1998). *Ship collision analysis,* A.A. Balkema, ISBN 90-5410-962-9, Rotterdam-Brookfield

Isaacs, R. (1965). *Differential games,* John Wiley & Sons, New York

LaValle, S. M. (2006). *Planning algorithms,* Camridge, ISBN 0-521-86205-1, New York

Levine, W.S. (1996). *The Control Handbook,* CRC Press, ISBN 0-8493-8570-9, Florida

Lisowski, J.; Rak A. & Czechowicz W. (2000). Neural network classifier for ship domain assessment. *Journal of Mathematics and Computers in Simulation*, Vol. 51, No. 3-4, 399-406, ISSN 0378-4754

Lisowski, J. (2001). Computational intelligence methods in the safe ship control process, *Polish Maritime Research*, Vol. 8, No. 1, 18-24, ISSN 1233-2585

Lisowski, J. (2002). Computational intelligence and optimisation methods applied to safe ship control process, *Brodogradnja*, Vol. 50, No. 2, 195-201, ISSN 502-141-274

Lisowski, J. (2004). Safety of navigation based on game theory – computer support algorithms of navigator decisions avoiding accidents at sea, *Journal of Shanghai Maritime University*, Vol. 104, No. 11, 75-85, ISSN 1672-9498

Lisowski, J. (2005a). Mathematical modeling of a safe ship optimal control process, *Polish Journal of Environmental Studies*, Vol. 14, No. I, 68-75, ISSN 1230-1485

Lisowski, J. (2005b). Game control methods in navigator decision support system, *Journal of Archives of Transport*, Vol. 17, 133-147, ISSN 0866-9546

Lisowski, J. (2007a). The dynamic game models of safe navigation, In: *Advances in marine navigation and safety of sea transportation*, Adam Weintrit, 23-30, Gdynia Maritime University-The Nautical Institute in London, ISBN 978-83-7421-018-8, Gdynia

Lisowski, J. (2007b). Application of dynamic game and neural network in safe ship control, *Polish Journal of Environmental Studies*, Vol. 16, No. 4B, 114-120, ISSN 1230-1485

Lisowski, J. (2008a). Computer support of navigator manoeuvring decision in congested waters. *Polish Journal of Environmental Studies*, Vol. 17, No. 5A, 1-9, ISSN 1230-1485

Lisowski, J. (2008b). Sensitivity of safe game ship control, *Proceedings of the 16th IEEE Mediterranean Conference on Control and Automation*, pp. 220-225, ISBN 978-1-4244-2504-4, Ajaccio, June 2008, IEEE-MED

Lisowski, J. (2008c). Optimal and game ship control algorithms avoiding collisions at sea, In: *Risk Analysis VII, Simulation and Hazard Mitigation*, Brebia C.A. & Beriatos E., 525-534, Wit Press, ISBN 978-1-84564-104-7, Southampton-Boston

Łebkowski, A. (2001). *Study and computer simulation of game positional control algorithm with consideration ship's speed changes in Matlab*, M.Sc. thesis, Gdynia Maritime University, Poland, (in Polish)

Nisan, N.; Roughgarden, T.; Tardos E. & Vazirani V.V. (2007). *Algorithmic Game Theory*, Cambridge University Press, ISBN 978-0-521-87282-9, New York

Nowak, A. S. & Szajowski, K. (2005). *Advances in Dynamic Games – Applications to Economics, Finance, Optimization, and Stochastic Control*, Birkhauser, ISBN 0-8176-4362-1, Boston-Basel-Berlin

Osborne, M. J. (2004). *An Introduction to Game Theory*, Oxford University Press, ISBN 0-19-512895-8, New York

Radzik, T. (2000). Characterization of optimal strategies in matrix games with convexity properties. *International Journal of Game Theory*, Vol. 29, 211-228, ISSN 0020-7226

Straffin, P. D. (2001). *Game Theory and Strategy*, Scholar, ISBN 83-88495-49-6, Warszawa (in Polish)

Pasmurow, A. & Zinoviev, J. (2005). *Radar Imaging and holography,* Institution of Electrical
        Engineers, ISBN 978-086341-502-9, Herts
Wierzbicki, A. (1977). *Models and sensitivity of control systems,* WNT, Warszawa (in Polish)

TOPIC AREA 2: Radar Applications

5

Wind Turbine Clutter

Beatriz Gallardo-Hernando, Félix Pérez-Martínez
and Fernando Aguado-Encabo
Technical University of Madrid (UPM) and Spanish Meteorological Agency (AEMET)
Spain

1. Introduction

The use of wind farms to generate electricity is growing due to the importance of being a renewable energy source. These installations can have over a hundred turbines of up to 120 m height each. Wind farm installations relatively near to radar systems cause clutter returns that can affect the normal operation of these radars. Wind turbines provoke clutter reflectivity returns with unpredictable Doppler shifts.

Wind turbines exhibit high radar cross sections (RCS), up to $1000m^2$ in some instances, and then, they are easily detected by radars. A typical wind turbine is made up of three main components, the tower, the nacelle and the rotor. The tower is a constant zero velocity return that can be suppressed by stationary clutter filtering. Unlike the tower, the turbine nacelle RCS is a function of the turbine yaw angle, and then, the radar signature will depend on this factor. Moreover, most wind turbines present curved surface nacelles which will scatter the energy in all directions and so the variability of the RCS is prominent. In addition, the rotor makes the blades move fast enough to be unsuppressed by conventional clutter filtering.

In this chapter, we will examine the characteristics of wind turbine clutter in great detail. For this purpose, we will use examples derived from real experimental data. After describing the experimental data gathered, we will perform several studies.

First of all, a complete statistical characterization of the experimental data will be accomplished. This statistical study will show the distinctive properties of this variety of clutter and then, it will give us clues for its correct detection, as every detection theory must rely on statistics. In this case we will study isolated turbines, so that the obtained characteristics will be accurate. After that, we will make an extensive frequency analysis. Different configurations will be studied, with variations such us the number of turbines, the yaw angle or the radar dwell time. This will show various wind turbine clutter situations that most affected radar systems have to deal with.

Finally, some mitigation techniques that have been published to date will be reviewed. Their main purposes, techniques and results will be analyzed and illustrated with descriptive examples.

2. Wind turbine clutter

2.1 Wind power

Wind power has proved to be one of the most profitable energy sources both in terms of economy and ecology. In fact, many countries have launched programs in order to deploy wind turbines as alternative sources of energy, trying to tackle the climate change as well as the increasing oil costs. As it can be seen in Fig. 1, wind energy production has been exponentially increasing (World Wind Energy Association, 2009) since the early 90s.

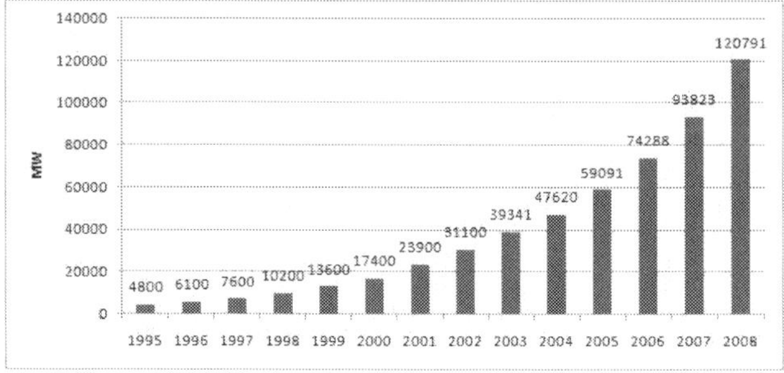

Fig. 1. Evolution of the wind power installed in the world.

Traditionally, Europe has leaded the wind energy market, with 70% of the sales by 2000 (Hatziargyriu & Zervos, 2001). However, the wind energy capacity has been promoted all over the world, and countries such as China or India are now using this technology to produce large amounts of electrical energy. USA is currently the largest wind power market, followed by Germany and Spain, Fig. 2 (World Wind Energy Association, 2009). With respect to penetration rates, this power provides 19% of the total energy consumed in Denmark, 11% in Spain and Portugal, 6% in Germany and 1% in USA (Thresher & Robinson, 2007). In terms of growth, world wind generation capacity more than quadrupled between 2000 and 2006. Wind farms will continue its expansion, as it is expected that within the next decades wind energy will occupy 20 % of the total annual power consumed both in Europe and USA (Thresher & Robinson., 2007). As a consequence, the impact of wind turbine clutter on radars is going be more and more important.

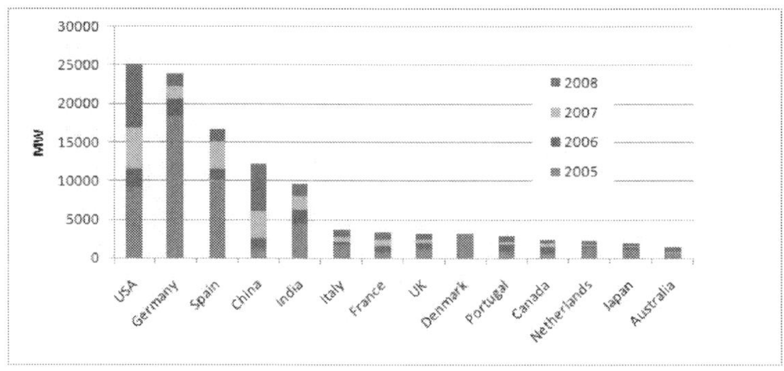

Fig. 2. Cumulative wind power capacity in the world.

2.2 Effects of wind turbines on radar systems

A typical wind turbine is made up of three components, the tower, the nacelle and the rotor. The tower means a constant zero velocity return than can be easily minimized by means of an appropriate clutter cancellation. Unlike the tower, the turbine nacelle radar cross section

(RCS) is a function of the turbine yaw angle and then, the radar signature will depend on this factor. More over, most wind turbines present curved surface nacelles which will scatter the energy in all directions and so, will increment the variability of the RCS (Poupart, 2003). Some studies conclude (Greving, 2008a) that the traditional RCS scheme is not applicable for objects on the ground and, therefore, it is not a useful parameter for the definition of safeguarding distances of wind farms to radars. Besides, the rotor makes the blades move fast enough to be unsuppressed by stationary clutter filtering, with maximum angular velocities between 30 and 50 rpm (Freris, 1992). To sum up, the main effects wind turbines have on radars are the following (Perry & Biss, 2007):

- The magnitude of the reflections from the wind turbine can cause radar receivers to be driven to saturation. In fact, typical tower heights reach 100 m, with blades from 30 to 50 m long, see Table 1 (Gamesa, 2009). Some studies address this problem and propose stealth technologies to mitigate this effect. The solutions involve both shaping and development of absorbing materials (Matthews et al., 2007)

Model	G58	G83	G90
Turbine rating (kW)	850	2000	2000
Blade length (m)	28.3	40.5	44
Tower height (m)	44-71	67-78	67-100
Rotation rate (rpm)	14.6-30.8	9-19	9-19
Max Tip speed (m/s)	91	80.5	87.5

Table 1. Typical Wind Turbine parameters.

- The rotation movement of blades cause Doppler shifts. The velocity of a blade depend on the distance from the centre, therefore, there is an increasing Doppler shift from the centre to the tip of the blade. This spectrum can fall within the limits of some radars or exceed them.

These effects result in various situations in different radars.

- For primary surveillance radars, air traffic control and air defence (Jackson, 2007), wind turbine effects include clutter, increased number of unwanted returns in the area of wind farms; desensitisation, reduced probability of detection for wanted air target; and a consequent loss of wanted target plotting and tracking. In conclusion, they provoke higher probability of false alarm and lower probability of detection.
- In weather radars (Vogt et al., 2008), the clutter, signal blockage and interference may cause the misidentification of thunderstorm features and meteorological algorithm errors such us false radar estimates of precipitation accumulation, false tornadic vortex and mesocyclone signatures and incorrect storm cell identification and tracking.
- Monopulse secondary radars performance is also affected by the presence of wind turbines (Theil & van Ewijk, 2007). The azimuth estimate obtained with the monopulse principle can be biased when the interrogated target emits its response when partially obscured by an large obstacle such as a wind turbine.

3. Radar signature of wind turbine clutter

3.1 Experimental data

In the experiments, made with the aid of the Spanish weather C-band radar network, we gathered data in normal and spotlight operation modes. In the first case, the aim is to calculate the Doppler spectrum in two ways: for each range gate, for a specific azimuth angle; and for each azimuth angle, for a specific range gate. This will show the variations of the wind turbine clutter Doppler spectrum as functions of range and azimuth angle. This spectrum is expected to have specific features which would aid the identification and mitigation of these clutter returns. In the second case, spotlight operation mode, the data are collected from a particular cell, known to experiment wind turbine clutter and so defined by specific range and azimuth angle. That is to say, the radar dish is stationary and a large and contiguous set of time series is collected. Thus, the information about temporal evolution of the amplitude of the signal and its Doppler spectrum can be easily extracted. These experiments allow us to do a detailed examination of the spectral characteristics and statistics of the wind turbine clutter signal.

All data were taken from a C-band weather radar near Palencia, Spain. Up to three different wind farms can be seen in a narrow sector between 30 and 45 km away from the radar. The main wind farm is composed by 54 wind turbines model G-58 (Gamesa, 2009), which provide an average power of 49300 kW. It was a clear day, so there weren't interfering weather signals. In the following figure we can distinguish the three wind farms.

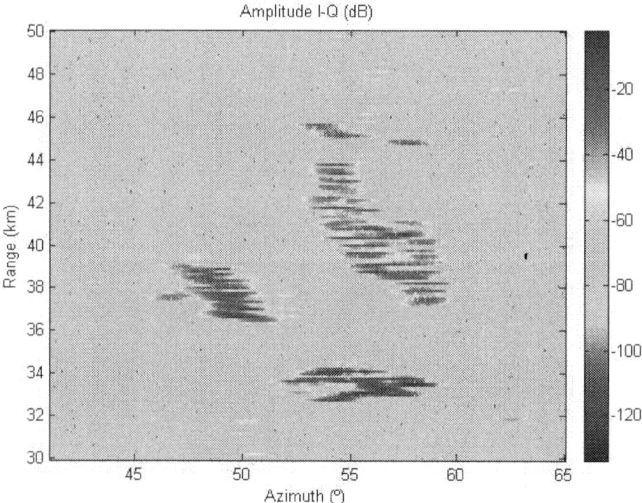

Fig. 3. PPI representation of the data under study.

The turbines layout within the wind farm let the radar resolve the different rows. It is usual to maintain a minimum distance between turbines (Jenkins, 1993) because when a wind turbine extracts energy from the wind it creates a wake of slower, more turbulent air which will negatively impact on the performance of adjacent turbines. This spacing is usually set from five to eight times the blade diameter, that is, about 200 m. Therefore, with a radar range resolution of 125 m it is possible to resolve different turbines in range. However, azimuth resolution does vary with the distance and most of the times two or more turbines will occupy the same resolution cell.

3.2 Scanning radar

By calculating the Doppler spectrum, defined as the power weighted distribution of radial velocities within the resolution volume of the radar (Doviak & Zrnic, 1984), for each azimuth angle, for a particular range gate, the spectral content versus the azimuth angle can be studied. I-Q radar data were gathered with the slowest antenna velocity, the lowest elevation angle (the most affected by the presence of wind farms) and the highest pulse repetition frequency (PRF). See Table 2 for detailed radar parameters. The spectral content of several range bins has been studied using a Short-time Fourier Transform (STFT) of partially overlapped time sectors to build a spectrogram. A Hamming window was used in order to diminish windowing effects.

An example has been represented in Fig. 4 (Gallardo-Hernando et al. 2008b). There was an isolated wind mill in the selected range gate, so, the spectrum is located at a very specific azimuth angle. This spectrum is extremely wide, as some of its components seem to be overlapped.

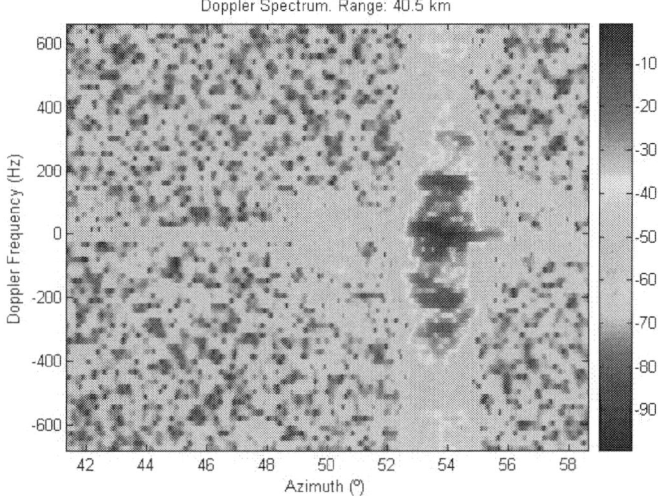

Fig. 4. Doppler spectrum versus azimuth angle.

Frequency	5500 MHz
Beam width	0.8°
Power	250 kW
Antenna gain	43 dBi
Pulse repetition frequency (PRF)	1300 Hz
Elevation angle	0.5°
Antenna velocity	12°/s
Pulse length	0.5 µs

Table 2. Scanning radar parameters.

The spectrum of two different turbines in the same range bin is plotted in Fig. 5. The Doppler frequency shift is different for each turbine for two reasons. First, the rotors can have different velocities of rotation. Second, although the turbines were rotating at the same velocity, the yaw angle could be different and so the radial velocity.

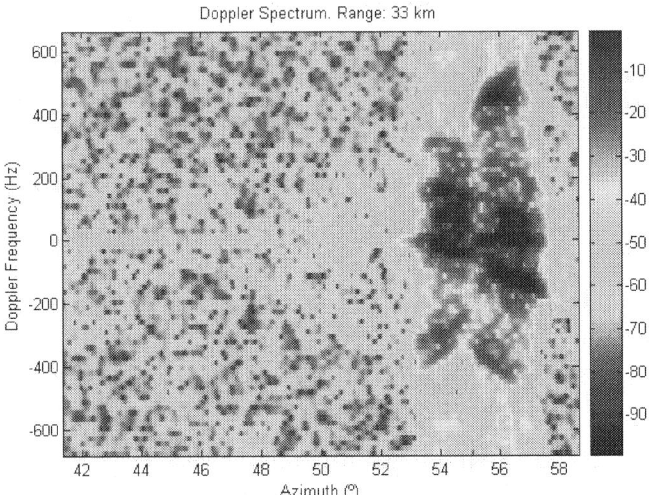

Fig. 5. Doppler spectrum of two adjacent wind turbines.

Fig. 6 shows a similar variation of the Doppler spectrum, now as a function of range. This spectrum is also extremely wide, and it obviously appears at every range gate with wind turbines located in.

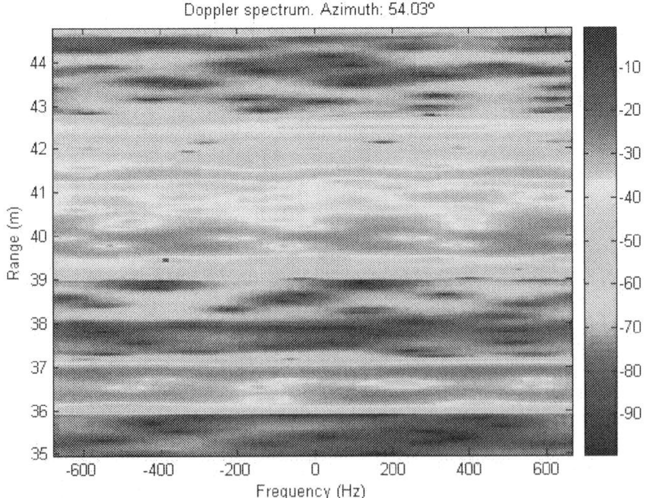

Fig. 6. Doppler spectrum calculated on the 54° azimuth bin.

3.2 Spotlight radar

In this operation mode, defined in (Isom et al., 2008), a large set of time series was collected while the radar dish was stationary and the azimuth angle defined to get data from the wind farm, 54.03°. Radar parameters are similar to those of Table 2, but for the radar antenna, zero velocity.

Fig. 7 shows the variation of the signal amplitude versus time. It shows a clear periodicity that is supposed to be caused by the motion of blades. Later on this periodicity will be studied, see section 3.3. The noise level seems to be 20 dB under the signal and the effect of target scintillation is clearly seen.

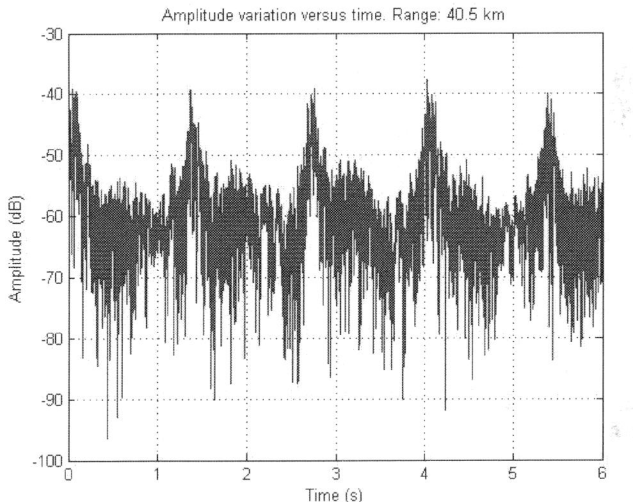

Fig. 7. Amplitude variation at 42.5 km bin versus time.

A frequency transform was made to calculate the Doppler spectrum. Fig. 8 shows the Doppler spectrogram in time, and the same periodicities seen in Fig. 7 seem to be here. The echo of the blades is characterized by short flashes, which occur when one blade is orthogonal with respect to the propagation direction of the transmitted signal (Gallardo-Hernando et al., 2008b). Once again, the spectrum is spread all over Doppler frequencies, and so, we cannot assure which components are moving towards or away from the radar.

Then, there are five very intense Doppler returns in Fig. 8. These flashes are separated approximately 1.33 seconds. By considering a three-blade wind turbine, this period means than one blade takes over 4 seconds to do an entire rotation of 360°, that is to say, the angular velocity of the blades is estimated to be 15 rpm. The reason why negative Doppler shifts (blades going down) are less powerful lies in the elevation angle, the differences of RCS between blade sides, and also in a possible shadowing of the radar beam. The aspect of these flashes is explained by the fact that the sum of the contributions of the different parts of a blade is coherent only when the blade is perpendicular to the line of sight. If there is no perpendicularity, the vector sum is destructive, as a consequence of the variability of the phase. Just in the blade tip the vectors are not enough to cancel the signal and a peak appears. This peak is visible in Fig. 8, and as the blade describes a circumference, a sinusoidal function appears in the spectrogram.

Although the blades tip velocities can be much higher than the maximum non ambiguous velocity of approximately 18m ·s⁻¹, the yaw angle involves a lower radial velocity.

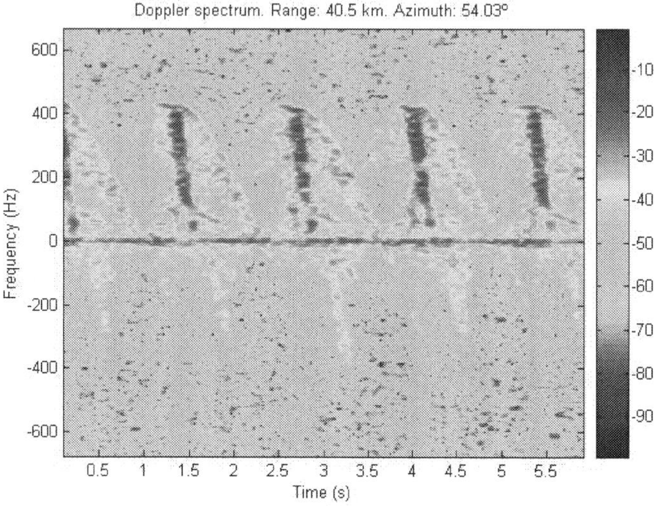

Fig. 8. Doppler spectrum at 42.5 km bin versus time.

In most cases the blade's energy returns are distributed over the entire Doppler frequency spectrum, there is a total ambiguity scheme, Fig. 9.

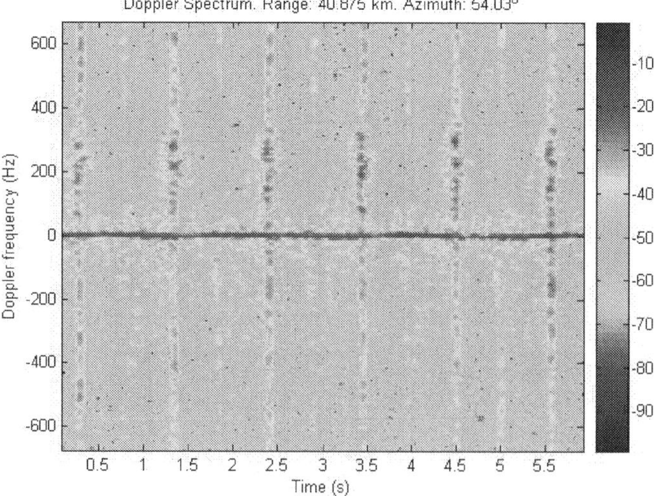

Fig. 9. Ambiguous WTC.

Focusing on the most powerful flashes, the wind turbine whose time behaviour has been represented in Fig. 9 rotates at 20 rpm. But, what is the meaning of the weaker flashes between them? There are several possibilities. First, they can correspond to the same blades on a different position than the perpendicular as they seem to have the same period. Second, there is another wind turbine in the same range bin with apparently the same rotation rate. Third, and more probable, they are the effect of the antenna side lobes. Then, for most wind farms and wind turbines, not even the time between flashes will be completely clean of clutter. More examples of spotlight WTC are plotted in Fig. 10.

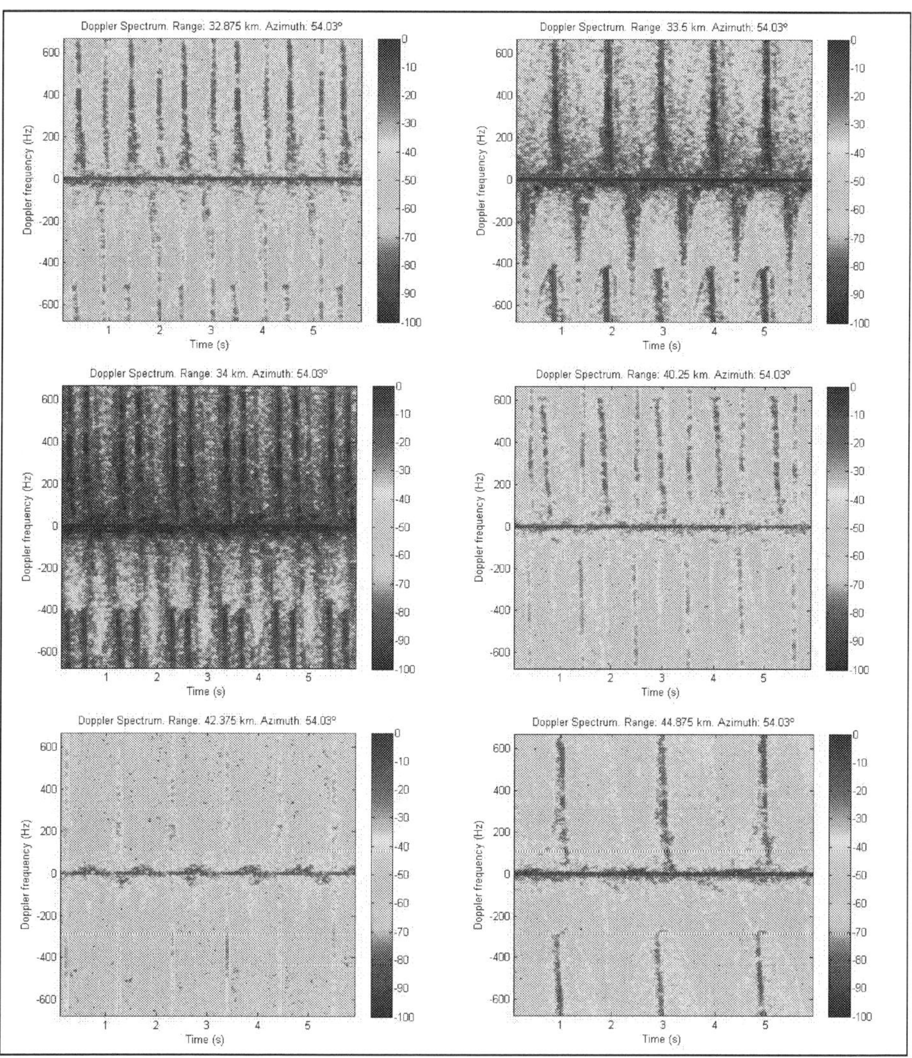

Fig. 10. Examples of WTC.

3.3 Statistics

The aim of the characterization of wind turbine clutter by means of statistical analysis is to model it as a stochastic process. Although this kind of clutter is not strictly stationary, it may exhibit some features that would allow an optimal detection of wind turbines for a latter mitigation.

<u>Doppler statistics</u>

This section focuses on detailing the dynamic behaviour of the Doppler spectrum as well as the relationship between amplitude and spectral qualities (Gallardo-Hernando et al., 2008b). As it can be seen in Fig. 11, the amplitude variations follow the behaviour of the Doppler centroid, defined as the centre frequency of the Doppler spectrum of the data. Its most significantly variations take place at the same time the amplitude maximums appear. The Doppler Bandwidth is centred on 200 Hz and has very small variations.

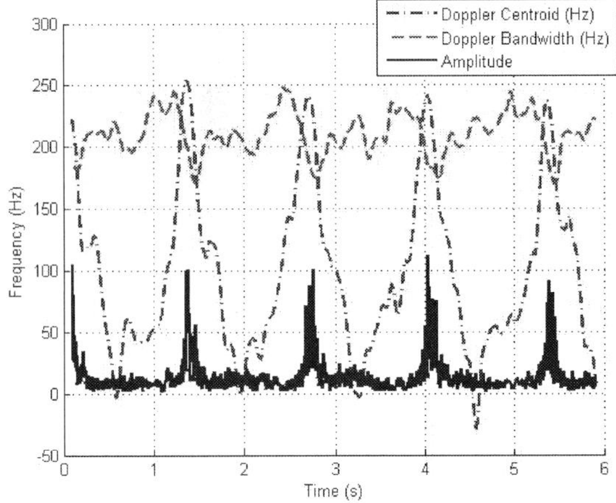

Fig. 11. Comparison of Doppler Centroid, Doppler Bandwidth and Amplitude of the signal versus time.

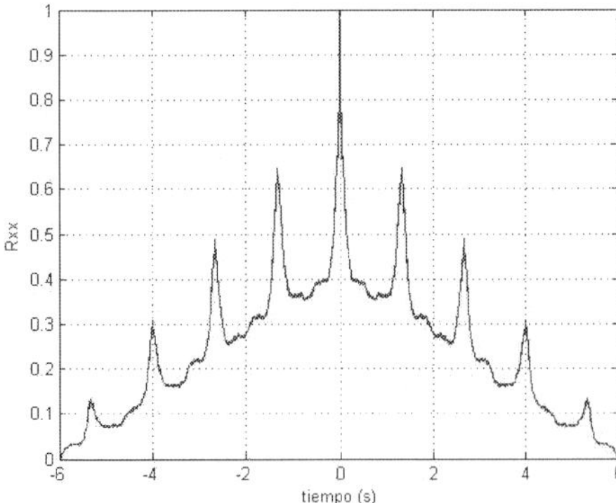

Fig. 12. Autocorrelation.

<u>Amplitude statistics</u>
The autocorrelation of the signal in time, Fig. 12, confirms the periodicities we mentioned before, as a main lobe appear at 1.15 seconds.
A modelling of the experimental amplitude probability density function has also been done. This empirical PDF has been fitted to the Log Normal, Weibull, Rayleigh, Gamma and K distribution. We employed the maximum-likelihood method in all the distributions except for K, where we implemented the method of moments. Fig. 13 shows the result of the fitting process of the experimental PDF to the different theoretic functions. The K distribution seems to provide the best fit. In order to determine the best fit, another technique has been used. The experimental and theoretic moments of the distributions have been calculated from the fitting resulting parameters and then compared. The experimental moments are better approximated by the K distribution, Fig. 14.

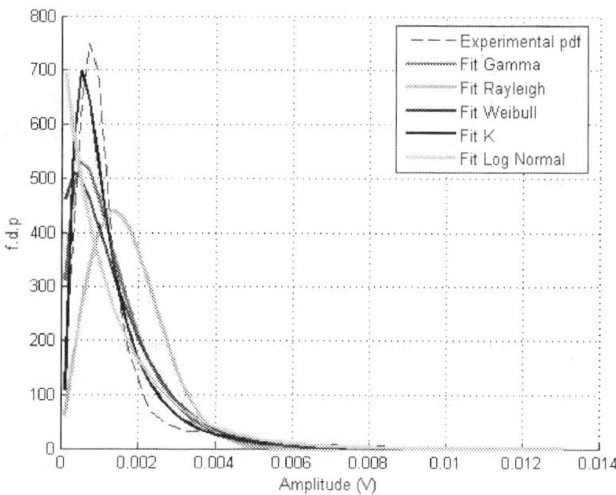

Fig. 13. Comparison of several distribution functions.

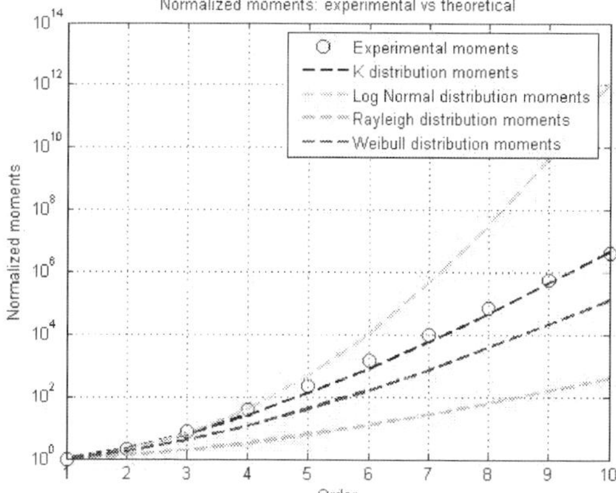

Fig. 14. Comparison of several distribution moments.

4. Mitigation concepts

As it has been shown in previous sections, wind turbine clutter is unpredictable. It can fluctuate from one scan to the following. The blades rotate at such a rate to produce Doppler shifts on the order of 70 or even 90 m·s⁻¹. These values can exceed the maximum non-ambiguous Doppler velocity of some radars and then, make more difficult WTC detection and mitigation.

This section summarizes some of the mitigation techniques that have been published to date. These state-of-the-art processing techniques have been categorized in function of the affected radar: primary air surveillance radars and meteorological radars.

Apart from processing techniques, stealth solutions are also being studied to reduce the problem (Matthews et al., 2007). These techniques try to develop radar absorbing materials as well as to design new wind turbines with reduced radar cross section, preserving the efficiency of turbines in terms of electricity production and construction costs. The main inconvenient of these solutions is that the materials employed might be only efficient for very specific radar frequency bands.

4.1 Air surveillance radars

Several techniques can be employed to minimise the effect of signal blocking and ghost target appearance which wind turbines can provoke. Some of them are listed below (Perry & Biss, 2007) and (Sergey et al., 2008):

- Moving Target Detection (MTD) Doppler processing would reduce the magnitude of the blade returns separating the blade spectrum into Doppler increments.
- Range Averaging Constant False Alarm Reduction (CFAR) processing. Wind turbines provoke the detection threshold to rise, and then, the shadow effect appears. CFAR would then be applied for each Doppler increment from MTD and then anomalous power bins would be substituted with average noise power.
- Increased System Bandwidth would allow detections between wind turbines by using a higher resolution clutter map.
- Plot and Track Filters would reduce false alarms on returns with non-aircraft profiles.
- Range and Azimuth Gating (RAG) maps would enable unique mitigation algorithms to be implemented only in wind farm areas, maintaining normal performance outside the wind farms.
- Sensitivity Time Control (STC) would minimize the radar sensitivity at short range in order to limit the return from the wind turbine while not affecting target detection and so, prevent the receptor to be driven to saturation.
- Enhanced target tracking techniques can be used after detection. Feature aided tracker (FAT) identifies features from signals and process them in a probabilistic manner. The tracker would incorporate special processing techniques such us adaptive logic, map aided processing, processing priorization, enhanced tracking filters or classification algorithms.

These techniques can be used all together and, theoretically, they would allow the detection of aircrafts in wind farm areas with similar results in terms of detection and false alarm probabilities than in areas free from wind turbine clutter.

4.2 Meteorological radars

Weather radars are one of the most affected radio systems by wind turbine clutter. This radar is a special type of primary radar intended to measure atmospheric volumetric targets:

large volumes of clouds and rain. The main distorting effects include reflections from the static parts, reflections from the rotating parts and shadowing or blocking (Greving & Malkomes, 2008b). These effects cause the meteorological algorithms to fail and give false radar estimates of precipitation and wind velocity, as it can be seen in Fig. 15.

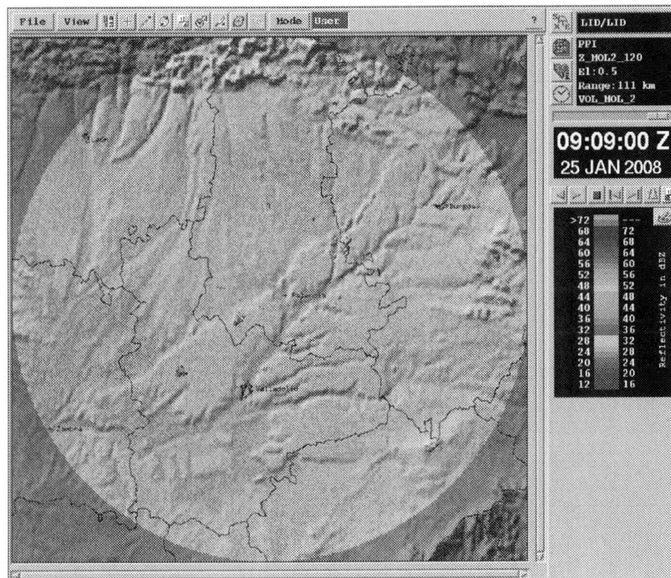

Fig. 15. PPI reflectivity image in a clear day. All the returns come from wind farms.

Prevention

Some of the mitigation efforts are focused on the prevention of this clutter (Donaldson et al., 2008). The assessment for new wind farms should be planned taking into account nearby weather radars by using line of sight calculations, standard 4/3 radio propagation model and echo statistics. But already built wind farms are still distorting weather radars, and then, specific processing is needed.

Interpolation

Wind turbine clutter spectrum usually exceeds most weather radar Doppler capacities. Current clutter filtering techniques are capable of removing the tower component effectively but the effects of the blade motion remains. The objective is to remove the blade components without distorting the desired weather signal. Some studies (Isom et al., 2007) propose interpolation techniques to mitigate WTC. These techniques use uncontaminated data to estimate the weather signal in bins which have been previously detected as contaminated. Results are plotted in Fig. 16. However large wind farms will cause an important loss of valuable weather information in their areas if the separation between turbines is narrower than twice the resolution distance, as none clean bins can be used.

Rain rate dependence

Interpolation has also been used in other works (Gallardo-Hernando et al.,2009) to show the dependence of this technique on rain rate variations. The study included simulated weather data as well as real clutter data retrieved from the radars described in section 3. The

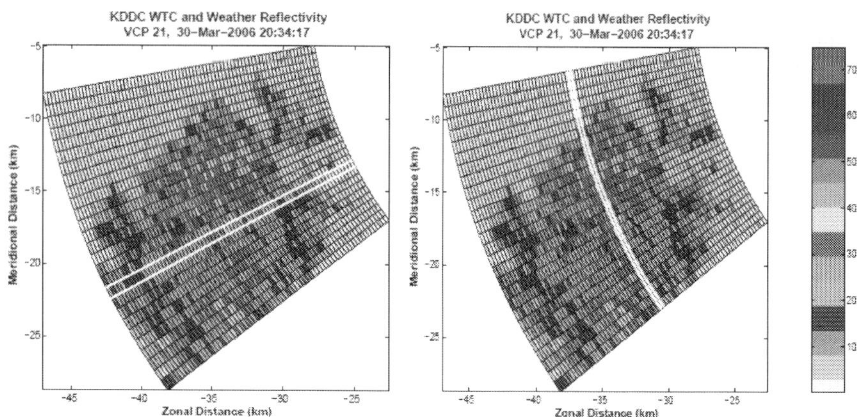

Fig. 16. Interpolation in PPI plots (Isom et al., 2007).

mitigation algorithm is based on the interpolation of correct values in the range bins which had been previously detected as contaminated by WTC. This interpolation is made in the direction of the velocity of the wind. The simulated rain Doppler spectrum was directly added to the WTC spectrum. Zero velocity clutter was previously removed. Fig. 17 shows the result of the addition of simulated rain spectrum to the WTC spectrum data in range plots. The estimated values of reflectivity, Fig. 17a, and velocity, Fig. 17b are drawn for all of the circumstances: WTC plus rain, simulated rain and corrected values. The simulated spectrum uses a rain intensity of 1 mm/h, which implies that the rain would be barely visible under the wind farms. The wind velocity does not vary with range. In reflectivity, the error drops from 32 to 4 dB. In velocity, the error drops from 23 to 0.5 m/s.

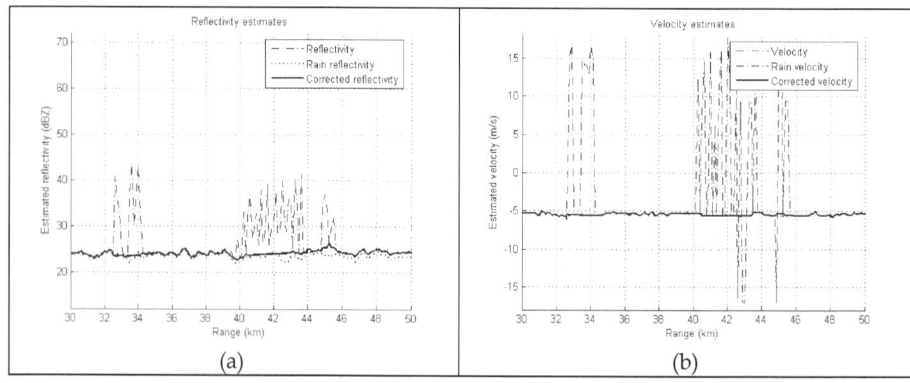

Fig. 17. Total, rain and corrected reflectivity and velocity values for R=1mm/h. In reflectivity, the error drops from 32 to 4 dB. In velocity, the error drops from 23 to 0.5 m/s.

Rain Doppler spectra were simulated from 1mm/h to 70 mm/h and used the same mitigation algorithm, with the previous detection of WTC contaminated range bins. The results are summarized in Fig. 18. Fig 18a shows the errors found in the reflectivity estimation, in dB before and after the algorithm. The error decreases almost exponentially as the rain intensity increases and there is a point where the error is almost the same using or not the algorithm. This happens when rain is much more powerful than clutter. Fig 18b shows the errors found in the velocity estimation before and after applying the algorithm. In this case at certain rain intensity WTC stops affecting the estimation of reflectivity, and the error is slightly greater when using the algorithm due to the evident loss of information.

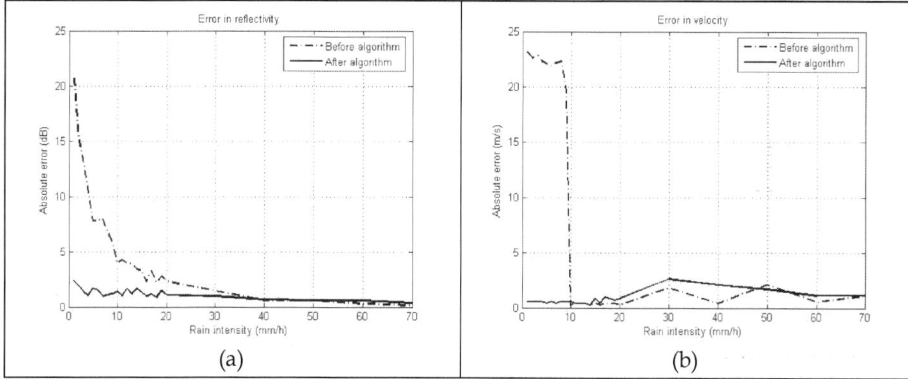

Fig. 18. Absolute error in reflectivity and velocity estimations.

Adaptive thresholding

As we have seen, these techniques require the previous detection of contaminated cells. However, it has also been shown that an adaptive detection can be applied in spotlight mode (Gallardo-Hernando et al., 2008c). This technique is based on the removal of flashes by means of adaptive thresholding. In Fig 19a a real WTC spotlight spectrum can be observed. Flashes are spread over the entire spectrum. Fig. 19b shows a spectrogram of the addition of real wind turbine spectrum and simulated 5mm/h variable velocity rain spectrum.

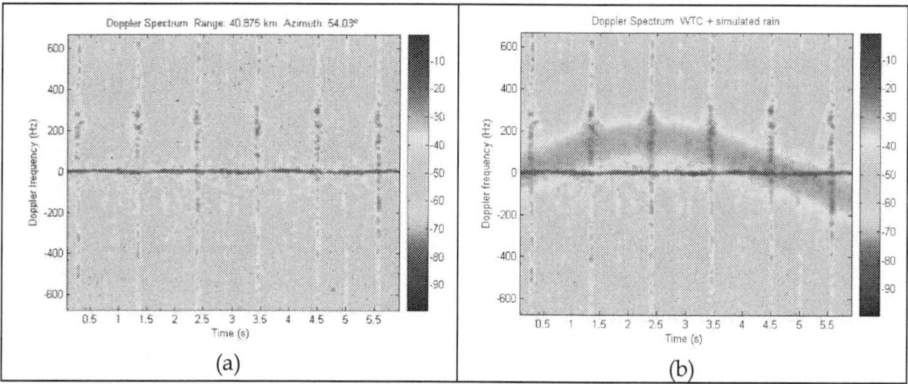

Fig. 19. WTC spotlight data spectrum and its addition to simulated weather data.

The election of an appropriate threshold has to be made regarding the significant changes in the amplitude of the signal, Fig. 20a. Finally, Fig. 20b shows the results after the flashes detection, removal and replacement with information of adjacent time bins.

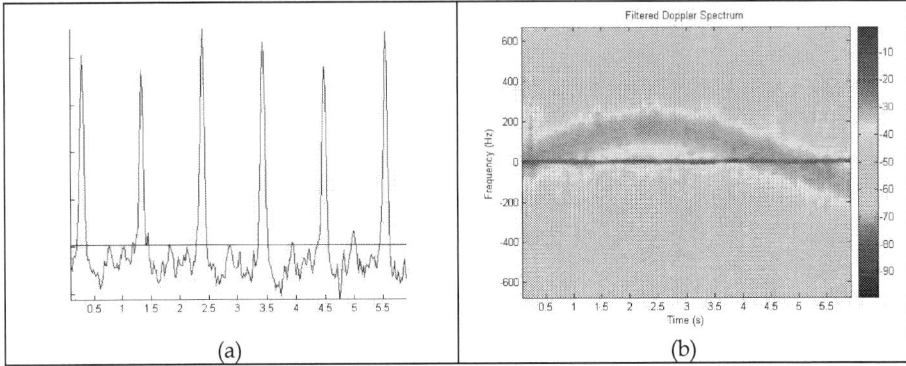

Fig. 20. Adaptive thresholding.

Tomographic techniques

In (Gallardo-Hernando et al., 2008a), an image processing technique to remove WTC in spotlight operation mode is presented. If Fig. 19b is considered as an image, it can be handled by means of specific image processing. The Radon transform of an image is calculated by integrating the intensity values among straight lines characterized by an angle and a distance. Therefore the vertical lines in the original image are going to be seen as $0°$ points in the Radon domain, as they only appear at $0°$ integrations. In particular, variations of $0.1°$ were used in a $-90°\leq\theta<90°$ interval. Fig. 21a shows the result of the transformation of Fig. 19b into the Radon domain. The six clutter flashes that appeared before are now six $0°$ points, whereas the rain is mostly near $90°$, as well as the ground clutter. WTC is now very easy to remove without distorting the weather information, in this case, values between $-5°\leq\theta<5°$ were filtered. Fig. 21b shows the results after the removal of the clutter points and the inverse transformation.

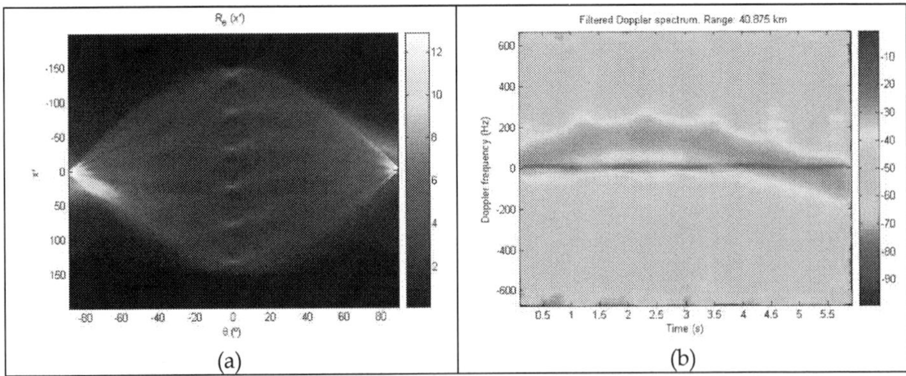

Fig. 21. Radon transformation and results after filtering.

<u>Adaptive Arrays</u>
Adaptive phased array radars have also been proposed as possible solution to WTC in weather radars (Palmer et al., 2008). These arrays offer the capability to obtain a signal that accurately represents the weather only scattering field. By carefully using the interfence of the radiation pattern, the WTC near the ground is rejected while the scattered energy of the weather above the ground is preserved.

5. Conclusion

In this chapter the main effects of wind turbines on the performance of radar systems have been explained. The radar signature of wind turbine clutter is unique and then, it requires a special treatment when developing mitigation techniques. WTC clutter remains spatially static, but it fluctuates continuously in time. In surveillance radars the return from wind turbines can be completely different from one scan to the following. In addition, apart from the powerful tower return, the movement of blades produces large Doppler shifts. Some of the latest mitigation techniques have been described; however, a more extensive study has to be accomplished. As future research, novel automatic detection techniques as well as accurate mitigation schemes in scanning radars have been planned to be developed.

6. References

Donaldson, N.; Best, C. & Paterson, B. (2008). Development of Wind Turbine Assessments for Canadian Weather Radars. *Proceedings of the Fifth European Conference on Radar in Meteorology and Hydrology, ERAD 2008*, pp. 1-4, Helsinki, Finland, July 2008. ISBN: 978-951-697-676-4.

Doviak, R. J., & Zrnic, D. S. (1984). Doppler Radar and Weather Observations, Academic Press, Inc., ISBN: 0-12-1420-X. Orlando, Florida. USA.

Freris, L. (1992). Inherit the Wind. *IEEE Review*, Vol. 38, No. 2, April 1992, pp. 155-159. ISSN: 0953-5683 .

Gallardo-Hernando, B.; Pérez-Martínez, F. & Aguado-Encabo, F. (2008a). Characterization Approach of Wind Turbine Clutter in the Spanish Weather Radar Network. *Proceedings of the Fifth European Conference on Radar in Meteorology and Hydrology, ERAD 2008*, pp. 1-5, Helsinki, Finland, July 2008. ISBN: 978-951-697-676-4.

Gallardo-Hernando, B.; Pérez-Martínez, F. & Aguado-Encabo, F. (2008b). Statistical Characterization of Wind Turbine Clutter in C-band Radars. *Proceedings of the 2008 International Conference on Radar*, pp. 360-364, Adelaide, Australia, September 2008. ISBN: 978-1-4244-2321-7.

Gallardo-Hernando, B.; Pérez-Martínez, F. & Aguado-Encabo, F. (2008c). La Energía Eólica y su Impacto en la Red de Radares Meteorológicos de la AEMet. XXIII Simposium Nacional de la Unión Científica Internacional de Radio, pp. 1-4, Madrid, Spain, September 2008. ISBN: 978-84-612-6291-5.

Gallardo-Hernando, B.; Pérez-Martínez, F. & Aguado-Encabo, F. (2009). Mitigation of Wind Turbine Clutter in C-band Weather Radars for Different Rainfall Rates. *Proceedings of the 2008 International Conference on Radar*, ISBN: 2-912328-55-1 pp. 1-6, Bordeaux, France, October 2009.

Gamesa. G58-850 Turbine http://www.gamesa.es/index.php/es/aerogeneradores

Greving, G. & Malkomes, M. (2008a). Application of the Radar Cross Section RCS for Objects on the Ground – Example of Wind Turbines. *Proceedings of the 2008*

International Radar Simposium, pp. 1-4, Wroclaw, Poland, May 2008. ISBN: 978-83-7207-757-8.

Greving, G.; & Malkomes, M. (2008b). Weather Radar and Wind Turbines – Theoretical and Numerical Analysis of the Shadowing Effects and Mitigation Concepts. *Proceedings of the Fifth European Conference on Radar in Meteorology and Hydrology, ERAD 2008*, pp. 1-5, Helsinki, Finland, July 2008. ISBN: 978-951-697-676-4.

Hatziargyriou, N.; & Zervos, A. (2001). Wind Power Development in Europe. *Proc. IEEE*, Vol. 89, no. 12, December 2001, pp. 1765-1782, ISSN: 0018-9219.

Isom, B. M.; Palmer, R. D.; Secrest, G. S.; Rhoton, R. D.; Saxion, D.; Winslow, J. L.; Reed, J.; Crum, T.; & Vogt, R. (2007). Characterization and mitigation of wind turbine clutter on the WSR-88D network. *Proceedings of the 33rd Conference on Radar Meteorology*, August 2007, paper 8B.8.

Isom, B. M. (2007b). Characterization and Mitigation of Wind turbine Clutter for the WSR-88D Network. University of Oklahoma.

Jackson, C. A. (2007). Wind farm characteristics and their effect on radar systems. *Proceedings of the 2007 IET International Conference on Radar Systems*, pp. 1-6, Edinburg, UK, October 2007, IET. ISBN: 978-0-86341-848-8

Jenkins, N. (1993): 'Engineering wind farms', Power Engineering Journal, Volume 7 Issue 2. April 1993. pp. 53-60. ISSN: 1479-8344

Matthews, J.; Pinto, J. & Sarno, C. (2007). Stealth Solutions to solve the radar-wind farm interaction problem. *Proceedings of the 2007 Antennas and Propagation Conference*, pp. 101-104, Loughborough, UK, April 2007, IEEE. ISBN: 1-4244-0776-1

Palmer, R. D.; Le, K. D. & Isom, B. (2008). Wind Turbine Clutter Mitigation using Fully Adaptive Arrays. *Proceedings of the 5th European Radar Conference*, pp. 356-359, Amsterdam, The Netherlands, October 2008. ISBN: 978-2-87487-009-5.

Perry, J. &Biss, A. (2007). Wind farm clutter mitigation in air surveillance radar. *Aerospace and Electronic systems Magazine*, IEEE, Vol. 22, No. 7, July 2007, pp. 35-40. ISSN: 0885-8985.

Poupart, G. (2003). Wind Farms Impact on Radar Aviation Interests. Final report. QinetiQ W/14/00614/00/REP DTI PUB URN 03/1294., 2003.

Sergey, L.; Hubbard, O; Ding, Z.; Ghadaki, H.; Wang, J. & Ponsford, T. (2008). Advanced mitigating techniques to remove the effects of wind turbines and wind farms on primary surveillance radars. *Proceedings of the 2008 IEEE Conference*, pp. 1-6, Rome, Italy, May 2008. ISBN: 978-1-4244-1538-0.

Theil, A. & van Ewijk, L.J. (2007). Radar Performance Degradation due to the Presence of Wind Turbines. *Proceedings of the 2007 IEEE Radar Conference*, pp. 75-80, Boston, USA, April 2007. ISBN: 1-4244-0284-0.

Thresher, R.; Robinson, M.; & Veers, P. (2007). To Capture the Wind. *IEEE Power energy Mag.* Vol. 5, No. 6, December 2007, pp. 34-46. ISSN: 1540-7977.

Vogt, R.; Reed, J.; Crum, T.; Snow, J.; Palmer, R.; Isom, B. & Burgess, D. (2008). Impacts of wind farms on WSR-88D operations and policy considerations. Preprints, *23rd Int. Conf. on Interactive Information Processing Systems (IIPS) for Meteorology, Oceanography, and Hydrology*, San Antonio, TX, Amer. Meteor. Soc., Paper 5B.7.

World Wind Energy Association (2009). "World Wind Energy Report 2008". Report. http://www.wwindea.org/home/images/stories/worldwindenergyreport2008_s.pdf. Retrieved on 16-May-2009.

6

Ground Penetrating Radar Subsurface Imaging of Buried Objects

Francesco Soldovieri[1] and Raffaele Solimene[2]

[1]Institute for Electromagnetic Sensing of the Environment, National Research Council,
[2]Department of Information Engineering, Second University of Naples,
Italy

1. Introduction

Ground Penetrating Radar (GPR) is a well assessed diagnostic instrumentation that finds application in a very large range of sectors where the aim is of achieving information about the presence, the position and the morphological properties of targets embedded in opaque media is c crucial task. Therefore, it is of interest in many fields which range from civil engineering diagnostics (Hugenschmidt & Kalogeropulos, 2009; Soldovieri et al. 2006), to archaeological prospecting (Conyers & Goodman, 1997; Piro et al., 2003; Orlando & Soldovieri, 2008), to cultural heritage management, to geophysics, only to quote a few examples.

Besides these classical fields of application, GPR methodology/technology is arising interest in new fields related to the security framework such as: through-wall imaging (Baranoski, 2008, Ahmad et al. 2003; Solimene et al., 2009), tunnel and underground facility detection (Lo Monte et al., 2009), borders surveillance.

Another field of recent interest is concerned with the exploration of planets; in particular, satellite platform radars are just investigating the Marsis subsurface (Watters et al., 2006) for the water search and other similar missions are under development for Lunar exploration.

GPR is based on the same operating principles of classical radars (Daniels, 1996). In fact, it works by emitting an electromagnetic signal (generally modulated pulse or Continuous Wave) into the ground; the electromagnetic wave propagates through the opaque medium and when it impinges on a non-homogeneity in the electromagnetic properties (representative of the buried target) a backscattered electromagnetic field arises. Such a backscattered field is then collected by the receiving antenna located at the air/opaque medium interface and undergoes a subsequent processing and visualization, usually as a 2D image.

Anyway, significant differences arise compared to the usual radar systems.

The first one is that while usually radar systems act in a free-space scenario, GPR works in more complicated scenarios with media that can have losses and exhibit frequency-dependent electromagnetic properties thus introducing dispersive effects.

Secondly, while the range of radars may be also of hundred of kilometres, in the case of GPR we have ranges, also in the most favourable cases, of some meters due to the limited radiated power and the attenuation of the signal.

Moreover, GPR resolution limits, which ranges from some centimetres to some metres (in dependence of the working frequency), usually are smaller as compared to the ones of the usual radar systems.

According to the working principles described above, the architecture of the radar system can be schematised very briefly by the following blocks:

- Electronic unit that has the task to drive and command the transmission and the reception of the antennas and to send the collected data to a monitor or to a processing unit;
- Two antennas that have the task of radiating the field impinging on the target and collecting the field backscattered by the target, respectively;
- A monitor/processing unit that has the task of the subsequent processing and visualization of the measurements collected by the receiving antenna.

From the schematization above, it emerges that a first advantage of the GPR instrumentation resides in the moderate cost and easiness of employ; in fact no particular expertise is required to collect the data. Secondly, the instrumentation is easily portable (unless very low frequencies are exploited thus increasing the physical size of the antennas) and allows to survey regions also of hundreds of square metres in reasonable time. Finally, the flexibility of the GPR system is ensured by the adoption of antennas (mostly portable) working at different frequencies and that can be straightforwardly changed on site.

It is worth noting that the necessities of probing lossy medium and of achieving resolution of the order of centimetres poses a very challenging task for the antennas deployed in the survey. On the other hand, as mentioned above, the portability of the system has to be ensured so that no complicated and large antenna systems can be employed. In particular, the trade-off between the necessity to have a large investigation range (that pushes to keep low the operating frequency) and the aim of achieving good spatial resolution makes the overall working frequency band exploited by GPR systems ranging from some tens of MHz to some GHz.

Therefore, the antennas designed for GPR applications have to comply with the following requirements: small dimension, ultra-wide-band behaviour and low levels of the direct coupling between the transmitting and receiving antennas.

The most widespread antennas employed for GPR surveys fall within the class of electrically small antennas such as dipole, monopole or bow-tie antennas.

Since these antennas are designed electrically small, the radiation pattern is broad and so low gain is achieved; keeping broad the radiation pattern allows to exploit synthetic aperture antenna which are particularly suitable to obtain focussing effect in the reconstructions. In addition, these simple antennas radiate a linear polarisation but also exhibit the drawback that relatively small frequency bandwidth is achieved. The working frequency band can be enlarged by loading techniques and the most-widely used technique is based on the application of resistive loading (Wu & King, 1965). The main disadvantage of resistive loading is the considerable reduction of the radiation efficiency due to the ohmic losses. In the last years, some approaches exploiting reactive elements to achieve the loading effect have been presented. A very recent development is concerned with the loading approaches based on the combination of resistive and capacitive loads; this offers the best performances in terms of trade off between small late-time ringing (increase of the frequency bandwidth) and high radiation efficiency (Lestari et al., 2004).

In addition, very recent advances are in course for adaptive (Lestari et al., 2005) and reconfigurable antennas, already exploited in communication field. In particular, the development and the implementation of RF/microwave switches have permitted the birth of a new concept of the antenna as a device that can dynamically adapt its behaviour to different measurement situations and operational contexts thanks to the change of its main parameters such as: input impedance, frequency band, radiation pattern. A design of an antenna for GPR purposes, with a geometry completely reconfigurable within the frequency band 0.3–1 GHz, and its performances compared to a reference bow-tie antenna are reported in (Romano et al., 2009). The structure has been "synthesized" by means of the total geometry morphing approach that represents the most structurally complicated but also versatile design method exploited to achieve the reconfigurability of an antenna.

Another class of antennas very widespread is the frequency independent one made up of spiral antennas (equiangular and conical). This kind of antennas are characterised by a geometry that repeats itself and in most cases the geometry (as in the case of spiral antennas) is completely described by an angle. Besides the frequency independent behaviour, spiral antennas are able to radiate a circular polarization; this is useful when the aim of the survey is to achieve information on objects that are elongated along a preferential direction. Another kind of frequency independent antennas is the Vivaldi antenna where the travelling wave energy is strongly bounded between conductors when the separation is very small, while becomes progressively weaker and more coupled with the radiation field when the separation increases. As main features, the Vivaldi antenna has an end-fire radiation mechanism, and the beamwidth is about the same in both planes E and H; the gain is proportional to the whole length and to the rate at which the energy is leaked away.

2. The measurement configuration and the radargram

Different configurations are adopted in GPR surveys and their choice is dictated by different applicative motivations such as: the type of investigation to be made; the kind of targets to be detected; the extent of the investigated region.

In general, all the reasons above push towards the adoption of the simplest acquisition mode for which the GPR system works in a monostatic or bistatic configuration. In the former case, the locations of the transmitting and the receiving antennas are coincident (or very close in terms of the radiated wavelength), whereas in the second case the transmitting and the receiving antennas are spaced by a fixed offset that is constant while they move along the survey line.

By moving the antenna system along a selected profile (line) above the ground surface a two-dimensional reflection profile (radargram) is obtained in which, for each location of the antenna system, a trace is achieved where the amplitude and the delay time of the recorded echoes (that can be related to the depth of the underground reflectors) is drawn (Daniels, 1996).

Such a radargram provides a rough information on the presence and the location of the targets but the actual shape of the target is blurred due to the effect of the propagation/scattering of the electromagnetic wave in the soil.

Such a statement is made clearer by considering the scattering by a point-object (an object small in terms of the probing wavelength) that is imaged as a hyperbola in the radargram. In fact, by denoting with x the position of the TX/RX system along the measurement line and with $(0,d)$ the position of the point-object , the two-way travel-time is given by

$$t = 2\sqrt{x^2 + d^2} \, / v \tag{1}$$

where v is the velocity of the electromagnetic wave in the soil. Therefore, the recorded data is represented as a hyperbola with a vertical axis and the apex at $(0, 2d/v)$. Figure 1 is a pictorial image of the consideration above.

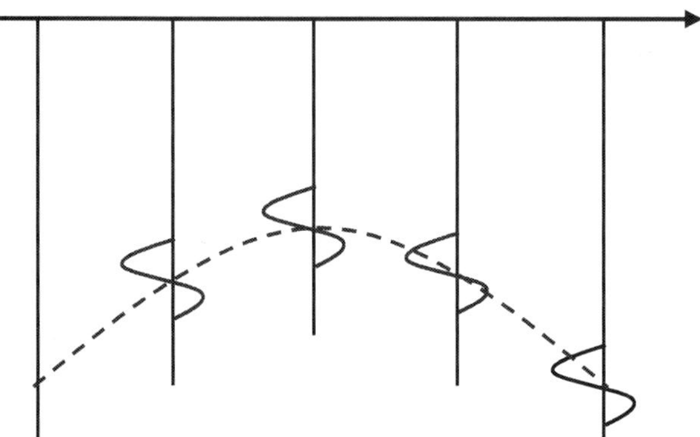

Fig. 1. Pictorial description of the radar response of a point-object.

As a result, the shape of the hyperbola depends on the electromagnetic properties of the investigated medium (electromagnetic velocity), the configuration (monostatic or bistatic) and on the depth of the scatterer.

Besides the configuration said above, others acquisition modes can be deployed in the survey. Among them, we recall the common midpoint (CMP) where the transmitting and receiving antennas are further spaced during the survey but leaving the same the midpoint between them. The adoption of this sounding configuration is particularly useful when one aims at determining the velocity of the electromagnetic wave in the medium (Huisman et al., 2003).

Further configurations are also exploited, but they have the drawback that are not particularly simple, as the multi-fold ones where one transmitting antenna is located at a fixed position while the backscattered field is collected by the receiving antenna moving at different locations. The same measurement configuration can be further complicated by repeating the measurement procedure for different location of the transmitting antennas.

All the configurations presented above are concerned with a reflection mode where both the transmitting and receiving antennas are from the same side with respect to the investigation domain (as for example in the case of subsurface prospecting).

A different mode can be considered for the case of the masonry structure as the trasillumination or transmission mode where the transmitting antenna and the receiving one are at the opposite sides of the structure under investigation.

3. GPR subsurface imaging

The aim of subsurface imaging is to obtain images of a scene which is buried in the soil in order to reconstruct non-homogeneities perturbing the background medium.

This can be achieved at different degree of difficulty depending on what one may want to retrieve. For example, one can be interested in detecting the buried scatterers "only" or even in determining their geometrical sizes or, finally, the nature of scatterers can be of concern.

From a mathematical point of view, subsurface imaging is a special case of inverse scattering problem where, at microwave frequencies, the scatterers are reconstructed in terms of a spatial map of their dielectric and/or conductivity features.

Accordingly, it is a non-linear and ill-posed problem whose solution generally requires to set up some optimization procedure (deterministic or stochastic) (Chew, 2005; Pastorino, 2007). Even though much work has been done in this field non-linear imaging schemes still suffer from reliability problems due to the false solutions (i.e. local minima of the cost functional to be minimized which can trap the optimization procedure) and are computationally demanding both as far as time and memory resources are concerned. As a matter of fact, these drawbacks hinder their use when imaging of a large (in terms of wavelength) spatial region is required and when the time to achieve imaging is a constraint.

Fortunately, in many practical contexts detecting and localizing the buried scatterers is the primary concern. This can be successfully achieved by adopting imaging algorithms based on simplified scattering models which neglect high order effects due to mutual interactions and which linearize thus the scattering equations. Indeed, the majority of the subsurface imaging algorithms are implicitly or explicitly based in the linear assumption. This is because, in the framework of scatterers localization and shape reconstruction linear imaging algorithms work well far beyond the limits of validity of the scattering linear models upon which they are founded (Pierri et al., 2006).

Here, we will consider only imaging algorithms based on linear models that are the most widely employed in practical scenarios. Such algorithms can be substantially categorized in two groups: *migration algorithms* and *inverse filtering algorithms*. Therefore, in the sequel we will describe and compare such algorithms. In particular, starting from the linearized scattering equations the way they are linked and the expected performance are derived under an unified framework.

Before attempting to pursue such a task it must be said that the imaging algorithms are only a part of the signal processing required in subsurface imaging. Most of the effort in signal processing concerns primarily with the reduction of the clutter. In fact, while system noise in GPR can be reduced by averaging, generally GPR data are heavily contaminated by clutter which affects the capability of detecting the buried scatterers. The strongest clutter mainly arises from the air/soil interface and different methods exist to mitigate such a contribution (Daniels, 2004). Connected with this question is the problem of subsurface dielectric permittivity and conductivity estimation. This problem is by its own of significant interest in several applicative fields, ranging from agriculture to monitoring of soil pollution and so on, and for this reason many estimation approaches have been developed. For example the dielectric permittivity can be exploited as an indicator of the soil surface water content (Lambot et al., 2008; Huisman et al., 2003).

However, for imaging purposes it is clear that if the electromagnetic properties of the soil were known one could estimate the field reflected from the air/soil interface and hence can completely cancel the corresponding clutter contribution. What is more, any imaging schemes requires the knowledge of the soil properties to effectively achieve "focusing"; otherwise blurred images where the scatterers appear smeared and delocalized.

4. Migration

Migration procedures essentially aim at reconstructing buried scattering objects from measurements collected above or just at the air/soil interface (from the air side). They were first conceived as a graphical method (Hagendoorn, 1954) based on high frequency assumption. After this approach has found a mathematical background based on the wave equation.

We illustrate the basic idea by assuming that the transmitting and the receiving antennas are just located at the air/soil interface where data are taken accordingly to a multimonostatic configuration (i.e., the receiver is at the same position as the transmitter while the latter moves in order to synthesize the measurement aperture). Moreover, for the sake of simplicity and according to the content of many contributions in the field of migration algorithms, we develop the discussion neglecting the air/soil interface. This means that we consider a homogeneous background scenario with the electromagnetic features of the subsurface. Moreover, the arguments are developed within a two-dimensional and scalar geometry so that the phenomena are described in terms of scalar fields obeying the two-dimensional and scalar wave equation.

Consider a point-like scatterer located in the object space at $\underline{r}_{sc} = (x_{sc}, z_{sc})$ and denote as $\underline{r}_O = (x_O, z_O)$ the observation variable, that is the positions where the scattered field is recorded. In particular, we assume $z_O = 0$ and $x_O \in [-X_M, X_M]$, where $\Sigma = [-X_M, X_M]$ is the synthesized measurement aperture (see Fig. 2).

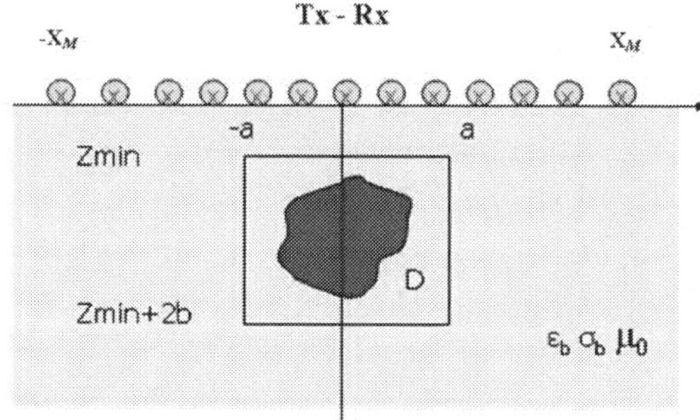

Fig. 2. Geometry of the subsurface prospecting problem

If $s_T(t)$ is the transmitted signal (ideally a delta impulse), the corresponding backscattered field (for the case at hand a B-scan measurement) is given by

$$s_R(\underline{r}_O, t) = s_T(t - 2\frac{|\underline{r}_O - \underline{r}_{sc}|}{v}) \qquad (2)$$

where v is the soil propagation speed assumed constant and the amplitude factor due to the propagation spreading has been neglected. Accordingly to the considerations said above (see eq. (1)), the backscattered signal in the $x_O - t$ data space appears as a diffraction hyperbola whose apex is in $(x, 2z/v)$ which can be translated in the $x - z$ image space by exploiting that $x = x_O$ and $z = vt/2$ (see Fig. 1).

The hyperbolic nature of the B-scan is due to the finite directivity of the antennas: migration aims at compensating for such a spreading by re-focalizing each segment of the hyperbola to its apex. To do this it is observed that the travel-time cannot be directly translated into depth because equal travel-times imply equal distances but the direction of arrival is not specified.

Hence, for each trace (i.e., A scan), the scatterer position should lie on a semicircle centred on the source-receiver position and whose radius is equal to the distance obtained by multiplying the travel-time by half the wave-speed in the soil. Accordingly, each $x_O - t$ data point is equally distributed over a semicircle in the image space (Hogendoorn, 1954; Bleinstein & Gray, 2001) so that all the semicircle intersect at $\underline{r}_{sc}$.

This graphical method is known as *Wave Interference Migration* (Gazdag & Sguazzero, 1984) sometimes also addressed in the literature as *A-scan-driven approach* (Marklein et al., 2002).

In order to estimate the travel-time, such a method requires that the hyperbolic curve be well discernable from other data features; this entails that a good Signal to Noise Ratio (SNR) is needed and that the scattering scenario is not much complex. In any case, an extended scatterer is considered as an ensemble of point scatterers which makes it clear that linearity of the scattering model is implicitly assumed.

A counterpart of *Wave Interference Migration* is the so-called *Diffraction Summation* (Gazdag & Sguazzero, 1984) also knew as pixel-driven approach (Marklein et al., 2002).

In this method the object space is divided in pixels and for each of them a diffraction hyperbola is constructed in the data (image) space. The reconstruction at each pixel is then obtained by summing all the traces (A-scans) that the synthetic hyperbola intersects in the data space.

This procedure can be implemented automatically and requires the evaluation of the following summation integral for each pixel (x, z)

$$R(x,z) = \int_{\Sigma} \int_{T} s_R(x_O,t)\delta(t - \frac{2}{v}\sqrt{(x-x_O)^2 + z^2})dx_O dt \tag{3}$$

where Σ and T are the measurement aperture and the interval of time during data are collected, respectively, and $R(x,z)$ is the corresponding migrated data.

If we denote as $S_R(x_O,\omega)$ the Fourier transform of $s_R(x_O,t)$ (by considering the temporal Fourier kernel $\exp(-j\omega t)$), then previous equation can be recast as

$$R(x,z) = \int_{\Sigma} \int_{\Omega} S_R(x_O,\omega)\exp(j\frac{2\omega}{v}\sqrt{(x-x_O)^2 + z^2})dx_O d\omega \tag{4}$$

with Ω being the adopted frequency bandwidth. Now, by putting $k = \frac{\omega}{v}$, eq. (4) can be rewritten in terms of integration in the wave-number k domain as

$$R(x,z) = \int_{\Sigma} \int_{\Omega_k} S_R(x_O,k)\exp(j2k\sqrt{(x-x_O)^2 + z^2})dx_O dk \tag{5}$$

where Ω_k now denotes the frequency band in the k domain and an unessential scalar factor has been omitted. Eq. (5) allows us to point out the equivalence between the *Diffraction Summation* and the *Range Migration Technique* presented in (Lopez-Sanchez & Fortuny-Guasch, 2000) and also to the *Synthetic Aperture Focusing Technique (SAFT)* (Markelein et al., 2002).

As can be seen, these methods are essentially based on a convolution in x_O and an integration in k. The convolution can be conveniently computed in the spatial Fourier domain as

$$R(x,z) = \int_{\Omega_{kx}} \int_{\Omega_k} f(k_x,k)S_R(k_x,k)\exp(-jk_x x)\exp(jk_z z)dk_x dk \qquad (6)$$

where Ω_{kx} is the selected frequency band in the spatial spectral k_x, $f(k_x,k) = \exp(j\pi/4)k\sqrt{2/(k_z^3\pi)}$ is an amplitude factor where $\sqrt{z}$ has been neglected, $k_z = \sqrt{4k^2 - k_x^2}$. $S_R(k_x,k)$ is the Fourier transform with respect to x_O of $S_R(x_O,k)$ (the spatial Fourier kernel $\exp(jk_x)$ is considered) whereas the Fourier transform of the exponential term $\exp(j2k\sqrt{x^2 + z^2})$ has been evaluated by means of the stationary phase method. Now, by substituting the frequency wavenumber k by the integration in k_z, eq. (6) can be recast as a double Fourier transform as

$$R(x,z) = \int_{\Omega_{kx}} \int_{\Omega_k} f(k_x,k)\frac{k_z}{\sqrt{k_x^2 + k_z^2}}S_R(k_x,k_z)\exp(-jk_x x)\exp(jk_z z)dk_x dk_z \qquad (7)$$

Eq. (7) has the computational advantage that it can be computed by Fast Fourier Transform (FFT) algorithms directly for each point in the object space but also requires data to be interpolated and re-sampled according to a rectangular grid in the k_x-k_z spatial spectral domain (Stolt, 1978).

When the amplitude factor is ignored, Eq. (7) becomes identical to the *Synthetic Aperture Radar (SAR)* imaging algorithm presented in (Soumekh, 1999) and also very similar to the *F-K Migration* as outlined in (Gilmore et al., 2006).

Interestingly, all the migration algorithms (a part the *F-K Migration*) have not been obtained by invoking some mathematical theory of the scattering phenomenon. Rather, they have been developed by simply adopting intuitive physically based arguments.

A more rigorous way to derive migration methods can be obtained by invoking on the wave equation (scalar for the case at hand) and by adopting the so-called "exploding source model". According to this model, the scattered field is thought as being radiated by a source at time $t=0$ embedded within a medium characterized by a wave velocity being half the actual one (this corresponds to consider the wave number of this equivalent medium twice the one of the subsurface medium). Therefore, outside the source region the scattered field is a solution of the following wave-equation (in the frequency domain)

$$\nabla^2 S_R(x,\omega) + 4k^2 S_R(x,\omega) = 0 \qquad (8)$$

with $\nabla \equiv (\frac{\partial}{\partial x},\frac{\partial}{\partial z})$ and where we still adopted $S_R(x,\omega)$ to denote the scattered field.

By Fourier transforming with respect to x one obtains that the field spatial spectrum at quota z is related to the one at quota z', with $z < z'$, as

$$S_R(k_x,\omega,z) = S_R(k_x,\omega,z')\exp[jk_z(z - z')] \qquad (9)$$

where k_z is the same as previously defined and only up-travelling waves (that is along the negative z direction) have been considered. Accordingly, the field in the objects space can be determined by forcing the boundary condition over the measurement line at $z=0$ and by a back-propagation procedure. Finally, a superimposition along the frequency ω returns

$$R(x,z) = \int_{\Omega_{kx}} \int_{\Omega} S_R(k_x,\omega)\exp(-jk_x x)\exp(jk_z z)dk_x d\omega \qquad (10)$$

This is the so-called *F-K Migration* which has been shown to be a generalization of the *Doppler compression technique* typical in SAR imaging which, in turn, can be considered a small angle (of view) approximation version (Cafforio et al., 1991). Moreover, it is not difficult to understand why this migration algorithm is also addressed in the literature as *Phase Shift Migration* (Gazdag & Sguazzero, 1984).

As can been seen, the *Phase Shift Migration* is very similar to the *Summation Diffraction* (in the frequency domain) and can be readily recast as a double Fourier integral in the spatial/spectral domain. However, it has been obtained by requiring less approximations on the underlying model. In particular, the amplitude spreading terms have not been ignored and the stationary phase method has not been employed. Indeed, the *Phase Shift Migration* derivation can be considered as the mathematical rationale supporting the intuitive discussion under which previous migration schemes have been developed.

Unfortunately, this mathematical model contains an implicit contradiction: while the field is back-propagated as a solution of a homogeneous wave equation the exploding source assumes it being radiated by a localized source. From a mathematical point of view, this contradiction manifests in the assumption of considering the scattered field as being composed of only up-going (towards the air/soil interface) waves. Therefore, it is expected that the field "extrapolation" in-depth in the object space works as long as the scatterers are reached. This becomes particularly apparent when the scatterers are buried within a dispersive and dissipative medium. In this case, while back-propagating the aperture field, if evanescent waves (which reflect the presence of the scatterers) are not filtered out, the migration rapidly degrades for points in the object plane beyond the scatterer (Oden et al., 2007).

By inverting the Fourier transformation with respect to k_x, eq. (10) can be rewritten as

$$R(x,z) = \int_\Sigma \int_\Omega s_R(x_O, z = 0, \omega) \frac{\partial}{\partial z} G^*(x - x_O, z, \omega) dx_O d\omega \tag{11}$$

where G^* is the complex conjugated Green's function. Migration scheme reported in eq. (11) is known as the *Rayleigh-Sommerfeld holography* which is a particular case (when data are collected over an infinite line) of the so-called *Generalized Holography* (Langenberg, 1987) which in turn is founded on the Porter-Bojarski integral equation.

The connection with the *Generalized Holography* is important because it establishes, in rigorous way, the relationship between the migrated field and the secondary sources, that is the ones arising from the interaction incident field-scatterers, and hence between the migrated field and the scatterers. This is especially true under scattering linear models (Langenberg, 1987). Note that in all the previous migration schemes the relationship between the migrated field and the scatterers has been implicitly assumed as an **ansatz**.

This question will be further addressed in the next section.

Finally, we observe that the time domain version of eq. (11) is nothing else that the well known *Kirchhoff Migration* (Berkhout, 1986). In fact, by Fourier transforming eq. (11) with respect to ω we roughly obtain

$$R(x,z) \cong \int_\Sigma \int_T \cos(-\hat{z}, \underline{r}_{Oc}) s_R(x_O, t) \frac{\partial}{\partial t} \delta(t - \frac{2}{v}\sqrt{(x - x_O)^2 + z^2}) dx_O dt \tag{12}$$

where $\cos(-\hat{z}, \underline{r}_{Oc})$ takes into account the angle between the unit normal vector at the measurement line and the vector $\underline{r}_{Oc} = \sqrt{(x - x_O)^2 + z^2}$.

5. Scattering equations and the Born approximation

In the previous section we described different migration algorithms and we showed that they are all very similar since one can pass from the different migration implementations by Fourier (spatial and/or temporal) transform operators.

However, the link between the migrated field and the scatterers to be reconstructed has not been clearly shown and remained only supported by intuitive arguments.

To cope with this question the equations describing accurately the scattering phenomenon are needed. Therefore, in this section, we just introduce the equations governing the scattering phenomena. Furthermore, we also shown as they simplify under the Born approximation for the case of penetrable scatterers (Chew, 1995).

Accordingly, the subsurface imaging problem is cast as an inverse scattering problem where one attempts to infer the electromagnetic properties of the scatterer from the scattered field measured outside the scatterer.

The statement of the problem is then the following: given an incident field, E_{inc}, that is the electromagnetic field existing in the whole space (the background medium) in absence of the scattering object and generated by a known sources, by the interaction with the object the scattered field E_s arises; from the knowledge of the scattered field some properties about the scatterer, either geometrical and/or structural, have to be retrieved.

Hence, it is mandatory to introduce the mathematical equations subtending the scattering phenomena to solve the above stated problem. To this end, we refer to a two-dimensional and scalar geometry as done in the previous section.

We consider a cylindrical dielectric object (that is invariant along the axis out-coming from the sheet) enclosed within the domain D illuminated by an incident field linearly polarized along the axis of invariance. The scattered field is observed over the domain Σ (not rectilinear necessarily). Moreover, we denote by $\varepsilon(\underline{r})$ e by $\varepsilon_b(\underline{r})$ the equivalent dielectric permittivity function of the unknown object and that of the background medium, respectively. In particular, the latter must not be necessarily constant (i.e., non-homogeneous background medium is allowed) but has to be known. The magnetic permeability is assumed equal the one of the free space μ_0 everywhere. The geometry of the problem is detailed in Fig. 2.

The problem, thus, amounts to retrieving the dielectric permittivity profile $\varepsilon(\underline{r})$ of the unknown object by the knowledge of the scattered field E_s.

It can be proven that the frequency domain mathematical relationship between the data and the unknown is furnished by the following scalar Helmoltz equation (Chew, 1995)

$$\nabla^2 E + k_b^2 E = -k_b^2 \chi(\underline{r}) E \tag{13}$$

with $E = E_{inc} + E_S$ is the total field, k_b is the subsurface (background) wave-number and $\chi(\underline{r}) = \varepsilon(\underline{r}) / \varepsilon_b - 1$ is the dimensionless contrast function. By adopting the Green's function method, eq. (13) leads to the pair of scalar integral equations (Colton & Kress, 1992)

$$E(\underline{r},\underline{r}_S;k_b) = E_{inc}(\underline{r},\underline{r}_S;k_b) + k_b^2 \int_D G(\underline{r},\underline{r}';k_b)E(\underline{r}',\underline{r}_S;k_b)\chi(\underline{r}')d\underline{r}' \qquad \underline{r} \in D$$

$$E_S(\underline{r}_O,\underline{r}_S;k_b) = k_b^2 \int_D G(\underline{r}_O,\underline{r};k_b)E(\underline{r},\underline{r}_S;k_b)\chi(\underline{r})d\underline{r} \qquad \underline{r}_O \in \Sigma$$

(14)

where G is the pertinent Green's function, $\underline{r}_O$ is the observation point and $\underline{r}_S$ is the source's position.

In accordance to the volumetric equivalence theorem (Harrington, 1961), the above integral formulation permits to interpret the scattered field as being radiated by secondary sources (the "polarization currents") which are just located within the scatterers spatial region.

The reconstruction problem thus consists of inverting the "system of equations (14)" for the contrast function. However, since (from the first equation) the field inside the scatterers depends upon the unknown contrast function, the relationship between the contrast function and the scattered field is nonlinear. The problem can be cast as a linear one if the first line equation is arrested at the first term of its Neumann expansion (Krasnov et al., 1976). In this way, $E \cong E_{inc}$ is assumed within the scatterer region and the so-called Born linear model is obtained (Chew, 1995). Accordingly, the scattering model now becomes

$$E_S(\underline{r}_O,\underline{r}_S;k_b) = k_b^2 \int_D G(\underline{r}_O,\underline{r};k_b)E_{inc}(\underline{r},\underline{r}_S;k_b)\chi(\underline{r})d\underline{r} \qquad \underline{r}_O \in \Sigma \qquad (15)$$

We just remark that, within the linear approximation, the internal field does not depend on the dielectric profile which is the same as to say that mutual interactions between different parts of the object and between different objects are neglected. In other words, this means to consider each part of the object as elementary scatterer independent on the presence of the other scatterers. It is worth noting that this is the same assumption as we made a priori while discussing about the migration algorithms.

Now, by considering a homogeneous background medium and a monostatic data collection ($\underline{r}_O = \underline{r}_S$), as done in the previous section, and by resorting to asymptotic arguments as far as the Green's function is concerned, eq. (15) can be rewritten as

$$E_S(\underline{r}_O;k_b) = k_b \int_D \frac{\exp(-2jk_b|\underline{r}_O-\underline{r}|)}{|\underline{r}_O-\underline{r}|}\chi(\underline{r})d\underline{r} \qquad \underline{r}_O \in \Sigma \qquad (16)$$

where a two-dimensional filamentary current with current $I(\omega) \propto 1/\omega$ has assumed as source of the incident field and a scalar factor has been omitted. If Σ is meant as in the previous section

$$E_S(x_O;k_b) = k_b \int_D \frac{\exp[-2jk_b\sqrt{(x_O-x)^2+z^2}]}{\sqrt{(x_O-x)^2+z^2}}\chi(x,z)dxdz \qquad (17)$$

Then by Fourier transforming the scattered field with respect to x_O and by exploiting the plane-wave spectrum of the Green's function we obtain

$$E_S(k_x,k_b) = \frac{k_b}{k_z}\hat{\chi}(k_x,k_z) \qquad (18)$$

with $k_z = \sqrt{4k_b^2 - k_x^2}$ is defined as in the previous section. Accordingly, when the spatial Fourier transformed data are rearranged in the $k_x - k_z$ spectral plane, the contrast function can be obtained as

$$\tilde{\chi}(x,z) = \int\limits_{\Omega_{k_x}} \int\limits_{\Omega_{k_z}} \frac{k_z}{\sqrt{k_x^2 + k_z^2}} E_S(k_x,k_z) \exp(-jk_x x) \exp(jk_z z) dk_x dk_z \qquad (19)$$

It is important to note that eq. (19) coincides just to the *F-K Migration* when integration in ω is replaced by the integration in k_z. Therefore, within the framework of the linear Born approximation we have established a clear connection between the migrated field and scatterers in terms of their contrast functions.

Similar results can be obtained for different kind of scatterers where other linear approximations can be adopted (Pierri et al., 2006; Solimene et al., 2008).

6. Inverse filtering imaging algorithms

As stated in the previous section, under the Born approximation and for a two-dimensional and scalar geometry, the scattering phenomenon may be modelled through a linear scalar operator

$$A : \chi \in X \to E_S \in \mathrm{E} \qquad (20)$$

where χ and E_S are the contrast function (the unknown of the problem) and the scattered field (the data of the problem), respectively.

Moreover, X and E represent the functional sets within we search for the contrast function and the one we assume the scattered field data belong to, respectively. In particular, we assume them as Hilbert spaces of square integrable functions, the first one of complex valued functions defined on the investigation domain D, whereas the second one of such functions supported over $\Lambda = \Sigma \times \Omega$. However, in general the data space depends on the adopted configuration, that is on the choice of the measurement domain and the strategy of illumination and observation.

It is worth remarking that the choice of X and E as Hilbert spaces of square integrable functions accommodates the fact that no a priori information are available on the unknown except that one on the finiteness of its "energy" as dictated by physical consideration. On the other side, it assures that E is "broad" enough to include the effect of uncertainties and noise on data.

Thus, the problem amounts to inverting equation (20) to determine the contrast function. Since the kernel of the operator in (20) is a continuous function on $X \times \mathrm{E}$, then the linear operator is compact (Taylor and Lay, 1980). As stated above this means that the inverse problem is an-illposed problem (Bertero, 1989). For compact and non-symmetric operator (as the one at hand) the singular value decomposition is a powerful tool to analyze and to solve the problem.

Therefore, we denote as $\{\sigma_n, u_n, v_n\}_{n=0}^{\infty}$ the singular system of operator A. In particular, $\{\sigma_n\}_{n=0}^{\infty}$ is the sequence of the singular values ordered in non-increasing way, $\{u_n\}_{n=0}^{\infty}$ and $\{v_n\}_{n=0}^{\infty}$ are orthonormal set of functions solution of the following shifted eigenvalue problems

$$Au_n = \sigma_n v_n$$
$$A^+ v_n = \sigma_n u_n \tag{21}$$

where A^+ is the adjoint operator of A [Taylor & Lay, 1980], which span the orthogonal complement of the kernel of A, $N(A)^\perp$, and the closure of the range of A, $\overline{R(A)}$, respectively.

A formal solution of equation (21) has the following representation (Bertero, 1989)

$$\chi = \sum_{n=0}^{\infty} \frac{\left\langle E_S, v_n \right\rangle_E}{\sigma_n} u_n \tag{22}$$

where $\left\langle \cdot, \cdot \right\rangle_E$ denotes the scalar product in the data space E.

By virtue of the compactness of A, $R(A)$ is not a closed set (Krasnov et al., 1987). This implies that the Picard's conditions is not fulfilled for any data functions (i.e., the ones having non null component orthogonal to $R(A)$) ; thus the solution may not exist and does not depend continuously on data (Bertero, 1989). This is just a mathematical re-statement of ill-posedness . From another point of view, we have to take into account that the actual data of the problem are corrupted by uncertainties and noise n , hence

$$\tilde{\chi} = \sum_{n=0}^{\infty} \frac{\left\langle E_S, v_n \right\rangle_E}{\sigma_n} u_n + \sum_{n=0}^{\infty} \frac{\left\langle n, v_n \right\rangle_E}{\sigma_n} u_n \tag{23}$$

Now, because of the compactness, the singular values tend to zero as their index increases. This implies that, the second series term (noise-related) in (23) in general does not converge. This leads to an unstable solution since even small error on data are amplified by the singular values close to zero.

The lack of existence and stability of solution can be remedied by regularizing the addressed ill-posed inverse problem (De Mol, 1992). For example, this can be achieved by discarding, in the inversion procedure, the "projections" of data on the singular functions corresponding to the "less significant" singular values, that is by filtering out the singular functions corresponding to the singular values which are below to a prescribed noise dependent threshold. This regularizating scheme is known as *Numerical Filtering* or *Truncated Singular Value Decomposition* (TSVD) and is the simplest one within the large class of windowing based regularizating algorithms (Bertero, 1989).

Accordingly, the finite-dimensional approximate stable solution is then given by

$$\tilde{\chi} = \sum_{n=0}^{N_T} \frac{\left\langle E_S + n, v_n \right\rangle_E}{\sigma_n} u_n \tag{24}$$

More in general, the basic idea of regularization theory is to replace an ill-posed problem by a parameter dependent family of well-posed neighbouring problems

$$\chi = A_\alpha^n E_S \tag{25}$$

with α being the so-called regularization parameter (in the TSVD this corresponds to the truncation index N_T in eq. (24)) and n the noise level, so that to establish a compromise between accuracy and stability (Groetsch, 1993). In particular, as $n \to 0$ also $\alpha \to 0$ and the regularized reconstruction must tends to the generalized inverse whose outcome is just shown in eq. (22).

The Tikhonov regularization scheme is a widespread regularization scheme which takes advantages from exploiting a priori information about the unknown (Tikhonov, 1977). In this case the inversion problem is cast as a constrained optimization problem

$$\tilde{\chi} = \min\{\|A\chi - E_S\|^2 + \alpha\|E_S\|^2\} \tag{26}$$

Here the constraint arises from the available a priori information and in general can be different from the energy constraint reported in the equation above.

The Landweber regularization scheme recasts the first kind integral equation to be inverted as an integral equation of second kind so that a well-posed problem is obtained (Groetsch, 1993). Accordingly, eq. (26) is recast as

$$\tilde{X} = A^+ E_S + (I - A^+ A E_S)\chi \tag{27}$$

and a solution is then obtained by an iterative procedure. In this case the regularization parameter is the number of iterations exploited in the iterative minimization N_I.

These regularization scheme can be compared if all are analyzed in terms of the operator properties. This can be done by expressing the different regularized reconstruction in terms of the singular system. By doing so, one obtains

$$\tilde{\chi} = \sum_{n=0}^{\infty} \frac{\sigma_n}{\sigma_n^2 + \alpha} \left\langle E_S, v_n \right\rangle_E u_n \tag{28}$$

for the Tikhonov regularization, and

$$\tilde{\chi} = \sum_{n=0}^{\infty} \frac{1 - (1 - \sigma_n^2)^{N_I}}{\sigma_n} \left\langle E_S, v_n \right\rangle_E u_n \tag{29}$$

for the Landweber method.

As can be seen, all this regularization methods result (but in a different) filtering of the unknown spectral expansion.

It is clear that, a part the computational convenience which can dictate the regularization algorithm to adopt, the key question is the choice of the regularization parameter. As remarked, this choice has to be done by accounting for the noise level, the mathematical features of the operator to be inverted and the available a priori information about the unknown. Different methods exist to select the regularization parameter. Such methods can explicitly exploit the knowledge of the noise level, (such as the Morozov discrepancy principle) or not (such as the generalized cross validation) (Hansen at al., 2008).

In inverse scattering problem, the scattered field has finite number of degrees of freedom (Solimene & Pierri, 2007) which in turn depends on the parameter of the configuration. Therefore, such an information can be exploited to select the regularization parameter.

The singular system formalism can be also employed to compare migration and inverse filtering. By looking at the scattering eq. (17), it is seen that the *Diffraction Summation* migrated field reconstruction, substantially, corresponds to achieve inversion by means of the adjoint operator, that is

$$\tilde{\chi} = A^+ E_S \tag{30}$$

whose expression in terms of the singular system is

$$\tilde{\chi} = \sum_{n=0}^{\infty} \sigma_n \langle E_S, v_n \rangle_E u_n \tag{31}$$

It is readily noted that migration allows to obtain a stable reconstruction because the singular values now appear at the numerator. However, it is not a regularization scheme in the sense of Tikhonov (Tikhonov & Arsenin, 1977) because even in absence of noise the actual unknown is not retrieved. From a practical point of view, this entails an *intrinsic* limit on the resolution achievable in the reconstructions regardless of the noise level. To explain this we observe that χ and its reconstructed version $\tilde{\chi}$ are linked by the following integral relationship

$$\tilde{\chi}(x,z) = \int_D K(x,z;x'z')\chi(x',z')dx'dz' \tag{32}$$

$K(x,z;x'z')$ being the so-called *model resolution kernel* (Dudley et al., 2000). Now, except for those measurement configuration that do not fulfill the so-called *golden-rule* (Langenberg, 1987), it can be shown that the scattering operator to be inverted is injective (this is rigorously true until data are collected continuously over E. Practical measurement set up requires data sampling).

This entails that, for each of the inverse filtering procedure described above, the model resolution kernel tends to a Dirac delta, that is, $K(x,z;x'z') \rightarrow \delta(x - x', z - z')$, when noise is absent and $\alpha \rightarrow 0$ (Dudley et al., 2000). This is not true for migration algorithms. Therefore, this means that, depending on the noise level, inverse filtering can achieve better resolution in the reconstructions.

This is shown in Fig. 3 where TSVD and F-K migration performances are compared.

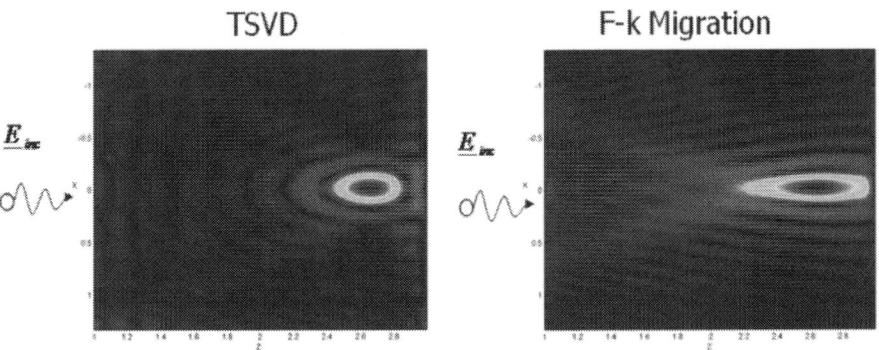

Fig. 3. Comparison between the reconstruction performances of TSVD and F-K migration for a point-target

However, as long as lossless medium is considered, as migration algorithms can be easily recast in the spatial Fourier domain they generally have a lower computational cost than inverse filtering. Indeed, in lossy scenarios, losses should be considered in the inversion procedure and this impairs the possibility of adopting only FFT based imaging algorithms (Meincke, 2001).

7. Reconstruction results

In order to give an idea of how subsurface reconstructions look like, in this Section we report a brief survey of reconstruction performances against synthetic data. The scatterd field data have been computed thanks to the Finite Difference Time Domain (FDTD) GPRMAX code (Giannopoulos, 2005) in time domain; these time-domain data are then Fourier transformed to pass to the frequency domain.

The TSVD inverse filtering approach based on the Born model is adopted to achieve the reconstructions; in particular, to emphasize the usefulness of linear reconstruction schemes test cases concerning applicative cases where the objects have electromagnetic properties very different from the ones of the soil (test cases outside of the Born regime) are considered.

The inversion algorithm assumes an investigation domain D whose extent ranges from 0.1m to 1.5m in depth and with the semi-horizontal extent a=0.75m. The soil has a relative dielectric permittivity equal to 9 and an electrical conductivity equal to 0.01 S/m. The measurement domain is located at the air/soil interface and has extent 1.5m. Over such a domain, 51 data are taken at a spatial step of 3cm. The working frequency Band Ω ranges from 300 to 900 MHz.

Different cases of a single object have been considered; in particular, we have considered the same geometry of the object, i.e., a circular cylindrical object of radius of 0.1 m and whose center is located at the depth of 0.6m. Three different values of the relative dielectric permittivity filling the object have been considered, i.e., 9.1 (low Born model error), 4 and 16.

The reconstruction results are given in terms of the modulus of the retrieved contrast function normalised with respect to its maximum in the investigation domain. The regions where the normalised modulus of the contrast function is significantly different from zero give information about the support of the targets.

Figure 4 depicts the reconstruction for the case of the target with relative dielectric permittivity equal to 9.1; the reconstruction permits to localize and reconstruct well the shape of the upper and lower sides of the circle. The features of the reconstruction agree with the theoretical expectations reported in (Persico et al. 2005) where the filtering effect of the regularized inversion algorithm consists in a low-pass filtering along the antenna movement direction (x-axis) and a band-pass filtering along the depth (z-axis). In this case, when we exploit the TSVD scheme for the inversion we retained in the summation (24) the singular values larger than 0.1 the maximum one, that is the TSVD threshold is equal to 0.1.

The second case is concerned with the circular target having a relative dielectric permittivity equal to 4 so that the Born model error is significant. Figures 5 and 6 depict the corresponding reconstructions when the TSVD threshold is equal to 0.1 and 0.01, respectively.

As can be seen, both the results permit to point out in a correct way the location and the shape of the upper side of the target. Moreover, by adopting a smaller value of the TSVD threshold (more singular values in the TSVD summation (24)), the achievable resolution

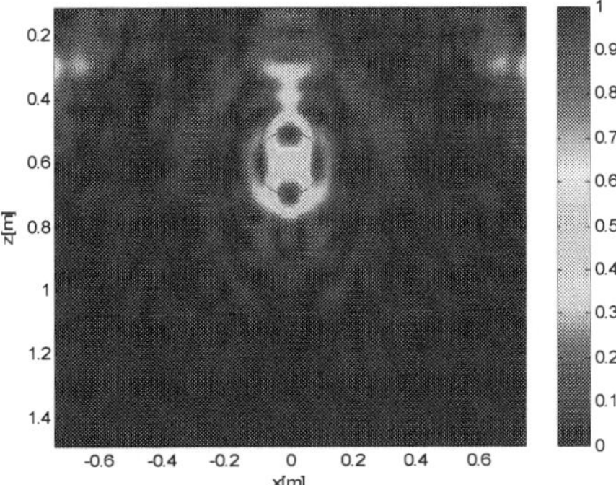

Fig. 4. TSVD reconstruction in the case of the target with dielectric permittivity equal to 9.1 (low model error) and TSVD threshold equal to 0.1

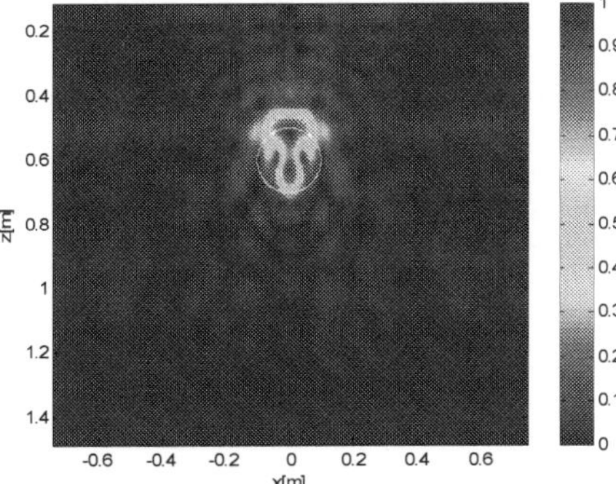

Fig. 5. TSVD reconstruction in the case of the target with dielectric permittivity equal to 4 (high model error) and TSVD threshold equal to 0.1

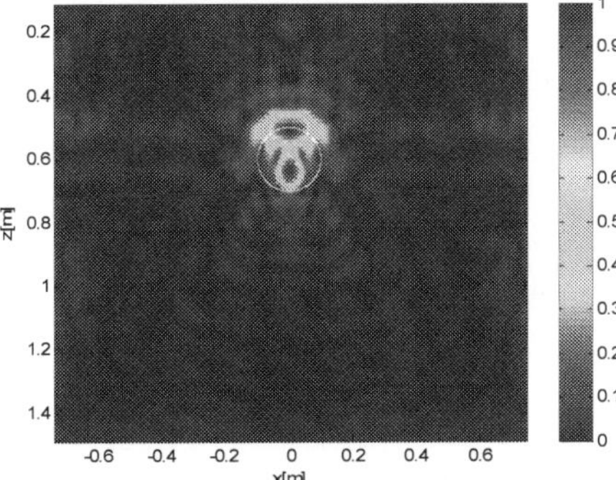

Fig. 6. TSVD reconstruction in the case of the target with dielectric permittivity equal to 4 (high model error) and TSVD threshold equal to 0.01

improves and both the upper and the lower sides of the scatterers are discernable in the reconstruction. In particular, for the lower side of the target, we observe that the reconstruction approach is able to detect it but does not provide to its actual position. This can be explained by noting that the velocity of the electromagnetic wave in the target is equal to v=15 cm/ns and different from the one assumed in the model equal to v=10 cm/ns. Therefore, the inversion model assumes a velocity smaller than the actual one inside the target and this leads to a reconstruction of the lower side at a depth smaller than the actual one.

Finally, we consider the case with the circular target having a relative dielectric permittivity equal to 16 (high Born model error). Figures 7 and 8 depict the tomographic reconstruction results when the TSVD threshold is equal to 0.1 and 0.01, respectively.

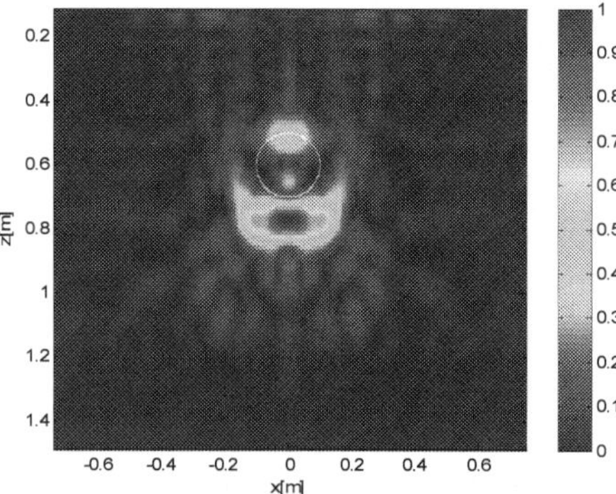

Fig. 7. TSVD reconstruction in the case of the target with dielectric permittivity equal to 16 (high model error) and TSVD threshold equal to 0.1

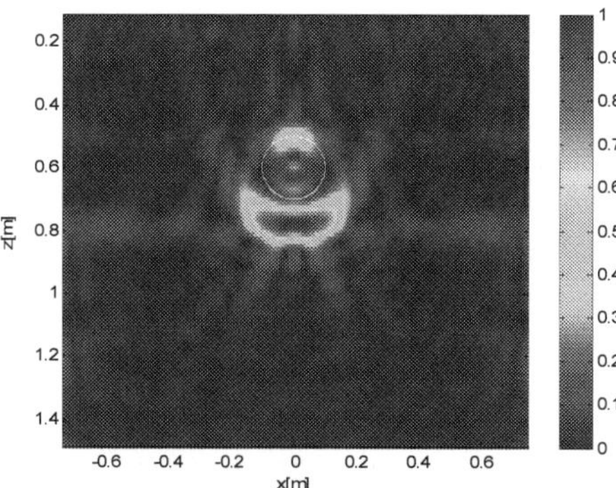

Fig. 8. TSVD reconstruction in the case of the target with dielectric permittivity equal to 16 (high model error) and TSVD threshold equal to 0.01

We can observe that both the results permit to point out in a correct way the location and the shape of the upper side of the target; also, by adopting a smaller value of the TSVD threshold (more singular values in the TSVD summation (24)), the reconstruction is slightly improved.

For the lower side of the target, we observe that the reconstruction approach is able to detect it but the reconstruction does not correspond to the actual target's shape. Similarly to the explanation given above, we note that the velocity of the electromagnetic wave in the target is now equal to v=7.5 cm/ns and different from the one assumed in the model equal to v=10 cm/ns. Therefore, the inversion model assumes a velocity larger than the actual one inside the target and this arises in a reconstruction more deeply located. Also, it is worth noting that for the smaller TSVD threshold, the reconstruction of the lower side has a shape more similar to the one of the target.

Finally, we present the case (see fig. 9) of a wrong choice of the TSVD threshold equal to 0.0001, in this case the necessity to improve the reconstruction (by lowering the TSVD threshold) is made it not possible by the effect of the uncertainties on data (in this particular case related only to numerical errors) that produce a reconstruction very difficult to be interpreted.

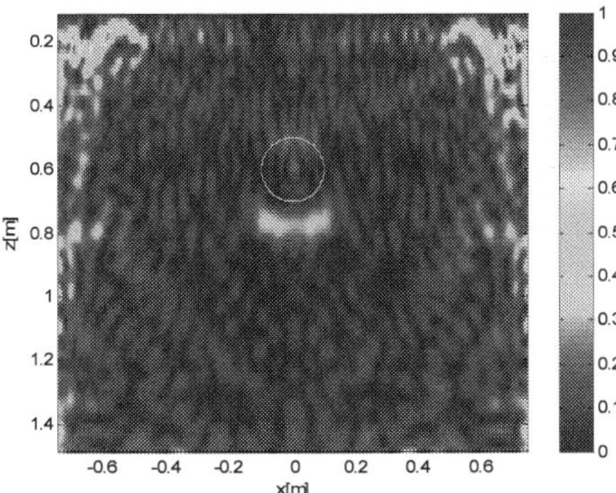

Fig. 9. TSVD reconstruction in the case of the target with dielectric permittivity equal to 16 (high model error) and TSVD threshold equal to 0.0001

8. Conclusions

This chapter has dealt with the imaging of the buried targets thanks to Ground Penetrating Radar. After a brief description of the operating principles of GPR we have focussed the attention on the data processing that is one of the main questions in order to achieve interpretable images of the subsurface.

In particular, we have described the main versions of the migration technique, usually exploited in data processing, and a relation between them has been pointed out. Then, we have presented a more recent data processing approach based on microwave tomography. Such a tomographic approach faces the inverse scattering problem and the adoption of a physics-based model allows us not only to obtain more interpretable reconstruction but also to analyse the performances of the reconstruction approach.

In particular, the performances of a microwave tomography based approach exploiting a simplified model of the electromagnetic scattering was investigated in detailed way. Such a linear inverse approach has revealed suitable in realistic applications where the aim is the geometry of the targets. Finally, the feasibility and limitations of this linear approach have been outlined by an analysis against synthetic data.

9. References

Ahmad, F., Amin, M. G. and Kassam, S. A. (2005). Syntethic aperture beamformer for imaging through a dielectric wall. *IEEE Trans. Aerosp. Electr. System.*, vol. 41, pp. 271–283.

Baranoski, E. J. (2008). Through-wall imaging: Historical perspective and future directions. *Journal of Franklin Institute*, Vol. 345, pp. 556–569.

Berkhout, A. J. (1986). Seismic Inversion in Terms of Pre-Stack Migration and Multiple Elimination. *IEEE Proc.* Vol. 74, pp. 415-427.

Bertero, M. (1989). Linear inverse and ill-posed problems. *Adv. in Electron. and Electron. Phys.* Vol. 45, pp. 1-120.

Bleinstein, N. and Gray, S. H. (2001) From the Hagedoorn Imaging Technique to Kirchhoff Migration and Inversion. *Geophys. Prospect.* Vol. 49, pp. 629-643.

Cafforio, C.; Prati, C. and Rocca E. (1991) SAR Data Focusing Using Seismic Migration Techniques. *IEEE Trans. Aerosp. Electr. Syst.* Vol. 27, pp. 194-207.

Chew, W. C. (2005). *Waves and fields in inhomogeneous media.* IEEE Press, New York.

Colton, D. and Kress, R. (1992). *Inverse Acoustic and Electromagnetic Scattering Theory.* Springer-Verlag, Berlin.

Conyers, L. B. and Goodman, D., (1997). *Ground Penetrating Radar: An Introduction for Archaeologists.* Alta Mira Press, Walnut Creek, London and New Delhi.

Daniels, D.J. (1996). *Surface-Penetrating Radar.* London, U.K.: Inst. Elect. Eng., 1996.

Daniels, D. J. (2004). *Ground Penetrating Radar.* IEE Radar, Sonar and Navigation Series, London.

De Mol, C. (1992). A critical survey of regularized inversion method. *Inverse Problems in Scattering Imaging ed M. Bertero and R. Pike*, Bristol Hilger, pp. 345-370.

Dudley, D.G.; Habashy, T.M. and Wolf, E. (2000). Linear inverse problems in wave motion: nonsymmetric first-kind integral equations. *IEEE Trans. Antennas Prop.* Vol. 48, pp. 1607-1617.

Gazdag, J. and Sguazzero P. (1984). Migration of Seismic Data. *IEEE Proc.* Vol. 72, pp. 1302-1315.

Giannopoulos, A. (2005). Modelling ground penetrating radar by GprMax, *Construction and Building Materials*, Vol. 19, pp. 755-762.

Gilmore, G.; Jeffry, I. and Lo Vetri J. (2006). Derivation and Comparison of SAR and Frequency-Wavenumber Migration Within a Common Inverse Scalar Wave Problem Formulation. *IEEE Trans. Geosc. Rem. Sen.* Vol. 44, pp. 1454-1461.

Groetsch, C. W. (1993). *Inverse Problems in the Mathematical Sciences.* Vieweg Verlag, Braunschweig, Germany.

Hagedoorn, J. G. (1954). A Process of Seismic Reflection Interpretation. *Geophys. Prospect.* Vol. 2, pp 85-127.

Hansen, P. C.; Nagy, J. G. and O'Leary, D. P. (2008) . *Deblurring Images: Matrices, Spectra and Filtering,* SIAM, New York.

Harrington, R. F. (1961). *Time-Harmonic Electromagnetic Fields.* McGraw-Hill, New York.

Hugenschmidt, J. and Kalogeropoulos, A. (2009). The inspection of retaining walls using GPR, *Journal of Applied Geophysics*, Vol. 67, pp. 335-344.

Huisman, J. A., Hubbard, S. S., Redman, J. D. and Annan, A. P. (2003). Measuring Soil Water Content with Ground Penetrating Radar: A Review, *Vadose Zone Journal*, Vol. 2, pp. 476-491.

Krasnov, M. L.; Kiselev, A. I. and Makarenko, G.I.(1976). *Integral Equations*. MIR, Moscow.

Lambot, S.; Slob, E. C.; Chavarro, D.; Lubczynski, M. and Vereecken, H. (2008). Measuring soil surface water content in irrigated areas of southern Tunisia using full-wave inversion of proximal GPR data. *Near Surface Geophys*. Vol. 16, pp. 403-410.

Langenberg, K. J. (1987). Applied Inverse Problems for Acoustic, Electromagnetic and Elastic Wave Scattering. *Basic Methods for Tomography and Inverse Problems ed P C Sabatier*. Bristol: Hilger, pp. 127-467.

Lestari, A. A.; Yarovoy, A. G. and Ligthart, L. P. (2004). RC-Loaded Bow-Tie Antenna for Improved Pulse Radiation. *IEEE Trans. Ant. Prop.*, Vol. 52, pp. 2555-2563.

Lestari, A. A.; Yarovoy, A. G. and Ligthart, L. P. (2005). Adaptive Wire Bow-Tie Antenna for GPR Applications, *IEEE Trans. Ant. Prop.*, Vol. 53, pp. 1745-1754.

Lo Monte, L.; Erricolo, D.; Soldovieri, F. and. Wicks, M. C. (2009). RF Tomography for Underground Imaging: the Effect of Lateral Waves. Proc. of International Conference on Electromagnetics in Advanced Applications (ICEAA), Torino, Italy.

Lopez-Sanchez, J. M. and Fortuny-Guasch, J. (2000). 3-D Radar Imaging Using Range Migration Techniques. *IEEE Trans. Antenn. Propag.* Vol. 48, pp. 728-737.

Markelein, R.; Mayer, K.; Hannemann, R.; Krylow, T.; Balasubramanian, K.; Langenberg, K. J. and Scmitz V. (2002). Linear and Nonlinear Algorithms Applied in Nondestructive Evaluation. *Inv. Probl.* Vol. 18, pp. 1733-1757.

Meincke, P. (2001). Linear GPR inversion for lossy soil and a planar air–soil interface. *IEEE Trans. Geosci. Remote Sens.* Vol. 39, pp. 2713–2721.

Oden, C. P.; Powers, H. M.; Wright, D. L. and Olhoeft, G. R. (2007). Improving GPR Image Resolution in Lossy Ground Using Dispersive Migration. *IEEE Trans. Geosc. Rem. Sens.* Vol. 45, pp. 2492-2500.

Orlando, L. and Soldovieri F. (2008). Two different approaches for georadar data processing: a case study in archaeological prospecting. *Journal of Applied Geophysics*, Vol. 64, pp. 1-13.

Pastorino, M. (2007). *Stochastic optimization methods applied to microwave imaging: A review.* *IEEE Trans. Ant. Prop.* Vol. 55, pp. 538-548.

Persico, R., Bernini, R. and Soldovieri, F. (2005). The role of the measurement configuration in inverse scattering from buried objects under the Born approximation. *IEEE Trans. Antennas and Propagation*, Vol. 53, pp. 1875-1887.

Pierri, R.; Liseno, A.; Solimene, R. and Soldovieri, F. (2006). Beyond physical optics SVD shape reconstruction of metallic cylinders. IEEE Trans. Anten. Propag. Vol. 54, pp. 655-665.

Piro, S.; Goodman, D. and Nishimura Y., (2003). The study and characterization of Emperor Traiano's villa (Altopiani di Arcinazzo, Roma) using high-resolution integrated geophysical surveys. *Archaeological Prospection*, Vol. 10, pp. 1-25.

Romano, N.; Prisco, G. and Soldovieri, F. (2009). Design of a reconfigurable antenna for ground penetrating radar applications. *Progress In Electromagnetics Research*, PIER 94, pp. 1-18.

Soldovieri, F.; Persico, R.; Utsi, E. and Utsi, V. (2006). The application of inverse scattering techniques with ground penetrating radar to the problem of rebar location in concrete. *NDT & E International*, Vol. 39, pp. 602-607.

Solimene, R. and Pierri, R. (2007). Number of Degrees of Freedom of the radiated field over multiple bounded domains. *Optics Letters*, Vol. 32, pp. 3113-3115.

Solimene, R.; Brancaccio, A.; Romano, J. and Pierri, R. (2008). Localizing thin metallic cylinders by a 2.5 D linear distributional approach: experimental results. *IEEE Trans. Anten. Propag.*, Vol. 56, pp. 2630-2637.

Solimene, R.; Soldovieri, F.; Prisco, G. and Pierri, R. (2009). Three-dimensional Through-Wall Imaging under ambiguous wall parameters. *IEEE Trans. Geosc. Rem. Sens.*, Vol. 47, pp. 1310-1317.

Soumekh, M. (1999). Synthetic Aperture Radar Signal Processing with MATLAB Algorithms. Wiley, New York.

Stolt, R. H. (1978). Migration by Fourier Transform. *Geophys.* Vol. 43 pp. 23-48.

Taylor, A. Lay, D. (1980). *Introduction to Functional Analysis.* Krieger, Malabar, Fla.

Tikhonov, A. N. and Arsenine, V. I. (1977). *Solution to ill-posed Problems.* Halsted, York.

Watters, T. R.; Leuschen C. J.; Plaut, J. J.; Picardi, G.; Safaeinili, A.; Clifford, S. M.; Farrell, W. M.; Ivanov, A. B.; Phillips, R. J. and Stofan, E.R. (2006). MARSIS radar sounder evidence of buried basins in the northern lowlands of Mars. *Nature*, Vol. 444, 905-908.

Wu, T. and King R.W. (1965). The cylindrical antenna with non-reflecting resistive loading. *IEEE Trans. Anten. Propag.*, Vol. 13, pp. 369–373.

7

Adaptive Ground Penetrating Radar Systems to Visualize Antipersonnel Plastic Landmines Based on Local Texture in Scattering / Reflection Data in Space and Frequency Domains

Yukimasa Nakano and Akira Hirose
Department of Electrical Engineering and Information Systems, The University of Tokyo
Japan

1. Introduction

Ground penetrating radars (GPRs) are widely used for buried object search, ruin investigation, groundwater assessment, and other various applications (Sato & Takeshita, 2000) (Moriyama et al., 1997). They are also expected to find nonmetallic landmines (Bourgeois & Smith, 1998) (Montoya & Smith, 1999) (Peters Jr. & Daniels, 1994) (Sato et al., 2004) (Sato, 2005) (Sato et al., 2006). Figure 1 shows the size and the structure of a plastic antipersonnel landmine. A near future target is to find or visualize antipersonnel landmines with a high distinction rate between landmines and other objects. The conventional metal detectors, based on electromagnetic induction, use so low frequency that the electromagnetic field penetrates through the soil very deep, and the false negative probability is very small. However, because of its long wavelength, the resolution is limited, and they can tell just whether inductive material exists or not. They cannot distinguish landmines from other metal fragments. GPRs employ much higher frequency. Then the resulting higher resolution will be helpful to discriminate landmines.

Currently, there are two methods to remove the plastic landmines. One is a metal detector, and the other is a rotary cutter. The former detects a blasting cap made of metal in the landmine. Because the cap is very small, we must set the sensitivity at a high level. Then, the positive fault rate is as high as about 99.9% (specificity=0.1%), resulting in a lot of time to remove the landmines. The latter, rotary cutter, looks like a bulldozer, bores the ground and tries to clear the landmines by exploding them. The problems in this method are necessity of additional removal by human because of impossibility of perfect clearance, necessity of sufficient areas for the approach of the rotary cutter, and land pollution by the exploded powder.

Accordingly, though these methods have certain merits, they have also demerits. Therefore, new landmine detection systems based on GPRs attract attention and are studied by a dozen of researcher groups/laboratories presently to solve the problem. Most of the proposed methods employ high-frequency and wide-band electromagnetic wave to visualize a plastic landmine itself instead of the metallic blasting cap. There are two types among the

Fig. 1. The size and the structure of a plastic antipersonnel landmines.

proposals. Some employ pulsed electromagnetic wave and others employ stepped-frequency continuous wave. We explain these two measurement methods in detail in the next section.

Unlike the metal detector which measures induced current, GPRs observe the reflected and scattered electromagnetic wave. In general, it is noted that the electromagnetic wave is reflected at boundaries between materials having different permittivity, and that the spatial resolution of the observation is almost the same as the wavelength. Therefore, it is possible to detect not only the metal but also the plastic body because the electromagnetic wave is reflected at the boundary of the soil and the plastic. In addition, the wide band electromagnetic wave has the possibility to observe the the accurate distance from the antenna to the target, physical property for electromagnetic wave and structural characteristics of the target. That is to say, a GPR system has a potential of detecting plastic landmines more strictly than the metal detector does.

However, high-frequency wave also induce a lot of clutter, which is caused by the roughness of the earth's surface and scattering substances other than the plastic landmines. Consequently, it is very difficult to extract significant features helpful for detecting the plastic landmines from the observed data by ignoring the clutter. Furthermore, it is also difficult to treat the extracted features effectively. A lot of processing methods were proposed so far. That is, we must resolve the following two steps to detect the plastic landmines. The first step is how to extract the features, and the second is how to treat the extracted features. To accomplish our goals, we must select or develop new effective methods. Previously we proposed an adaptive radar imaging system to visualize plastic landmines using complex-valued self-organizing map (CSOM) (Hara & Hirose, 2004) (Hara & Hirose,

2005). With the system, we observe reflection and scattering to obtain a complexamplitude two-dimensional image at multiple frequencies. In the resulting 3-dimensional (2-dimensional (space) × frequency) data, we extract local texture information as a set of feature vectors, and feed them to a CSOM for adaptive classification of the 3-dimensional texture (Hirose, 2006) (Hirose, 2003). By using the system, we could visualize antipersonnel plastic landmines buried shallowly underground. We also constructed a preliminary object identifier, which is a type of associative memory that learns the feature of the plastic-landmine class with adaptation ability to various ground conditions (Hirose et al., 2005).

However, the system requires a long observation time because it employs mechanical scan. Long observation time is one of the most serious problems in high-resolution imaging systems. Some methods to overcome the problem have been investigated (Kobayashi et al., 2004) (Shrestha et al., 2004).

We then developed a portable visualization system with an antenna array to reduce the observation time (Masuyama & Hirose, 2007). The array has 12×12 antenna elements, resulting in about 144 pixels. The element aperture size is 28mm×14mm, which determines the spatial resolution. In texture evaluation and adaptive CSOM classification, a higher resolution leads to a better performance. We recently proposed a resolution enhancement method using a special antenna-selection manner in combination with elimination of direct coupling and calibration of propagation pathlength difference (Masuyama et al., 2007).

However, even with such resolution enhancement, the visualization performance is still worse than that obtained with the first mechanical-scanning system. The resolution is still insufficient, and the mutual coupling between antenna elements are not completely ignorable.

In this chapter, we propose two techniques to improve the visualization ability without mechanical scan, namely, the utilization of SOM-space topology in the CSOM adaptive classification and a feature extraction method based on local correlation in the frequency domain. In experimental results, we find that these two techniques improve the visualization performance significantly. The local-correlation method contributes also to the reduction of tuning parameters in the CSOM classification.

The organization of this chapter is as follows. In Section 2, we explain the merits and demerits of three plastic landmine detection systems that utilize the electromagnetic technology. Then we show the processing flow of our plastic landmine detection system based on the CSOM and show the conventional and proposal methods in Section 3. In Section 4, we show the experimental results with the observed data. Finally, we summarize and conclude this chapter in Section 5.

2. Conventional technology

As mentioned before, there are three methods which utilize the electromagnetic technology to detect plastic landmines. One is the metal detector based on induction current, another is the pulse radar using the electromagnetic wave, the other is stepped-frequency radar using the electromagnetic wave. In the following subsections, we briefly explain the characteristics of these methods respectively.

2.1 Metal detector

The fundamental principle of the metal detector is to sense the mutual interaction between a coil of the detector and target conductors using low frequency electromagnetic field induced by the coil. The used frequency is from about f = 100kHz to 1MHz. Because the frequency is very low, the metal detector is highly tolerant against inhomogeneous ground and the depth of the buried landmines. Additionally, it is very easy to manufacture a highly sensitive metal detector because it consists of some simple analog oscillation circuits. Consequently, the metal detector is very suitable for landmine detection, where we expect not to miss landmine at all, and this method is widely used for the current landmine detection.

However, it is impossible for the metal detector to detect targets other than conductors and difficult to search a target with a high spatial resolution because of the long wavelength. In addition, we cannot obtain the time and/or frequency response which shows the features originating at the targets because the bandwidth is very narrow. The most critical problem is the very high probability of the positive fault which is caused by the metal scattering substance. In particular, as the landmines are buried in a field of combat, there are a large amount of metal fragments around buried landmines, and according to a report, the positive fault is about 99.9%. Therefore, it takes long time to remove landmines perfectly, which is a very big problem. In addition, when we use multiple metal detectors simultaneously, improper signals are often caused by the mutual coupling among closely located detectors.

2.2 Pulse GPR

A pulse GPR observes the time response of the electromagnetic-wave pulse irradiated toward the ground. The time response represents the depth of a scatterer. When we sweep an antenna in two dimension horizontally, we can obtain the three dimensional data. Besides, as the electromagnetic wave is reflected at the boundary of materials that have different permittivity, it is possible to observe the reflection wave from the plastics that forms a landmine, not only the metal blasting cap. The pulse GPRs have another merit. As the pulse have a wide bandwidth, as wide as that of the stepped-frequency range mentioned below, we can observe not only the time response but also the frequency response through the Fourier transform, and these data may show characteristics of the target. Regarding the measuring time, we can conduct the measurement more speedily in comparison with the stepped-frequency GPRs.

However, as a maximum frequency component of the pulse radar is usually about f = 6GHz at the most. That is, the shortest free-space wavelength is about 5cm. Then the pulse GPRs cannot observe sufficient amount of characteristics of plastic landmines whose size is typically the same as the wavelength. Besides, the ground surface is very rough, and the soil, including various scatterers, often causes serious clutter. Therefore, it is very difficult to obtain clear images. To solve this problem we need to utilize sharper pulse which consists of a wideband wave with a high power. However, as a high-peak pulse is distorted by the nonlinearity in transmitter circuits and switched antennas, the problem cannot be solved sufficiently.

2.3 Stepped-frequency GPR

Stepped-frequency GPRs observe the reflected continuous wave at a wide-range frequency points. This method does not need to output strong power instantaneously. Then, the electromagnetic wave has little influence on the nonlinearity of the circuits and the switches. As the results, stepped-frequency GPR accomplishes higher SN ratio than the pulse GPR. In addition, it is easier for the stepped-frequency radar to observe the high frequency wave and select bandwidth freely than a pulse radar system. Besides, we can obtain the time response, like what pulse radar provide, through the inverse Fourier transform of the observed frequency domain data.

As above, stepped-frequency GPRs enable to accomplish high SN ratio at the high frequency with a wide band, and obtain not only the time response but also frequency domain feature very effectively. It is true that even the stepped-frequency GPR has a drawback. Namely, it takes too long time to measure the scattering because of the time required for frequency sweeping. However, this problem will be solved by inventing new appropriate devices in the near future.

Then we can expect a higher precision with the system utilizing the stepped-frequency GPR than the conventional systems. To achieve this purpose, there are two important points we should consider carefully. One is to extract useful features from the obtained data, and the other is to fully utilize the features. However, a perfect technique has not been suggested yet.

In the next section, we show the details of our CSOM-based signal processing published in our previous paper (Nakano & Hirose, 2009).

3. System construction

3.1 Overall construction

Figure 2 shows the processing flow in our plastic landmine visualization system. We describe the components briefly.

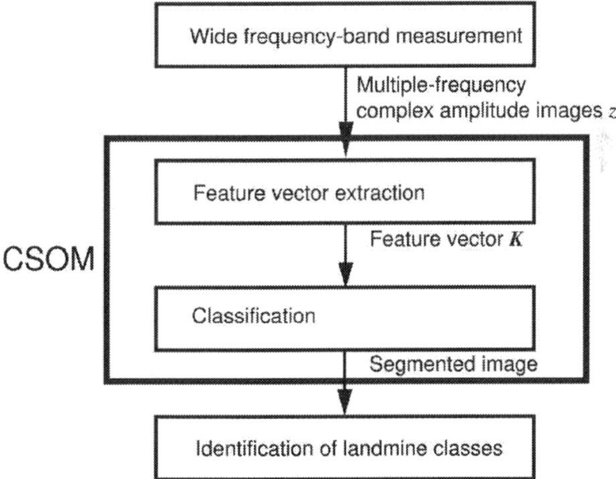

Fig. 2. The overall processing flow (Nakano & Hirose, 2009).

First, we operate our system at a stepped-frequency mode to obtain wideband electromagnetic reflection and/or scattering images at multiple frequency points. The details of the system is given in Ref.(Masuyama & Hirose, 2007). We acquire complex amplitude images at 10 observation frequency points from 8 to 11.6GHz at an interval of 0.4GHz.

Next, we generate a spatially segmented image by using a CSOM that classifies local texture adaptively. The classification consists of two steps. In the first step, we extract feature vectors representing local complex-amplitude textural quantity in a local window that sweeps all over the image. As shown in Fig.3, we prepare a sweeping window in each frequency image at a synchronizing real-space location. We calculate correlations between pixel values in the window in terms of real-space relative distance and frequency-domain distance. We assume that the correlation values represent the texture at around the pixel at the window center, and we put the values at the center pixel as the textural feature. In the second step, we classify the extracted feature vectors adaptively by using a CSOM (Hara & Hirose, 2004). Then we color pixels correspondingly with the resulting classes to generate a segmented spatial image.

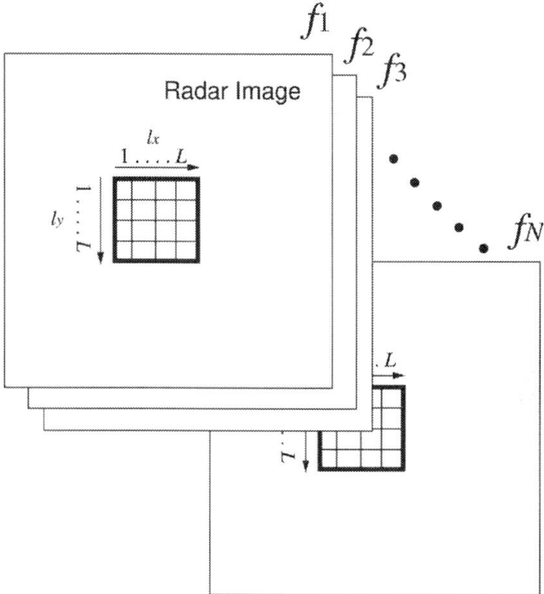

Fig. 3. The scanning window for local textural feature extraction (Nakano & Hirose, 2009).

Lastly, we identify landmine classes included in the segmented image. We use an associative memory that learns respective features of landmine classes and other objects such as metal fragments, stones and clods. We prepare a set of teacher features for the learning beforehand by observing land pieces where we know the landmine locations.

In this paper, we propose two novel methods in the CSOM processing. One is a dynamics in the feature vector classification, and the other is a feature vector extraction method. The former is utilization of SOM-space topology in the CSOM adaptive classification by introducing a ring CSOM, and the latter is the extraction of local correlation in the frequency domain.

3.2 Utilization of SOM-space topology in the CSOM adaptive classification

As the first proposal, instead of the conventional K-mean algorithm, we employ a SOM dynamics that utilizes SOM-space topology in the CSOM adaptive classification. Figure 4 shows the CSOM structure, which forms a ring in the CSOM space.

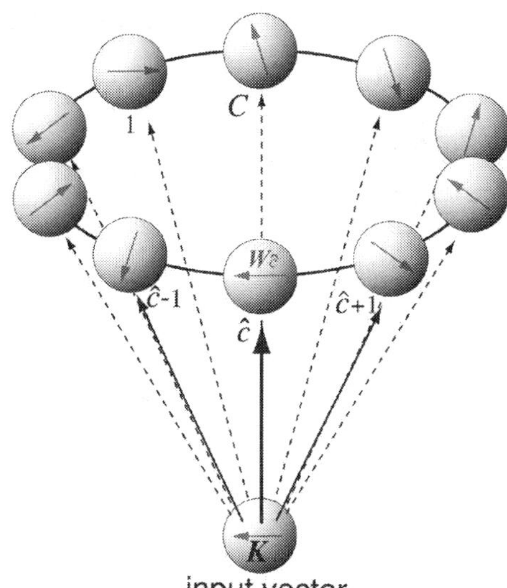

Fig. 4. The ring-CSOM structure. Balls represent reference vectors in the CSOM, and the directions of the arrows show the vector values, among which the winner $W_{\hat{c}}$ and the two neighbors $W_{\hat{c}\pm1}$ change in the self-organization.

In the CSOM in our previous system, we classified the feature vectors by using the K-mean clustering algorithm, which is the simplest SOM dynamics, as (Hara & Hirose, 2004)

$$W_{\hat{c}}(t+1) \quad = \quad W_{\hat{c}}(t) + \alpha(t)(K - W_{\hat{c}}(t)) \tag{1}$$

$$\alpha(t) \quad = \quad \alpha(0)\left(1 - \frac{t}{TMAX}\right) \tag{2}$$

$W_{\hat{c}}(t)$: reference vector of the winner
K : input feature vector
t : iteration number in self-organization
$TMAX$: maximum iteration number
$\underline{\alpha(t)}$: self-organization coefficient
where the winner $W_{\hat{c}}(t)$ is the reference vector nearest to K among all the reference vectors $W_1, W_2, \ldots, W_c, \ldots, W_C$. We update $W_{\hat{c}}$ iteratively by presenting K sequentially. In the new system, we also introduce the self-organization of reference vectors at the winner neighbors $(c \pm 1)$ in the SOM space shown in Fig.4 as

$$W_{\hat{c}\pm1}(t+1) \quad = \quad W_{\hat{c}\pm1}(t) + \beta(t)(K - W_{\hat{c}\pm1}(t)) \tag{3}$$

$$\beta(t) \quad = \quad \beta(0)\left(1 - \frac{t}{TMAX}\right) \tag{4}$$

where $\beta(t)$ is another self-organization coefficient for the neighboring classes, which is usually smaller than $\alpha(t)$. The classes $\hat{c} \pm 1$ are neighbors of the winner class $\hat{c}$ in the CSOM space. The CSOM space is used only to determine the winner neighbors, whereas the winner is determined in the K space.

The reason of the modification is as follows. In the previous method, we used the K-means algorithm (1), which is the simplest dynamics in the SOM. Because we have only about 10 classes in the adaptive vector quantization in the landmine visualization system, we paid less attention to the SOM-space topology. Nevertheless, we could obtain sufficiently meaningful classification results (Hara & Hirose, 2004).

However, with the present portable visualization system with a lower spatial resolution at the antenna array (Masuyama & Hirose, 2007), the failure probability in the classification became slightly higher than the first laboratory system. We sometimes fail to extract texture features sufficiently because of the decrease in resolution. As described later, in such an insufficient extraction case, we found that only a small number of classes were used in the vector quantization.

We therefore propose the utilization of the SOM-space topology in the CSOM adaptive classification, to activate most of the prepared reference vectors, by introducing additional self-organization at neighbor class vectors. In this paper, we change the values of only the two adjacent-class vectors $W_{c \pm 1}$ as shown in (3). The neighbor vector number is small because the number of the totally prepared classes is small, i.e., only about 10. The structure of the CSOM should also be suitable for the small size, namely, one-dimensional as shown in Fig.4.

3.3 Frequency-domain local correlation method

The second proposal is to adopt frequency-domain local correlation in the texture feature extraction. We modify the point on which we put stress based on a new concept in extracting features in the frequency domain.

Figure 5(a) illustrates the conventional method, in which the feature vector K is calculated for complex pixel values $z(l_x, l_y, f)$ as

$$K = [M, K_s, K_f] \tag{5}$$

$$M = \frac{1}{L^2} \sum_{l_x=1}^{L} \sum_{l_y=1}^{L} z(l_x, l_y, f_b) \tag{6}$$

$$K_s = [K_s(0,0), K_s(1,0), K_s(0,1), K_s(1,1)] \tag{7}$$

$$K_s(i,j) = \frac{1}{L^2} \sum_{l_x=1}^{L} \sum_{l_y=1}^{L} z(l_x, l_y, f_b) z^*(l_x + i, l_y + j, f_b) \tag{8}$$

$$K_f = [K_f(f_1), \cdots, K_f(f_N)] \tag{9}$$

$$K_f(f_n) = \frac{1}{L^2} \sum_{l_x=1}^{L} \sum_{l_y=1}^{L} z(l_x, l_y, f_b) z^*(l_x, l_y, f_n) \tag{10}$$

where M, K_s, and K_f are the mean, real-space-domain correlations, and frequency-domain correlations, respectively. Real-space discrete coordinate l_x and l_y determine pixel positions in the local window as shown in Fig.3.

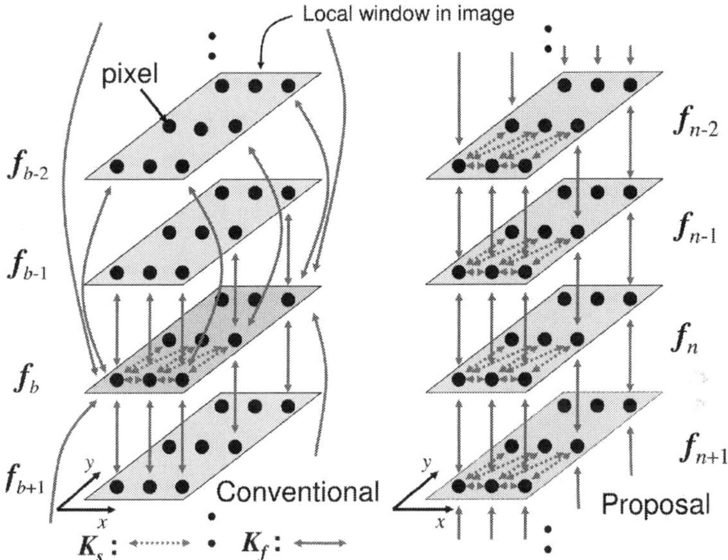

Fig. 5. (a)Conventional and (b)proposed textural feature extraction methods based on correlation in real-space and frequency domains (Nakano & Hirose, 2009).

We prepare a local observation window of $L \times L$ pixels to extract the local textural feature in the window by calculating correlations between pixel values. In (6)–(10), M is the average of pixel values in the window at a base frequency f_b, which we select among the 10- frequency observation points in advance. The vector K_s in (7) is the local correlations in the $L \times L$ real-space window at f_b, while K_f is the correlations between pixel values at f_b and other frequencies f_n at identical positions.

The effectiveness of K_f (f_n) in (10) as a feature vector was suggested by the following frequency-dependent interference. A plastic landmine usually has parallelism among its ceiling, bottom, and air gap inside, if any, which causes interference, whose spectral profile is periodic in the frequency domain. That is to say, we will observe a resonance at integral multiple of a certain frequency periodically in the frequency domain, resulting in a specific peak at certain f_n in K_f (f_n). We intended to capture this phenomenon in (10). However, we found in our series of experiments that we normally observe only a single peak in the 8–12GHz band. If we expect multiple frequency peaks, we have to expand the observation bandwidth. However, very high-frequency electromagnetic wave cannot penetrate ground so deep.

Then we have changed our direction. Note that, in the spatial texture case described above, we paid attention to local correlation caused by the Markovian property. In the same way, also in the frequency domain, we decided to calculate the local correlation to observe the frequency space texture in a simple manner.

Figure 5(b) illustrates our proposal, namely the frequency-domain local correlation method, to extract the frequency-domain feature. We define our new K_f as

$$K_f \;=\; [K_f(f_1), \cdots, K_f(f_{N-1})] \tag{11}$$

$$K_f(f_n) \;=\; \frac{1}{L^2} \sum_{l_x=1}^{L} \sum_{l_y=1}^{L} z(l_x, l_y, f_n) z^*(l_x, l_y, f_{n+1}) \tag{12}$$

where K_f is the feature vector representing the correlation coefficients between the data at adjoining frequency points. This method enables us to eliminate the base frequency f_b, which means that we do not need to choose a special frequency. To extend this f_b-free advantage further, we also modify M and $K_s(i, j)$ slightly as

$$M \;=\; \frac{1}{L^2 N} \sum_{l_x=1}^{L} \sum_{l_y=1}^{L} \sum_{n=1}^{N} z(l_x, l_y, f_n) \tag{13}$$

$$K_s(i,j) \;=\; \frac{1}{L^2 N} \sum_{l_x=1}^{L} \sum_{l_y=1}^{L} \sum_{n=1}^{N} z(l_x, l_y, f_n) z^*(l_x + i, l_y + j, f_n) \tag{14}$$

That is, M and K_s are averaged for the all used frequency data.

The frequency-domain local correlation method is suitable for the processing in this system. Instead of the radar cross section, we use the texture of complex amplitude when we distinguish plastic landmines and other objects such as metal fragments and stones. If we can obtain a sufficiently high resolution in real-space and frequency domains, we should take into account the correlation between one pixel and another at a certain distance. However, when the system has only a low resolution, it is more effective to concentrate on the correlation between neighboring pixels, in which we can expect useful correlation information.

Additionally, in the proposed method, it is a great merit that we do not need the base frequency f_b, which was required in the conventional method. Previously we have a number of possible f_b. As presented below, it is a problem that a different f_b results in a different segmentation image. The new method is free from this problem because we have only one way to construct K.

4. Experiments and results

Table 1 shows the parameters used in the following experiments. We have determined the values of $\alpha(0)$ and $\beta(0)$ empirically. We bury a mock plastic landmine under ground iteratively. We change the burial situation every time, including the ground surface and underground. The surface-roughness amplitude is about 2cm peak-to-peak. In any case, the landmine is buried at around the center of the observation area.

Figure 6(a) shows an experimental result (Result 1). The numbers show the observation frequencies. The upper blue maps show the amplitude data, while the lower color maps show the phase data. Scales of amplitude and phase are shown at the top. The position in every map corresponds to the position in real space. As mentioned above, we use these complex amplitude data obtained at the 10 frequency points.

Figure 6 (b) shows segmented images generated with the previous method. The numbers are base frequencies f_b used respectively. We can choose feature vectors K in 10 ways because there are 10 possible f_b. Each gray level indicates one of the 10 classes. We can find a segmented area at the buried plastic landmine position at f_b=8GHz and 8.8GHz. However, we cannot at other f_b.

Target (Plastic landmine)	
Size	78mm^{ϕ}, 40mm high
Burial depth	$2 \sim 3\text{cm}$
System	
Antenna height	$2 \sim 3$ cm above ground
Window size	$L = 4$
Frequency number	$N = 10$
Class number	$C = 10$
Initial learning coefficients	$\alpha(0) = 0.4$
	$\beta(0) = 0.1$
Maximum learning iteration	$TMAX = 10$

Table 1. Parameters of target and system (Nakano & Hirose, 2009).

Figure 6 (c) shows the result of segmentation by utilizing the SOM-space topology in the CSOM adaptive classification. We find that there are more classes used in the classification, i.e., 10 classes in most cases, than that in the case of the previous method. We can confirm that we can classify the landmine area appropriately at most f_b. For example, also at 8.4GHz and 9.2GHz, we are successful in the segmentation this time. These results reveal that we can improve the performance of classification by the utilization of the SOM-space topology in the CSOM.

Figure 6 (d) shows the segmentation result obtained by the frequency-domain local correlation method as well as the utilization of SOM-space topology. As mentioned before, there is only one manner to extract feature vectors K in this proposed method because we have no f_b. Here we show four result examples for various initial reference vectors in the CSOM since the result of the CSOM may depend on the initial reference-vector values. In all the cases, the landmine area is segmented correctly. We confirm a high robustness of the present method with the two proposal.

Figure 7(a) shows a measurement result (Result 2) in a different situation from that of Fig.6(a). The landmine classification seems more difficult in this case than that of Fig.6 because the calibration of direct coupling components (Masuyama et al., 2007) is somewhat sensitive to noise, occasionally resulting in insufficient compensation of antenna-selection-mode dependent amplitude.

Figure 7(b) shows the segmented images obtained with the previous method. We can classify the landmine area only when f_b=9.2GHz. We completely failed in the segmentation at other f_b. Figure 7(c) shows a result by utilizing the SOM-space topology in the CSOM. We can segment the landmine area only at 9.2GHz again.

Figure 7 (d) shows the results obtained by employing the two proposed methods. It is confirmed that we can classify the landmine area perfectly. We show four results for various initialization again. The results indicate that we can segment landmine areas stably.

In addition, we recognize that more classes are used for the classification in Fig.7(d) than in Fig.7(c) despite we use the same dynamics for the classification in the CSOM. For this reason, we can extract more characteristic feature quantities with the frequency-domain local correlation method than that with the previous one.

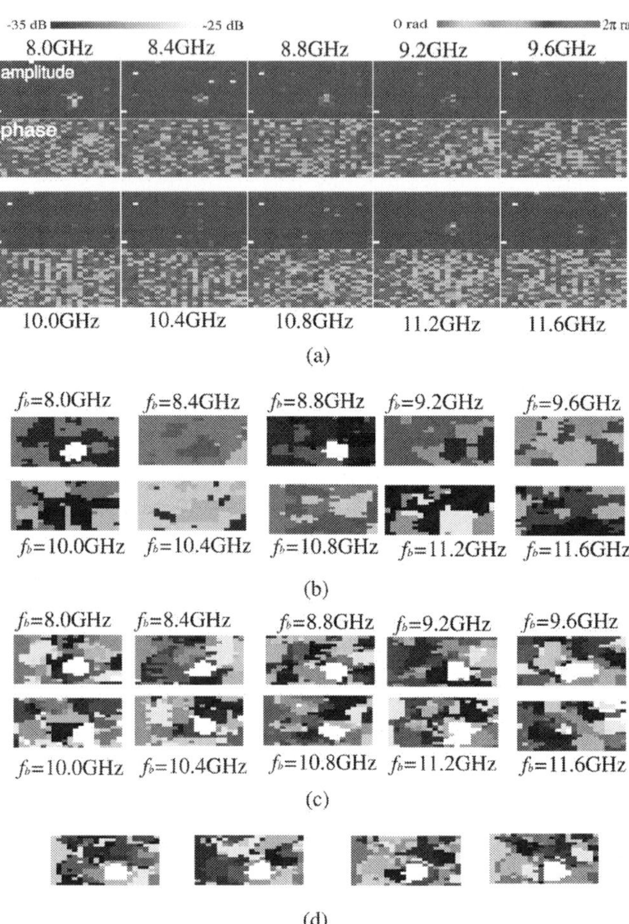

Fig. 6. Experimental results 1. (a)Amplitude and phase images at 10 frequency points, (b)classification results with the previous method. Numbers denote base frequency f_b, (c)classification results with utilization of SOM-space topology in the CSOM. Numbers denote base frequency. (d)Classification results with utilization of SOM-space topology in the CSOM and frequency-domain local correlation method. The four images show the results with various initial reference vectors in the CSOM (Nakano & Hirose, 2009).

5. Summary

In this chapter, first we explained the ground-penetrating radars (GPRs) which are studied currently as a new technology for the antipersonnel plastic landmine detection. In this field, researchers usually choose a measurement type from the pulse GPR or the stepped frequency GPR. Though both of these methods have merits and demerits, a stepped-frequency GPR has an advantage in the high ability to extract features over a pulse GPR.

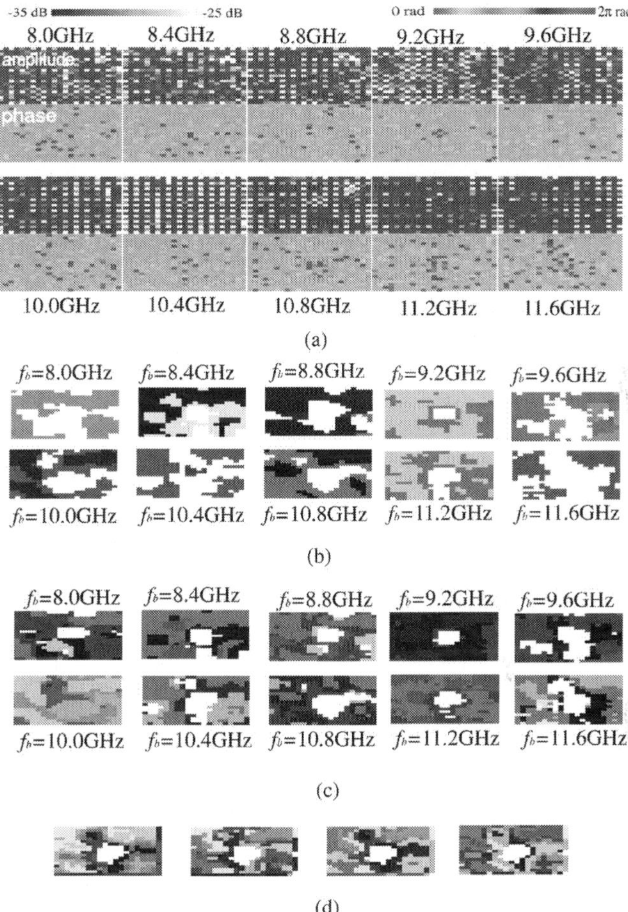

Fig. 7. Experimental results 2. (a)Amplitude and phase images at 10 frequency points, (b)classification results with the previous method. Numbers denote base frequency f_b, (c)classification results with utilization of SOM-space topology in the CSOM. Numbers denote base frequency. (d)Classification results with utilization of SOM-space topology in the CSOM and frequency-domain local correlation method. The four images show the results with various initial reference vectors in the CSOM (Nakano & Hirose, 2009).

Next, we described two techniques, which is based on the stepped-frequency GPR, to improve the performance of the GPR system to visualize plastic landmines. One is to utilize SOM-space topology in the CSOM adaptive classification to stabilize the classification process. Unlike the K-means algorithm, we can use most of the prepared classes in the learning vector quantization. The other technique is to employ local correlation as the feature vector components in the frequency domain. It extracts complex texture information better and, at the same time, eliminates the base frequency, which had to be chosen by the user as one of the tuning parameters. Experimental results demonstrated better and stable visualization of the plastic landmine.

6. References

Bourgeois, J. M. & Smith, G. S. (1998). A complete electromagnetic simulation of the separatedaperture sensor for detecting buried land mines, *IEEE Trans. Antenna and Propagation* 46(10): 1419–1426.

Hara, T. & Hirose, A. (2004). Plastic mine detecting radar system using complex-valued selforganizing map that deals with multiple-frequency interferometric images, *Neural Networks* 17(8-9): 1201–1210.

Hara, T. & Hirose, A. (2005). Adaptive plastic-landmine visualizing radar system: effects of aperture synthesis and feature-vector dimension reduction, *IEICE Transactions on Electronics* E88-C(12): 2282–2288.

Hirose, A. (ed.) (2003). *Complex-Valued Neural Networks: Theories and Applications*, World Scientific Publishing Co. Pte. Ltd.

Hirose, A. (ed.) (2006). *Complex-Valued Neural Network*, Heidelberg: Springer-Verlag.

Hirose, A., Toh Jiayun, A. & Hara, T. (2005). Plastic landmine identification by multistage association, *IEICE Tech. Rep.* (NC2004-156).

Kobayashi, T., Feng, X.&Sato, M. (2004). Agpr for landmine detection using an array antenna, *International Symposium on Antennas and Propagation (ISAP) Sendai.*

Masuyama, S. & Hirose, A. (2007). Walled LTSA array for rapid, high spatial resolution, and phase sensitive imaging to visualize plastic landmines, *IEEE Transactions on Geoscience and Remote Sensing* 45(8): 2536–2543.

Masuyama, S., Yasuda, K. & Hirose, A. (2007). Removal of direct coupling in a walled-LTSA array for visualizing plastic landmines, *International Symposium on Antennas and Propagation (ISAP) 2007 Niigata*, pp. 1238–1241.

Montoya, T. P. & Smith, G. S. (1999). Land mine detection using a ground-penetrating radar based on resistively loaded vee dipoles, *IEEE Trans. Antenna and Propagation* 47(12): 1795–1806.

Moriyama, T., Nakamura, M., Yamaguchi, Y. & Yamada, H. (1997). Radar polarimetry applied to the classification of target buried in the underground,*Wideband Interferometric Sensing and Imaging Polarimetry*, Vol. 3210 of *Proc. of SPIE*, pp. 182–189.

Nakano, Y. & Hirose, A. (2009). Improvement of plastic landmine visualization performance by use of ring-csom and frequency-domain local correlation, *IEICE Transactions on Electronics* E92-C(1): 102–108.

Peters Jr., L. & Daniels, J. J. (1994). Ground penetrationg radar as a surface environmental sensing tool, *Proceedings of the IEEE*, Vol. 82,No.12, pp. 1802–1822.

Sato, M. (2005). Dual sensor alis evaluation test in afghanistan, *IEEE Geoscience and Remote Sensing Society Newsletter* pp. 22–24.

Sato, M., Hamada, Y., Feng, X., Kong, F.-N., Zeng, Z. & Fang, G. (2004). GPR using an array antenna for landmine detection, *Near Subsurface Geophysics* 2: 7–13.

Sato, M., Takahashi, K., Feng, X. & Kobayashi, T. (2006). Stepped-frequency handheld demining dual sensor alis, *Proceeding of 11th International Conference on Ground Penetrating Radar*, p. UXO.10.

Sato, M. & Takeshita, M. (2000). Estimation of subsurface fracture roughness by polarimetric borehole radar, *IEICE Trans. Electron.* E83-C(12): 1881–1888.

Shrestha, S. M., Arai, I. & Tomizawa, Y. (2004). Landmine detection using impulse ground penetrating radar, *Internationall Symposium on Antennas and Propagation (ISAP) Sendai.*

8

Application of Radar Technology to Deflection Measurement and Dynamic Testing of Bridges

Carmelo Gentile
Politecnico di Milano, Dept. of Structural Engineering
Italy

1. Introduction

Vibration testing of bridges and large structures is generally performed by using piezoelectric or force-balanced accelerometers since these sensors are very accurate and relatively inexpensive. Although accelerometers have been – and still are – extensively and successfully used, their drawbacks are well-known: (1) the sensors must be installed at selected locations that are representative of the structure motion and access may be difficult and often dangerous; (2) the installation and wiring are the most time-consuming tasks during the tests; (3) the use in permanent monitoring systems is prone to the typical failures of any system provided with cables and electrical circuits; and (4) accelerometers do not provide a direct measurement of displacement, something that could positively affect the development of the Structural Health Monitoring (SHM) in operational conditions.

Within this context, the development of innovative non-contact systems for vibration measurement is very attractive and especially applications of laser-based systems (Cunha & Caetano 1999, Kaito *et al.* 2005) are reported in the literature. Other recent investigations suggest the application of Global Positioning Systems (GPS, Nickitopoulou *et al.* 2006, Meng *et al.* 2007) or image analysis and vision systems (Lee *et al.* 2006, Silva *et al.* 2007). Furthermore, a non-contact system using microwaves was described by Farrar *et al.* (1999) and used to measure the vibration response of the well-known I-40 bridge over Rio Grande river (Farrar & Cone 1995); this sensor did not provide any *range resolution*, i.e. was not capable to detect different targets in the scenario illuminated by the microwave beam.

Recent progresses in radar techniques and systems have favoured the development of a microwave interferometer, potentially suitable to the non-contact vibration monitoring of large structures (Pieraccini *et al.* 2004, Bernardini *et al.* 2007, Gentile & Bernardini 2008, Gentile 2009). The main characteristic of this new radar system – entirely designed and developed by Italian researchers – is the possibility of simultaneously measuring the static or dynamic deflections of several points of a structure, with sub-millimetric accuracy. For the radar vibrometer, each discontinuity of a structure – such as the "corner zones" corresponding to the intersection of girders and cross-beams in the deck of bridges – represents a potential source of reflection of the electromagnetic waves generated by the radar; in such cases, an echo can be generated and the corner zones act as a series of virtual sensors. In addition to its non-contact feature, the sensor provides other advantages including a wide frequency range of response, portability and quick setup time.

The radar sensor detects the position and the displacement of target points placed at different distances from the equipment by using two well-known radar techniques: the Stepped-Frequency Continuous Wave (SF-CW, Wehner 1995) and the Interferometric techniques (Henderson & Lewis 1998).

This chapter first describes the basic principles implemented in the new sensor, its technical characteristics and the results of the laboratory tests carried out to evaluate the accuracy and the intrinsic performance of the equipment. Subsequently, the application to field test of different full-scale bridges is summarized and reviewed. The field tests include the measurement of vibration responses on structural elements very difficult to access by using conventional techniques, such as stay cables (Gentile 2009). During the presented experimental tests, the microwave interferometer and conventional accelerometers were simultaneously used; hence, the analysis of acquired signals included extensive comparison between the time series recorded by the radar and the accelerometers and the comparison clearly exemplifies the major advantages and drawbacks of the radar equipment. Furthermore, resonant frequencies and mode shapes of the investigated vibrating systems, that were identified from the radar signals, are compared to the corresponding quantities estimated from the accelerometer's records.

2. Radar techniques implemented in the microwave interferometer

The most peculiar and important characteristic of a conventional radar is its ability to determine the range (i.e. the distance) by measuring the time for the radar signal to propagate to the target and back. Although the name *radar* is derived from radio detection and ranging, a radar is capable of providing more information about the target than its name would imply. Typical applications (see e.g. Skolnik 1990) include the evaluation of the radial velocity, the angular direction, size and shape of the target.

The main functions of the new radar sensor herein described are the simultaneous detection of the position and deflection of different targets placed at different distances from the sensor. This performance is obtained by implementing two well-known radar techniques:

1. the Stepped-Frequency Continuous Wave technique (Wehner 1995), employed to detect the positions of different targets placed along the radar's line of sight;
2. the Interferometric technique (Henderson & Lewis 1998), implemented to compute the displacement of each target, by comparing the phase information of the back-scattered electromagnetic waves collected at different times.

2.1 The Stepped Frequency – Continuous Wave (SF-CW) technique

The usual radar waveform to determine the range is the short pulse. The shorter the pulse, the more precise is the measurement of the range because the range resolution Δr is related to the pulse duration τ by the following relationship:

$$\Delta r = \frac{c\tau}{2} \qquad (1)$$

where c is the speed of light in free space. For a signal of duration τ, it can be shown (see e.g. Marple 1987) that the time-bandwidth product satisfies the equality $\tau B = 1$, where B is the equivalent bandwidth in Hz. Hence, the range resolution Δr may be expressed as:

$$\Delta r = \frac{c}{2B} \tag{2}$$

Eqs. (1) and (2) show that a better range resolution (corresponding to a smaller numerical value of Δr) can be obtained by either decreasing τ or increasing B. Instead of using short-time pulses, SF-CW radars exhibit a large bandwidth by linearly increasing the frequency of successive pulses in discrete steps, as shown in Fig. 1; hence, a SF-CW radar has a narrow instantaneous bandwidth (corresponding to individual pulse) and attains a large effective bandwidth:

$$B = (N-1)\Delta f \tag{3}$$

with a burst of N electromagnetic pulses, generally named tones, whose frequencies are increased from tone to tone by a constant frequency increment Δf.
In a SF-CW radar, the signal source dwells at each frequency $f_k = f_o + k\Delta f$ $(k=0,1,2, ..., N{-}1)$ long enough to allows the received echoes to reach the receiver. Hence, the duration of each single pulse (T_{tone}) depends on the maximum distance (R_{max}) to be observed in the scenario:

$$T_{tone} = \frac{2R_{max}}{c} \tag{4}$$

The number N of tones composing each burst can be computed as:

$$N = \frac{2R_{max}}{\Delta r} \tag{5}$$

a)

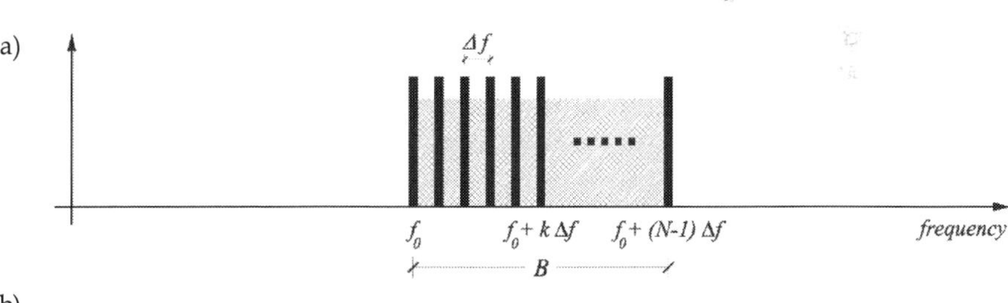

b)

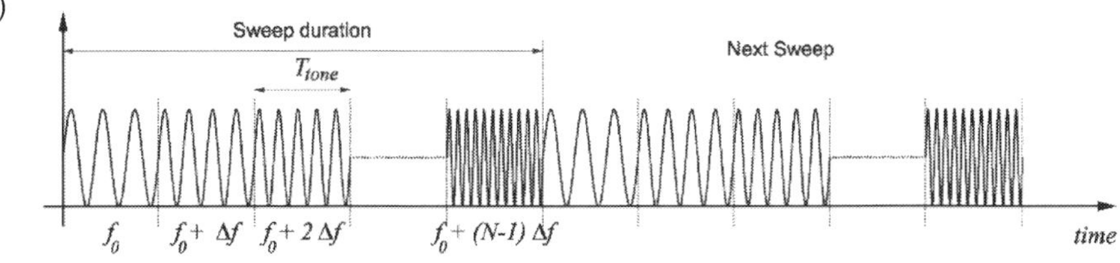

Fig. 1. Representation of SF-CW waveform in: a) frequency domain and b) time domain

The time scheduling (4) permits the SF-CW radar to receive the response of the furthest target before transmitting the following tone. On the other hand, eqs. (4)-(5) clearly highlight that the maximum sampling rate of the scenario f_{sample} depends on R_{max} and Δr. Specifically, accounting for eqs. (4)-(5), f_{sample} can be expressed as:

$$f_{\text{sample}} \cong \frac{1}{N T_{\text{tone}}} = \frac{c}{2 N R_{\max}} = \frac{c \Delta r}{4 R_{\max}^2} \tag{6}$$

Eq. (6) clearly shows that:

a. the maximum sampling rate decreases as the maximum measured distance increases (since the system has to wait for a longer time to receive the echo of the furthest target);

b. the maximum sampling rate increases as the range resolution increases.

The use of SF continuous waveform exhibits several advantages:

1. SF-CW radars can reach the same far distance of a pulse radar by transmitting lower power. In turn, low transmitted power generally allows SF-CW radars to be included in the Short-Range Device category as a license-free use equipment;

2. SF modulated radars can transmit and receive signals with precise frequency control by using Direct Digital Synthesizer (DDS), an innovative up-to-date device for generating SF waveforms.

SF waveforms produce a synthetic profile of scattering objects through the procedure summarized in Fig. 2. At each sampled time instant, both in-phase (I) and quadrature (Q) components of the received signals are acquired so that the resulting data consist of a vector of N complex samples, representing the frequency response measured at N discrete frequencies. By taking the Inverse Discrete Fourier Transform ($IDFT$), the response is reconstructed in the time domain of the radar: each complex sample in this domain represents the echo from a range (distance) interval of length $c/2B$. The amplitude range profile of the radar echoes is then obtained by calculating the magnitude of the $IDFT$ of acquired vector samples.

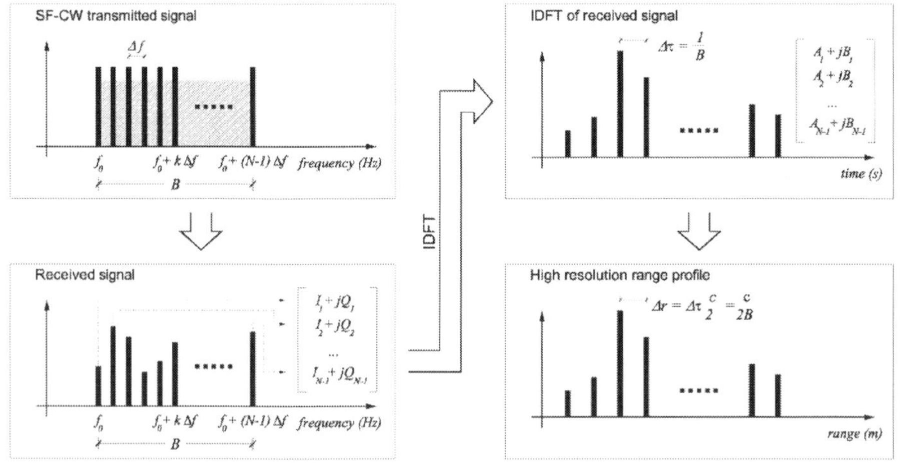

Fig. 2. Evaluation of high-resolution range profile from SF-CW waveform

The amplitude range profile gives a one-dimensional (1-D) map of the scattering objects in the space illuminated by the antenna beam, as a function of their relative distance from the sensor.

The concept of range profile is better illustrated in Fig. 3, where an ideal range profile is shown, as obtained when the radar transmitting beam illuminates a series of targets at different distances and different angles from the axis of the system. The peaks in the plot of Fig. 3 correspond to points with good electromagnetic reflectivity: the sensor can simultaneously detect the displacement or the transient response of these points. Fig. 3 also shows that the radar has only 1-D imaging capabilities, i.e. different targets can be individually detected if they are placed at different distances from the radar. Hence, measurement errors may arise from the multiplicity of contributions to the same range bin, coming from different points placed at the same distance from the radar but not lying on the same axis (Gentile *et al.* 2008).

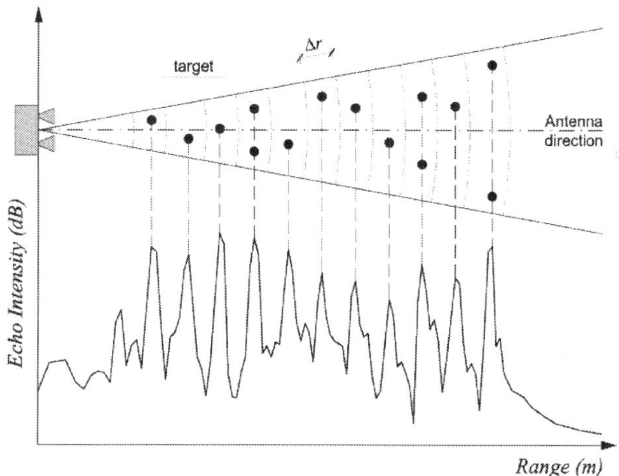

Fig. 3. Idealization of a radar image profile (range profile)

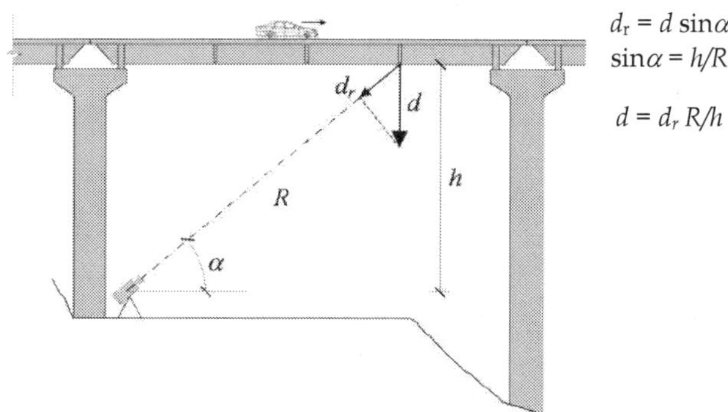

Fig. 4. Radial displacement vs. actual (i.e. vertical) displacement

2.2 The Interferometric technique

Once the range profile has been determined at uniform sampling intervals $\Delta t=1/f_{sample}$, the displacement response of each range bin is evaluated by using the phase interferometry (Henderson & Lewis 1998); according to this technique, the displacement of a scattering object is evaluated by comparing the phase information of the electromagnetic waves reflected by the object in different time instants.

Since two images obtained at different times exhibit phase differences, depending on the motion of the scatterers along the direction of wave propagation, the radial displacement d_r (i.e. the displacement along the radar line-of-sight) is simply computed from the phase shift $\Delta\vartheta$ as:

$$d_r = -\frac{\lambda}{4\pi}\Delta\vartheta \qquad (7)$$

where λ is the wavelength of the electromagnetic signal.

It is worth underlining that the interferometric technique, represented by eq. (7), provides a measurement of the radial displacement of all the range bins of the structure illuminated by the antenna beam; hence, the evaluation of the actual displacement requires the knowledge of the direction of motion. For many bridges (simple or continuous spans, frame or truss bridges), the displacement under traffic loads can be assumed as vertical and it can be easily evaluated by making straightforward geometric projections, as shown in Fig. 4.

3. Description and technical characteristics of the microwave interferometer

The radar techniques described in the previous section have been implemented in an industrially engineered micro-wave interferometer by IDS (Ingegneria Dei Sistemi, Pisa, Italy); the new sensor, named IBIS-S (*Image By Interferometric Survey of Structures*), consists of a sensor module, a control PC and a power supply unit (Fig. 5).

The sensor module is a coherent radar (i.e. a radar preserving the phase information of the received signal) generating, transmitting and receiving the electromagnetic signals to be processed in order to compute the displacement time-histories of the investigated structure. This unit, weighting 12 kg, includes two horn antennas for transmission and reception of the electromagnetic waves and is installed on a tripod equipped with a rotating head (Fig. 5), allowing the sensor to be orientated in the desired direction.

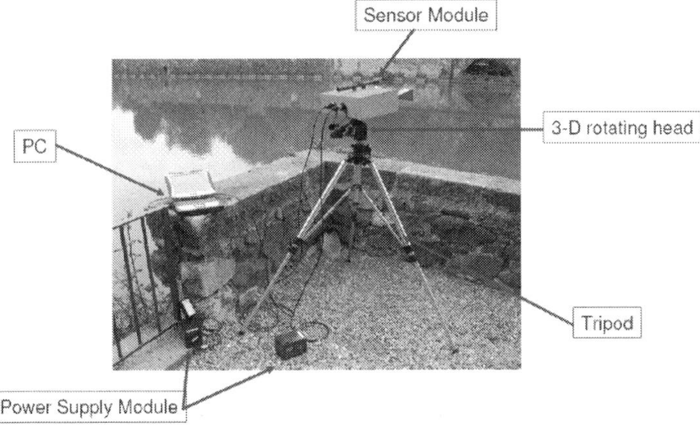

Fig. 5. View of the IBIS-S microwave interferometer

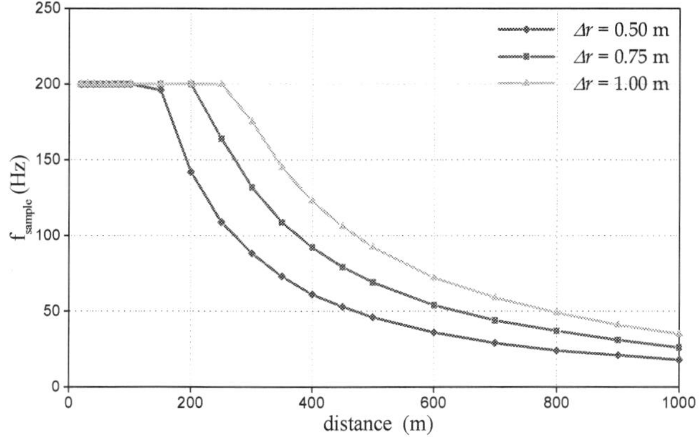

Fig. 6. Sampling rate vs. maximum distance for three different values of Δr

The radio-frequency section radiates at a central frequency of 16.75 GHz with a maximum bandwidth of 300 MHz; hence, the radar is classified as K_u-band, according to the standard radar-frequency letter-band nomenclature from IEEE Standard 521-1984.

The sensor unit is connected to the control PC by means of a standard USB 2.0 interface; the control PC is provided with a specific software for the management of the system and is used to configure the acquisition parameters, to manage and store the measurements and to show the displacements in real time.

The power supply unit, consisting of 12V battery packs, provides the power to the system for 5-6 hours.

The IBIS-S interferometric radar was designed to provide:

1. minimum range resolution Δr of 0.50 m, so that two targets can still be detected individually if their relative distance is greater than or equal to 0.50 m;
2. maximum sampling frequency of the scenario f_{sample} of 200 Hz, which is an excellent performance since the significant frequency content of displacement time-histories is generally in the frequency range 0–20 Hz for a civil engineering structure. In addition, sampling interval Δt=0.005 s is in principle well suitable to provide a good waveform definition of the acquired signals.

As a consequence of the radar techniques implemented in the sensor, the maximum operating distance depends on f_{sample} and Δr (see eq. 6). The dependence of sampling rate on the maximum distance is shown in Fig. 6, for three different distance resolutions. The inspection of Fig. 6 reveals that, for a range resolution of 0.5 m, the sampling rate drops off for distances greater than 150.0 m while, for a range resolution of 1.0 m, the sampling rate starts to decrease for distances greater than 300.0 m and reaches the value of 35 Hz for a range of 1000.0 m.

4. Accuracy and validation of the radar technique

4.1 Laboratory test

Unlike other non-contact techniques of deflection measurement, that are characterized by an accuracy generally ranging between 1.0-4.0 mm (image-based techniques) and 1.0 cm (GPS), sub-millimetric accuracy has in principle to be expected from the design specification on the components of the microwave interferometer. This performance was verified in various laboratory tests, before using the radar in the field on full-scale structures. Among these tests, the free-vibration response of a simple mass-spring system was measured (Bernardini *et al.* 2007).

The test set-up was arranged by installing the mass-spring system, modified by adding a small and light passive radar reflector (corner reflector), in front of the radar sensor at a distance of 7.0 m. Fig. 7 shows a sketch of the test set-up and a photograph of the oscillator equipped with the corner reflector. The control PC of the sensor was configured to measure targets up to a distance of 50.0 m and with a scenario sampling frequency of 50 Hz.

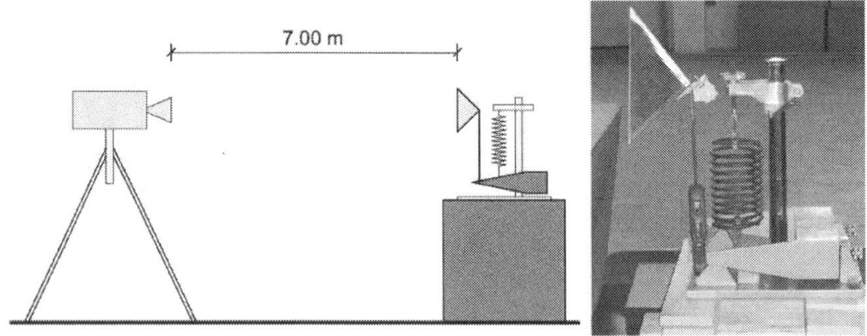

Fig. 7. Mass-spring system tested in laboratory and test set-up

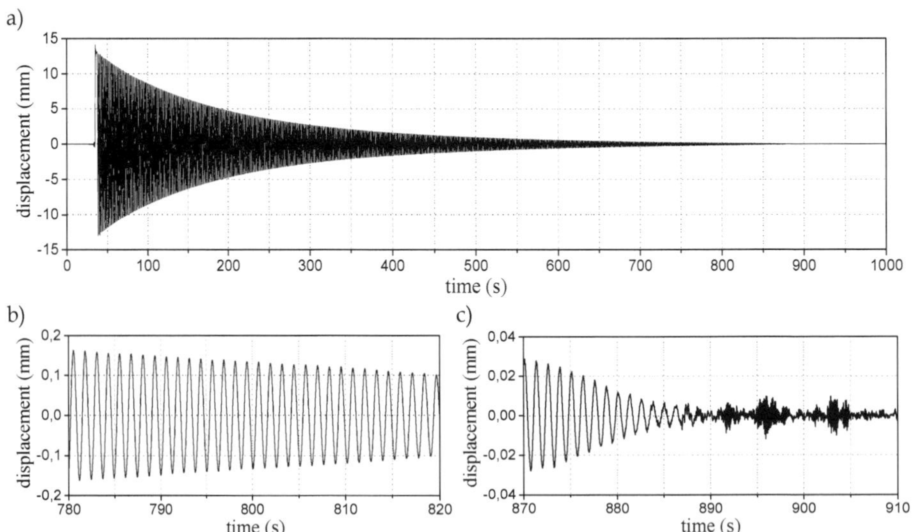

Fig. 8. a) Radar-based measurement of the displacement of a mass-spring system in laboratory test; b) zoom of the measured displacement in time interval 780–820 s; c) zoom of the measured displacement in time interval 870–910 s

Fig. 8a shows the free-damped displacement measured by the radar sensor in 1000 s of observation and the measured time-history corresponded perfectly to what expected for a lightly damped single-degree-of-freedom system. In order to better illustrate the characteristics of the measured response, Figs. 8b and 8c show temporal zooms of the displacement time-histories in the low-amplitude range. Fig. 8b clearly shows that the damped harmonic motion is very well described when its amplitude ranges between 0.1 and 0.2 mm. A similar performance appears in Fig. 8c, corresponding to the end of the free-damped motion: the measurement seems to exhibit excellent quality until the amplitude of the displacement exceeds 0.01–0.02 mm.

The inspection of Figs. 8a-c clearly highlights that – at least in a laboratory test and at a distance of 7.0 m – the accuracy of the sensor is better than 0.02 mm. It is worth underlining that other similar tests are currently in execution for distance ranging from 10.0 to 100.0 m.

4.2 Ambient vibration test of a reinforced concrete bridge

The radar equipment was first used on site during the ambient vibration test (AVT) of a reinforced concrete bridge crossing the river Adda (Gentile & Bernardini 2008), between the towns of Capriate and Trezzo (about 50 km north-east of Milan, Italy).

Plan and elevation of the bridge are shown in Fig. 9. The deck has a total length of 113.3 m and consists of two variable-depth balanced cantilevers (47.9 m long), connected by one simply-supported drop-in girder. Each balanced cantilever consists of a three-cell concrete box girder, while the central girder consists of a concrete slab supported by 4 girders and 3 cross-beams. The total width of the deck is 10.10 m.

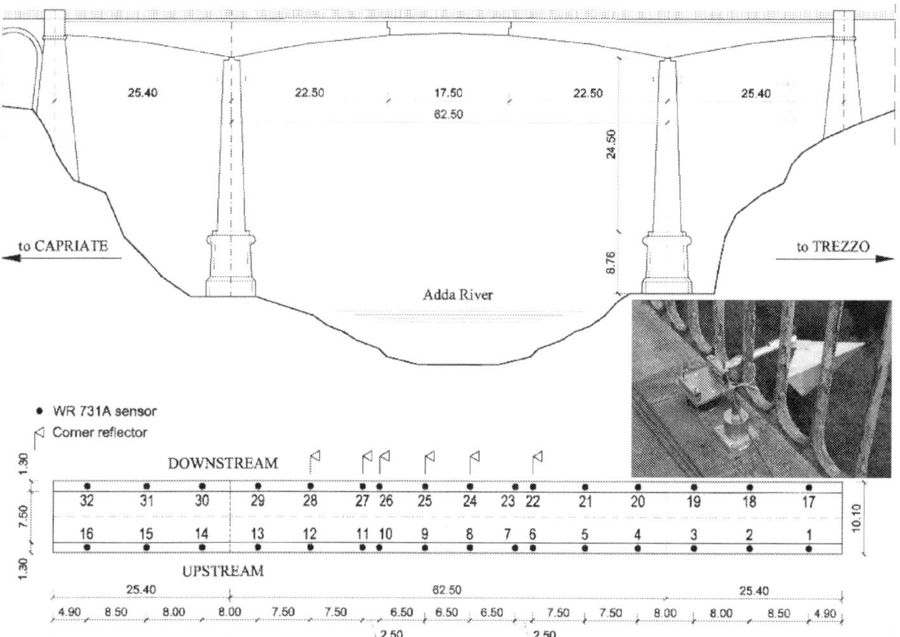

Fig. 9. Elevation and plan of the Capriate bridge, and sensor layout during the bridge tests

In this test, the ambient vibration response of the bridge was measured by simultaneously using the radar sensor and conventional accelerometers. In order to provide accurate comparison between the signals acquired from the different sensors, simple passive radar reflectors were placed as close as possible to the accelerometers (Fig. 9).

Two series of AVTs were carried out and the response of the bridge was measured at selected points using WR-731A sensors, each with a WR-P31 power unit/amplifier. These sensors, allowing acceleration or velocity responses to be recorded, were used as conventional accelerometers in the first series of tests to identify the dynamic characteristics of the bridge; velocity time-histories were recorded during the second series of tests, when the microwave interferometer and the WR-731A sensors were simultaneously used.

The objective of the tests was two-fold. First, the agreement between the time-histories evaluated from the radar and the ones recorded by conventional sensors was extensively investigated (over a time window of 3000 s) in correspondence of several target surfaces (Fig. 9); more specifically, the velocity time-histories directly recorded by the WR 731A sensors were compared to the ones computed by deriving the displacements obtained by the IBIS-S sensor. Subsequently, resonant frequencies and mode shapes of the bridge, identified from the radar signals, were compared to the corresponding quantities estimated from the accelerometer's records.

An example of comparison between radar and conventional signals is given in Figs. 10a-b; the figures refer to the velocities simultaneously recorded at test points TP22 and TP27 over a short time period (12 s) and clearly show an excellent agreement between the data obtained from radar and conventional sensors. A similar agreement was obtained for all corner reflectors during 3000 s of simultaneously acquired time window, provided that the deflection response exceeds 0.01–0.02 mm.

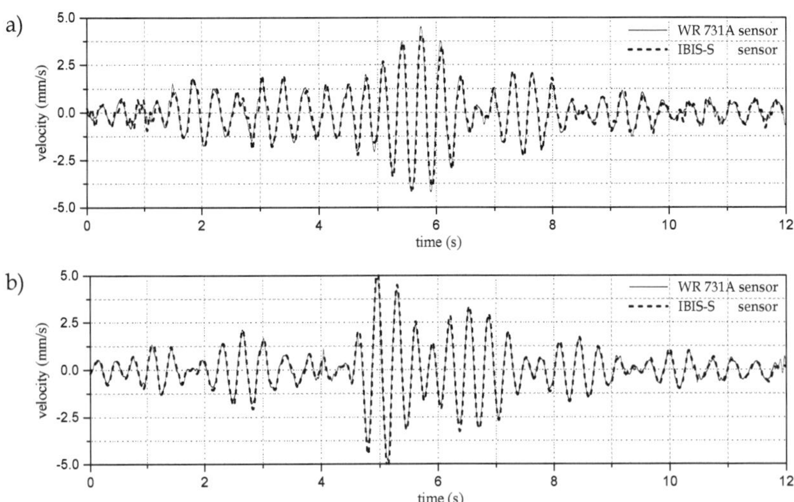

Fig. 10. Capriate bridge: comparison between the velocity time-histories collected by WR 731A and IBIS-S sensors at: a) TP22; b) TP27

As a consequence of the time-histories agreement, resonant frequencies and mode shapes provided by the radar sensor turned out to be as accurate as those obtained with traditional accelerometers (Gentile & Bernardini 2008), as shown in Fig. 11.

Finally, it is worth underlining that the microwave interferometer exhibits a remarkable stability in long term functioning on site (required for effective employment in AVT or continuous dynamic monitoring) and its use is relatively simple.

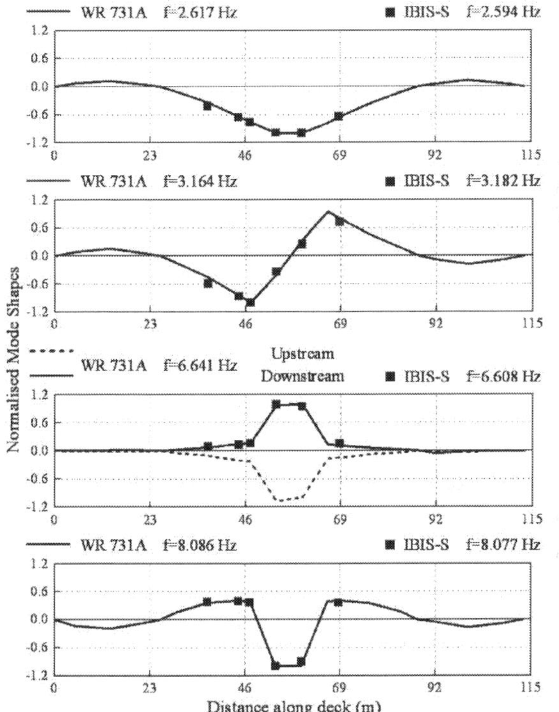

Fig. 11. Capriate bridge: comparison between natural frequencies and mode shapes identified from WR 731 and IBIS-S sensors data

4.3 Static test of a steel-composite bridge

In order to perform direct validation of the deflection measured by the radar and to assess the performance of the non-contact radar technique without the use of corner reflectors, extensive static tests were carried out on some spans of a steel-composite bridge. Steel and steel-composite bridges are much more reflective to electromagnetic waves than the concrete ones; furthermore, the deck generally includes a large number of corner zones, provided by the intersection of girders and cross-beams.

Since the static tests with the radar equipment were conducted taking profit of the simultaneous execution of standard reception tests of the investigated bridge, experimental data were collected by simultaneously using the radar technique and conventional techniques, with validation purposes.

The investigated bridge belongs to a motorway intersection recently completed in the neighbourhood of Forlanini Avenue, Milan, that is the main road linking the city centre to the city airport of Linate. The new infrastructure, shown in Fig. 12, includes two viaducts over-passing Forlanini Avenue.

The south-side viaduct is a continuous span steel-concrete composite bridge, consisting of 8 spans; the intermediate spans are generally 50.0 m long, while the end-spans are 38.0 m long, for a total length of 376.0 m. The structure consists of ladder beams with cantilevers; hence, the cross-section (Fig. 13) is characterised by two main longitudinal girders with transverse cross-beams, at a longitudinal spacing of 4.17 m. The cross-beams are extended beyond the girder to form cantilevers spanning 4.15 m. The girders are 2.55 m high while the floor beams are 1.00 m high. Girders and floor beams have wide flanges supporting a reinforced concrete slab, 25.0 cm thick. The total width of the deck is 18.0 m for three traffic lanes and two lateral emergency lanes.

Fig. 12. Aerial view of the new viaduct over Forlanini Avenue, Milan, Italy

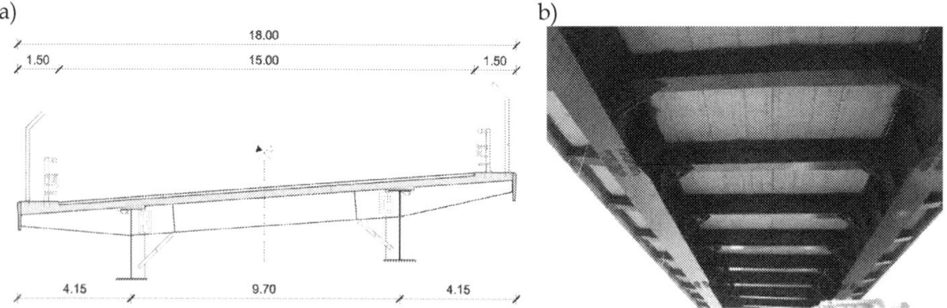

Fig. 13. a) Typical cross-section of the bridge over-passing Forlanini Avenue; b) Bottom view of the bridge deck

As it is usual in reception tests of new bridges, the focus of the test program was the measurement of vertical deflections of the bridge under live load. Vehicles of known weight were located at selected points of the viaduct and vertical deflections were measured at the centre of the girders of loaded spans by using traditional Linear Variable Differential Transformer (LVDT) Schaewitz extensometers.

The source of the live load for the test program was a series of 12 two-axle trucks, weighting between 415 kN and 440 kN. The test vehicles were placed according to four different arrangements to provide four live load cases. For all live load configurations, the test vehicles were positioned to simultaneously load two spans of the viaduct and longitudinally with the rear axle centred on the mid-span (Fig. 14). Since the deck is characterized by a significant transverse slope (Fig. 13), the position of the vehicles was transversely non-centred between the two main girders in order to experimentally evaluate the torsion effects. (Fig. 14a).

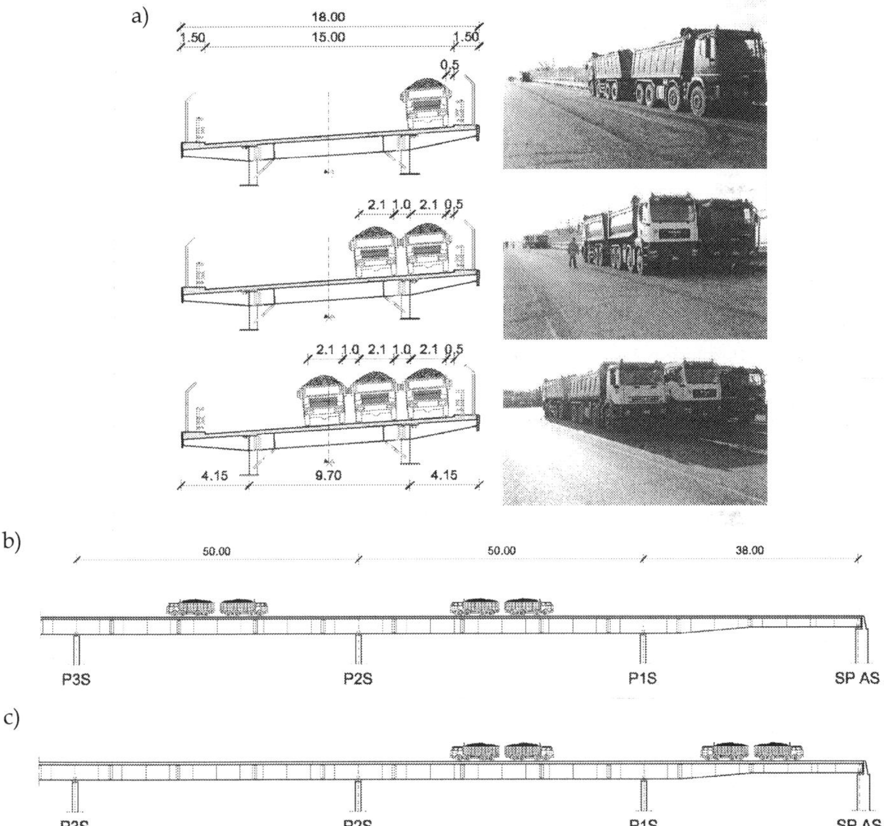

Fig. 14. Bridge over-passing Forlanini Avenue: a) transverse positioning of the test vehicles; b) longitudinal scheme of load condition LC1; c) longitudinal scheme of load condition LC2

The results of two load configurations will be presented and discussed in the following. These two load cases, shown in Fig. 14, are herein after referred to as LC1 (test vehicles loading the two spans between piers P3S and P1S on the west-side of the structure, Fig. 14b) and LC2 (test vehicle loading the two end spans on the west-side, Fig. 14c).
In all load conditions, the radar has been configured to measure targets up to a distance of 150.0 m, with a scenario sampling frequency of 10 Hz. Fig. 15 shows the position of the extensometers on span P2S-P3S and IBIS-S sensor during load configuration LC1. Since the

deck includes a large number of corner zones, provided by the intersection of girders and cross-beams, the exterior position of the microwave interferometer (Fig. 15) has to be preferred, in order to avoid the possible occurrence of multiple contributions to the same range bin coming from different reflective zones placed at the same distance from the radar (Gentile *et al.* 2008).

Fig. 16 shows the range profile of the scenario detected in LC1, projected along the longitudinal axis of the bridge. The analysis of the results provided by the microwave interferometer begins with the inspection of the range profile; this inspection, performed on site, allows to verify that the sensor positioning provides a correct image of the scenario. The radar image of Fig. 16 exhibits several peaks clearly marking the relative distance from the sensor of the transverse cross-beams reflecting the electromagnetic waves. It should be noticed that the peaks of Fig. 16 identify with excellent accuracy the cross-beams, provided that the distance between the radar and the axis of P3S pier (2.40 m, see Fig. 15) is properly accounted for.

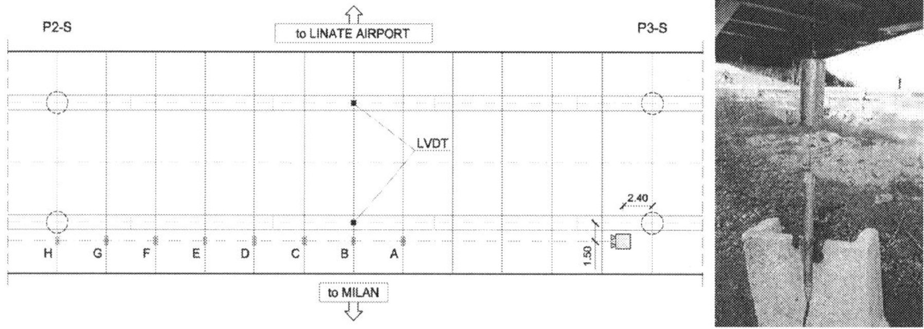

Fig. 15. Bridge over-passing Forlanini Avenue: radar and LVDT position in load condition LC1

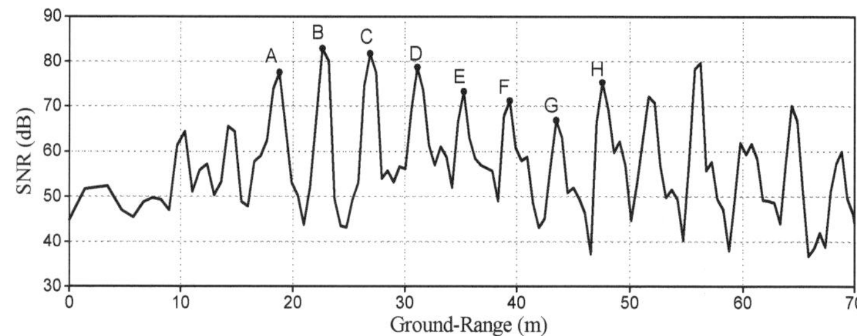

Fig. 16. Ground-Range profile in load condition LC1

It is further observed that the areas of the cross-beams corresponding to the peaks of Fig. 16 are in principle placed along the radar's line of sight (as it is schematically shown in Fig. 15).

Fig. 17 shows an example of displacement time-histories corresponding to different range bins and hence to different positions along the deck. It is observed that all deflections exhibit similar evolution and the time windows corresponding to successive entrance and motion (150-400 s, 700-950 s and 1050-1250 s) of the test vehicles along the bridge are clearly identified in Fig. 17; in addition, as it has to be expected, deflection decreases from mid-span (curve B in Fig. 17) to pier (curve H in Fig. 17). Fig. 17 also compares the deflection obtained by the radar at mid-span (region B of Fig. 15) to the one directly measured by the neighbouring extensometer; it has to be noticed that the radar-based measurement slightly exceeds the conventional measurement, conceivably as a consequence of the torsion behaviour of the deck.

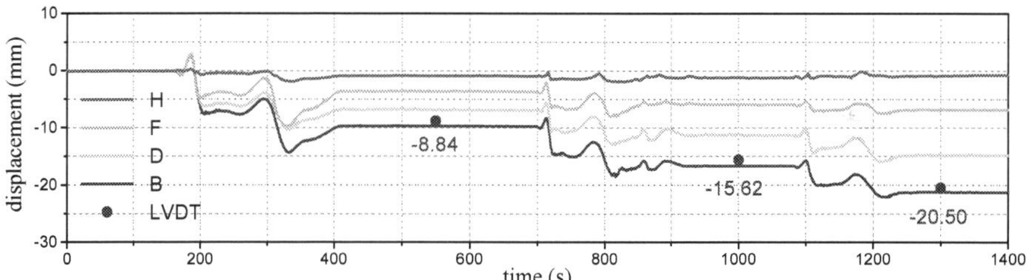

Fig. 17. Load Condition LC1: progress in time of the vertical displacements measured by the radar technique and comparison with the extensometer measurement

In load condition LC2, the radar position was moved as shown in Fig. 18, in order to have significant echo from all the transverse cantilevers and to obtain the deflected elastic curve of the whole span P1S-P2S. The corresponding range profile (Fig. 19) allows to clearly identify the cross-beams and confirms that all cantilevers of span P1S-P2S provide a sufficient echo.

Again the progress in time of the deflections measured by the radar clearly corresponds to the different phases of the load condition (Fig. 20) and good agreement was found between the results provided by radar and extensometer; in this case, as shown in Fig. 20, the difference between conventional and radar measurement is larger than in LC1 (Fig. 17), due to either the torsion behaviour of the deck or the slightly curved geometry of the investigated span.

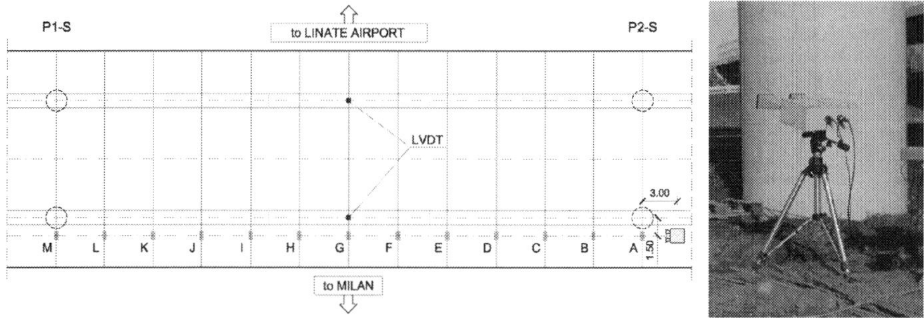

Fig. 18. Bridge over-passing Forlanini Avenue: radar and LVDT position in load condition LC2

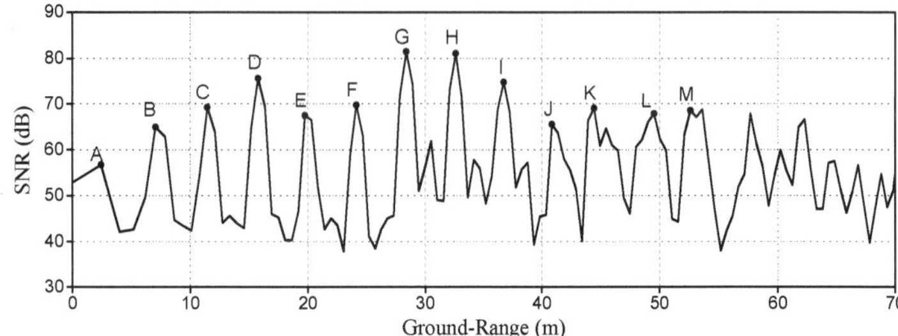

Fig. 19. Ground-Range profile in load condition LC2

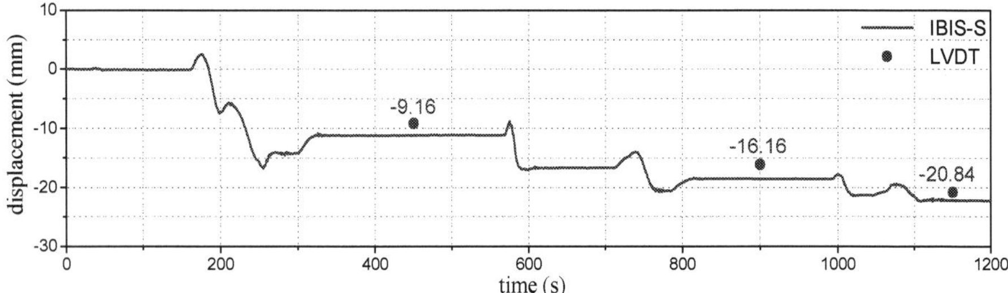

Fig. 20. Load Condition LC2: progress in time of the vertical displacements measured by the radar technique and comparison with the extensometer measurement

5. Dynamic measurements on cable stays

The application of the radar technique to the measurement of cable vibrations seems especially promising in order to perform systematic dynamic assessment of stay cables in a simple and quick way.

It is worth underlining that dynamic measurements on stay cables are generally aimed at: (1) evaluating the amplitude of cable vibrations; (2) identifying the local natural frequencies and damping ratios; (3) evaluating the cable forces and monitoring the changes in these forces over time; (4) investigating potential fatigue problems in the cables.

When a linear correlation exists between the mode order and the corresponding natural frequency of a stay cable (Robert *et al.* 1991), the tension force can be obtained from cable's natural frequencies using the taut string theory (see e.g. Irvine 1981, Caetano 2007). For the tension members that deviate from a taut string, still the cable forces can be predicted by using the identified natural frequencies with reference to more advanced formulations (accounting for the effects of both the sag and the bending stiffness on the dynamic behaviour of cables, see e.g. Casas 1994, Mehrabi & Tabatabai 1998). Subsequently, the knowledge of cable forces is used to check the correct distribution of the internal forces in the bridge at the end of construction, while monitoring the possible changes in stay cable forces over time may provide an efficient method for Structural Health Monitoring. For example, a significant drop in the tension force of a cable with simultaneous increases in the forces of the neighbouring cables may be a clear indication of a loss of cross section (or slippage at the anchorage) for the cable exhibiting the force drop.

When compared to other techniques of remote sensing, the microwave interferometry exhibits several advantages:
1.   low power transmitted;
2.   higher accuracy;
3.   possibility of simultaneously measuring the dynamic response of all cables belonging to an array.

In addition, the possible issues that may occur in the application of the radar technique to bridges and large structures practically cannot affect the radar the survey of an array of cables; peculiarly:

a.   the typical position of the sensor in the survey of an array of cables is inclined upward, as schematically shown in Fig. 21a; hence, the only targets encountered along the path of the electromagnetic waves are the stays itself so that 1-D imaging capability is perfectly adequate to the test scenario;
b.   it can assumed that the in-plane motion of the cable is orthogonal to its axis, so that the actual deflection d can be expressed as:

$$d = \frac{d_r}{\cos\left[\pi/2 - (\alpha_c + \alpha_s)\right]} \tag{8}$$

where α_c and α_s are the slope of the cable and of the sensor, respectively (Fig. 21a). In other, words, the prior knowledge of the direction of motion is available for cable systems so that it is possible to evaluate the actual displacement from the radial one.

Furthermore, since it is quite easy to predict the scenario under the radar beam (Fig. 21b), the inspection of the range profile allows to quickly verify that the sensor positioning provides a correct image of the scenario.

The accuracy and operational simplicity of the radar techniques in vibration survey of stay-cables arrays has been to date verified on two cable-stayed bridges (Gentile *et al.* 2008, Gentile 2009): some results obtained on the cable-stayed bridge crossing the Oglio river between the municipalities of Bordolano and Quinzano (northern Italy) are herein after presented and discussed.

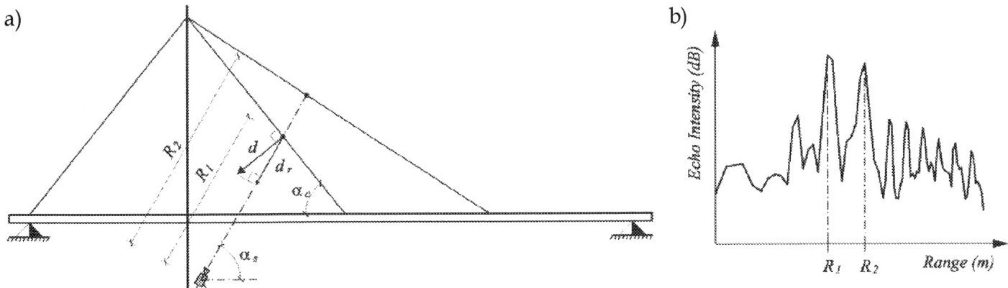

Fig. 21. a) Radial displacement versus actual (in-plane) displacement of a stay cable; b) Typical range profile expected for an array including two cables

The dynamic characteristics of the investigated cable-stayed bridge were well-known since ambient vibration measurements were carried out on the bridge in Spring 2004 by the Vibration Laboratory of L'Aquila University (Benedettini & Gentile 2008), using Sprengnether servo-accelerometers. During this test, 10 global modes of the bridge were identified in the frequency range 0–10 Hz and also the dynamic response of one cable (S_{2U} in Fig. 22) was recorded.

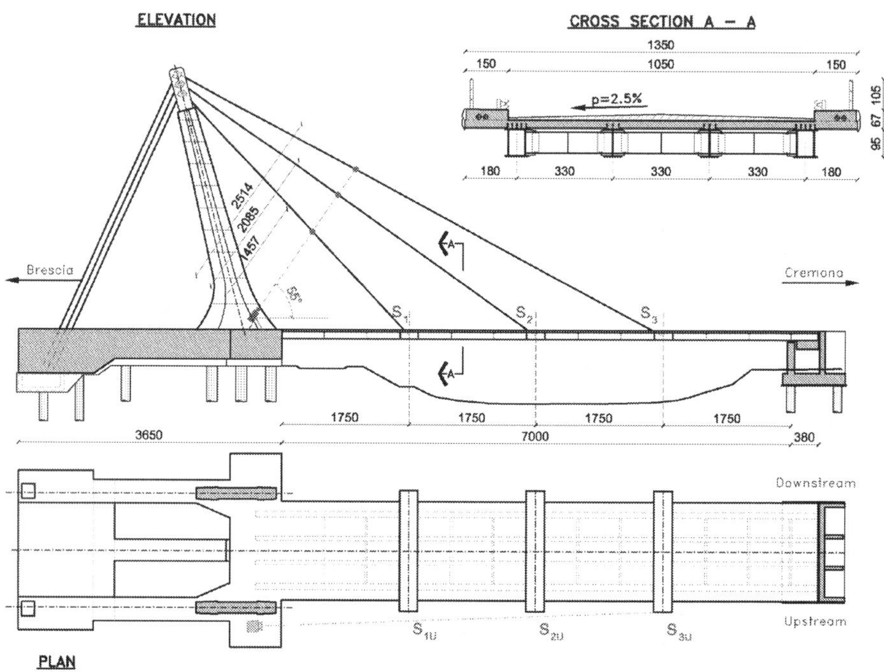

Fig. 22. Elevation, plan and cross-section of the investigated cable-stayed bridge. Position of the radar sensor during the test of stay cables S_{1U}, S_{2U} and S_{3U} (upstream side)

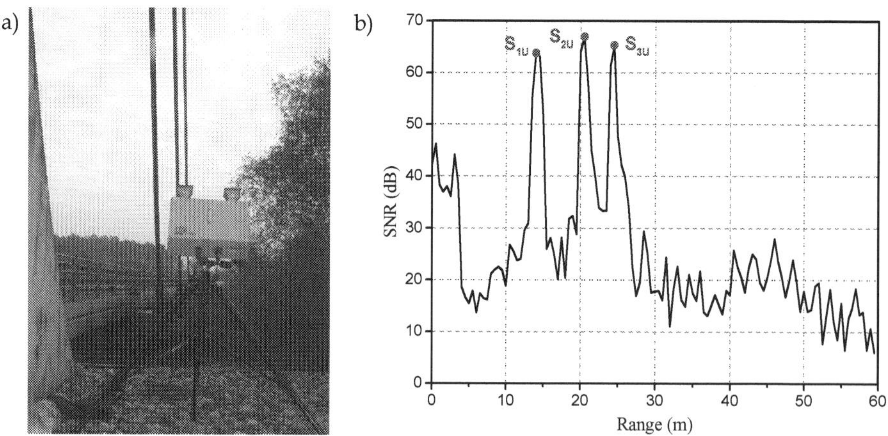

Fig. 23. a) View of the radar position during the test of stay cables S_{1U}-S_{3U} (upstream side); b) Range profile of the test scenario

Elevation and plan views of the bridge and typical cross-section are presented in Fig. 22. The cross-section (Fig. 22) consists of 4 steel girders framed by floor beams; girders and floor beams are all composite with a 30.0 cm concrete slab. The steel girders are 95 cm high, with the outer girders (spaced 9.90 m centre to centre) being of box section while the inner ones are wide flange sections. The girders are framed by floor beams 5.83 m spaced; the floor beams providing the anchorage of the stays are of box section while the other ones are wide flange sections (Fig. 22). The total width of the deck is 13.50 m for two traffic lanes and two pedestrian walkways; the suspended span is 70.0 m long. The cast-in-place concrete towers are 35.65 m high and each consists of an inclined, varying width, concrete leg bearing an upper steel device providing the anchorage for the stay cables.

In the test of array including cables S_{1U}-S_{3U} on the upstream side (Fig. 22), the radar sensor was placed on the basement of the upstream-side tower, as shown in Figs. 22 and 23a. The range profile of the test scenario is presented in Fig. 23b and exhibits three well defined peaks, clearly identifying the position in range of the cables.

3000 s of radar data were acquired at a rate of 200 Hz. Figs. 24a-b show the auto-spectral densities (ASD) associated to the ambient response of stay cables S_{1U} and S_{3U}, respectively. The ASD plots in Fig. 24 are a synthesis of the frequency content present on each cable and allowed the identification of several local resonant frequencies, marked with the vertical lines, in the frequency range of analysis (0–25 Hz).

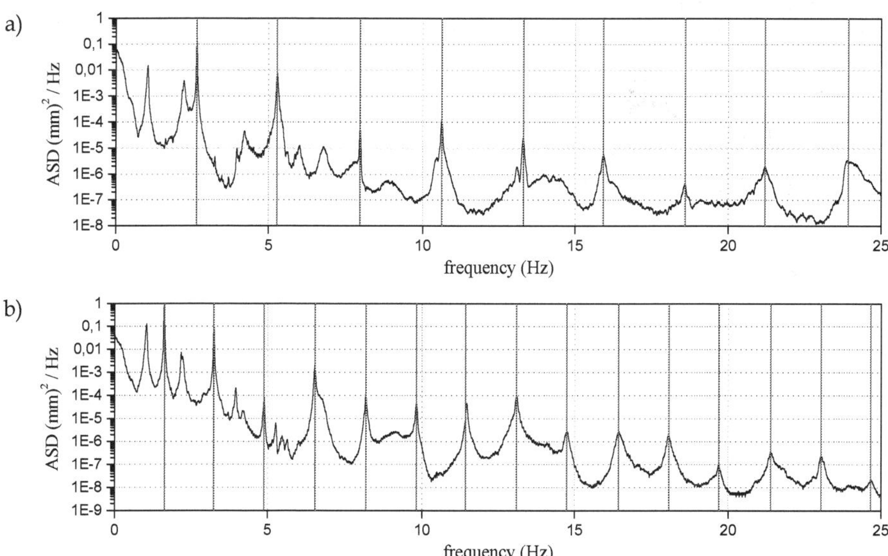

Fig. 24. Auto-spectra of the displacement data measured by the radar on cables: a) S_{1U}; b) S_{3U}

Furthermore, Fig. 25 presents a direct comparison between: (1) the auto-spectrum of the acceleration measured on stay cable S_{2U} by a conventional accelerometer in the test of Spring 2004 and (2) the auto-spectrum of the acceleration obtained from the radar sensor (and computed by deriving the displacement data).

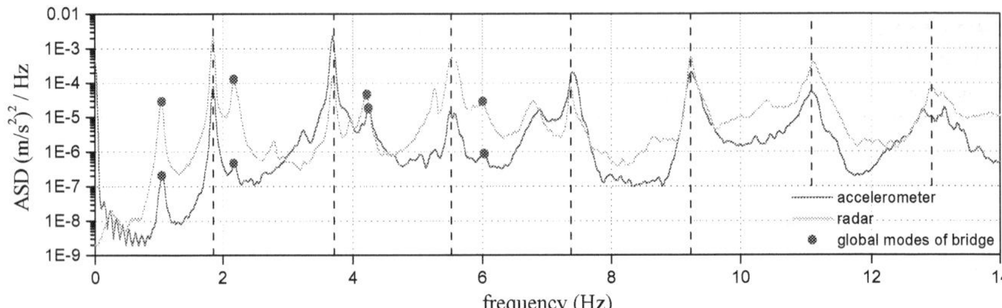

Fig. 25. Comparison between the auto-spectra of the acceleration data of cable S_{2U} obtained from conventional and radar sensor

From the inspection of the spectral plots in Figs. 24a-b and 25, one can observe that:

1. the response of cables S_{1U} and S_{3U} (Fig. 24) in the frequency range 0-25 Hz is characterized by several equally spaced and well-defined peaks. Since a linear correlation exists between the mode order and the corresponding natural frequency of the cables, the tension forces computed from cable's natural frequencies using the taut string theory (see e.g. Irvine 1981, Casas 1994, Caetano 2007) turned out to be practically equal to the design values;

2. the values of the first seven natural frequencies of stay cable S_{2U}, identified on the basis of the auto-spectra obtained using conventional and radar measurement systems are virtually coincident (1.84, 3.70, 5.53, 7.37, 9.24. 11.1 and 12.93 in Fig. 25);

3. some global natural frequencies of the bridge (corresponding to peaks of the ASDs placed at 1.06, 2.18, 4.25 and 6.03 Hz) are also apparent in Figs. 24-25 and are equal to the ones identified by (Benedettini & Gentile 2008) in the AVT of the bridge.

6. Conclusions

An innovative radar sensor, developed for remote (non contact) measurements of deflections on bridges and large structures in both static and dynamic conditions, has been presented. The new sensor exhibits various operational advantages with respect to contact sensors and, at the same time, provides a direct measurement of displacement, which is of utmost importance in the in-service monitoring of large structures. Furthermore, based on the design specifications of the system components, the displacements sensitivity is expected to range between 1/100 and 1/10 mm.

The accuracy of the sensor has been evaluated in laboratory tests carried out on a simple mass-spring system; the free-vibration tests carried out on the oscillator confirm that the displacement sensitivity is better than 2/100 mm; in addition, the laboratory tests indicated both an excellent quality for the measured displacements and a good operating stability of the equipment.

The use of the radar interferometer in static and dynamic tests of two full-scale bridges exhibited excellent stability, when employed on site for long time intervals.

Based on the analysis of the results of the above tests, including (a) the inspection of the displacement responses obtained from the radar sensor and (b) the comparison with the data recorded by conventional sensors (LVDT extensometers and accelerometers), the following conclusions can be drawn:

1. the radar-based deflection measurements obtained during live-load static tests seem as accurate as that obtained with the conventional extensometers;
2. the velocity time-histories evaluated from the radar sensor (and obtained by deriving the displacements) generally exhibit very good agreement with the ones recorded by conventional sensors;
3. an excellent agreement was found between the modal parameters (resonant frequencies and mode shapes) identified from radar data and from data obtained with conventional accelerometers.

Furthermore, the radar technique turned out to be especially suitable to vibration survey of stay cables since it allows to simultaneously measure the dynamic response of several stay-cables and provides measurements of high level of accuracy in terms of identification of natural frequencies.

Notwithstanding the good results obtained by using the microwave sensor, it has to be recalled that two main issues may occur in the application of radar techniques to bridges and large structure:

a. as a consequence of the 1-D imaging capabilities of the radar sensor, measurement errors may arise when different points of a structure are placed at the same distance from the radar;
b. the radar provides a measurement of the variation of the target position along the sensor's line of sight (i.e. the radial displacement); hence, the evaluation of the actual displacement requires the prior knowledge of the direction of motion.

7. Acknowledgements

The author would like to acknowledge: (1) the support provided by the Italian Ministry of University and Research (MIUR), within a National Project funded in 2006, and by IDS (Ingegneria Dei Sistemi, Pisa, Italy), owner and developer of the microwave interferometer used in the tests; (2) Dr. Giulia Bernardini (IDS) and the Local Administrations owners of the tested bridges: without their involvement and co-operation, this research would not have been possible; (3) Prof. Francesco Benedettini (L'Aquila University) for the kind concession of the data recorded in Spring 2004 and presented in Fig. 25; (4) the help provided by Marco Antico, Marco Cucchi and Dr. Nicola Gallino (Laboratory of Vibrations and Dynamic Monitoring of Structures, Politecnico di Milano) in conducting the field tests.

8. References

Benedettini, F. & Gentile, C. (2008): F.E. modelling of a cable-stayed bridge based on operational modal analysis, *Proceedings of 26th Int. Modal Analysis Conf.* (IMAC-XXVI).

Bernardini, G. et al. (2007). Microwave interferometer for ambient vibration measurements on civil engineering structures: 1. Principles of the radar technique and laboratory tests, *Proceedings of Int. Conf. on Experimental Vibration Analysis of Civil Engineering Structures* (EVACES'07), 143-152.

Caetano, E. (2007). *Cable vibrations in cable stayed bridges*, SED 9, IABSE.

Casas, J.R. (1994). A combined method for measuring cable forces: the cable-stayed Alamillo Bridge, Spain, *Structural Engineering International* , 4(4), 235–240.

Cunha A. & Caetano E. (1999). Dynamic measurements on stay cables of cable-stayed bridges using an interferometry laser system, *Experimental Techniques*, 23(3), 38-43.

Farrar, C.R., Darling, T.W., Migliori, A. & Baker, W.E. (1999). Microwave interferometers for non-contact vibration measurements on large structures, *Mech. Syst. Signal Process.*, 13(2), 241-253.

Farrar, C.R. & Cone, K.M., (1995). Vibration testing of the I-40 bridge before and after the introduction of damage, *Proceedings of 13th Int. Modal Analysis Conference* (IMAC-XIII).

Gentile, C. & Bernardini, G. (2008). Output-only modal identification of a reinforced concrete bridge from radar-based measurements, *NDT&E International*, 41(7), 544-553.

Gentile, C., Bernardini, G. & Ricci, P.P. (2008). Operational modal analysis of a cable-stayed bridge from conventional and radar-based measurements, *Proc. Eurodyn 2008*.

Gentile, C. (2009). Radar-based measurement of deflections on bridges and large structures: advantages, limitations and possible applications, *Proc. SMART'09*.

Irvine, M. (1981). *Cable Structures*, MIT Press.

Henderson, F.M. & Lewis, A.J. (Eds.) (1998). *Manual of Remote Sensing. Principles and Applications of Imaging Radar*, Wiley & Sons, 3rd Ed.

Kaito, K., Abe, M. & Fujino, Y. (2005). Development of a non-contact scanning vibration measurement system for real-scale structures, *Structure & Infrastructure Engng.* 1(3), 189-205.

Lee, J.J. & Shinozuka, M. (2006). A vision-based system for remote sensing of bridge displacement. *NDT&E International*, 39(5), 425-431.

Marple, S.L. Jr. (1987). *Digital spectral analysis with applications*, Prentice-Hall.

Mehrabi, A.B. & Tabatabai, H. (1998). Unified finite difference formulation for free vibration of cables. *J. Structural Engineering*, ASCE, 124(11), 1313-1322.

Meng, X., Dodson, A.H. & Roberts, G.W. (2007). Detecting bridge dynamics with GPS and triaxial accelerometers, *Engineering Structures*, 29(11), 3178-3184.

Nickitopoulou, A., Protopsalti, K. & Stiros, S. (2006). Monitoring dynamic and quasi-static deformations of large flexible engineering structures with GPS: Accuracy, limitations and promises, *Engineering Structures*, 28(10), 1471-1482.

Pieraccini, M. et al. (2004). Highspeed CW step-frequency coherent radar for dynamic monitoring of civil engineering structures, *Electron. Lett.*, 40(14), 907-908.

Robert, J., Bruhat, D., & Gervais, J.P. (1991). Mesure de la tension de câbles par méthode vibratoire, *Laboratoire des Ponts et Chaussés*, 173(3), 109-114.

Silva, S., Bateira, J. & Caetano, E. (2007). Development of a vision system for vibration analysis, *Proceedings of Int. Conf. on Experimental Vibration Analysis of Civil Engineering Structures* (EVACES'07), 113-121.

Skolnik, M.I. (Ed.) (1990). *Radar Handbook*, McGraw-Hill.

Wehner, D.R. (1995). *High-resolution radar* , 2nd Ed., Norwood, MA: Artech House.

9

Radar Systems for Glaciology

Achille Zirizzotti, Stefano Urbini, Lili Cafarella and James A.Baskaradas
Istituto Nazionale di Geofisica e Vulcanologia
Italy

1. Introduction

This chapter deals with radar systems, measurements and instrumentation employed to study the internal core and bedrock of ice sheets in glaciology . The Earth's ice sheets are in Greenland and Antarctica. They cover about 10% of the land surface of the planet. The total accumulated ice comprises 90% of the global fresh water reserve. These ice sheets, associated with the ocean environment, provide a major heat sink which significantly modulates climate.

Glaciology studies aim to understand the various process involved in the flow (dynamics), thermodynamics, and long-term behaviour of ice sheets. Studies of large ice masses are conducted in adverse environmental conditions (extreme cold, long periods of darkness). The development of remote sensing techniques have played an important role in obtaining useful results. The most widely used techniques are radar systems, employed since the 1950s in response to a need to provide a rapid and accurate method of measuring ice thickness. Year by year, polar research has become increasingly important because of global warming. Moreover, the discovery of numerous subglacial lake areas (water entrapped beneath the ice sheets) has attracted scientific interest in the possible existence of water circulation between lakes or beneath the ice (Kapitsa et al., 2006; Wingham et al., 2006; Bell et al., 2007). Recent studies in radar signal shape and amplitude could provide evidence of water circulation below the ice (Carter 2007, Oswald & Gogineni 2008).

In this chapter the radar systems employed in glaciology, radio echo sounding (RES), are briefly described with some interesting results. RES are active remote sensing systems that utilize electromagnetic waves that penetrate the ice. They are used to obtain information about the electromagnetic properties of different interfaces (for example rock-ice, ice-water, seawater-ice) that reflect the incoming signal back to the radar. RES systems are characterized by a high energy (peak power from 10 W to 10 KW) variable transmitted pulse width (about from 0.5 ns to several microseconds) in order to investigate bedrock characteristics even in the thickest zones of the ice sheets (4755 m is the deepest ice thickness measured in Antarctica using a RES system). Changing the pulse length or the transmitted signal frequencies it is possible to investigate particular ice sheet details with different resolution. Long pulses allows transmission of higher power than short pulses, penetrating the thickest parts of the ice sheets but, as a consequence, resolution decreases. For example, the GPR system, commonly used in geophysics for rock, soil, ice, fresh water, pavement and structure characterization, employs a very short transmitted pulse (0.5 ns to 10 ns) that allow detailing of the shallow parts of an ice sheet (100-200 m in depth) (Reynolds 1997).

Consequently, in recent years, GPR systems are also employed by explorers to find hidden crevasses on glaciers for safety.

RES surveys have been widely employed in Antarctic ice sheet exploration and they are still an indispensable tool for mapping bedrock morphologies and properties of the last unexplored continent on Earth. The advantage of using these remote sensing techniques is that they allow large areas to be covered, in good detail and in short times using platforms like aeroplanes and surface vehicles.

2. Glaciological radar

2.1 Summary of physical and electromagnetic properties of ice

The difference in electromagnetic properties at an interface between two different media can reflect radio waves. Air-ice, ice-bedrock surfaces and internal layers are targets for RES systems, forming interfaces that reflect a transmitted radar signal back to the receiver antenna. The radar technique is widely used in glaciology to obtain information about electromagnetic properties of the reflecting layers and their position (Bogorodsky, 1985). RES data can give information on ice thickness, bedrock depth, position of internal ice layers, and water content of bedrock interfaces.

The electromagnetic properties of materials are described by the relative complex electrical permittivity ε_c, the relative magnetic permeability μ_r (magnetic permeability $\mu = \mu_r \cdot \mu_o$ with $\mu_o = 4\pi \cdot 10^{-7}$ H/m) and the conductivity σ. Conductivity and permittivity depend on bulk physical state and chemical composition (for example: porosity, water content) while the relative permeability is always $\mu_r = 1$ for non-magnetic materials.

The propagation of an electromagnetic wave in materials is characterized by its attenuation and velocity. In particular, radio waves are easily transmitted in air (good dielectric, low attenuation) but they propagate very little in sea water (poor dielectric, high attenuation). Table 1 reports some dielectric properties of the materials of the glaciological environment. Electromagnetic wave propagations is described using Maxwell equations. Compact solution, in general, can be obtained defining the relative complex permittivity $\varepsilon_c = \varepsilon'$ - j ε'' (where $\varepsilon'' = \sigma/\omega\varepsilon_o$) or $\varepsilon_c = \varepsilon'(1\text{-}j\sigma/\omega\varepsilon_o\varepsilon')$ where $\varepsilon_o = 8.85 \cdot 10^{-12}$ F/m is the vacuum permittivity. Based on their electromagnetic properties, materials can be separated into isulating ($\sigma << \omega\,\varepsilon_o\varepsilon'$, the complex permittivity is mainly real) and conductive media ($\sigma >> \omega\varepsilon_o\varepsilon'$ the complex permittivity is mainly imaginary).

Material	Relative permittivity	Conductivity mS m^{-1}	Velocity [10^8 m/s]	Attenuation [dB m^{-1}]
Air	1	0	3.0	0
Distilled water	80	0.01	0.33	-0.002
Fresh water	80	0.5	0.33	-0.1
Salt water	80	3000	0.1	-1000
Ice	3.2	0.03	1.68	-0.01
Rock	4÷12	100-0	0.86-1.5	25-0

Table 1. Electromagnetic properties of the materials of the glaciological environment

The complex permittivity of the ice in glaciers and ice sheets depends on several factors including crystal orientation fabrics, density, impurity concentration (internal layers of dust from major volcanic events are always present in glaciers) or sea salt (acidity), water content and temperature.

Several ice core drilling, under the framework of climatological projects in Greenland and Antarctic base stations, allow an assessment of ice composition and its main physical characteristics (Stauffer 1991, Epica 2004). As regards the polar ice temperature, the upper and lower layers have constant heat sources (air and bedrock respectively). The temperature profile in the ice sheet changes with depth due to the heat conduction of ice. From the ice surface, the ice temperature increases up to a value that depends on the ice thickness, pressure and geothermal heat contributions. At Concordia Station (Dome C, Antarctica) the mean ice surface temperature is -55°C (annual average temperature of the site) while it reaches -3°C at 3225 m of ice depth (Epica 2004, Laurent A., 2003). Observing physical properties and DiElectric Profile (DEP) directly on the ice core obviously represent the best way to acquire this kind of information (Wolff 2004). These measurements are usually made in the LF frequency spectrum at a fixed temperature and must be extended to the frequencies used and temperature (Corr & Moore, 1993). An alternative approach for investigating ice properties is using doped ice in the laboratory. A possible disadvantage of this approach is that the artificial ice properties may not be the same as those of natural ice. The relative permittivity measurements on doped ice grown in laboratory, ranges between 3.2 to 3.13 over a wide temperature range and for a wide radar frequency range (Fujita et al. 2000). In the same conditions, the conductivity measured in ice containing acids (impurities) varies from 0.03 mS/m (@ -10°C f < 1 GHz) to 0.0008 mS/m (@ -80°C f<600MHz). Based on these results the ice can be considered as an insulating media (low absorption) for frequencies greater than 600 kHz ($\sigma << \omega\varepsilon_0\varepsilon'$); the ice conductivity is virtually independent of frequency below about 600 MHz. As a consequence, electromagnetic waves velocity in artificial ice ranges between 167.7 to 169.5 m/µs (Fujita et al. 2000) while the in situ measurements, determined by different methods, show a wider range of values due to the fact that natural ice is inhomogeneous.

2.2 Introduction to RES principles

Electromagnetic wave velocity (at radio frequencies) in ice can be approximated (for an insulating low absorption medium) as:

$$v = \frac{c}{\sqrt{\varepsilon'}}$$

(1)

where c is the speed of light $3 \cdot 10^8$ m/s. Accurate determination of radar wave velocity in ice is fundamental to derive the depth of the various reflecting interfaces, and is established from radar measurements with the simple equation $h = v \cdot t/2$ where t is the two-way traveling time of the radar echoes. Bearing in mind that electromagnetic waves have different velocities through firn and snow layers (213 m/µs), this may introduce an error in the order of 10% of the firn layer thickness (Rees 1992).

In addition to laboratory measurements, electromagnetic wave velocity can be estimated in many different ways using radar data. These can be divided into direct and indirect methods. Lowering a probe into a borehole with source and receiving sensors measuring speeds at several intervals is a direct way to obtain velocity information.

Wide Angle Reflection and Refraction or Common Mid Point (WARR or CMP) are another indirect techniques for measuring wave velocity and are widely employed in seismic reflection procedures. They consist measurements of arrival time of the reflections at several source-receiving distances. It is possible to reconstruct the hyperbola describing the arrival times of a reflected event. Sometimes, when a single antenna (transmitting-receiving) or fixed geometry between transmitting and receiving antenna (as in airborne assemblies) are used it is not possible to carry out a WARR/CMP profile. In the absence of any other kind of information, recognized diffractions (hyperbolas) from particular points along the profiles could be used to estimate velocity with a similar procedure to the previous method (Daniels 1996, Reynolds 1997).

RES radar equation: dry and wet bedrock ice interfaces are important in subglacial exploration and hydrological studies on ice sheets. The physical condition of the ice interfaces can be estimated from the RES measurement by studying the reflectivity coefficient of the radar equation. The radar equation allows the estimation of signal power of received echoes (Skolnik 1990, Borogosky 1985). In the case of a coherent flat reflector, neglecting noise and scattering due to inhomogeneities and multiple reflections between layers, it can be expressed as:

$$p_r = \frac{p_t G_a^{2} \lambda^2}{(4\pi)^2 \cdot r^2} \cdot g_f \cdot l_p \tag{2}$$

where p_t and p_r are the transmitted and received power, G_a is the antenna gain, λ is the RES signal wavelength, r is the range from transmitting antenna, the square term is the spherical divergence of the electromagnetic signal power in a far field condition.

The gain term g_f depends on the refractive effect due to the change in the relative permittivity from air to ice an airborne RES system:

$$g_f = \frac{\frac{h}{r}+1}{\frac{h}{r}+n_{12}} \tag{3}$$

where h is the aeroplane flight altitude, r is the radar range in the ice and n_{12} is the refractive index between the two media air and ice. This term provides little dB gain in the radar equation, for example @ h=300m and r=300 m g_f=2.2 dB (g_f< 5 dB for r<3000 m).

The power loss term l_p in decibel (L_p) can be expressed as the sum of several contributions

$$L_p = 2T_{ai} + R_{ir} + 2L_i \tag{4}$$

where the subscripts a, i and r are for air, ice and rock respectively. In this equation L_i is the ice absorption of the electromagnetic wave, and T and R are the losses due to the transmission and reflection coefficients at the different interfaces (Peters et al., 2005) where T and L are considered twice because of the double passage of the reflected signal in the medium. Recently, a linear trend of ice absorption (in decibels) with depth has been observed with a constant absorption rate that differ according to variation of chemical content and temperature of ice (Raymond 2009, Jacobell et al 2009).

Reflection and transmission coefficients: in order to investigate the different properties of reflecting surfaces, the simple case of homogeneous, horizontal surfaces can be considered. For example, the electromagnetic power reflected by the subglacial homogeneous, horizontal ice-water interface depends only on the reflection coefficients, i.e. on the physical nature of the two media. In table 2 the power losses due to the reflection R or transmission T coefficients at the relative interfaces are summarized. The reflection coefficient ρ at the surface separating two media 1 and 2 is:

$$\rho = \frac{\sqrt{\varepsilon_{c1}} - \sqrt{\varepsilon_{c2}}}{\sqrt{\varepsilon_{c1}} + \sqrt{\varepsilon_{c2}}} \tag{5}$$

The power loss due to reflection R is equal to $|\rho|^2$ while the power loss due to transmission T is given by T= 1 - R (Reynolds 1997).

Reflection/ Transmission losses	Power loss [dB]
T_{ai} (air – ice)	-0.35
R_{ai} (air - ice)	-11.0
R_{iw} (ice – water)	-3.5
R_{ir} (ice – rock $\varepsilon'=4\div12$)	-11.2 ÷ -25

Table 2. Power loss due to the reflection R or transmission T coefficients.

RES wave attenuation: The dielectric absorption of radar waves can occur due to conduction or relaxation processes in the ice. Conduction causes electrons to move away from their nuclei, losing energy from the electric field into the host material. On the other hand relaxation causes energy to be lost through the oscillation of water molecules. Dielectric absorption is increased in impure ice which is characterized by enhanced relative permittivity and electrical conductivity (positively related to temperature).
In an insulating homogeneous medium, the amplitude of electric field E decay in the z direction of the medium as:

$$E = E_0 e^{-\alpha \cdot z} \tag{6}$$

where is the attenuation coefficient of incident electromagnetic wave of amplitude E_0 @ z=0. In the hypothesis of an insulating medium ($\sigma << \omega\varepsilon_0\varepsilon'$)$\alpha$ can be approximated to:

$$\alpha = 8.686 \sqrt{\frac{\mu}{\varepsilon_0\varepsilon}} \frac{\sigma}{2} \tag{7}$$

expressed in dB/m. A low value of conductivity σ corresponds with a low value of attenuation coefficient α indicating a low absorption medium. From RES measurements the α term can be evaluated using the radar equation. Considering a linear trend of ice absorption it is possible to solve the radar equation obtaining a constant absorption rate for the ice. In Antarctica α can assume values from 9 to 27 dB/km in different zones with different ices characteristics and physical condition (Macgregory et al. 2007, Jacobel et al. 2009). This means a total absorption of the radar signal from about 36 dB to 108 dB on 4km of ice thickness. Also considering geometrical spreading and other attenuation (included in the radar equation) a sensitivity of about -130dBm is required in RES receivers.
Resolution and detectability: Two reflectors adjacent to each other in space (transverse to the radar beam) or in time (parallel to the radar beam) can be detected if their distance is greater than the resolution.
Vertical resolution depends mainly on the length of the transmitted pulse; the shorter is the pulse (large bandwidth) the more precise the range measurement. Using long pulse or continuous wave radar with phase or frequency modulation it is possible to achieve the resolution of short pulse with the advantage of a high transmitted power signal (pulse

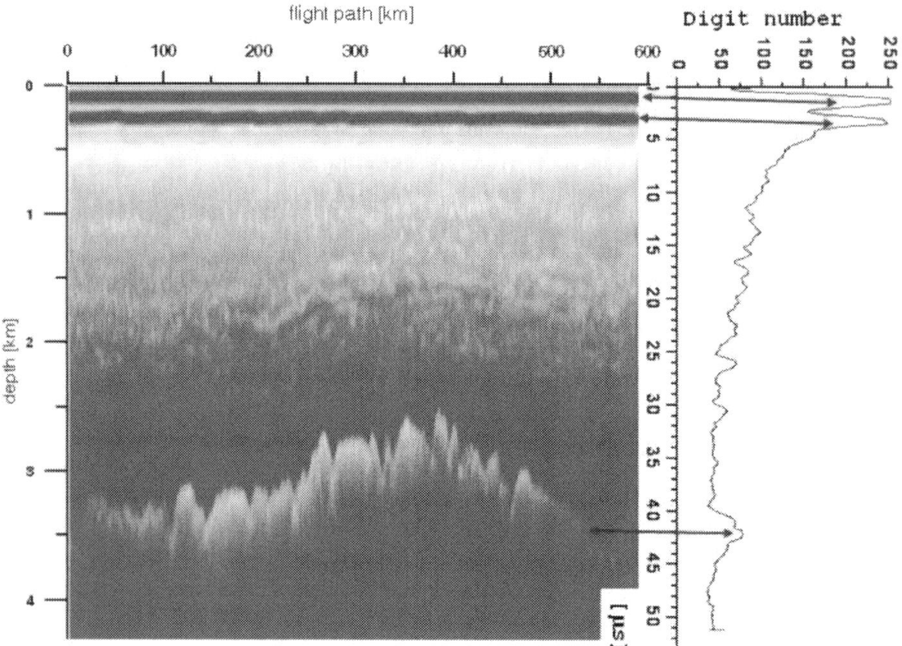

Fig. 1. Radargram examples in colour mode (on the left) and O-scope mode (on the right) Transmitted pulse ice-surface and bedrock reflection are indicated by arrow.

compressed or CW radar in Sholnik, 1990). Vertical range resolution in ice can be calculated using $\Delta h = v \cdot \tau/2$ where τ is the pulse length of the radar and v is the ice wave velocity. In the case of very short pulses the lower limits depend on the signal frequency, (theoretically equal to approximately one–quarter of radar signal wavelength (Reynold, 1997). For example, in the case of GPR systems (where less than one cycle of the wave is transmitted) the resolution depends only on the wavelength of the transmitted signal in the analysed medium.

Horizontal resolution is the capability to detect two adjacent reflectors in space. It depends mainly on the area illuminated by the antennas systems (beamwidth) or it depends on the minimum distance between two acquired radar traces (pulse repetition rate and the speed of horizontal movement for aeroplane or ground-based radar). Both of these are usually greater than the first Fresnel zone in the RES systems in use.

3. Radar technology for glaciology

The first indication that snow and ice can be penetrated by high frequency radio signal was observed at Admiral Byrd's base, Little America (Antarctica) in 1933. Pilots reported that radar altimeters were erroneous over ice and the U.S. Army started an investigation that discovered that polar ice and snow are transparent to VHF and UHF band radiation. Following the investigation results, Waite and Schmidt, in 1957 demonstrated that a radar altimeter could be used to measure the thickness of polar glaciers (Waite & Schmidt, 1961). In 1963 at Cambridge University's Scott Polar Research Institute (SPRI) the first VHF radar system specifically for radio echo sounding was developed and subsequently around the world various national institutions involved in Arctic and Antarctic polar researches started developing and using RES systems. These first RES systems implemented short pulse envelope radar operating from 30 MHZ to 150 MHz and were used in Greenland and Antarctica. The acquired radar trace was displayed on an oscilloscope. The received radar trace represented on the right of fig. 1 is still named an "O-scope" view from "oscilloscope" while the radargram on the left is a colour map of trace amplitude. The recording system was a motor-driven 35 mm camera with adjustable film velocity, synchronized to the surface speed. Position determination was obtained initially by observation of discernible landmarks by system operator and subsequently by inertial navigation systems using motion-sensing devices.

Technological improvements made these RES systems more powerful. Today digital acquisition of radar traces allows a higher number of recorded traces in fast digital mass storage devices, allowing wide area investigations with improved horizontal resolution. Signal processing of the acquired traces significantly improve dynamic range and signal to noise level. Moreover differential GPS measurements allow precise determination of the geographical position of the acquired traces, reducing uncertainties to ±5 m and ±10 cm in post-processing.

In 1990 the first coherent RES system was developed by the University of Kansans and tested in Antarctica (Raju 1990, Cogineni 1998). A coherent radar detects both the amplitude and phase of the radar signal. Coherent systems allow pulse compression with a net gain in the received signal to noise ratio. This increased sensibility of the received signal level permits a reduction in the transmitted peak power. Furthermore a coherent integration of moving airborne platform forms a synthetic aperture radar (SAR) which improves along-track (horizontal) resolution (Peters et al., 2005).

Multi channel synthetic-aperture radar (SAR) systems for bedrock imagining have been used in Greenland to produce maps of basal backscatter over an area extending 6 km × 28 km. Interferometric processing of data from this 8-channel system also permitted the discrimination of off-nadir scattering source providing further insight into basal topography (Plewes et al. 2001, Paden et al. 2005; Allen et al. 2008).

Different type of radar have been developed to focus on ice sheet details. The following categorisations have been based on the transmitted radar waveform.

Pulsed radar: Historically the first RES systems developed were all short-pulse-type radar systems. A short sine wave pulse of constant amplitude is transmitted with operating frequencies ranging from 30 MHz to 600 MHz. An envelope (logarithmic) receiver detects the received echoes as peaks of amplitude from the ambient noise level (see O-scope radar trace in fig 1). These systems typically used incoherent receivers that only detected the power (or amplitude) of the received radar signal. Incoherent radar is suitable for range determination and certain echo strength analyses. In general, echo amplitudes statistics are the primary means of determining the level of reflection and scattering.

FM-CW (Frequency Modulated Continuous Wave) radar: A FM-CW radar transmits a continuous waveform in which the carrier frequency increases linearly with time. The modulation bandwidth, which determines range resolution, is the difference between the start and stop frequencies of the transmitted signal. In an FM-CW radar a sample of transmitted signal is mixed with the received signal to generate a beat signal. The frequency of transmitted and received signals will be different because of time delay, which is directly proportional to the range, associated with signal propagation to the target and back. The beat signal contains the range, amplitude and phase information of the target. These RES systems working at high frequency C band and X band (Arcone 1997) were designed for cryopheric research, to profile frozen lakes and temperate glacier (Arcone 2000) in Alaska. Airborne and ground based UHF radar systems were used to map the accumulation rate variability in Greenland (Kanagaratnam et al, 2004)

Impulse radar: Pulsed radar in the UHF frequency ranges is limited in penetration depths on temperate glaciers due to scattering and absorption of electromagnetic waves in warm ice. As seen in paragraph 2.1, ice conductivity increases with temperature and decrease at lower frequencies. For these reasons, to increase depth penetration in temperate glaciers, a radar working at frequencies lower than 10 MHz was developed. In order to increase the vertical resolution, the pulse length was reduced, transmitting approximately one cycle (monopulse radar) at a centre frequency determined by the antenna length. These radars use resistively loaded dipole antennae (Wu 1965), coupled with an avalanching transistor circuit (Cook, 1960), and more recently, using MOSFETs (metal oxide semiconductor field effect transistors) (Conway 2009) to increase the stability of the transmitted power, making it possible to generate high-voltage pulses. These systems are similar to GPR radar systems with a higher transmitted power and working at lower frequencies.

Ground Penetrating Radar (GPR): GPR is a widely used radar system to explore various media from a few to several hundred meters in depth using two methods: discrete (fixed points measurement) or continuous profile to investigate a media subsurface. Fixed point measurement is generally used to measure ice properties (for example WARR or CMP velocity measurements) or to carry out a multi-fold acquisition (Reynolds 1997). Continuous profile is used to maximize the productivity of a survey. Commercial GPR systems, commonly used in geophysics for rock, soil, ice, fresh water, pavement and structure characterization, employ a very short transmitted pulse (about from 0.3 ns to 300 ns) to study the shallow portion of terrain in detail. GPRs have also been used for ice probing. Since the 1970's, surface-based surveys have been performed on permafrost and frozen soils, glacial ice, sea ice, sedimentary environments, rock, riverbeds, and lakebed deposits, snow accumulations for internal layering detection. The low transmitted energy of the GPR signal does not allow it to reach deeper ice levels. For this reason GPR systems are used to analyze shallow parts of ice sheets as a useful tool that avoids deeper internal layer detection and for finding buried crevasses on glaciers for safety (see next paragraph).

GPR radar systems, similarly to impulse radar, employ resistively loaded dipole antenna arrangements which produce a 1-2 cycle wavelet in the HF-UHF (3-3000 MHz) frequency range to obtain the required short pulse that gives GPR its good depth resolution. GPR antennas are the most important part of the GPR system. In fact, because of the wide spectrum of the transmitted waveform GPR needs broadband antennas. Moreover, to cover the range of working frequencies, GPR systems employ several different antennas each working at a particular frequency range. Commercially available or special purpose antennas are either strings of resistors or resistively coated and flared dipoles, either unshielded or in compact transmit-receive units with a conductive shielding to mitigate above ground radiation (Reynolds 1997).

4. Radar applications in Glaciology

4.1 Ice depth and bedrock elevation.

Since the first RES system were used in ice-sheet surveys to estimate ice thickness, the measurements were controlled by seismic and gravity based measurements. A RES system is able to achieve an echo reflection from the ice surface and from the ice-bedrock interface. Calculating the difference in time (or depth) between them it is possible to estimate the ice thickness to a high degree of accuracy. Bedrock elevation can then be obtained adding the ice surface elevation (obtained from radar satellite measurements (EARS1 and other available Digital Elevation Models) to the ice thickness. Ice thickness measurement is the most widespread application of radar systems. Most of this dataset comes from airborne radar investigations undertaken by the United States National Science Foundation NSF, the Scott Polar Research Institute, United Kingdom (SPRI), the British Antarctic Survey, the Alfred Wegener Institute, Germany, the Technical University of Denmark (TUD) and, in recent years, the Italian Antarctic Program (PNRA).

RES measurements have been extensively used to investigate the grounding line (GL is the point where continental ice, flowing toward the open sea, starts to float) of many outlet Antarctic glaciers. Bedrock gate shapes together with ice velocity information enables estimation of ice mass discharge into the sea. This estimate plays an important role in the mass balance calculation for the Antarctic ice cap modulated by climatological changes.

Another application, when very detailed RES survey are indispensable, is for ice core site selection. In recent years, ice coring in the thickest areas of ice sheets has allowed the reconstruction of about 1 million years of the climate history of our planet. This analysis clearly revealed how human activities have changed the percentage of greenhouse gases and how this has influenced the recent climate. Precise information on bedrock depth, topography, and internal layering behavior are as important as surface topography and ice flow speed when choosing the best drilling point.

Examples of this kind of application are the definition of EPICA, (Dome C, Antarctica) and TALDICE (Talos Dome, Antarctica) ice drilling points (Bianchi et al. 2003). The first allowed reconstruction of climate history for about 900 thousand years B.P. reaching 3270 m of ice cored. The second was less deep (1620 m) and allowed reconstruction of climate history of about 250 thousands year but, due to a higher rate of snow accumulation, with increased detail. This site was also chosen in order to improve knowledge of the response of near coastal sites to climate changes. In figure 2 the ice thickness data is used to construct a three-dimensional view of the main bedrock morphology of the area. As shown in the figure, the bedrock topography consists of a mountainous region with NW-SE trending ridges and depressions interposed between higher relief to the NE and a deeper basin to the south. The thickness of the ice sheet on the area ranges from 1000 to 2000 m.

Focusing under the topographic summit (Fig. 2), bedrock morphology and ice layer slope made it clear that the original drilling position (black triangle) could have led to a wrong ice core interpretation. Based on RES data (Urbini et al 2006), the point was so shifted to ID1 (black square) which is 5 km from original point (TD summit).

RES technique has been used over wide areas in Antarctica and Greenland with the aim of geomorphological studies. Detailed bedrock maps permitted hypotheses on the main geological features of a continent buried under thousands of meters of ice. An example (Fig. 3) of this type of study is the analysis of the Aurora trench, close to the Dome C region. Data was collected during the 1999 and 2001 Italian Antarctic Expeditions (Zirizzotti, 2009),

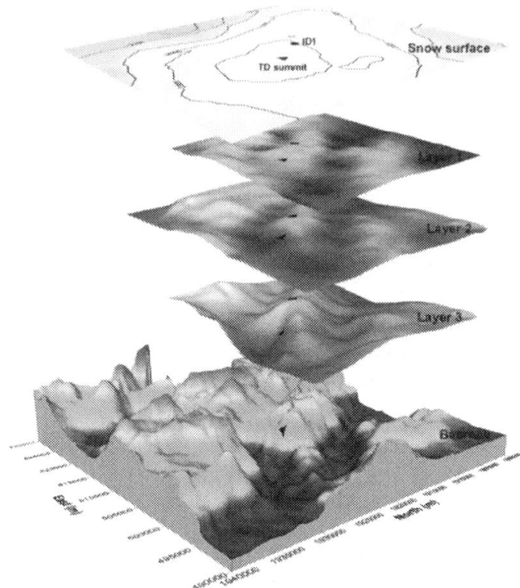

Fig. 2. Talos Dome flight paths (on the left) and bedrock and ice surface from RES measurements (on the right).

(about 6000 km of radar traces were acquired). The radar data was used to determine ice thickness and bedrock topography for the entire area; in addition, analysis of the shape and amplitude of bottom reflections allowed the detection of sub-glacial "lake" mirror features in 30 radar tracks (Tabacco et al., 2003). The deepest point in the Aurora trench (Fig. 3), located at 118.328° E; 76.054°S, has a bed elevation of -1549 m and an ice thickness of 4755 m. This location must be included among the thickest ice cover areas ever discovered in Antarctica (Cafarella, et al. 2006).

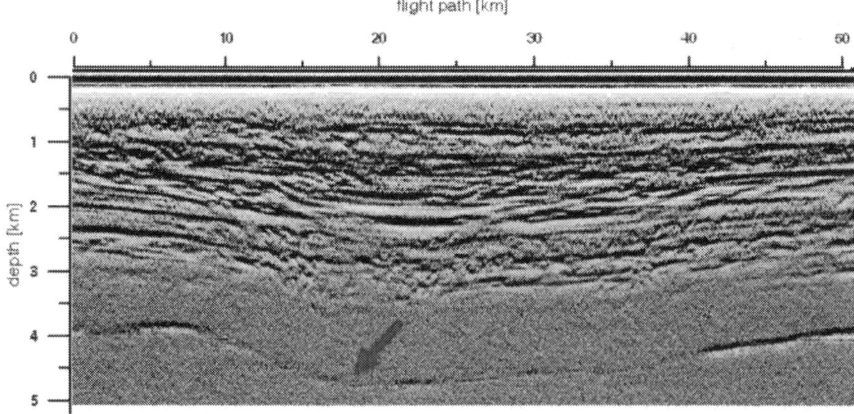

Fig. 3. The Aurora trench deepest point

4.2 Interglacial layers

Reflections from internal ice sheet layers have played a very important role in glaciological exploration. Since the first RES observations were collected, a large number of reflections in the range between the surface and the bedrock appeared on radargrams. They usually show great reflectivity variation, high correlation with the main bedrock morphologies and increasing smoothness towards the surface. Many sources for these internal reflections have been identified: volcanic dust, acidity variation, changes in the size or shape of air bubbles within the ice, variation in ice crystal orientation and density, palaeo-surfaces (like buried wind crusts), layers of liquid water, etc. These layers are important because their continuity is assumed to represent an isochronous layer in the time-depth-age relationship in the two way travelling time of the radar echo. In other words it is assumed that the arrival times of the strongest internal reflections represents a specific event (or short time period), and that layers yielding a strong radar reflection are isochronous (Vaughan et al. 1999). As a consequence, variations in snow thickness between these reference layers provide information on snow accumulation variability. This kind of information is very important in many glaciological issues, like for example the assessment of an ice-drill location (at the same depth, sites where snow accumulation is less, correspond to older deposits). In this context, RES and GPR surveys have been used to investigate the same area at different scale resolutions. RES is mainly used to provide information about the main layer distribution over great distances and depths neglecting shallow small scale details (like wind driven redistribution). On the contrary, GPR is mainly used to analyze the problem of ice-atmosphere interaction in superficial snow redistributions (mega-dune areas, erosion and accumulation driven by surface morphology and prevalent wind direction along the maximum slope direction).

In figure 4 a GPR radargram (antenna frequency 200 MHz, investigated depth 70 m) is reported. It shows a very high variation in snow accumulation over relatively short distances due to ice-atmosphere interaction.

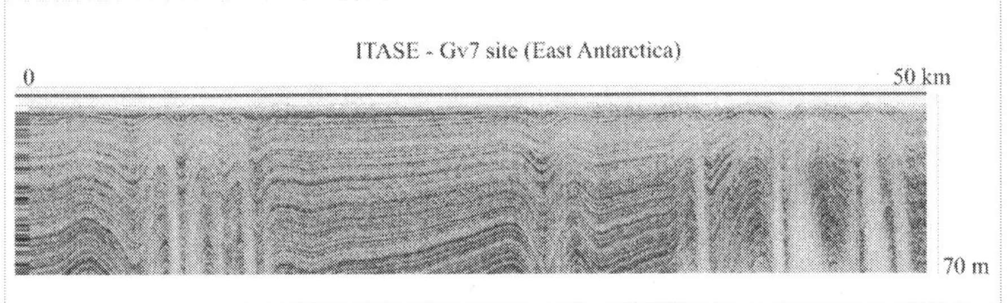

Fig. 4. GPR radargram of internal layers

4.3 Subglacial lake exploration

The earliest hypothesis of the existence of lakes beneath the Antarctic ice sheet was formulated studying airborne radio-echo soundings collected during the 1960's and 1970's (Drewry, 1983).

During the last ten years the characteristics of these subglacial lakes have been thoroughly investigated by means of geophysical instruments. Today we know that more than 150 lakes exist beneath (3 km) the Antarctic ice sheets (Siegert et al. 2005). They are mainly located beneath the main ice divides and the total volume of water stored in these lakes is between 4000 and 12000 km³ (Dowswell & Siegert, 1999).

Subglacial lakes occur essentially because of three factors: the pressure beneath an ice sheet reduces the melting point of ice; the ice sheet insulates the base from the cold temperatures observable on the surface; and finally, the ice base is kept warm by geothermal heating.

RES technique is generally used to identify subglacial lakes, but a strong signal alone is not enough. Strong radar reflection from the ice sheet base could be ascribed both to water-saturated basal sediment or to subglacial lakes. Lake identification is possible if other conditions occur. The identification of flat, horizontal reflectors with almost constant echo strength surrounded by sharp edges is necessary (fig. 6). Moreover the reflection strength from ice-water and ice-rock interfaces must differ by more than 10 dB

Approximately 81% of the detected lakes (on the right of fig. 5) lie at elevations less than a few hundred meters above sea level whereas the majority of the remaining lakes are perched at higher elevations.

The largest, and most well documented among the subglacial lakes, is without doubt lake Vostok (map of fig. 5). Lake Vostok was identified in 1996 by Russian and British scientists

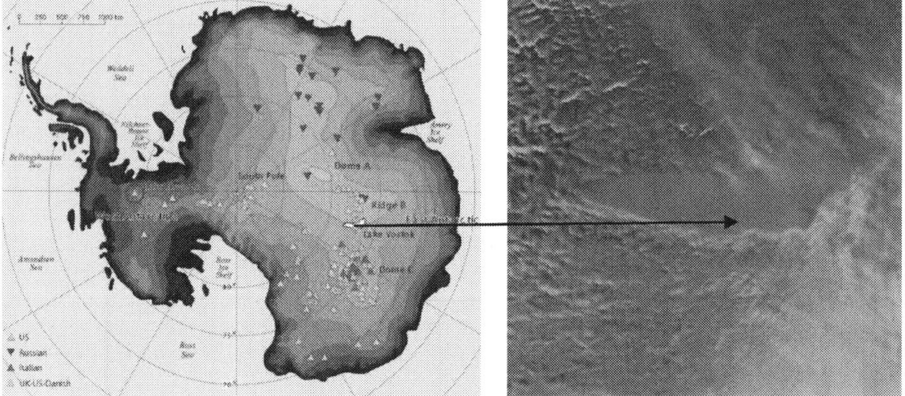

Fig. 5. On the left, Antarctic lake exploration (catalogued lakes). On the right a satellite images of Vostok lake and its position on the catalogued lakes map.

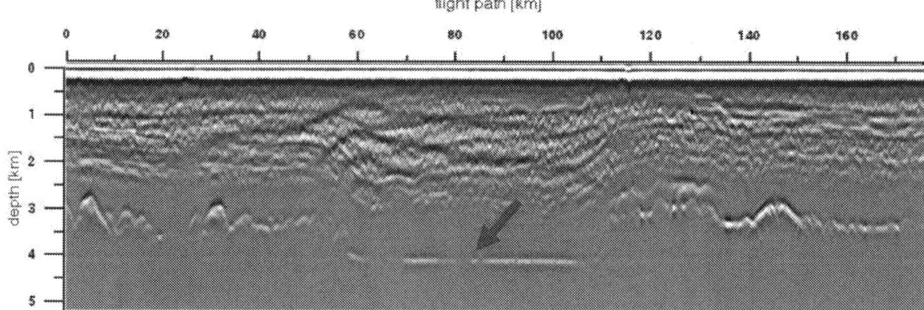

Fig. 6. The Vostok lake radar signal.

(Kapitsa et al., 1996) who integrated data survey from seismic, airborne ice-penetrating radar and spaceborne altimetric observations (on the left of fig. 5). The horizontal extent of the lake is estimated from the flat surface (0.01 degrees) observed in the ERS-1 ice surface altimetry. The 4 kilometer-thick ice sheet floats on the lake , just as ice sheets become floating ice shelves at the grounding line. Lake Vostok is about 230 km long and has an area of about 10000 km^2, similar to that of lake Ontario. A maximum water depth of 510 m was measured using seismic methods, and mean water depth is about 200 m. The estimated volume of lake Vostok is about 2000 km^3. The other known subglacial lakes are one to two orders of magnitude smaller than this.

Recently, scientific observations indicate the existence of a subglacial hydrodynamic network in which liquid water can flow and can be stored in river and lake systems as happens on the other continents. Recently, some peculiar studies on the amplitude of radar signals received from the ice-bedrock interface have been proposed in order to discriminate wet or dry bedrock conditions. Some interesting results emerged from these studies and maps of dry/wet interfaces have been proposed by different authors helping to improve knowledge of the existence of water circulation beneath kilometers of ice (Jacobel 2009).

4.4 Crevasse exploration

Moving over a glacial environment involve many kinds of risk. Recently, there are ever more studies that need to travel or cover long distances across the Antarctic continent or an Alpine glacier. If we consider exploration and logistic teams moving on the ice surface, the higher risk could be ascribed to the presence of buried crevassed area. Many accidents occurred and many lives were lost because of it. In Antarctica scientific traverse programme (as ITASE i.e.) involved the use of tractor as Caterpillar and Pisten Bully towing heavy sledge for thousands kilometres to climb glaciers in order to reach the ice plateau that cover the Antarctic continent. Passing through an unknown snow bridge hiding crevasses, (10-20 m wide and hundred meters deep) with 18 tons heavy trucks is not courageous, but it is crazy. Obviously, the gap left by the discovery of a crevasse represents a perfect target for a radar investigation. Figures 7 show some examples of "lucky" token.

Fig. 7. Past and future of Antarctic exploration problems.

The shallow depth (25-50 m of depth) of investigation and the necessary high resolution, turned the choice of instrumentation to the GPR. An helicopter coupled with a very precise

GPS could give an accurate crevasses position map,in which geometry, thickness and condition of the snow bridges can be evaluated. Figures 8 shows how crevasses appear in radargrams.

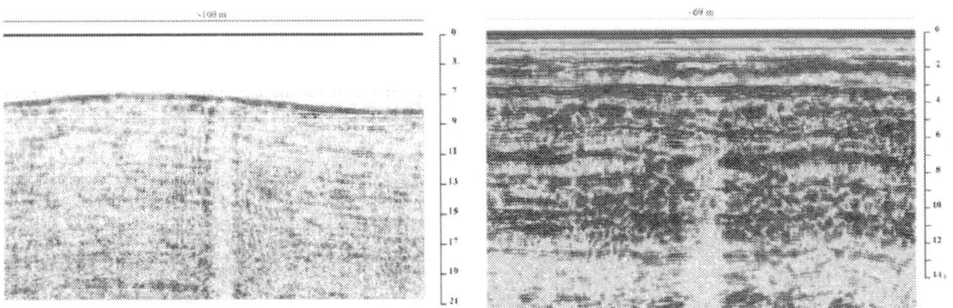

Fig. 8. GPR radargrams showing crevasses

Note the difference in electromagnetic response of the ice corresponding to a low (on left) and high accumulation area (on right) and how the condition of the snow bridge is well described by the status of snow layering.

5. References

Allen, C; Paden, J; Dunson, D; Gogineni, P; (2008). Ground-based multi-channel synthetic-aperture radar for mapping the ice-bed interface, *IEEE Radar Conference, Rome, Italy*, (1428-1433).

Arcone, SA; Yankielun, NE; Chacho, Jr EF. (1997) Reflection profiling of arctic lake ice using microwave FM-CW radar," *IEEE Transactions on Geoscience and Remote Sensing*, 35(2), pp. 436-443.

Arcone, SA; Yankielun, NE. (2000). 1.4 GHz radar penetration and evidence of drainage structures in temperate ice: Black Rapids Glacier, Alaska, U.S.A. *Journal of Glaciology*, 46(154), (477-490).

Bell, R. ; Studinger, M. ; Shumman C.A. ; Fahnestock M.A.; Joughin I. (2007). Large subglacial lakes in East Antarctica at the onset of fast-flowing ice streams", *Nature*, vol. 445, (904-907).

Bianchi, C.; Cafarella, L.; De Michelis, P.; Forieri, A.; Frezzotti, M.; Tabacco, E.; Zirizzotti, A. ; (2003), Radio Echo Sounding (RES) investigations at Talos Dome (East Antarctica): bedrock topography and ice thickness, *Annals of geophysics*, 46, 6, (1265-1271).

Bogorodsky, V.V.; C.R. Bentley and P.E. Gudmandsen; (1985). *Radioglaciolog"*, Reidel Publishing Company.

Cafarella, L.; Urbini, S.; Bianchi, C.; Zirizzotti, A.; Tabacco, I.E.; Forieri, A.; (2006). Five subglacial lakes and one of Antarctica's thickest ice covers newly determined by Radio Echo Sounding over the Vostock-Dome C region, *Polar Research*, 25, no.1, (69-73).

Carter, S.P.; Blankenship, D.D.; Peters, M.E.; Young, D.A.; Holt, J.W.; Morse D.L.;(2007). Radar-based subglacial lake classification in Antarctica", *Geochem. Geophys. Geosyst.*, Vol. 8, doi: 10.1029/ 2006GC001408.

Conway, H; Smith, B; Vaswani, P; Matusuoka, K; Rignot, E; Claus, P. (2009). A low-frequency ice-penetrating radar system adapted for use from an airplane: test results from Bering and Malaspina Glaciers, Alaska, USA. *Annals of Glaciology* 50 (51) (93–97).

Cook, J.C. (1960). Proposed monocycle-pulse VHF radar for airborne ice and snow measurement. *Commun. Electr.*, 5 l: (588-594) [also AIEE Trans., 79].

Corr, H.; Moore, J.C.; Nicholls, K.W. (1993). Radar absorption due to impurities in Antarctic ice", *Geophys. Res. Lett.*, vol. 20, num.11, (1071-1074).

Daniels, D. J. (1996) *Surface-penetrating Radar* Institution of Electrical Engineers, ISBN 0852968620, 9780852968628.

Dowdeswell, J.A.; Siegert ,M.J. (2002). The physiography of modern Antarctic subglacial lakes, *Global and Planetary Change*, 35, (221-236).

Drewry, D.J. (Editor), (1983), *Antarctica: Glaciological and Geophysical Folio* Scott Polar Research Institute, Cambridge.

Eisen, O.; Wilhelms, F.; Seinhage, D.; Schander, J. (2006). Instruments and Methods Improved method to determine radio-echo sounding reflector depths from ice-core profiles of permittivity and conductivity", *J. Glaciology.*, vol. 52, no. 177.

EPICA community members (2004). Eight glacial cycles from an Antarctic ice core, *Nature*, vol. 429, (623-628).

Fujita, S.; Matsuoka, T.; Ishida, T.; Matsuoka, K.; Mae, S. (2000). A summary of the complex dielectric permittivity of ice in the megahertz range and its applications for radar sounding of polar ice sheets", *Physics of Ice Core Records*, (185-212).

Gogineni, S.; Chuah, T.; Allen, C; Jezek, K; Moore, R.K. (1998). An improved coherent radar depth sounder. Journal of Glaciology Vol. 44:148, (659-669).

Jacobel, R.W.; Welch, B.C.; Osterhouse, D.; Pettersson, R.; Macgregor, J.A. (2009). Spatial variation of radar-derived basal conditions on Kamb Ice Stream, West Antarctica. *Annals of Glaciology* 50(51), (10–16).

Kanagaratnam, P; Gogineni, SP; Ramasami, V; Braaten, D. (2004). A wideband radar for high-resolution mapping of near-surface internal layers in glacial ice," *IEEE Transactions on Geoscience and Remote Sensing*, 42(3), (483-490).

Kapitsa, A.P.; Ridley, J.K.; Robin, G.D.; Siegert, M.J; Zotikov I.A. (2006). A large deep freshwater lake beneath the ice of central East Antarctica" *Nature*, vol. 381, (684-686). Laurent, A. (2003). EPICA Dome C DRILLING SEASON. *2002-2003 Field Season Report.*

MacGregor, J.A.; Winebrenner, D.P.; Conway, H.; Matsuoka, K.; Mayewski, P.A. Clow, G.D. (2007). Modeling englacial radar attenuation at Simple Dome, West Antarctica, using ice chemistry and temperature data, *J. Geophysics Res.*, 112, F03008, doi:10.1029/2006JF000717.

Oswald, G.K.A.; Gogineni, S.P. (2008). Recovery of subglacial water extent from Greenland radar survey data, *J. Galcyology*, vol. 54, num. 184, (94-106).

Paden, J.D.; Allen, C.T.; Gogineni, S.; Jezek, K.C.; Dahl-Jensen, D.; Larsen, L.B. (2005). Wideband measurements of ice sheet attenuation and basal scattering", *Geoscience and Remote Sensing Letters IEEE* 2, 2, (164 – 168).

Peters, M.E.; Blankenship, D.D., Morse D.L. (2005). Analysis techniques for coherent airborne radar sounding: Application to West Antarctic ice streams, *J. Geophys. Res.*, 110, B06303, doi: 10.1029/2004JB003222, 2005.

Plewes, L.A.; Hubbard B. (2001). A review of the use of radio-echo sounding in glaciology Progress in Physical Geography, vol. 25, num. 2, (203-236).

Raju, G.; Xin, W.; Moore, R. K. (1990). Design, development, field operations and preliminary results of the coherent Antarctic radar depth sounder (CARDS) of the University of Kansas, USA. *J. Glaciology*, vol. 36,123, (247-258).

Rees, W.G.; Donovan, R.E. (1992). Refraction correction for radio-echo sounding of large ice masses. *Journal of Glaciology* 32, (192-194).

Reynold J.M. (1997). *An introduction to Applied and Enviromental geophysics*. Chichester: Wiley.

Raymond, C.F.; Catania, G.A.; Nereson, N.; van der Veen, C.J. (2006) Bed radar reflectivity across the north margin of Whillans Ice Stream, West Antarctica, and implications for margin processes. *Journal of Glaciology*, 52, 176, (3-10)

Siegert, M.J.; Carter, S.P.; Tabacco, I.E.; Popov, S.; Blankenship, D.D. (2005). A revised inventory of Antarctic subglacial lakes. *Antarctic Science*, 17(3), (453-460).

Skolnik M. (1990). *Radar Handbook* McGraw-Hill,1990, 2nd edn.

Stauffer, B. (1991). The Greenland Icecore Project (GRIP). Communications. *The Journal of the European Science Foundation*, 24, (12-13).

Tabacco, I.E.; Passerini, A.; Corbelli, F.; Gorrnan M. (1998). Determination of the surface and bed topography at Dome C, East Anatrctica. *Journal of Glaciology*, 39, (495-506).

Tabacco, I.E.; Forieri, A.;Vedova, A.D.; Zirizzotti, A.; Bianchi, C. (2003) . Evidence of 13 new subglacial lakes in the Dome C-Vostok area, *Terra Antarct.* 8, (175–179).

Urbini, S; Cafarella, L; Zirizzotti, A; Bianchi, C; Tabacco, I ; Frezzotti, M. (2006). A Location of a new ice core site at Talos Dome (East Antarctica). *Annals of Geophysics* 49, 4-5 (1133-1138).

Vaughan, D.G.; Bamber J.L.; Giovinetto M.; Russell J.; Cooper A.P.R. (1999). Reassessment of Net Surface Mass Balance in Antarctica. *J. Climate*, 12, (933–946).

Waite, AH; Schmidt; SJ. (1961). Gross errors in Height Indication from Radio Altime- ters Operating Over Thick Ice or Snow. *Institute of Radio Engineers International Convention Record*, 5, (38-53)

Wingham, D.; Siegert, J.; Shepherd, A.; Muir A.S. (2006). Rapid discharge connects Antarctic subglacial lakes. *Nature*, vol. 440, (1033-1036).

Wolff, E. (2004). EPICA Dome C Core EDC99 Dielectric Profiling Data. IGBP PAGES/World Data Center for Paleoclimatology Data Contribution. Series # 2004-037, NOAA/NGDC, Paleoclimatology Program, Boulder CO, USA, 2004.

Wu, R.; King W.. P. (1965). The cylindrical antenna with nonre- flecting resistive loading," *IEEE Trans. Antennas Propagat.*, vol. 13, (369-373).

Zirizzotti, A.;. Baskaradas, J.A; Bianchi, C.; Sciacca, U.; Tabacco, I.E.; Zuccheretti, E. (2008), Glacio RADAR system and results, *IEEE Radar Conference, 2008. RADAR '08.*, 1-3, 26, doi: 10.1109/RADAR.2008.472099

TOPIC AREA 3: Radar Functional Chain and Signal Processing

10

Multisensor Detection in Randomly Arriving Impulse Interference using the Hough Transform

Chr. Kabakchiev[1], H. Rohling[2], I. Garvanov[3],
V. Behar[4], and V. Kyovtorov[3]
[1]Faculty of Mathematics & Informatics - Sofia University,
[2]Technical University Hamburg-Harburg,
[3]Institute of Information Technologies - BAS,
[4]Institute for Parallel Processing – BAS
[1,3,4]Bulgaria
[2]Germany

1. Introduction

In this chapter, several advanced detection algorithms for Track-Before-Detect (TBD) procedures using the Hough Transform (HT) are proposed and studied. The detection algorithms are based on the scheme described in (Carlson et al., 1994) to use the Hough transform for simultaneous target detection and trajectory estimation. The concept described in (Carlson et al., 1994) accepts that a target moves within a single azimuth resolution cell, and the distance to the target is estimated for several last scans forming the $(r\text{-}t)$ data space. The Hough transform maps all points from the $(r\text{-}t)$ space into the Hough space of patterns. The association with a particular pattern is done by thresholding the Hough parameter space with a predetermined threshold. In order to enhance the target detectability in conditions of Randomly Arriving Impulse Interference (RAII), a CFAR processor is proposed to be used for signal detection in the $(r\text{-}t)$ space instead of the detector with a fixed threshold as it is suggested in (Carlson et al., 1994). The results obtained show that such a Hough detector works successfully in a noise environment. In real-time and realistic applications, however, when the two target parameters (range and azimuth) vary in time, the usage of the Polar Hough transform (PHT) is more suitable for radar applications because the input parameters for the PHT are the output parameters of a search radar system. Such a Polar Hough detector combined with a CFAR processor is proposed for operation in RAII conditions. The results obtained by simulation illustrate the high effectiveness of this detector when operating in strong RAII situations. Finally, the TBD-PHT approach is applied to the design of a multi-channel Polar Hough detector for multi-sensor target detection and trajectory estimation in conditions of RAII. Three different structures of a nonsynchronous multi-sensor Polar Hough detector, decentralized with track association (DTA), decentralized with plot association (DPA) and centralized with signal association (CSA), are considered and analyzed. The detection probabilities of the three multi-sensor Hough detectors are evaluated using the Monte Carlo approach. The results obtained show that the detection probability of the centralized detector is higher than that of the decentralized detector. The DPA Hough detector is close to the potential of the most effective multi-sensor CSA Hough detector. The target measurement errors in the (r-t) space mitigate the operational efficiency of multi-sensor Hough detectors. The needed operational efficiency requires the appropriate sampling of the Hough parameter space.

In recent years, the mathematical methods for extraction of useful data about the behavior of observed targets by mathematical transformation of the received signals have been widely used for design of a set of highly effective algorithms for processing of radar information. As a result, the more precise estimates of moving target parameters can be obtained in very dynamic radar situations. Actually, the target trajectories are estimated based on several radar plots. In the process of trajectory estimation, it is very important to use all current information for the detected target - amplitude and spectrum of the signals reflected from targets, target geometric dimensions, coordinates etc. According to the classical method, a target trajectory is estimated by determining of an existing kinematical dependence between few measurements of target coordinates. An optimization problem is solved where the optimization criterion is minimization of the distance between the two target coordinates, expected and measured. As a rule, different modifications of both methods, Kalman filter and Bayesian estimation, are used in these algorithms. However, the real-time implementation of these algorithms demand serious computational resources because the number of optimization problems, solved in the process of trajectory estimation, increases exponentially with the number of trajectories and the measurement density in the surveillance area. Recently, another modern approach is often used for trajectory estimation. In recent years the two mathematical transforms, Hough and Radon, become increasing attention. The use of these transforms makes it possible to map two-dimensional images containing straight lines into the space of possible straight line parameters, where each straight line in an image corresponds to the peak in the parameter space, which has coordinates equal to the respective straight line parameters. For that reason, these mathematical transformations are very attractive for applications related to detection of straight lines in images. Such areas of applications are, for example, image processing, computer vision, seismic studies and etc. The idea to use the standard Hough transform (HT) for joint target detection and trajectory estimation on the background of white Gaussian noise was firstly introduced in (Carlson et al., 1994). According to this concept a target is assumed to move within a single azimuth resolution cell, and the target range is estimated in each scan. The data stored for several last scans forms the matrix hereinafter called as the (r-t)-data space. The HT maps all points from the (r-t) space where the target is detected into the Hough space of straight line parameters. The association with a particular straight line is done by estimating the quantity of information extracted from the signals received from the target with coordinates associated with this line.

In order to enhance target detectability in RAII conditions, a CFAR processor can be used for signal detection in the (r-t) space instead of a detector with a fixed threshold proposed in (Carlson et al., 1994). It is well known that different CFAR processors can be applied to signal detection in a complex noise environment (Finn & Johnson, 1968; Rohling, 1983; Gandhi & Kassam, 1988; Goldman, 1990; Himonas, 1994). The adaptive CFAR processors for signal detection in conditions of RAII are studied in (Kabakchiev & Behar, 1996, Behar et al., 2000, Garvanov et al., 2003; Garvanov, 2003). In this chapter, it is assumed that the noise amplitude is a Rayleigh distributed random variable and therefore the noise power is an exponentially distributed variable. Different Hough detectors that employ CFAR processors such as *Cell Averaging* (CA), *Excision* (EXC), *Binary Integration* (BI), *Excision with Binary Integration* (EXC BI), *Adaptive Post detection Integration* (API), *K-stage Detector*, *Order Statistic* (OS) for signal detection in the (r-t) space are studied and compared in (Behar et al., 1997; Behar & Kabakchiev, 1998; Kabakchiev et al., 2005; Doukovska, 2005; Doukovska & Kabakchiev, 2006; Garvanov et al., 2006; Garvanov et al., 2007; Doukovska, 2007; Doukovska et al., 2008). The structure of these Hough detectors includes the following operations - CFAR signal detection in the area of observation and the HT of the target range measurements from the observation area into the Hough parameter space, binary integration of data in the parameter space and, finally, linear trajectory estimation. All these CFAR Hough detectors have been studied in cases when a target moves in the same azimuth direction at a constant velocity. The results obtained in (Kabakchiev, Garvanov, Kyovtorov et al., 2005; Kabakchiev & Kyovtorov et al., 2005; Behar et al., 2007; Kyovtorov, 2007) show that different CFAR processors work successfully in combination with a Hough detector in conditions of RAII, and allow to evaluate the parameters of the targets.

In real radar applications, however, when the two target parameters, range and azimuth, vary in time, the PHT can be successfully used because in that case the input parameters of the PHT are two polar coordinates of a target – range and azimuth. Such advanced structures of the TBD using the PHT (TBD-PHT) have been developed and studied in (Garvanov et al., 2006; Garvanov et al., 2007). The PHT is analogous to the standard HT and performs all the data collected for several previous scans into a single large multi-dimensional polar data map. The general structure of an adaptive Polar Hough detector with binary integration is similar to that of a standard Hough detector. The only difference between them is that the PHT uses (*range-azimuth-time*) space while the standard HT employs (r-t) space. The detection probability of a Polar Hough detector is calculated by Brunner's method as for a standard Hough detector. The use of the PHT instead of the standard HT allows detecting target trajectories in real situations when targets move at variable speeds along arbitrary linear trajectories.

The TBD approach that applies the HT to multi-sensor detection in conditions of intensive RAII is proposed in (Garvanov, 2007; Kabakchiev, 2007; Kabakchiev, 2008; Garvanov, 2008, Garvanov et al., 2008). As usual, the fusion center of a decentralized system applies binary integration of the data received from each sensor. In such a system, at the first stage, radars produce local decisions by the TBD-PHT processing and at the second stage - all the local decisions are transferred from radars into the fusion node where coordinates and time are associated in the Global Observation Space (GOS). The centralized system, however, firstly associates data with common coordinates and time received from sensors and then performs them by the TBD-PHT processing. In this context, two variants of a centralized asynchronous net with association of signals or signal detections are developed and analyzed. The algorithm with association of signals includes two stages. At the first stage, the signals received from sensors are non-coherently accumulated in the signal matrixes of the fusion centre, because the size of signal matrixes and their cells are the same for the entire radar net. At the second stage the accumulated signals are transferred into the GOS. The algorithm with association of signal detections firstly accumulates decisions for signal detection after CFAR processing in each sensor, secondly - transfers detections in the GOS. These different types of multi-sensor TBD-PHT processors that operate in the presence of RAII have been developed and studied in (Garvanov et al., 2007; Kabakchiev et al., 2007; Kabakchiev et al., 2007; Kabakchiev et al., 2008; Garvanov et al., 2008).

The expressions for calculating the probability characteristics, i.e. the probability of target detection, trajectory estimation and the false alarm probability, are derived under the assumption that both target coordinates (range and azimuth) and both parameters of the Hough parameter space (ρ and θ) are measured with or without errors. The results obtained show that the detection probability of multi-sensor centralized TBD-PHT processors is higher than that of the decentralized detectors.

The performance evaluation of multi-sensor TBD-PHT processors has been carried out by Monte-Carlo simulations in MATLAB computing environment. The DPA based Hough detector is close to the potential of the most effective multi-sensor CSA Hough detector. The target coordinate measurement errors in the (r-t) space mitigate the operational efficiency of multi-sensor Hough detectors. The needed operational efficiency requires the appropriate sampling of the Hough parameter space.

The chapter includes the following paragraphs - abstract, introduction, Single-channel Hough detector in condition of RAII, Performance analysis of a conventional single-channel Hough detector with a CFAR processor, Performance analysis of a single-channel polar Hough detector with a CFAR processor, Multi-sensor (multi-channel) polar Hough detector with a CFAR processor, performance analysis of a multi-sensor polar Hough detector with a CFAR processor and finally conclusion.

2. Single-channel Hough detector in a local RAII environment

2.1 Conventional Hough detector

The basic concept of using the HT to improve radar target detection in white Gaussian noise is firstly introduced in (Carlson et al., 1994). According to this concept, it is assumed that a target moves in a straight line within in a single azimuth resolution cell. The structure of a Hough detector proposed by Carlson is shown in Fig. 1. The Hough detector estimates trajectory parameters in the Hough parameter space that constitute a straight line in the (r-t) space.

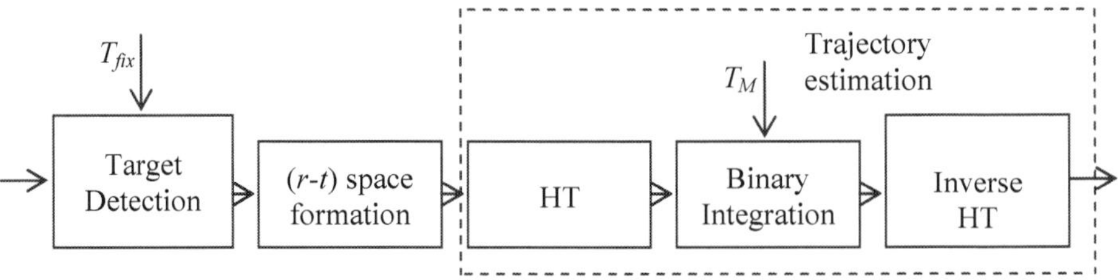

Fig. 1. Structure of a conventional Hough detector with two fixed thresholds

Naturally, two or more large interference spikes in the signal data plane can also constitute a straight line and can create false alarms (false tracks). The control of false alarms in the Hough detector begins with setting an appropriate threshold T_{fix} for signal detection and formation of the (r-t) space. The (r-t) space is divided into cells, whose coordinates are equal to the range resolution cell number - in the range and to the scan number in the history - in the time. The HT maps points from the observation space termed as the (r-t) space, into curves in the Hough parameter space called as the (ρ-θ) space, by:

$$\rho = r\cos\theta + t\sin\theta \qquad (1)$$

Here r and t are the measured distance to the target and time, respectively. The mapping can be viewed as the sampling of θ in the range of 0° to 180° and then the calculating of the corresponding parameter (ρ).
The result of transformation is a sinusoid with magnitude and phase depending on the value of the point in the (r-t) space. Each point in the Hough parameter space corresponds to one straight line in the (r-t) space with two parameters (ρ,θ). Each of the sinusoids corresponds to a set of possible straight lines through the point. If a straight line exists in the (r-t) space, by means of the Hough transform it can be viewed as a point of intersection of sinusoids defined by the Hough transform. The parameters ρ and θ define the linear

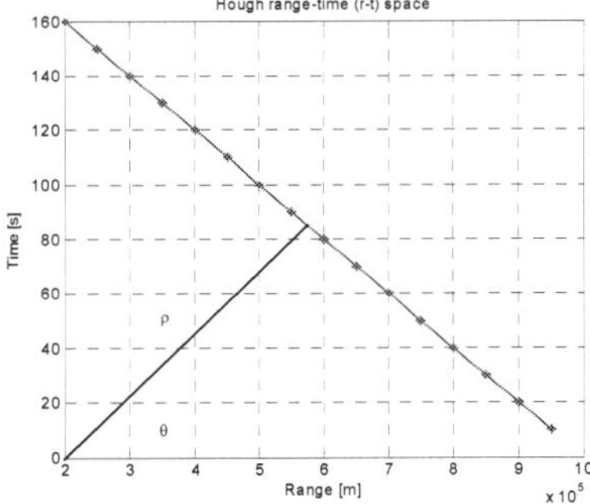

Fig. 2. Range-time (r-t) space

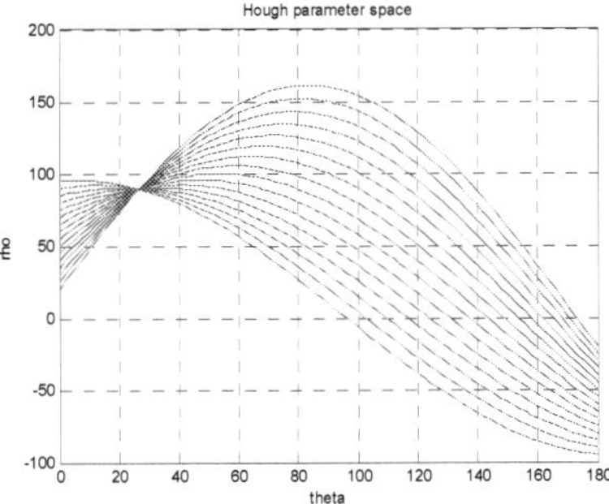

Fig. 3. Hough parameter (ρ-θ) space

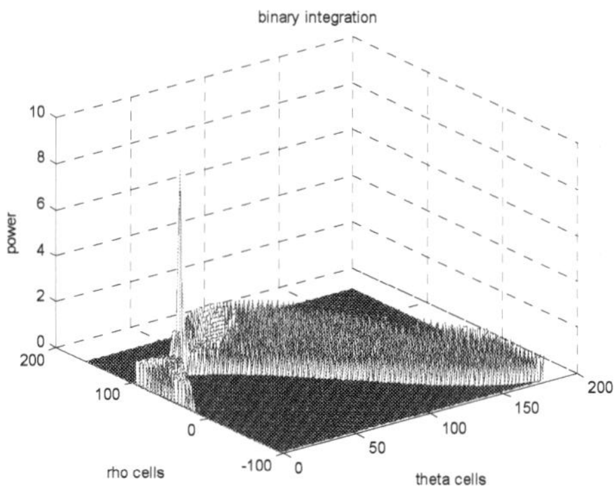

Fig. 4. The output of the Hough detector with binary integration

trajectory in the Hough parameter space, which could be transformed back to the $(r\text{-}t)$ space showing the current distance to the target. Figures 2, 3 and 4 illustrate the $(r\text{-}t)$ space, the Hough parameter space and the output signal of the Hough detector with binary integration in case of two closely moving target.

2.2 Conventional Hough detector with a range CFAR processor

It is proved that different CFAR processors used for signal detection in the $(r\text{-}t)$ space on the homogeneous background of unknown intensity and in the presence of randomly arriving impulse interference with known parameters improve the detection performance. In such CFAR processors, it is usually assumed that the noise amplitude is a Rayleigh distributed variable and the power, therefore, is an exponentially distributed variable. As shown in (Kabakchiev & Behar, 1996; Doukovska, 2006), such CFAR processors combined with a conventional Hough detector can improve the detection probability characteristics. In (Doukovska et al., 2006; Garvanov et al., 2007), the performance of a Hough detector with a fixed threshold is compared with the performance of Hough detectors with two-dimensional CFAR processors - CFAR BI (binary integration), EXC CFAR BI (excision and binary integration) and API CFAR (adaptive post integration). In Fig. 5, the detection probability of these Hough detectors is plotted as a function of the signal to-noise ratio (SNR).

The comparison analysis of these detectors shows that the Hough detector with a CFAR BI processor is the most preferable, because has the relatively good detection characteristics in conditions of RAII and can be implemented at the least computational cost. For that reason this Hough detector is considered and analyzed in this chapter. The structure of the CFAR BI processor is shown in Fig. 6.

In a CFAR pulse train detector with binary integration, the binary integrator counts "L" decisions (Φ_l) at the output of a CFAR pulse detector. The pulse train detection is declared if this sum exceeds the second digital threshold M. The decision rule is:

$$\begin{cases} H_1: & if\ \sum_{l=1}^{L}\Phi_l \geq M \\ H_0: & otherwise \end{cases} \tag{2}$$

where L is the number of pulse transmissions, $\Phi_l=0$ – if no pulse is detected, and $\Phi_l=1$ – if pulse detection is indicated.

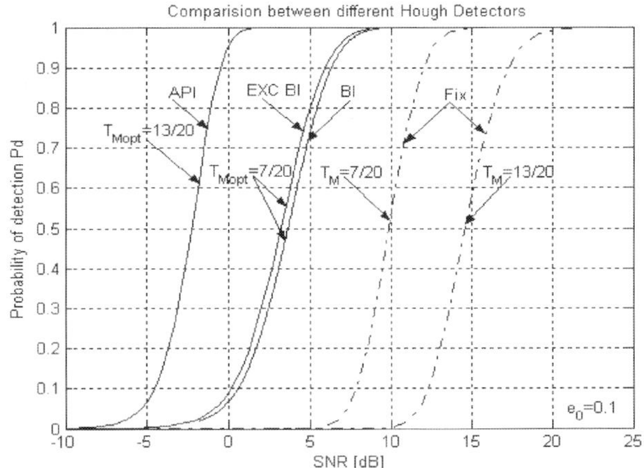

Fig. 5. Target detection probability for Hough detectors with API, BI and EXC BI CFAR processors

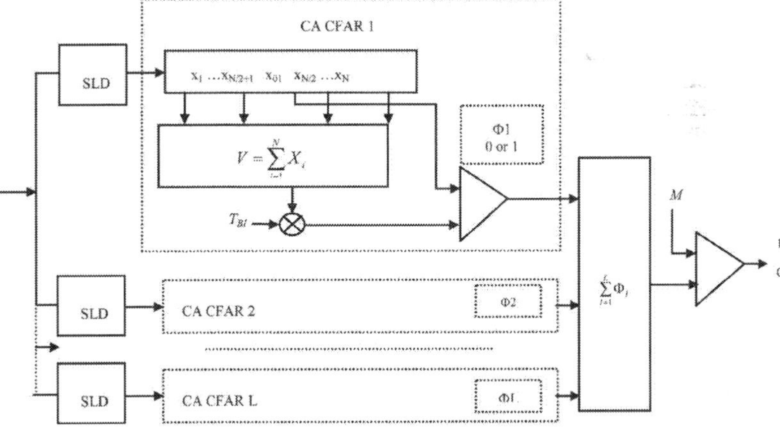

Fig. 6. Structure of a CFAR BI processor

In a Hough detector with a CFAR BI processor, the two-dimensional $(r\text{-}t)$ space of binary data is formed at the output of the CFAR processor, as a result of N_{SC} radar scans. According to (Carlson et al., 1994), the Hough transform is applied to coordinates of such cells in the $(r\text{-}t)$ space, where the detection is indicated. In this way the Hough parameter space is formed. Each cell from the Hough parameter space is intersected by a limited set of sinusoids obtained by the Hough transform. If the number of intersections in any of cells exceeds a fixed threshold (T_M), both target and linear trajectory detections are indicated. The procedure of detection is repeated in the Hough parameter space cell by cell.

The general structure of such an adaptive Hough detector with binary integration of data in the Hough parameter space that can be used in real search radar is shown in Fig. 7.

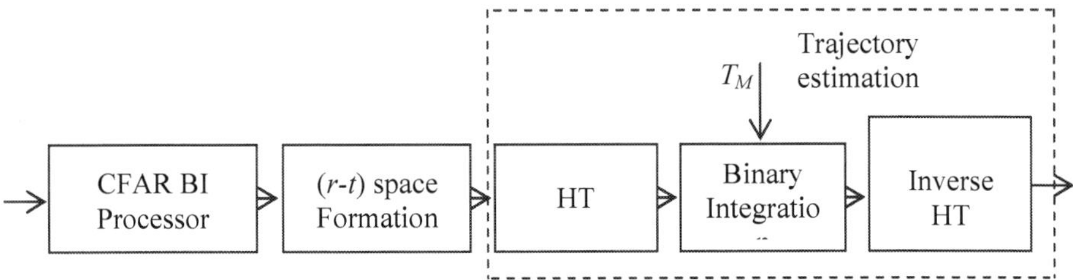

Fig. 7. Structure of an adaptive Hough detector with adaptive detection threshold in (r-a) space

The analysis of the performance of the Hough detector with a CFAR BI processor is done in (Behar et al., 1997; Garvanov, et al., 2007; Doukovska, 2007).

2.3 Polar Hough detector with a CFAR processor

In real radar applications, the estimated target coordinates are given in the polar coordinate system (distance and azimuth). In order to employ the Hough detector with a structure shown in Fig. 1, the polar target coordinates must be firstly transformed into the Cartesian coordinate system and further performed using the Hough transform. Such transformation of coordinates, however, additionally complicates the signal processing. For that reason, the detection algorithm employing the polar Hough transform is very comfortable for trajectory and target detection because the input parameters for the polar Hough transform are the output parameters of the search radar. Another important advantage is the signal processing stability when the target changes its speed and moves at different azimuths. According to (Garvanov et al. 2006, Garvanov et al. 2007), the polar Hough transform is defined by two polar coordinates, distance and azimuth - (r, a). In such a way, the polar Hough transform represents each point of a straight line in the form:

$$\rho = r\cos(a - \theta), \ 0 < (a - \theta) \le \pi \tag{3}$$

where r and a are the polar target coordinates (distance and azimuth), θ is the angle and ρ is the smallest distance to the origin of polar coordinate system.

The general structure of such a polar Hough detector with binary integration of data in the Hough parameter space that can be used in real search radar is shown in Fig. 8.

As a result of N_S radar scans, the polar coordinate data map that contains the data for two targets moving with variable speeds and cross trajectories is formed (Fig. 9).

A single (ρ, θ) point in the parameter space corresponds to a single straight line in the (r-a) data space with ρ and θ values. Each cell in the Hough parameter space is intersected by a limited set of sinusoids generated by the polar Hough transform (Fig. 10). If the number of intersections in any cell of the Hough parameter space exceeds a fixed threshold (T_M), both target and linear trajectory detections are indicated (Fig. 11). The procedure of detection is repeated in the polar Hough parameter space cell by cell. The analysis of the performance of CFAR processor with binary integration (CFAR BI) used in combination with a polar Hough detector is done in (Garvanov, et al., 2006; Garvanov et al., 2007).

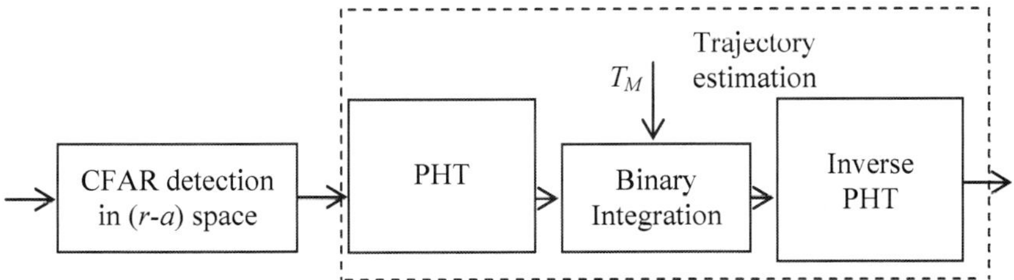

Fig. 8. Structure of a polar Hough detector with a CFAR processor

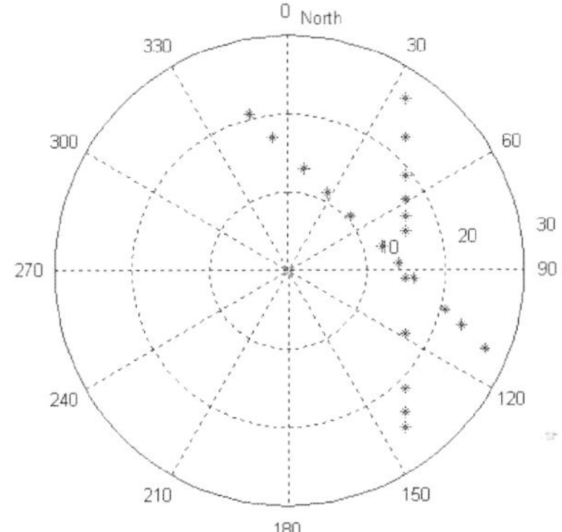

Fig. 9. Illustration of target trajectory in Cartesian coordinate system

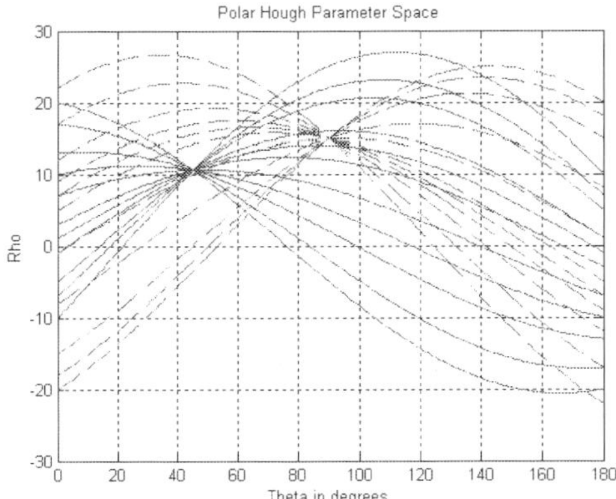

Fig. 10. Polar Hough parameter space

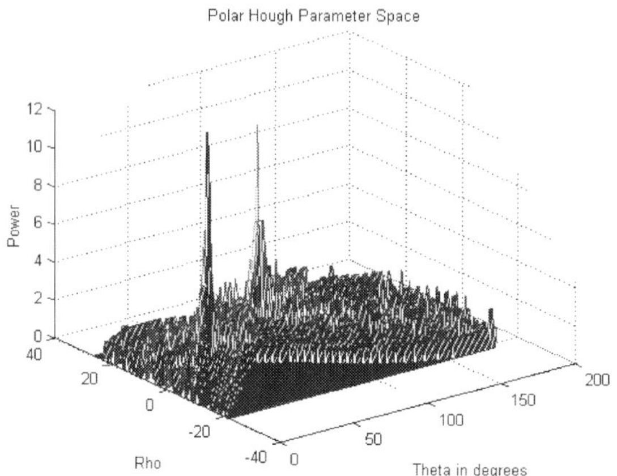

Fig. 11. Binary integration in Polar Hough parameter space

3. Performance analysis of a conventional single-channel Hough detector with a CFAR processor

3.1 Signal and interference model

The Bernoulli model (Poisson model) describes a real radar situation when the impulse noise comes from a single impulse-noise source (Kabakchiev & Behar, 1996). According to this model, in each range resolution cell the signal sample may be corrupted by impulse noise with constant probability P_I. Therefore, the elements of the reference window are drawn from two classes. One class represents the interference-plus-noise with probability P_I. The other class represents the receiver noise only with probability $(1-P_I)$. According to the theorem of total probability, the elements of a reference window (N) are independent random variables distributed with the following PDF:

$$f(x_i) = \frac{(1-P_I)}{\lambda_0}\exp\left(\frac{-x_i}{\lambda_0}\right) + \frac{P_I}{\lambda_0(1+I)}\exp\left(\frac{-x_i}{\lambda_0(1+I)}\right), \quad i=1,...,N \tag{4}$$

In the presence of the desired signal in the test resolution cell the signal samples are independent random variables distributed with the following PDF:

$$f(x_{0l}) = \frac{(1-P_I)}{\lambda_0(1+S)}\exp\left(\frac{-x_{0l}}{\lambda_0(1+S)}\right) + \frac{P_I}{\lambda_0(1+I+S)}\exp\left(\frac{-x_{0l}}{\lambda_0(1+I+S)}\right), \quad l=1,...,L \tag{5}$$

where S is the signal-to-noise ration, I is the interference-to-noise ratio and λ_0 is the average power of the receiver noise.

The probability of occurrence of a random pulse (L) in each range resolution cell can be expressed as $P_I = F_j.t_c$, where F_j is the average pulse repetition frequency and t_c is the transmitted pulse duration. It must be noted that if the probability P_I is small $(P_I < 0.1)$, the size of a reference window N is large, and $N.P_I$ =const, then the Bernoulli model may be approximated with a Poisson model of impulse noise.

3.2 Probability characteristics

CFAR BI processor

The detection probability of a CFAR BI processor, which uses the pulse train binary detection rule M-out-of-L, is evaluated as in (Kabakchiev & Behar, 1996). The probability of pulse train detection is evaluated by:

$$P_D^{CFAR\ BI} = \sum_{l=M}^{L} C_L^l P_D^{CA\ CFAR^l} \left(1 - P_D^{CA\ CFAR}\right)^{L-l} \tag{6}$$

where $P_D^{CA\ CFAR}$ P_D is the probability of pulse detection for CA CFAR detector in the presence of random impulse noise. This probability is calculated as follows:

$$P_D^{CA\ CFAR} = \sum_{i=0}^{N} C_N^i P_I^i \left(1 - P_I\right)^{N-i} \left\{ \frac{P_I}{\left(1 + \frac{(1+I)T}{1+I+S}\right)^i \left(1 + \frac{T}{1+I+S}\right)^{N-i}} + \frac{1 - P_I}{\left(1 + \frac{(1+I)T}{1+S}\right)^i \left(1 + \frac{T}{1+S}\right)^{N-i}} \right\} \tag{7}$$

where T is a predetermined scale factor keeping a constant false alarm rate (P_{FA}). The probability of false alarm of a CFAR BI processor is evaluated by (7), setting $s=0$.

Hough detector

All decisions for signal detection in N range resolution cells obtained for N_{SC} scans form the two-dimensional (r-t) space. The total false alarm probability in the Hough parameter space is equal to one minus the probability of no false alarm occurrence in any Hough space cell. For independent Hough cells this probability is calculated as (Carlson, 1994):

$$P_{FA}^{Hough} = 1 - \prod_{N_{nm}=T_M}^{\max(N_{nm})} \left[1 - P_{fa}^{nm}\right]^{W(N_{nm})} \tag{8}$$

where P_{fa}^{nm} is the cumulative false alarm probability for a cell (n, m), $\max(N_{nm})$ is the accessible Hough space maximum and $W(N_{nm})$ is the number of cells from the Hough parameter space whose values are equal to N_{nm}, T_M is a linear trajectory detection threshold. The cumulative probability of target detection in the Hough parameter space P_D^{Hough} cannot be written in the form of a simple Bernoulli sum. As a target moves with respect to the radar, the SNR of the received signal changes depending on the distance to the target and the probability for one pulse of detection $P_D^{CFAR\ BI}(j)$ changes as well. Then the probability P_D^{Hough} can by calculate by Brunner's method. For N_{SC} scans of radar the following is valid:

$$P_D^{Hough} = \sum_{i=T_M}^{N_{SC}} P_D^{CFAR\ BI}(i, N_{SC}) \tag{9}$$

The probability characteristics of both modifications of a Hough detector, standard and polar, are the same because they calculated in the same way and the two observation spaces, (r-t) and (r-a), are identical.

3.3 Simulation analysis

The performance of a single channel Hough detector with a CFAR processor can be studied using the Monte Carlo approach. Here, a new simulation algorithm for Monte Carlo analysis of the Hough detector with a CFAR processor is described (Behar, et al., 2007). The detection performance is evaluated in terms of the probability of detection calculated as a function of the detector parameters. Let consider radar that provides range, azimuth and elevation as a function of time. Time is sampled by the scan period, but resolution cells sample range, azimuth and elevation. The trajectory of a target moving within the same "azimuth-elevation" resolution cell is a straight line specified by several points in the $(r\text{-}t)$ space. In the $(r\text{-}t)$ space, the target trajectory can be also specified by another two parameters - the distance of the normal ρ from the origin to the line, and the angle θ between the normal ρ and the x-axis. The corresponding (ρ, θ) parameter space is sampled into ρ and θ dimensions. When the primary detection threshold is crossed in any (r, t) cell, unity is added to (ρ, θ) cells intersecting the corresponding sinusoidal curve in the parameter space. In this way, each cell in the Hough parameter space accumulates detection decisions, which increases its value due to the intersection of several sinusoids. The secondary detection threshold, applied to the accumulated value in each cell of the $(\rho\text{-}\theta)$ parameter space, indicates a detection of a target trajectory. The point $(\hat{\rho},\hat{\theta})$ where the secondary threshold is exceeded specifies the target detected trajectory. The simulation algorithm of the Hough detector with a CFAR BI processor, developed for Monte Carlo analysis, includes the following stages:

1. The $(r\text{-}t)$ space is sampled. The following data is needed – the range resolution cell (δR), scan time (t_{sc}), and the number of scans (N_{sc}). The sampled $(r\text{-}t)$ space is of size $[N \times M]$, where $N = N_{sc}$, and $M = \dfrac{R_k - R_n}{\delta R}$.

2. The hypothesis matrix $(IndTr)$ is formed as follows:

$$\begin{cases} IndTr(i,j) = 1, \ j = Vt_{SC}i/\delta R \\ IndTr(i,j) = 0, \ j \neq Vt_{SC}i/\delta R \end{cases} \tag{10}$$

The number of nonzero elements $K_{t\,\arg et}$ in the hypothesis matrix $IndTr$ equals the number of all the target positions in the $(r\text{-}t)$ space:

$$K_{t\,\arg et} = \sum_{i=1}^{N}\sum_{j=1}^{N} IndTr(i,j)/\ IndTr(i,j) \neq 0 \tag{11}$$

3. The process of target detection in each cell of the $(r\text{-}t)$ space is simulated. The detection is carried out by a CFAR algorithm (Behar, 1997). As a result, the following matrix who's each element indicates whether the target is detected or not in the corresponding cell of the $(r\text{-}t)$ space, is formed:

$$Det^{q}(i,j) = \begin{cases} 1, t\arg et\ is\ \det ected \\ 0, t\arg et\ is\ not\ \det ected \end{cases} \tag{12}$$

where q is the simulation cycle number.

4. The $(\rho\text{-}\theta)$ parameter space is sampled. It is a matrix of size $[K \times L]$. The parameters K and L are determined by the number of discrete values of the θ parameter, which is sampled in the interval (θ_1, θ_2) by sampling step $\Delta\theta$, and the size of the $(r\text{-}t)$ space.

$$K = 2\sqrt{N^2 + M^2}; \quad L = \frac{\theta_2 - \theta_1}{\Delta\theta} \tag{13}$$

5. All the nonzero elements of the matrix Det^q are performed using the Hough transform. In such a way, the $(r\text{-}t)$ space is mapped into the $(\rho\text{-}\theta)$ parameter space resulting into the matrix $\{Ht\}^q_{K,L}$.

6. Target trajectory detection is simulated. This is done by comparing the value of each element of the parameter space, i.e. of the matrix $\{Ht\}^q_{K,L}$, with the fixed threshold T_M. It means that the decision rule "T_M out of N_{sc}" is applied to each element in the parameter space. According to this criterion, the linear target trajectory specified as a point $(\hat{\rho}, \hat{\theta})$ in the Hough parameter space is detected if and only if the value $Ht^q(\hat{\rho}, \hat{\theta})$ exceeds the threshold T_M.

$$DetHo^q(i, j) = \begin{cases} 1, & Ht^q(i, j) > T_M \\ 0, & \text{otherwise} \end{cases} \tag{14}$$

7. In order to estimate the probability characteristics, steps 3-6 are repeated N_q times where N_q is the number of Monte Carlo runs.

The false alarm probability in the $(r\text{-}t)$ space is estimated as:

$$\hat{P}_{fa} = \frac{1}{(NM - K_{target})N_q} \sum_{i=1}^{N} \sum_{j=1}^{M} \sum_{l=1}^{N_q} \{Det^q(i, j)/ IndTr(i, j) \neq 1\} \tag{15}$$

The target detection probability in the $(r\text{-}t)$ space is estimated as:

$$\hat{P}_d = \frac{1}{K_{target} + N_q} \sum_{i=1}^{N} \sum_{j=1}^{M} \sum_{l=1}^{N_q} \{Det^q(i, j)/ IndTr(i, j) = 1\} \tag{16}$$

The false alarm probability in the $(\rho\text{-}\theta)$ space is estimated as:

$$\hat{P}_{FA} = \frac{1}{KLN_q} \sum_{i=1}^{K} \sum_{j=1}^{L} \{DetHo^q(i, j)/ i \neq I, j \neq J\} \tag{17}$$

The probability of trajectory detection in the $(\rho\text{-}\theta)$ space is estimated as:

$$\hat{P}_D = \max_{I,J} \sum_{l=1}^{N_q} DetHo^q(i, j) \tag{18}$$

4. Performance analysis of a single-channel polar Hough detector with a CFAR processor

Simple examples illustrate the advantages of the PHT in situations, when the target moves with a non-uniform velocity along arbitrary linear trajectories. It is the case when the polar

Hough transform is very convenient for joint target and trajectory detection (Garvanov, et al., 2006; Garvanov, et al., 2007). The input data of the CFAR processor is simulated using the average noise power of 1, the signal-to-noise ratio of 20dB, the probability of impulse noise appearance of 0.1 and the average interference-to-noise ratio of 30dB (Fig. 12). The signal received from a target is corrupted by the random impulse interference and the receiver noise. After CFAR processing, the output matrix includes 1 or 0 depending on that whether the signal is detected or not detected (Fig. 13). The heavy interference environment is characterized by the number of false alarms.

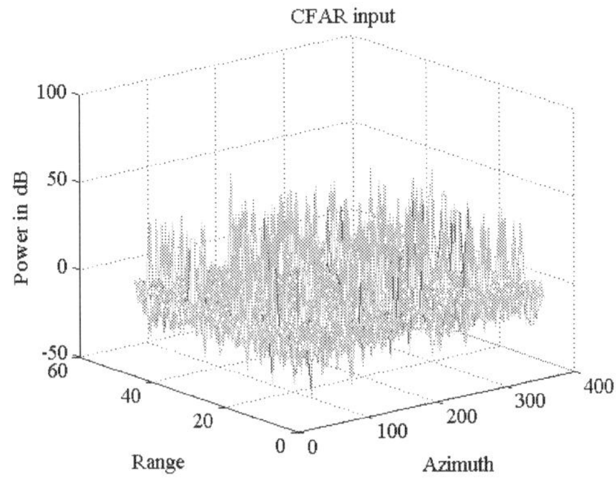

Fig. 12. Range-azimuth matrix including noise, RAII, and signal. It is the input for the CFAR processor

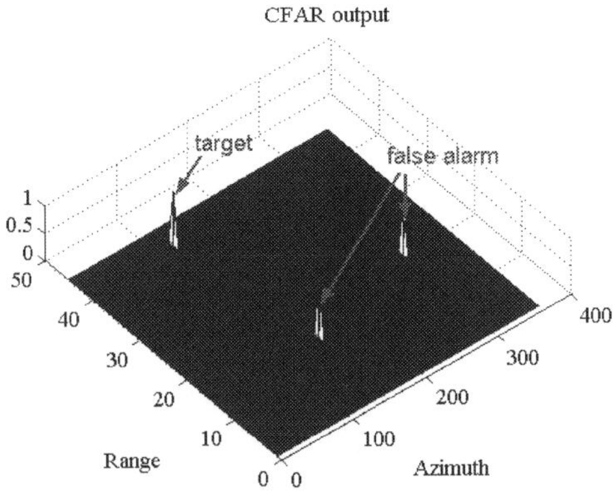

Fig. 13. Range-azimuth matrix after CFAR processing. It includes 0 and 1

The target changes its velocity and moves at different azimuths. After N_{SC} radar scans, as a result of the CFAR processing the matrix of target detections in the range-azimuth plane is formed (Fig. 14). The matrix includes the true points of the target trajectory and false alarms. This information fills the range-azimuth space, which is the input for the polar Hough transform. The positions of a target in the polar coordinate system after N_{SC} radar scans are shown on Fig. 15. This result is equivalent to Fig. 14.

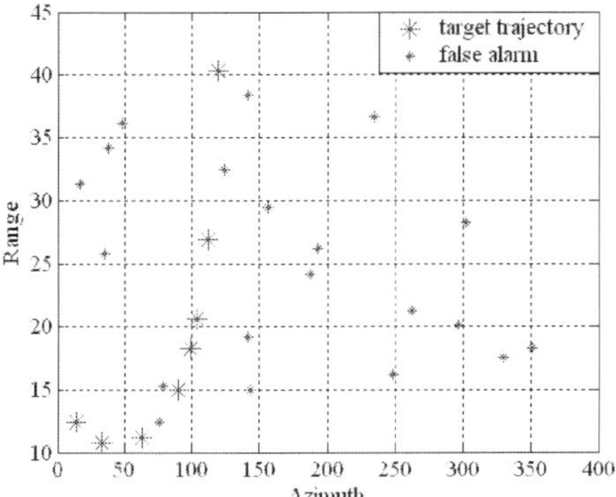

Fig. 14. An observation of a target in the range-azimuth plane after N_{SC} radar scans

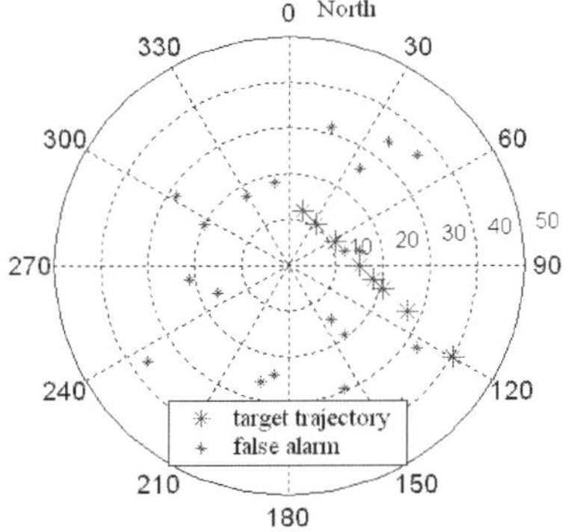

Fig. 15. An observation of a target in polar coordinate system after N_{SC} radar scans

The polar Hough transform maps points (targets and false alarms) from the observation space (polar data map) into curves in the polar Hough parameter space, termed as the (ρ-θ) space (Fig. 16). The results of transformation are sinusoids with unit magnitudes. Each point in the polar Hough parameter space corresponds to one line in the polar data space with parameters ρ and θ. A single ρ-θ point in the parameter space corresponds to a single straight line in the *range-azimuth* data space with these ρ and θ values. Each cell from the polar parameter space is intersected by a limited set of sinusoids obtained by the polar Hough transform. The sinusoids obtained by the transform are integrated in the Hough parameter space after each of radar scans (Fig. 17).

In each cell of the Hough parameter space is performed binary integration and comparison with the detection threshold. If the number of binary integrations (BI) in the polar Hough parameter space exceeds the detection threshold, target and linear trajectory detection is indicated. Target and linear trajectory detection is carried out cell by cell in the entire polar Hough parameter space. In order to compare the effectiveness of the two detectors, the CFAR BI detector (Fig. 18 and 19) and the Hough detector with a CFAR BI processor (Fig. 20 and 21) their performance is evaluated for the case of RAII using the two methods, analytical and Monte Carlo.

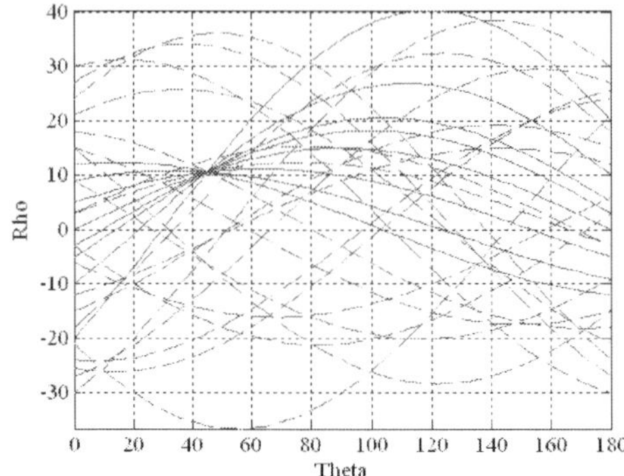

Fig. 16. Hough parameter space showing the sinusoids corresponding to the data point from Fig. 14

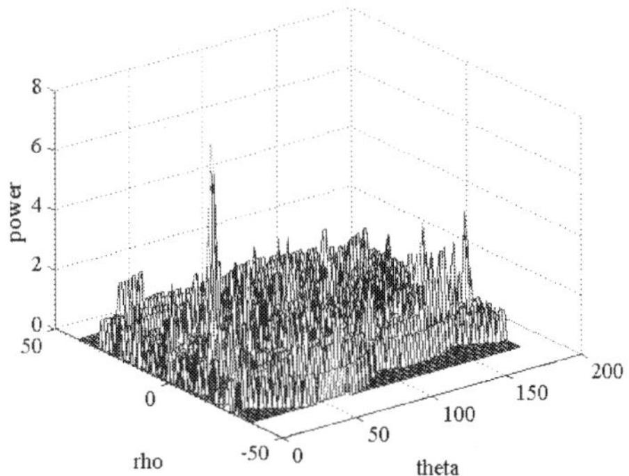

Fig. 17. Binary integration of data in Hough parameter space for example shows on Fig. 15

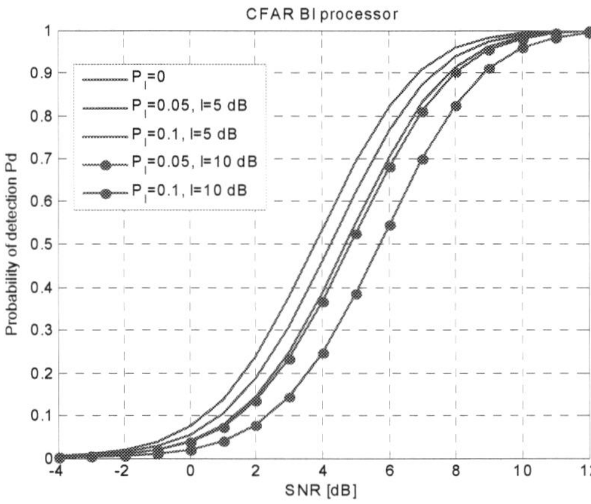

Fig. 18. Probability of detection of CFAR BI processor (analytically calculated)

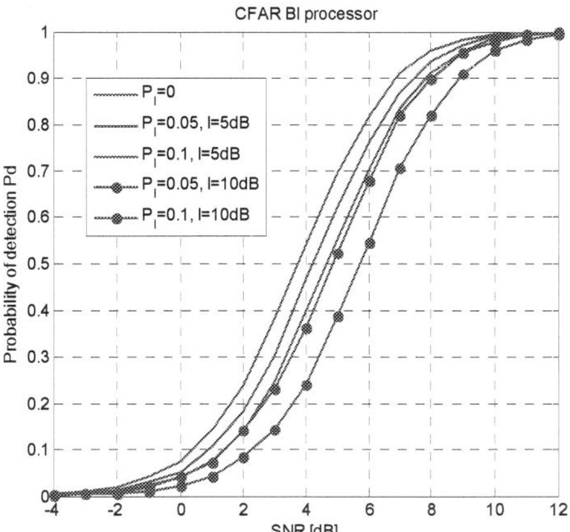

Fig. 19. Probability of detection of CFAR BI processor (Monte Carlo simulation)

The effectiveness of both detectors is expressed in terms of the detection probability, which is calculated as a function of the signal-to-noise ratio (SNR). The probability of detection is calculated using the same parameters for both detectors. These parameters are: the probability of false alarm - 10^{-4}, the decision rule in the polar data space is "10-out-of-16", the decision rule in the Hough parameter space is "7-out-of 20", the interference-to-noise ratio is I=5,10dB, and the probability of interference appearance is P_I=0; 0.05 and 0.1 (Garvanov, 2003; Doukovska, 2005; Doukovska, 2006; Doukovska, 2007; Garvanov, 2007; Doukovska, 2008).

Analysis of the graphical results presented in Fig. 18-21 shows that the calculations of the probability of detection using the two different approaches analytical and Monte Carlo produce the same results. This provides reasons enough to use the simulation method

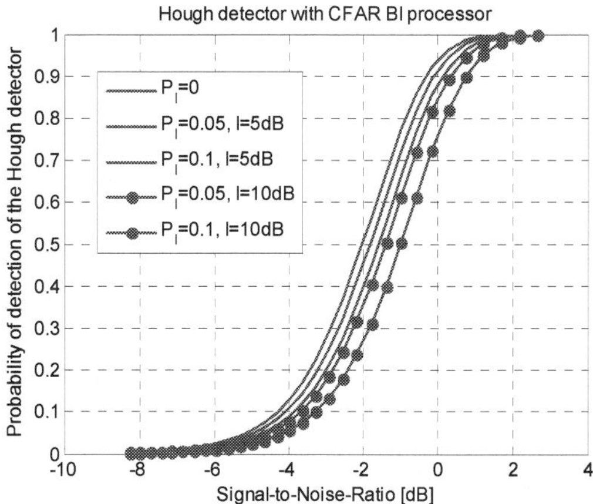

Fig. 20. Probability of detection of CFAR BI with Hough (analytically calculated)

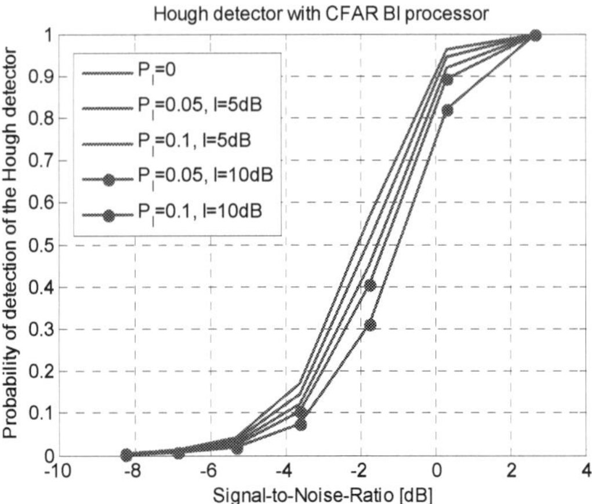

Fig. 21. Probability of detection of CFAR BI with Hough (Monte Carlo simulation)

described in the previous section for analysis of Hough detectors with the other CFAR processors. It can be also concluded that the combination of the two detectors, CFAR and Hough, improves the joint target and trajectory detectability in conditions of RAII.

5. Multi-sensor (multi-cannel) polar Hough detector with a CFAR processor

The target detectability can be additionally improved by applying the concept of multi-sensor detection. The important advantage of this approach is that both the detection probability and the speed of the detection process increase in condition of RAII (Garvanov, 2007; Kabakchiev, 2007; Garvanov, 2008; Kabakchiev, 2008). The speeding of the detection process and the number of channels are directly proportional quantities. The usage of multiple channels, however, complicates the detector structure and requires the data association, the universal timing and the processing in the universal coordinate system. Three radar systems with identical technical parameters are considered in this chapter. Three variants of a multi-sensor Hough nonsynchronous detector in a system with three radars are developed and studied. The first of them is the decentralized (distributed) Hough detector with track association (DTA), whose structure is shown in Fig. 22.

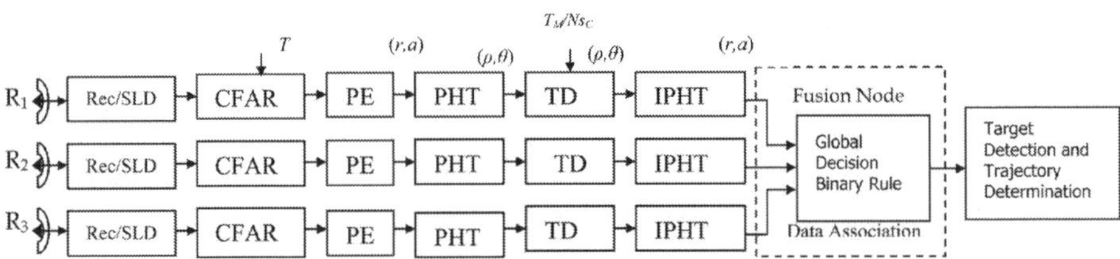

Fig. 22. Decentralized (distributed) Hough detector, with track association (DTA)

The structure of DTA Hough detector shown in Fig. 22 consists of three single-channel detectors described in Section 4. The final detection of a target trajectory is carried out after association of the output data of channels, where a target trajectory was detected or not detected. The signal processing carried out in each channel includes optimum linear filtration, square law detection (SLD), CFAR detection, plot extraction (PE), PHT, inverse PHT, data association in the fusion node and finally, target detection and trajectory estimation. In each channel, the range-azimuth observation space is formed after N_{SC} radar scans. After CFAR detection, using the PHT, all points of the polar data space, where targets are detected, are mapped into curves in the polar Hough parameter space, i.e. the $(\rho\text{-}\theta)$ parameter space. In the Hough parameter space, after binary integration of data, both target and linear trajectory are detected (TD) if the result of binary integration exceeds the detection threshold. The polar coordinates of the detected trajectory are obtained using the inverse PHT of $(\rho\text{-}\theta)$ coordinates. Local decisions are transferred from each channel to the fusion node where they are combined to yield a global decision. In many papers, the conventional algorithms for multi-sensor detection do not solve problems of data association in the fusion node, because it is usually assumed that the data are transmitted without loses. Here, the details related to the data association and the signal losses in the fusion node are reviewed and the problems of the centralized signal processing in a signal processor are considered. Here are provided the size of signal matrixes and their cells; range and azimuth resolution; and data association. Signal detection in radar is done in range-azimuth resolution cells of definite geometry sizes. Detection trajectory is done in the Hough parameter space with cells of specified sizes. The global decision in the fusion node of a radar system is done in result from association of signals or data in a unified coordinate system. The unified coordinate system of the range-azimuth space predicts cell's size before association of data received from different radars. The radar system is synchronized by determining the scale factor used in a CFAR processor, the size of the Hough parameter space and the binary decision rule of the Hough detector.

Unlike the decentralized structure of a multi-sensor Hough detector, in the DPA Hough detector the process of data association is carried out in the global $(r\text{-}t)$ space of a Hough detector. The global $(r\text{-}t)$ space associates coordinates of the all detected target (plots) in radars, i.e. associates all the data at the plot extractor outputs, as shown in Fig. 23.

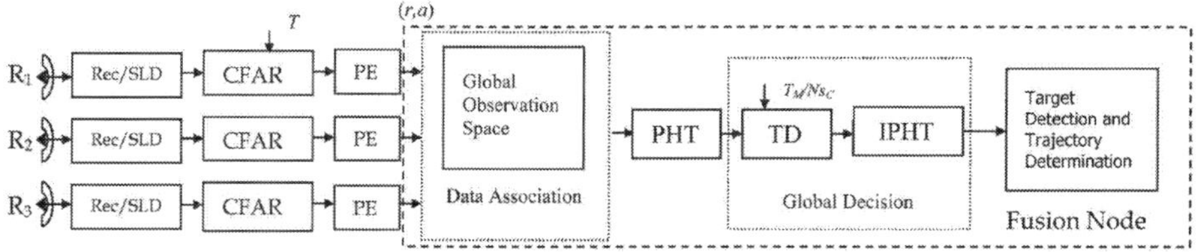

Fig. 23. Decentralized Hough detector, with plot association (DPA)

It can be seen that the decentralized plot association (DPA) Hough detector has a parallel multi-sensor structure. In each channel, the local polar observation space, i.e. (r,a) are the polar coordinates of detected targets, is formed. All coordinate systems associated with radars are North oriented, and the earth curvature is neglected. At the first stage the local polar observation spaces of radars are associated to the Global Coordinate system resulting into the Global polar observation space. At the second stage, the polar Hough transform is applied to the global observation space and then the trajectory detection is performed in each cell of the Hough parameter space. The polar coordinates of the detected trajectory are obtained using the inverse polar Hough transform (IPHT) applied to the Hough parameter space.

All factors, such as technical parameters of radar, coordinate measurement errors, rotation rate of antennas and etc. are taken into account when sampling the Hough parameter space. The probability characteristics of such a system are better that those of the decentralized Hough detector.

The third structure of a multi- sensor Hough detector called as the centralized Hough detector is the most effective for target trajectory detection (Fig. 24). In this multi-sensor detector data association means association of signals processed in all channels of a system. The effectiveness of a centralized Hough detector is conditioned by the minimal information and energy losses in the multi-sensor signal processing. However, a centralized Hough detector requires the usage of fast synchronous data buses for transferring the large amount of data and the large computational resources.

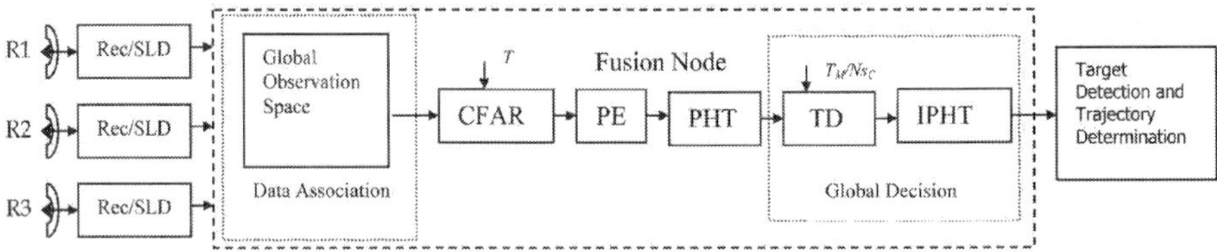

Fig. 24. Centralized Hough detector, with signal association (CSA)

In such a multi-sensor Hough detector, the received signals are transferred from all receive/transmit stations (MSRS) to the fusion node. The global polar observation space is formed after N_{SC} radar scans. After CFAR detection, using the polar Hough transform (PHT), all points of the global polar observation space, where targets are detected, are mapped into curves in the polar Hough parameter space, i.e. the $(\rho\text{-}\theta)$ parameter space. The global target and linear trajectory detection is done using the binary decision rule "M-out-of -N_{SC}". The polar coordinates of the detected trajectory are obtained using the inverse polar Hough transform (IPHT) applied to the Hough parameter space.

6. Performance analysis of a multi-sensor polar Hough detector with a CFAR processor

The first example, given in this section, illustrates the advantages of a three-radar system that operates in the presence of randomly arriving impulse interference. The three radars have the same technical parameters as those in (Carlson et al., 1994; Behar et al. 1997; Behar & Kabakchiev, 1998; Garvanov, 2007; Kabakchiev, 2007; Kabakchiev, 2008; Garvanov, 2008). The radar positions form the equilateral triangle, where the lateral length equals 100km. The performance of a multi-sensor polar Hough detector is evaluated using Monte Carlo simulations. The simulation results are obtained for the following parameters:
-      Azimuth of the first radar - 45⁰;
-      Target trajectory - a straight line toward the first radar;
-      Target velocity – 1 Mach;
-      Target radar cross section (RCS) - 1 sq. m;
-      Target type - Swerling II case;

- Average SNR is calculated as $S=K/R^4 \cong 15dB$, where $K=2.07*10^{20}$ is the generalized power parameter of radar and R is the distance to the target;
- Average power of the receiver noise - $\lambda_0=1$;
- Average interference-to-noise ratio for random interference noise - $I=10dB$;
- Probability of appearance of impulse noise – $P_I=0.033$;
- Size of a CFAR reference window - $N=16$;
- Probability of false alarm in the Hough parameter space - $P_{FA}=10^{-2}$;
- Number of scans - $N_{SC}=20$;
- Size of an observation area – 100 x 30 (the number of range resolution cells is 100, and the number of azimuth resolution cells is 30);
- Range resolution - 1 km;
- Azimuth resolution -2°;
- Size of the Hough parameter space - 91 x 200 (θ cells -91, and ρ cells – 200);
- Sampling in θ - 2°;
- Sampling in ρ - 1km;
- Binary detection threshold in the Hough parameter space - $T_M=2\div20$.

The performance of the three multi-sensor polar Hough detectors, centralized (CSA), decentralized (DTA) and decentralized (DPA), are compared against each other. The detection performance is evaluated in terms of the detection probability calculated for several binary decision rules applied to the Hough parameter space. The simulation results are plotted in Fig. 25. They show that the detection probability of a centralized detector is better than that of a distributed detector. It can be seen that the detection probability of the two types of detectors, centralized and decentralized, decreases with increase of binary decision rules (T_M/N_{SC}). The maximum detection probability is obtained when the binary decision rule is 7/20.

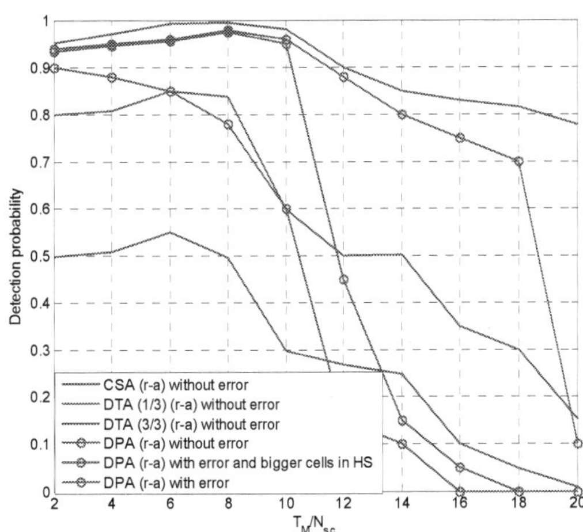

Fig. 25. Detection probability of the three multi-sensor Hough detectors - centralized (CSA), decentralized (DTA) and decentralized (DPA) detectors for different binary rules in Hough parameter space

These results are in accordance with the results obtained for a single-channel Hough detector (for a single radar) operating in the interference-free environment as in (Finn & Johnson, 1968; Doukovska, 2005). The detection probability is low, because the input average SNR≅15dB and it is commensurate with value of INR=10dB. The results have shown that the detection probability of the TBD Polar Hough Data Association detector is between the curves of detector with binary rules in distributed Hough detector 1/3 - 3/3.

It is apparent from Fig. 25 that the potential curve of a decentralized (DPA) Hough detector (Fig. 23) is close to the potential curve of the most effective multi-sensor centralized Hough detector (Fig. 24). It follows that the effective results can be achieved by using the communication structures with low-rate-data channels. The target coordinate measurement errors in the (r-t) space mitigate the operational efficiency of multi- sensor Hough detectors. The needed operational efficiency requires the appropriate sampling of the Hough parameter space.

The second example is done in order to compare the effectiveness of the two Hough detectors, single-channel and three-channel decentralized (DPA), operating in conditions of RAII. The effectiveness of each detector is expressed in terms of the probability of detection calculated as a function of the signal-to-noise ratio. The detection probability of these Hough detectors is presented in Fig. 26. The detection probability is plotted for the case when the random pulse appearance probability is in range from 0 to 0.1 (P_l).

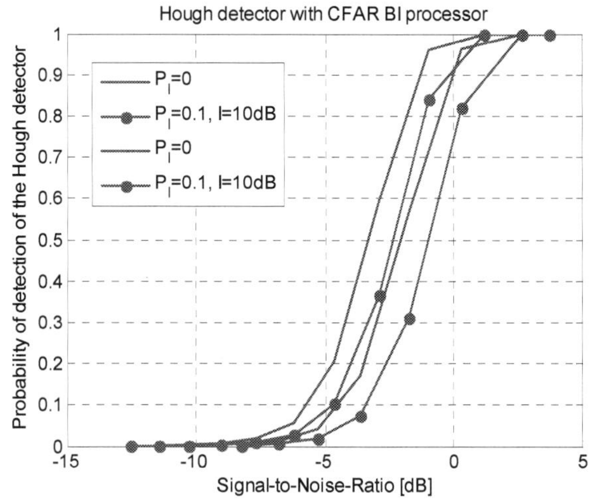

Fig. 26. Detection probability of two Hough detectors, single-channel (dash line) and three-channel decentralized with plot association (solid line) with Monte Carlo simulation analysis

It is obvious from the results obtained that in conditions of RAII a multi-sensor Hough detector is more effective than a single-channel one. The higher effectiveness is achieved at the cost of complication of a detector structure.

7. Conclusions

In this study, a new and more suitable modification of the Hough transform is presented. The polar Hough transform allows us to employ a conventional Hough detector in such real situations when targets move with variable speed along arbitrary linear trajectories and clutter and randomly arriving impulse interference are present at the detector input. The polar Hough transform is very comfortable for the use in search radar because it can be directly applied to the output search radar data. Therefore, the polar Hough detectors can be attractive in different radar applications.

It is shown that the new Hough detectors increase probabilities, detection and coincidence, when the target coordinates are measured with errors.

Three different structures of a multi-sensor polar Hough detector, centralized (CSA) and decentralized (DPA), are proposed for target/trajectory detection in the presence of randomly arriving impulse interference. The detection probabilities of the multi-sensor Hough detectors, centralized and decentralized, are evaluated using the Monte Carlo approach. In simulations, the radar parameters are synchronized in order to maintain a constant false alarm rate. The results obtained show that the detection probability of the centralized detector is higher than that of the decentralized detector.

The results obtained shows that the required operational efficiency of detection can be achieved by using communication structures with low-rate-data channels. The target coordinate measurement errors in the $(r\text{-}t)$ space mitigate the operational efficiency of multi-sensor Hough detectors. The needed operational efficiency requires the appropriate sampling of the Hough parameter space.

The proposed multi-sensor Hough detectors are more effective than conventional single-channel ones due to the usage of the Hough transform for data association. This operation increases the effectiveness of trajectory detection in the presence of randomly arriving impulse interference.

8. Acknowledgment

This work was partially supported by projects: IIT-010089/2007, DO-02-344/2008, BG051PO001/07/3.3-02/7/17.06.2008, MU-FS-05/2007, MI-1506/2005.

9. References

Behar V., Chr. Kabakchiev, Hough detector with adaptive non-coherent integration for target detection in pulse jamming, In: *Proc. of IEEE 5-th inter. symp., ISSSTA'98*, Sun City, South Africa, vol. 3, 1998, pp. 1003-1007, ISBN:0-7803-4281-X.

Behar V., Chr. Kabakchiev, L. Doukovska, Adaptive CFAR PI Processor for Radar Target Detection in Pulse Jamming, *Journal of VLSI Signal Processing*, vol. 26, 2000, pp. 383 – 396, ISSN:0922-5773.

Behar V., L. Doukovska, Chr. Kabakchiev, Target Detection and Estimation using the Hough Transform, *Lecture Notes in Computer Science,* Springer-Verlag Berlin Heidelberg, LNCS 4310, pp. 525-532, 2007, ISBN:978-3-540-70940-4.

Behar V., L. Doukovska, Chr. Kabakchiev, H. Rohling, Comparison of Doppler and Hough Target Velocity Estimation Techniques, *Proc. of the International Radar Symposium – 2007*, Cologne, Germany, 2007, pp. 157-162.

Behar V., B. Vassileva, Chr. Kabakchiev, Adaptive Hough detector with binary integration in pulse jamming, *Proc. of Int. Conf. ECCTD'97*, Budapest, 1997, pp. 885-889, ISBN:963-420-5232.

Carlson B., E. Evans, S. Wilson, Search Radar Detection and Track with the Hough Transform, *IEEE Trans., vol. AES - 30.1.1994*, Part I, pp. 102-108; Part II, pp. 109-115; Part III, pp. 116-124, ISSN:0018-9251.

Doukovska L., Hough Detector with Binary Integration Signal Processor, *Comptes rendus de l'Academie bulgare des Sciences*, vol. 60, №5, 2007, pp. 525-533, ISSN:0861-1459.

Doukovska L., Hough Transform in the CFAR Algorithms in Presence of Randomly Arriving Impulse Interference, *PhD Thesis, Institute of Information Technologies*, Bulgarian Academy of Sciences, Bulgaria, 2005, Thesis leader: prof. Chr. Kabakchiev, (in Bulgarian).

Doukovska L., Chr. Kabakchiev, Performance of Hough Detectors in Presence of Randomly Arriving Impulse Interference, *Proc. of the International Radar Symposium – 2006*, Krakow, Poland, 2006, pp. 473-476, ISBN: 978-83-7207-621-2.

Doukovska L., Chr. Kabakchiev, V. Kyovtorov, I. Garvanov, Hough Detector with an OS CFAR Processor in Presence of Randomly Arriving Impulse Interference, *Proc. of the 5-th European Radar Conference - EuRAD*, Amsterdam, Holland, 2008, pp. 332-335., ISBN: 978-2-87487-009-5

Doukovska L., V. Behar, Chr. Kabakchiev, Hough Detector Analysis by means of Monte Carlo Simulation Approach, *Proc. of the International Radar Symposium – 2008*, Wroclaw, Poland, 2008, pp. 103-106, ISBN: 978-83-7207-757-8

Finn H. M., R. S. Johnson, Adaptive detection mode with threshold control as a function of spatially sampled clutter estimation, *RCA Review*, 29, 3, 1968, pp. 414-464.

Gandhi, P. P., S. A. Kassam, Analysis of CFAR processors in nonhomogeneous background, IEEE Trans., vol. AES-24, No 4, 1988, pp. 443-454.

Garvanov I., Methods and algorithms for keeping constant false alarm rate in the presence of pulse jamming, *PhD Thesis, Institute of Information Technologies, Bulgarian Academy of Sciences*, Bulgaria, 2003, Thesis leader: prof. Chr. Kabakchiev, http://www.iit.bas.bg/staff_en/I_Garvanov/Dissertation_en.pdf.

Garvanov I., Chr. Kabakchiev, Radar Detection and Track Determination with a Transform Analogous to the Hough Transform, *Proc. of the International Radar Symposium – 2006*, Krakow, Poland, 2006, pp. 121-124, ISBN: 978-83-7207-621-2

Garvanov I., Chr. Kabakchiev, Radar Detection and Track in Presence of Impulse Interference by using the Polar Hough Transform, *Proc. of the European Microwave Association*, Vol. 3, March 2007, pp. 170-175, ISBN 88-8492-324-7

Garvanov I., Chr. Kabakchiev, Sensitivity of API CFAR Detector Towards the Change of Input Parameters of Pulse Jamming, *Proc. of the International Radar Symposium 2004*, Warsaw, Poland, 2004, pp. 233-238.ISSN: 0885-8985

Garvanov I., Chr. Kabakchiev, Sensitivity of Track before Detect Multiradar System toward the Error Measurements of Target Parameters, *Cybernetics and Information Technologies*, Volume 7, № 2, 2007, pp. 85-93. ISSN: 1311-9702

Garvanov I., Chr. Kabakchiev, H. Rohling, Detection Acceleration in Hough Parameter Space by K-stage Detector", Proc. of the Int. Conf. *Numerical Methods and Applications – 2006*, Borovets, Bulgaria, LNCS 4310, 2007, pp. 558-565. ISBN:978-3-540-70940-4

Garvanov I., Chr. Kabakchiev, L. Doukovska, V. Kyovtorov, H. Rohling, Improvement in Radar Detection Through Window Processing in the Hough Space, *Proc. of the International Radar Symposium – 2008*, Wroclaw, Poland, 2008, pp. 139-144, ISBN: 978-83-7207-757-8

Garvanov I., V. Behar, Chr. Kabakchiev, CFAR Processors in Pulse Jamming, *Proc. of the Int. Conf. "Numerical Methods and Applications – 02"*, Lectures Notes and Computer Science –LNCS 2542, 2003, pp. 291-298, ISBN:978-3-540-70940-4

Goldman H., Performance of the Excision CFAR detector in the presence of interferers, *IEE Proc.*, vol. 137, Pt.F, (3), 1990, pp. 163-171, ISSN 0956-375X

Himonas S., CFAR Integration Processors in Randomly Arriving Impulse Interference, *IEEE Trans.*, vol. AES-30, No 3, 1994, pp. 809-816.

Kabakchiev Chr., V. Behar, CFAR Radar Image Detection in Pulse Jamming, *IEEE Fourth Int. Symp. ISSSTA'96*, Mainz, Germany, 1996, pp 182-185.

Kabakchiev Chr., V. Behar, Techniques for CFAR Radar Image Detection in Pulse Jamming, *26-th Europ. Microwave Conf. EuMC'96*, Prague, Czech Republic, 1996, pp 347-352.

Kabakchiev Chr., I. Garvanov, H. Rohling, Netted Radar Hough Detector in Randomly Arriving Impulse Interference, *Proc. of the IET International Conference on Radar Systems, RADAR 2007*, UK, 2007, CD ROM 7a.1, pp.5, ISSN: 0537-9989, ISBN: 978-0-86341-849-5

Kabakchiev Chr., I. Garvanov, L. Doukovska, Excision CFAR BI detector with Hough transform in Randomly Arriving Impulse Interference, *Proc. of the International Radar Symposium-2005*, Berlin, Germany, 2005, pp. 259 – 264.

Kabakchiev Chr., I. Garvanov, L. Doukovska, V. Kyovtorov, Comparison between Two Multi-Sensor Fusion Techniques for Target Height Finding, *Proc. of the International Radar Symposium – 2007*, Cologne, Germany, 2007, pp. 809-813.

Kabakchiev Chr., I. Garvanov, L. Doukovska, V. Kyovtorov, H. Rohling, Data Association Algorithm in Multiradar System, *Proc. of the 2008 IEEE Radar Conference, Rome*, Italy, 2008, pp. 1771-1774. ISSN: 1097-5659, ISBN: 1-4244-1593-

Kabakchiev Chr., I. Garvanov, L. Doukovska, V. Kyovtorov, H. Rohling, Data Association Algorithm in TBD Multiradar System, *Proc. of the International Radar Symposium – 2007*, Cologne, Germany, 2007, pp. 521-525.

Kabakchiev Chr., I. Garvanov, V. Kyovtorov Height Finding Based on Networks of Radar Systems, *Proc. of the International Radar Symposium - 2005*, Berlin, Germany, 2005, pp. 433-438.

Kabakchiev Chr., V. Kyovtorov, Doukovska L., I. Garvanov, TBD approaches though HT for multi-antennae target elevation measurement, *International Radar Symposium – IRS'09*, Hamburg, Germany, 2009 (on review).

Kyovtorov V., Detection and Assessment of Target Coordinates in a Radar Sensor Network, *PhD Thesis, Technical University – Sofia*, Bulgaria, 2007, Thesis leader: prof. Chr. Kabakchiev, (in Bulgarian).

Rohling H., Radar CFAR Thresholding in Clutter and Multiple Target Situations, *IEEE Trans.*, vol. AES-19, No 4, 1983, pp. 608-621.

11

Tracking of Flying Objects on the Basis of Multiple Information Sources

Jacek Karwatka
Telecommunications Research Institute
Poland

1. Introduction

In the paper, a new approach to the estimation of the localization of flying objects on the basis of multiple radiolocation sources has been proposed. The basic purpose of the radiolocation is the detection and tracking of objects. The main task of a tracking system is the estimation and prediction of localization and motion of tracked objects. In the classical tracking systems, the estimation of localization and motion of tracked objects is based on the series of measurements performed by a single radar. Measurements obtained from other sources are not taken directly into account. It is possible to assess the estimate of a tracking object on the basis of the different sources of information, applying, for example, least squares method (LSM). However, such solution is seldom applied in practice, because necessary formulas are rather complicated. In this paper, a new approach is proposed. The key idea of the proposed approach is so called matrix of precision. This approach makes possible tracking not only on the basis of radar signals, but also on the basis of bearings. It makes also possible the tracking of objects on the basis of multiple information sources. Simplicity is the main attribute of proposed estimators. Their iterative form corresponds well with the prediction-correction tracking model, commonly applied in radiolocation. In the paper numerical examples presenting advantages of the proposed approach are shown.
The paper consists of seven parts. Introduction is the first one.
In the second part the idea of the matrix of precision is presented and it is demonstrated how it can be used to uniformly describe the dispersion of measurement. Traditionally, the measurement dispersion is described by a matrix of covariance. Formally, the matrix of precision is an inverse of the matrix of covariance. However, these two ways of description are not interchangeable. If an error distribution is described by the singular matrix of precision then the corresponding matrix of covariance does not exist, more precisely, it contains infinite values. It means in practice that some components of a measured vector are not measurable, in other words, an error of measurement can be arbitrarily large. The application of the matrix of precision makes possible an uniform description of measurements taken from various sources of information, even if measurements come from different devices and measure different components. Zero precision corresponds to components which are not measured.
In the third part, the problem of estimation of stationary parameters is formulated. Using the matrix of precision, the simple solution of the problem is presented. It appears that the

best estimate is a weighted mean of measurements, where the weights are the matrices of the precision of measurements. It has been proved that the proposed solution is equivalent to the least squares method (LSM). Additionally it is simple and scalable.

In the fourth part of this paper the problem of the estimation of states of dynamic systems, such as a flying aircraft, is formulated. Traditionally, for such an estimation the Kalman filter is applied. In this case the uncertainty of measurement is described by the error matrix of covariance. If the matrix of precision is singular, it is impossible to determine the corresponding matrix of covariance and utilize the classical equation of Kalman filter. In the presented approach this situation is typical. It appears that there is such a transformation of Kalman filter equations, that the estimation based on the measurements with error described by the singular matrix of precision, can be performed. Such a transformation is presented and its correctness is proved.

In the fifth part, numerical examples are presented. They show the usability of the concept of matrix of precision.

In the sixth part, the summary and conclusions are shown, as well as the practical application of presented idea is discussed. The practical problems which are not considered in the paper are also pointed out.

2. The concept of precision matrix

Traditionally in order to describe the degree of the dispersion distribution the covariant matrix is used. There exist another statistics which can be used to characterize dispersion of the distribution. However, the covariance matrix is the most popular one and in principle the precision matrix is not used. Formally the precision matrix is the inverse of covariance matrix.

$$\mathbf{W}_x = \mathbf{C}_x^{-1} \tag{1}$$

The consideration may take much simpler form if the precision matrix is used. Note that precision matrix is frequently used indirectly, as inverse of covariance matrix. For example, in the well-known equation for the density of multi-dimensional Gauss distribution:

$$\sqrt{\frac{\left|\mathbf{C}^{-1}\right|}{(2\pi)^k}}\, e^{\left(-\frac{(\mathbf{x}-\mathbf{m_x})^T \mathbf{C}^{-1}(\mathbf{x}-\mathbf{m_x})}{2}\right)} \tag{2}$$

The equation (1) can not be used if the covariance matrix is singular. As long as the covariance (precision) matrix is not singular, the discussion which statistics is better to describe degree of the dispersion distribution is as meagre as discussion about superiority of Christmas above Easter. Our interest is in analysis of extreme cases (e.g. singular matrix).

At first, the singular covariance matrix will be considered. Its singularity means that some of the member variables (or their linear combinations) are measured with error equal to zero. Further measurement of this member variable makes no sense. The result can be either the same, or we shall obtain contradiction if the result would be different. The proper action in such a case is modifying the problem in such a manner that there is less degrees of freedom and the components of the vector are linearly independent. The missing components computation is based on the linear dependence of the variables. In practice, the presence of the singular covariance matrix means that the linear constraints are imposed on the components of the measured vectors and that the problem should be modified. The second possibility (infinitely accurate measurements) is impossible in the real world.

In distinction from the case of the singular covariance matrix, when the measurements are described by the singular precision matrix, measurements can be continued. The singularity of the precision matrix means that either the measurement does not provide any information of one of components (infinitive variation) or there exist linear combinations of member variables which are not measured. This can results from the wrong formulation of the problem. In such a case, the system of coordinates should be changed in such a manner that the variables which are not measured should be separated from these which are measured. The separation can be obtained by choosing the coordination system based on the eigenvectors of precision matrix. The variables corresponding to non-zero eigenvalues will then be observable.

There is also the other option. The singular precision matrix can be used to describe the precision of measurement in the system, in which the number of freedom degrees is bigger then the real number of components, which are measured. Then all measurements may be treated in a coherent way and the measurements from various devices may be taken into account. Thus all measurements from the devices which do not measure all state parameters can be included (i.e. direction finder). In this paper we are focusing on second option.

Each measuring device uses its own dedicated coordinate system. To be able to use the results of measurements performed by different devices, it is necessary to present all measured results in the common coordinate system. By changing coordinate system the measured values are changed as well. The change involves not only the digital values of measured results, but also the precision matrix describing the accuracy of particular measurement. We consider the simplest case namely the linear transformation of variables. Let $\mathbf{y}$ denote the vector of measurements in common coordinate system and $\mathbf{x}$ denote the vector of measurements in the measuring device dedicated coordinate system. Let $\mathbf{x}$ and $\mathbf{y}$ be related as follows:

$$\mathbf{y} = \mathbf{Ax} + \mathbf{b} \tag{3}$$

In this case the covariance of measurement changes according to formula:

$$\mathbf{C}_y = \mathbf{AC_x A}^T \tag{4}$$

The precision matrix as inverse of covariance matrix changes according to formula:

$$\mathbf{W_y} = \mathbf{C_y^{-1}} = \mathbf{A}^{T-1}\mathbf{W_x A}^{-1} \tag{5}$$

Consider the case when the transformation $\mathbf{A}$ is singular. In this case, we can act in two ways. If the precision matrix $\mathbf{W_x}$ is not singular it can be inverted and the covariance matrix $\mathbf{C_x}$ can be find. Using the formula (4) and the obtained result $\mathbf{C_y}$ it is possible to invert back, to obtain the precision matrix $\mathbf{W_y}$. The second method consists in increasing transformation $\mathbf{A}$ in such way to be invertible. It is done by writing additional virtual lines to the transformation matrix $\mathbf{A}$. The new transformation matrix is now as follows:

$$\mathbf{A}' = \begin{bmatrix} \mathbf{A} \\ \mathbf{A_v} \end{bmatrix} \tag{6}$$

Additional rows of this matrix should be chosen in such way that the obtained precision matrix has block-diagonal form:

$$\mathbf{W}_y' = \begin{bmatrix} \mathbf{W}_A & \mathbf{0} \\ \mathbf{0} & \mathbf{W}_v \end{bmatrix} \tag{7}$$

it means that additional virtual lines of transformation $\mathbf{A}'$ should be chosen to fulfill the following relationship:

$$\mathbf{A}_v^{T-1} \mathbf{W}_x \mathbf{A}_v^{-1} = \mathbf{0} \tag{8}$$

In such a way, the additional virtual lines do not have any influence on the computation result of precision matrix in the new coordinate system.

Consider the specific case. Let the measuring device measures target OX and OY positions independently, but with different precision. Measurement in the direction of the axis OX is done with accuracy $w_{OX} = 0.25^2$ and in the direction of axis OY measured precision is $w_{OY} = 0.2^2$. Then the precision matrix has a form:

$$\mathbf{W}_x = \begin{bmatrix} 0.25^2 & 0 \\ 0 & 0.2^2 \end{bmatrix} \tag{9}$$

We are interested in accuracy of the new variable which is a linear combination of these variables. Let $y = \mathbf{A}x$ where:

$$\mathbf{A} = \begin{bmatrix} -1 & 2 \end{bmatrix}$$

Matrix $\mathbf{A}$ is not invertible. We add an additional row so that the matrix $\mathbf{A}'$ is invertible and formula (8) is fulfilled. General form of the matrix fulfilling this condition is of the form:

$$\mathbf{A}' = \begin{bmatrix} -1 & 2 \\ 50n & 16n \end{bmatrix} \tag{10}$$

We chose $n = 1$. In that case:

$$\mathbf{A}'^{-1} = \frac{1}{116} \begin{bmatrix} -16 & 2 \\ 50 & 1 \end{bmatrix} \tag{11}$$

The precision matrix in the new coordinate system has the following form:

$$\mathbf{W}_y = \mathbf{A}'^{-1T} \mathbf{W}_x \mathbf{A}'^{-1} = \frac{1}{116} \begin{bmatrix} 1 & 0 \\ 0 & 0.0025 \end{bmatrix} \tag{12}$$

We ignore additional variables which have been added to make $\mathbf{A}'$ invertible. Finally the accuracy of the measurement of variable y is $1/116$.

The same result can be obtained using the first method. Then the covariance matrix is of the form:

$$\mathbf{C}_x = \begin{bmatrix} 4^2 & 0 \\ 0 & 5^2 \end{bmatrix} \tag{13}$$

The value of the covariance of variable y according to formula (4) is:

$$\mathbf{C}_y = \begin{bmatrix} -1 & 2 \end{bmatrix} \begin{bmatrix} 16 & 0 \\ 0 & 25 \end{bmatrix} \begin{bmatrix} -1 \\ 2 \end{bmatrix} = 116 \tag{14}$$

While the second method is more complicated it is nevertheless more general, and permits to find the precision matrix in the new coordinate system even if the precision matrix is singular.

In a case when it is necessary to make nonlinear transformation of variables $\mathbf{y} = \mathbf{f}(\mathbf{x})$, the linearization of transformation is proposed. Instead of matrix $\mathbf{A}$ the Jacobian matrix of transformation $\mathbf{f}$ is used:

$$\mathbf{J} = \begin{bmatrix} \dfrac{\partial \mathbf{f}}{\partial \mathbf{x}} \end{bmatrix} \tag{15}$$

In order to find the value of partial derivative, it is necessary to know point $\mathbf{x}_0$ around which linearization is done.

3. Estimation of stationary parameters

In engineering it is often necessary to estimate certain unknown value basing on several measurements. The simplest case is, when all measurements are independent and have the same accuracy. Then, the intuitive approach i.e. using the representation of results as arithmetic average of measurements is correct (it leads to minimal variation unbiased estimator). The problem is more complicated if the measured value is a vector and individual measurements have been performed with different accuracy. Usually, to characterize measurements accuracy the covariance matrix of measurement error distribution is used. Application of precision matrix leads to the formulas which are more general and simple. Further the precision matrix describes the precision of measurements when the number of degrees of freedom is smaller than the dimension of the space in which the measurement has been done. For example, direction finder measures the direction in the three dimensional space. The direction finder can be used as a radar with an unlimited error of measured distance. In this case, the covariance matrix of the measurement does not exist (includes infinite values). But there exists the precision matrix (singular). Information included in the precision matrix permits to build up optimal (minimal variation unbiased) linear estimator.

3.1 Formulation of the problem

Consider the following situation. The series of n measurements $\mathbf{x}_i$ of a vector $\mathbf{x}$ value is done under the following assumptions:
1. There is no preliminary information concerning the value of the measured quantity $\mathbf{x}$.
2. The measured quantity is constant.
3. All measurements are independent.
4. All measurements are unbiased (expected value of the measured error is zero).
5. All measurements can be presented in common coordinate system and the precision matrix $\mathbf{P}_i$ which characterizes the measurement error distribution in these coordinates, can be computed.

The problem is to find the optimal linear estimator of unknown quantity $\mathbf{x}$ and to find its accuracy. By optimal estimator, we understand the unbiased estimator having minimal covariance. By estimation of precision we understand presenting its covariance (or precision) matrix.

3.2 The solution

The first step consists in presenting all measurement results in the common coordinate system. The precision matrices must be recalculated to characterize the measurement error distribution in the common coordinate system. The method of the recalculation the precision (covariance) matrixes in the new coordinated system have been presented before. If all measurements have been done in the same coordinate system, the first step is omitted.

The unbiased minimal variation estimator of unknown quantity $\mathbf{x}$ is a weighed average of all measurements $\mathbf{x}_i$, where the weights are the precision matrices $\mathbf{W}_i$ of the individual measurements. The precision matrix $\mathbf{W}_{\bar{\mathbf{x}}}$ of such estimator is equal to the sum of the precision matrixes $\mathbf{W}_i$ of all measurements.

$$\hat{\mathbf{x}} = \frac{\sum \mathbf{W}_i \mathbf{x}_i}{\sum \mathbf{W}_i} \tag{16}$$

$$\mathbf{W}_{\hat{\mathbf{x}}} = \sum \mathbf{W}_i \tag{17}$$

In formula (16) the results of the individual measurements are represented as column vectors. Formulas (16-17) also can be presented in the iterative form (18-21) :

$$\hat{\mathbf{x}}_1 = \mathbf{x}_1 \tag{18}$$

$$\mathbf{W}_{\hat{\mathbf{x}}_1} = \mathbf{W}_1 \tag{19}$$

$$\hat{\mathbf{x}}_{i+1} = \frac{\mathbf{W}_{\hat{\mathbf{x}}_i} \hat{\mathbf{x}}_i + \mathbf{W}_{i+1} \mathbf{x}_{i+1}}{\mathbf{W}_{\hat{\mathbf{x}}_i} + \mathbf{W}_{i+1}} \tag{20}$$

$$\mathbf{W}_{\hat{\mathbf{x}}_{i+1}} = \mathbf{W}_{\hat{\mathbf{x}}_i} + \mathbf{W}_{i+1} \tag{21}$$

Proof:

The presented formula represents the particular case of the general estimator of state $\mathbf{x}$ of the linear system, in which information about $\mathbf{x}$ is accessible via $\mathbf{y} = \mathbf{A}\mathbf{x} + \mathbf{e}$, where $\mathbf{y}$ is the vector of the measured results, $\mathbf{A}$ is observation which is the projection of the system state on the measured vector, $\mathbf{e}$ is random noise having average value and covariance $E\mathbf{e}\mathbf{e}^T = \mathbf{C}$, which represents the measurement errors. In such case, the best linear unbiased estimator has the following form:

$$\hat{\mathbf{x}} = \mathbf{C}^{-1}\mathbf{A}\left(\mathbf{A}^T\mathbf{C}^{-1}\mathbf{A}\right)^{-1}\mathbf{y} \tag{22}$$

In our case, the formula is radically simplified. All measurement results are presented in the same coordination system, therefore the observation Matrix has the following form:

$$A = \begin{bmatrix} 1 & 0 & ... & 0 \\ 0 & 1 & ... & 0 \\ ... & ... & ... & ... \\ 0 & 0 & ... & 1 \\ & & .... & \\ 1 & 0 & ... & 0 \\ 0 & 1 & ... & 0 \\ ... & ... & ... & ... \\ 0 & 0 & ... & 1 \end{bmatrix} \qquad (23)$$

Since all measurements are independent, the covariance matrix has the block-diagonal form:

$$C = \begin{bmatrix} C_1 & 0 & ... & 0 \\ 0 & C_2 & ... & 0 \\ ... & ... & ... & ... \\ 0 & 0 & ... & C_n \end{bmatrix} \qquad (24)$$

$$C^{-1} = \begin{bmatrix} C_1^{-1} & 0 & ... & 0 \\ 0 & C_2^{-1} & ... & 0 \\ ... & ... & ... & ... \\ 0 & 0 & ... & C_n^{-1} \end{bmatrix} = \begin{bmatrix} W_1 & 0 & ... & 0 \\ 0 & W_2 & ... & 0 \\ ... & ... & ... & ... \\ 0 & 0 & ... & W_n \end{bmatrix} \qquad (25)$$

After simplification and changing of the multiplication order (symmetrical matrix) we obtain:

$$\hat{x} = C^{-1}A\left(A^T C^{-1} A\right)^{-1} y = \sum W_i \left(\sum W_j\right)^{-1} y_i = \sum W_i y_i \left(\sum W_j\right)^{-1} \qquad (26)$$

As we can see, it is another form of formula (16). Its interpretation is easy to read and easy to remember. The optimal linear estimator represents the weighed average of all measurements, when weighs are the precision matrices of individual measurements.

4. Modified Kalman filter

In many cases it is necessary to estimate the changing state vector of the system. The value of the vector changes between successive measurements. Such situation exists in radiolocation when the position of moving target is tracked. Restricting ourselves to linear systems, the solution of this problem is known as the Kalman filter. In the Kalman filter theory, the model of system is presented as the pair of equations:

$$x_{i+1} = Fx_i + q_{i+1} \qquad (27)$$

$$z_{i+1} = Hx_{i+1} + r_{i+1} \qquad (28)$$

The first equation describes dynamics of system in the discrete time domain. Vector q models the influence of internal noise. It is assumed that expected value q is zero and it is independent from the vector of state value $\mathbf{x}$, as well as of the values of vector $\mathbf{q}$ at the previous iterations:

$$E\mathbf{q}_i = \mathbf{0} \tag{29}$$

$$\underset{i \neq j}{E\mathbf{q}_i \mathbf{q}_j} = \mathbf{0} \tag{30}$$

$$E\mathbf{q}_i \mathbf{q}_i^T = \mathbf{Q} \tag{31}$$

The covariance of internal noise $\mathbf{q}$ is described by matrix $\mathbf{Q}$. The second equation describes output $\mathbf{z}$ of the system which depends on actual state of the vector $\mathbf{x}$. The vector $\mathbf{r}$ models the influence of noise on the measuring process. Further it is assumed that the expected value $\mathbf{r}$ is equal to zero and it is independent from the value of the state vector $\mathbf{x}$, internal noise vector $\mathbf{q}$, as well as of the vector $\mathbf{r}$ from the previous iterations:

$$E\mathbf{r}_i = \mathbf{0} \tag{32}$$

$$\underset{i \neq j}{E\mathbf{r}_i \mathbf{r}_j} = \mathbf{0} \tag{33}$$

$$E\mathbf{r}_i \mathbf{r}_i^T = \mathbf{R} \tag{34}$$

It is assumed that we have estimate of the value $\hat{\mathbf{x}}_i$, accuracy of which is described by covariance matrix $\mathbf{P}_i$ and as well as the new measurement $\mathbf{z}_{i+1}$ which is accurately described by the covariance matrix $\mathbf{R}_{i+1}$. How taking into consideration the last measurement can we obtain the estimated value of the state vector at the time $i+1$? The solution of this problem leads to the classic Kalman filter equation. Kalman filter can be presented in the iteration form:

new estimate =state prediction + correction
correction = gain(new measurement – measurement prediction)*

The equation of the state prediction has the following form:

$$\hat{\mathbf{x}}_{i+1|i} = \mathbf{F}\hat{\mathbf{x}}_i \tag{35}$$

The covariance matrix of the state prediction has the following form:

$$\mathbf{P}_{i+1|i} = \mathbf{F}\mathbf{P}_i\mathbf{F}^T + \mathbf{Q}_{i+1} \tag{36}$$

The equation of the measurement prediction has the form:

$$\hat{\mathbf{z}}_{i+1|i} = \mathbf{H}\hat{\mathbf{x}}_{i+1|i} \tag{37}$$

The Kalman gain is defined as:

$$\mathbf{K}_{i+1} = \mathbf{P}_{i+1|i}\mathbf{H}^T \left[\mathbf{H}\mathbf{P}_{i+1|i}\mathbf{H}^T + \mathbf{R}_{i+1} \right]^{-1} \tag{38}$$

The new estimate of the state has the following form:

$$\hat{\mathbf{x}}_{i+1} = \hat{\mathbf{x}}_{i+1|i} + \mathbf{K}_{i+1}\left(\mathbf{z}_{i+1} - \hat{\mathbf{z}}_{i+1|i}\right) \tag{39}$$

The covariance matrix of new estimate of the state has the following form:

$$\mathbf{P}_{i+1} = \left[\mathbf{1} - \mathbf{K}_{i+1}\mathbf{H}\right]\mathbf{P}_{i+1|i} \tag{40}$$

Equations (27-40) constitutes the description of the classic Kalman filter. The Kalman filter can be considered as the system in which the input consists of string of measurements $\mathbf{z}_i$ together with the covariance matrix $\mathbf{R}_i$ which characterizes the accuracy of the i-th measurement and with the output being new estimate of the state vector $\hat{\mathbf{x}}_i$ together with the covariance matrix $\mathbf{P}_i$ which describes accuracy of this estimate. In the case when the precision of measurements is described by the singular precision matrix $\mathbf{W}_i$, it is impossible to obtain corresponding covariance matrix $\mathbf{R}_i$ and the classic Kalman filter equation can not be used (covariance matrix is required). Such situation happens for instance when the movement of the object based on the bearings is tracked. The bearing can be treated as a degenerated plot in which precision of distance measurement is equal zero. Thus bearing is the plot described by the singular precision matrix. Determination of the covariance matrix is impossible. However, it is not necessary. The only equation in which the covariance matrix $\mathbf{R}$ is used is equation (38). The covariance matrix in this equation is inverted by indirect way. A part of the equation (38) in which matrix $\mathbf{R}$ exists has form:

$$\left[\mathbf{HPH}^T + \mathbf{R}\right]^{-1} \tag{41}$$

It is easy to prove that:

$$\mathbf{A}\left[\mathbf{R}^{-1} + \mathbf{A}^{-1}\right]\mathbf{R} = \left[\mathbf{A} + \mathbf{R}\right] \tag{42}$$

We invert both sides:

$$\mathbf{R}^{-1}\left[\mathbf{R}^{-1} + \mathbf{A}^{-1}\right]^{-1}\mathbf{A}^{-1} = \left[\mathbf{A} + \mathbf{R}\right]^{-1} \tag{43}$$

Hence:

$$\left[\mathbf{HPH}^T + \mathbf{R}\right]^{-1} = \mathbf{R}^{-1}\left[\mathbf{R}^{-1} + \left[\mathbf{HPH}^T\right]^{-1}\right]^{-1}\left[\mathbf{HPH}^T\right]^{-1} \tag{44}$$

It can be written as:

$$\left[\mathbf{HPH}^T + \mathbf{R}\right]^{-1} = \mathbf{W}\left[\mathbf{W} + \left[\mathbf{HPH}^T\right]^{-1}\right]^{-1}\left[\mathbf{HPH}^T\right]^{-1} \tag{45}$$

Finally, the equation (38) has form:

$$\mathbf{K}_{i+1} = \mathbf{P}_{i+1|i}\mathbf{H}^T\mathbf{W}_{i+1}\left[\mathbf{W}_{i+1} + \left[\mathbf{HP}_{i+1|i}\mathbf{H}^T\right]^{-1}\right]^{-1}\left[\mathbf{HP}_{i+1|i}\mathbf{H}^T\right]^{-1} \tag{46}$$

The only problem which can appear here is the possible singularity of the matrix $\mathbf{HP}_{i+1|i}\mathbf{H}^T$.

This is the covariance matrix of the measurement prediction. If this matrix is singular, it means that we have certainty about measurement result of some parts and act of filtration have no sense as it has been discussed before. During normal filtration process such situation never happens. In such transformed form of Kalman filter the measurements which accuracy is described by the precision matrix can be used as the input, even if this matrix is singular. The price which is to pay is the necessity to twice invert the matrix which dimension corresponds to the dimension of the observation vector.

5. Examples

5.1 Estimation of the static system.

In order to demonstrate the proposed technique, the following example is used. The position of tracking target is determined using three measurements. The first measurement comes from the radar. The second one comes from the onboard GPS device. The third one comes from the direction finder. The radar is a device which measures three coordinates: slant range (R [m]), azimuth (β [rad]), and elevation angle (ϕ [rad]). For our simulation we assume that standard deviation of particular errors are: $\begin{bmatrix} 1500 & 0.001 & 0.001 \end{bmatrix}$ ([m rad rad])

i.e. the radar works as an altimeter precisely measuring elevation and azimuth while the slant range is not precisely measured. As far as GPS device is concerned we know that this device measures position with accuracy of up to 100 m and this measurement can be presented in any Cartesian coordinate system. The altitude is not measured by GPS. The standard deviation of bearing error for direction finder is [0.001] ([rad]). We know also very precisely the location of the direction finder. For this example the simulation has been done. We have chosen Cartesian coordinate system located at the radar as the global coordinate system. Furthermore, we have assumed that the tracking target is at the point $\mathbf{x} = \begin{bmatrix} 30000 & 40000 & 12000 \end{bmatrix}$. The radar position is then at $\mathbf{R} = \begin{bmatrix} 0 & 0 & 0 \end{bmatrix}$. Bearing finder is in the point $\mathbf{N} = \begin{bmatrix} 70000 & 10000 & 0 \end{bmatrix}$. We obtain following measurements results:

$$\mathbf{z}_{Rad} = \begin{bmatrix} 50771 & 0{,}64363 & 0{,}23388 \end{bmatrix} \tag{47}$$

$$\mathbf{z}_{GPS} = \begin{bmatrix} 30029 & 39885 \end{bmatrix} \tag{48}$$

$$\mathbf{z}_{Nam} = \begin{bmatrix} -0{,}92733 \end{bmatrix} \tag{49}$$

Basing on these results we want to obtain the estimator of $\mathbf{x}$. The first step is to present the results of measurements in the common coordinate system. In our case it is XYZ coordinate system connected to the radar. Now, some values should be attached to missing variables. If the transformation is linear, we can assign them any values. If the transformation of variables is nonlinear, we linearize around an unknown point. The values of the missing variables are estimated based on results of other measurements. After assigning the missing variables we obtain:

$$\mathbf{z}_{Rad} = \begin{bmatrix} 50771 & 0.64336 & 0.23388 \end{bmatrix} \tag{50}$$

$$\mathbf{z}_{GPS} = \begin{bmatrix} 30029 & 39885 & 11766 \end{bmatrix} \tag{51}$$

$$\mathbf{z}_{Nam} = \begin{bmatrix} 51363 & -0.92733 & 0.23113 \end{bmatrix} \tag{52}$$

Transforming to new coordinate system, we obtain following results:

$$\mathbf{z}_{Rad}^{XYZ} = \begin{bmatrix} 29638 & 39507 & 11766 \end{bmatrix} \tag{53}$$

$$\mathbf{z}_{GPS}^{XYZ} = \begin{bmatrix} 30029 & 39885 & 11766 \end{bmatrix} \tag{54}$$

$$\mathbf{z}_{Nam}^{XYZ} = \begin{bmatrix} 30001 & 39997 & 11766 \end{bmatrix} \tag{55}$$

In order to determine the precision matrix we invert the corresponding covariance matrices:

$$\mathbf{W}_x = \mathbf{C}_x^{-1} \tag{56}$$

The accuracy of the GPS device is 100 m. Therefore we consider 100 as the standard deviation of error distribution.

$$\mathbf{W}_{GPS} = \mathbf{C}_{GPS}^{-1} = \begin{bmatrix} 100^2 & 0 \\ 0 & 100^2 \end{bmatrix}^{-1} = \begin{bmatrix} 0.0001 & 0 \\ 0 & 0.0001 \end{bmatrix} \tag{57}$$

The standard deviation of the direction finder is 0.001 rad. The variance and precision are accordingly:

$$\mathbf{W}_{Nam} = \mathbf{C}_{Nam}^{-1} = \begin{bmatrix} 0.001^2 \end{bmatrix}^{-1} = 10^6 \tag{58}$$

After inserting zeros to the rows and columns corresponding to missing variables we obtain the following precision matrices:

$$\mathbf{W}_{Rad} = \begin{bmatrix} 0.(4)e-6 & 0 & 0 \\ 0 & 10^6 & 0 \\ 0 & 0 & 10^6 \end{bmatrix} \tag{59}$$

$$\mathbf{W}_{GPS} = \begin{bmatrix} 0.0001 & 0 & 0 \\ 0 & 0.0001 & 0 \\ 0 & 0 & 0 \end{bmatrix} \tag{60}$$

$$\mathbf{W}_{Nam} = \begin{bmatrix} 0 & 0 & 0 \\ 0 & 0 & 0 \\ 0 & 0 & 10^6 \end{bmatrix} \tag{61}$$

When applying the linear transformation of variables: $\mathbf{y} = \mathbf{Tx} + \mathbf{a}$ the covariance matrix changes according to the formula: $\mathbf{C}_y = \mathbf{TC}_x\mathbf{T}^T$. In the case of the nonlinear transformation: $\mathbf{y} = \mathbf{Y}(\mathbf{x})$, we replace the matrix $\mathbf{T}$ by the Jacobian matrix $\mathbf{J}$, which is the linearization of this transformation.

$$J = \begin{bmatrix} \dfrac{\partial y_1}{\partial x_1} & \cdots & \dfrac{\partial y_1}{\partial x_n} \\ \cdots & \cdots & \cdots \\ \dfrac{\partial y_n}{\partial x_1} & \cdots & \dfrac{\partial y_n}{\partial x_n} \end{bmatrix}_{x=\hat{x}} \tag{62}$$

$\mathbf{C_y} = \mathbf{J}\mathbf{C_x}\mathbf{J}^T$. Similarly we obtain: $\mathbf{W_y} = \mathbf{J}^{-1^T}\mathbf{W_x}\mathbf{J}^{-1}$. In our case, the radar and the direction finder (after inserting the missing variables) operate in the spherical coordinates which are related to global system by the set of formulas:

$$X = X_0 + R\cos(\varphi)\sin(\beta) \tag{63}$$

$$Y = Y_0 + R\cos(\varphi)\cos(\beta) \tag{64}$$

$$Z = Z_0 + R\sin(\varphi) \tag{65}$$

The Jacobian matrix of such transformation has the following form:

$$J = \begin{bmatrix} \cos(\varphi)\sin(\beta) & R\cos(\phi)\cos(\beta) & -R\sin(\varphi)\sin(\beta) \\ \cos(\varphi)\cos(\beta) & -R\cos(\varphi)\sin(\beta) & -R\sin(\varphi)\cos(\beta) \\ \sin(\varphi) & 0 & R\cos(\varphi) \end{bmatrix} \tag{66}$$

Transforming variables from the local coordinate system of the measuring device to global system, we obtain as follows:
For the radar:

$$\mathbf{W}_{Rad}^{XYZ} = \mathbf{J}_{\mathbf{z}_{Rad}}^{-1}{}^T \mathbf{W}_{Rad} \mathbf{J}_{\mathbf{z}_{Rad}}^{-1} = \begin{bmatrix} 26.998 & -18.659 & -5{,}2424 \\ -18.659 & 16.124 & -6{,}9881 \\ -5{,}2424 & -6.9881 & 36.713 \end{bmatrix} e-5 \tag{67}$$

In case of the GPS device we do not have any change of variables. We are only adding the zero values corresponding to additional variable to the precision matrix:

$$\mathbf{W}_{GPS}^{XYZ} = \begin{bmatrix} 0.0001 & 0 & 0 \\ 0 & 0.0001 & 0 \\ 0 & 0 & 0 \end{bmatrix} \tag{68}$$

For the direction finder:

$$\mathbf{W}_{Nam}^{XYZ} = \mathbf{J}_{\mathbf{z}_{Nam}}^{-1}{}^T \mathbf{W}_{Nam} \mathbf{J}_{\mathbf{z}_{Nam}}^{-1} = \begin{bmatrix} 14.4 & 19.201 & <\text{1e-15} \\ 19.201 & 25.604 & <\text{1e-16} \\ <\text{1e-15} & <\text{e-16} & <\text{1e-32} \end{bmatrix} e-5 \tag{69}$$

Finally, we obtain the position estimator:

$$\mathbf{x}^* = \frac{\mathbf{z}_{Rad}^{XYZ}\mathbf{W}_{Rad}^{XYZ} + \mathbf{z}_{GPS}^{XYZ}\mathbf{W}_{GPS}^{XYZ} + \mathbf{z}_{Nam}^{XYZ}\mathbf{W}_{Nam}^{XYZ}}{\mathbf{W}_{Rad}^{XYZ} + \mathbf{W}_{GPS}^{XYZ} + \mathbf{W}_{Nam}^{XYZ}} \tag{70}$$

resulting in:

$$\mathbf{x}^* = \begin{bmatrix} 30008 & 39973 & 11908 \end{bmatrix} \tag{71}$$

Note that utilization of the measurements from GPS device and direction finder, which do not measure the height, actually results in some improvement of altitude estimation. This is due to the fact that the variables which provide the radar measurements in XYZ coordinates are not independent, but they are strongly correlated. The linear combination of them is measured with high accuracy. Thus, by improving the estimation of XY variables we also improve the estimation of the variable in Z direction.

5.2 Estimation in the dynamic system

The next example presents the application of the proposed technique to the estimation of the parameters of a dynamic system. It is the most typical situation when the racking target is moving. Our purpose is the actualization of motion parameters of tracking target based on three measurements. We assume that the tracking is two dimensional. The following parameters are tracked: x - the position on axis OX; vx - the motion speed on axis OX; y - the position on axis OY; vy - the motion speed on axis OY. We consider the flight altitude z as fixed and equal 2000. Finally, we have the following data:
At time:

$$T0 = 100 \tag{72}$$

Based on previous measurements, we obtain estimation of the initial parameters of tracked target motion:

$$\mathbf{x}_0 = \begin{bmatrix} x \\ vx \\ y \\ vy \end{bmatrix} = \begin{bmatrix} -259.5 \\ 126.7 \\ 60075 \\ -114.3 \end{bmatrix} \tag{73}$$

The covariance matrix of this estimation has the form:

$$\mathbf{C}_{\mathbf{x}_0} = \begin{bmatrix} 600^2 & 0 & 0 & 0 \\ 0 & 20^2 & 0 & 0 \\ 0 & 0 & 600^2 & 0 \\ 0 & 0 & 0 & 20^2 \end{bmatrix} \tag{74}$$

At time:

$$T1 = 108 \tag{75}$$

The tracked target has been detected by the radar R1 which is located at the beginning of our coordinate system

$$\mathbf{R1} = \begin{bmatrix} 0 \\ 0 \\ 0 \end{bmatrix} \tag{76}$$

The parameters of the detection are:

$$\mathbf{z}_{R1} = \begin{bmatrix} 59025 \\ 0.1169 \\ 0.2122 \end{bmatrix} \tag{77}$$

We assume that the radar measurement is unbiased. The covariance matrix of the measurement error has the following form:

$$\mathbf{C}_{R1} = \begin{bmatrix} 55^2 & 0 & 0 \\ 0 & 0.08^2 & 0 \\ 0 & 0 & 0.15^2 \end{bmatrix} \tag{78}$$

The next position measurements of the tracking target took place at the time:

$$T2 = 113.8 \tag{79}$$

We obtain the bearing of tracked target. The direction finder was located at the point:

$$\mathbf{N2} = \begin{bmatrix} 10000 \\ 40000 \\ 100 \end{bmatrix} \tag{80}$$

The bearings value is:

$$\mathbf{z}_{N2} = \begin{bmatrix} -0.3853 \end{bmatrix} \tag{81}$$

The covariance measurement error is:

$$\mathbf{C}_{N2} = \begin{bmatrix} 0.05^2 \end{bmatrix} \tag{82}$$

At the near time:

$$T3 = 114 \tag{83}$$

We obtain bearing from another direction finder which was located an point:

$$\mathbf{N3} = \begin{bmatrix} 30000 \\ 50000 \\ 120 \end{bmatrix} \tag{84}$$

The bearing value is:

$$\mathbf{z}_{N3} = \begin{bmatrix} -1.2433 \end{bmatrix} \tag{85}$$

The covariance measurement error is:

$$\mathbf{C}_{N3} = \begin{bmatrix} 0.05^2 \end{bmatrix} \tag{86}$$

Traditionally, to solve such a problem, the theory of Kalman filter is used. In this theory two assumptions are usually accepted. First that all measurements are carried out by the same device. Second, that the measurements are done in regular time intervals. None of the above assumptions is satisfied in our example. To address the issue of different measuring device, we assume that the measurements are always presented in the common global coordinate system, and their dispersion is described by the precision matrix which can be singular. Using the modified formulas of Kalman filters (46), the estimation can be done, even if the precision matrix is singular. In order to address the issue of the irregular time intervals of the measurements, the Kalman filter has been modified to consider the changes of time interval between individual measurements. It means that the matrix $\mathbf{F}$ describing dynamics of the tracked process and matrix $\mathbf{Q}$ described covariance of the internal noise of tracked process depend on time Δt which elapses from the last measurement. We assume the linear model of the tracked target movement in which velocity does not have systematic changes. Therefore the equation of motion has the following form:

$$\begin{cases} x(t) = x_0 + vx_0 * t \\ vx(t) = vx_0 \\ y(t) = y_0 + vy_0 * t \\ vy(t) = vy_0 \end{cases} \tag{87}$$

We introduce the following state vector of Kalman filter:

$$\mathbf{X} = \begin{bmatrix} x \\ vx \\ y \\ vy \end{bmatrix} \tag{88}$$

Therefore, the dynamic matrix $\mathbf{F}(\Delta t)$ has the following form:

$$\mathbf{F}(\Delta t) = \begin{bmatrix} 1 & \Delta t & 0 & 0 \\ 0 & 1 & 0 & 0 \\ 0 & 0 & 1 & \Delta t \\ 0 & 0 & 0 & 1 \end{bmatrix} \tag{89}$$

We also assume that the internal noise of the process has additive character and it is the source of change of motion velocity. In such a case, the velocity variation is increasing linearly with time:

$$\mathbf{Q}_v = Q \begin{bmatrix} \Delta t & 0 \\ 0 & \Delta t \end{bmatrix} \tag{90}$$

The internal noise related to velocity affects the position through the dynamic process described by matrix $\mathbf{F}(\Delta t)$. It can be proved that in this case the covariance matrix $\mathbf{Q}(\Delta t)$ of the process internal noise has the following form:

$$\mathbf{Q}(\Delta \mathbf{t}) = Q \begin{bmatrix} \dfrac{\Delta t^3}{3} & \dfrac{\Delta t^2}{2} & 0 & 0 \\[2mm] \dfrac{\Delta t^2}{2} & \Delta t & 0 & 0 \\[2mm] 0 & 0 & \dfrac{\Delta t^3}{3} & \dfrac{\Delta t^2}{2} \\[2mm] 0 & 0 & \dfrac{\Delta t^2}{2} & \Delta t \end{bmatrix} \tag{91}$$

The parameter Q provides the information on how much the velocity of the motion changes during one second in the medium square sense. In the simulations we assume that $Q = 1$. Additionally we assume that the observation vector has the following form:

$$\mathbf{z} = \begin{bmatrix} x \\ y \end{bmatrix} \tag{92}$$

Therefore, the observation matrix $\mathbf{H}$ has a form:

$$\mathbf{H} = \begin{bmatrix} 1 & 0 & 0 & 0 \\ 0 & 0 & 1 & 0 \end{bmatrix} \tag{93}$$

The direction finder, as well as the radar, make measurements at spherical coordinate system. To present measurements results at the global reference system we use formulas (63-65) described in the first example. The necessary elements have been already discussed. The algorithm of the estimation of the motion parameters can be described on the basis of measurements coming from different sources. Utilizing previous measurements we know the estimation $\hat{\mathbf{x}}_0$ of the tracked target motion parameters in time t_0 and the covariance matrix $\mathbf{p}_0$, describing the precision of this estimate. When at the time $t_1 > t_0$ the next measurement occur, the following steps are taken:

- Calculate the matrices $\mathbf{F}(\Delta t)$ and $\mathbf{Q}(\Delta t)$ according to formulas (89, 91)
- Calculate the prediction of the state vector $\hat{\mathbf{x}}_{t1|t0}$ using formula (35)
- Calculate the covariance matrix $\mathbf{P}_{t1|t0}$ of the prediction of the state using formula (36)
- Calculate the prediction of detection $\hat{\mathbf{z}}_{t1|t0}^{XYZ}$ at the global coordinate system
- Calculate the prediction of detection $\hat{\mathbf{z}}_{t1|t0}^{Rad}$ at the local radar (direction finder) coordinate system
- Expand the measurements vector $\tilde{\mathbf{z}}_{t1}^{Rad}$. The values which has not been measured are replaced by predicted values from the previous measurements.
- Create the precision matrix $\mathbf{W}_{t1}^{Rad}$ of the measurements at the local radar (direction finder) coordinate system. The values which have not been measured have zero precision.
- Transform the expanded vector of measurements $\tilde{\mathbf{z}}_{t1}^{Rad}$ and corresponding precision matrix $\mathbf{W}_{t1}^{Rad}$ to global coordinate system $\tilde{\mathbf{z}}_{t1}^{XYX}$ and $\mathbf{W}_{t1}^{XYZ}$.

- Using formula (39-40, 46) calculate the new estimate of the motion parameters $\hat{\mathbf{x}}_{t1}$ and their corresponding precision matrix $\mathbf{P}_{t1}$.

Note that tracking is done in two dimensional space. The altitude Z is not tracked and is fixed as Z=2000. If the altitude is required in computation like for example in formula (97) we take Z=2000m.

For the first measurements we have accordingly:

$$\Delta t = 108 - 100 = 8 \tag{94}$$

$$\hat{\mathbf{x}}_{108|100} = \mathbf{F}(8)\hat{\mathbf{x}}_{100} \approx \begin{bmatrix} 754 \\ 127 \\ 59161 \\ -114 \end{bmatrix} \tag{95}$$

The covariance matrix of the prediction:

$$\mathbf{P}_{108|100} = \mathbf{F}(8)^T \mathbf{P}_{100}\mathbf{F}(8) + \mathbf{Q}(8) \tag{96}$$

The prediction of the detection at the global coordinate system XYZ:

$$\hat{\mathbf{z}}_{108|100}^{XYZ} = \begin{bmatrix} 754 \\ 59161 \\ 2000 \end{bmatrix} \tag{97}$$

In the last formula we use information on tracking target altitude 2000m. Now, we consider the system of the radar R1. The prediction of the detection at the radar local coordinate system:

$$\hat{\mathbf{z}}_{108|100}^{Rad} = \begin{bmatrix} 59200 \\ 0.013 \\ 0.034 \end{bmatrix} \tag{98}$$

The radar measures all coordinates. It is not necessary to increase measuring vector and to use the prediction.

$$\tilde{\mathbf{z}}_{108}^{Rad} = \begin{bmatrix} 59025 \\ 0.117 \\ 0.212 \end{bmatrix} \tag{99}$$

The dispersion of the radar measurements is described by the precision matrix of the form:

$$\mathbf{W}_{108}^{Rad} = \begin{bmatrix} 1/55^2 & 0 & 0 \\ 0 & 1/0.8^2 & 0 \\ 0 & 0 & 1/0.15^2 \end{bmatrix} \tag{100}$$

After transformation to the global coordinate system XYZ we obtain:

$$\tilde{\mathbf{z}}_{108}^{XYZ} = \begin{bmatrix} 6732.7 \\ 57306 \\ 2000 \end{bmatrix} \tag{101}$$

The precision matrix is transformed according to formula:

$$\mathbf{W}_{108}^{XYZ} = \mathbf{J}_{\hat{z}}^{-1T} \mathbf{W}_{108}^{Rad} \mathbf{J}_{\hat{z}}^{-1} \tag{102}$$

where $\mathbf{J}_{\hat{z}}$ is Jacobian matrix of transformation (66) calculated at the point $\hat{\mathbf{z}}_{108}^{Rad}$.

Because we are tracking only the components XY, we ignore the last row in the measurement vector $\tilde{\mathbf{z}}_{108}^{XYZ}$, as well as the last row and the last column in the precision matrix $\mathbf{W}_{108}^{XYZ}$.

Now, we can determine Kalman gain $\mathbf{K}$ and the state estimate $\hat{\mathbf{x}}_{108}$, as well as the corresponding covariance matrix $\mathbf{P}_{108}$:

$$\mathbf{K}_{108} = \mathbf{P}_{108|100} \mathbf{H}^T \mathbf{W}_{108} \left[\mathbf{W}_{108} + \left[\mathbf{HP}_{108|100} \mathbf{H}^T \right]^{-1} \right]^{-1} \left[\mathbf{HP}_{108|100} \mathbf{H}^T \right]^{-1} \tag{103}$$

$$\hat{\mathbf{x}}_{108} = \hat{\mathbf{x}}_{108|100} + \mathbf{K}_{108} \left(\tilde{\mathbf{z}}_{108}^{XYZ} - \hat{\mathbf{z}}_{108|100} \right) \tag{103}$$

$$\mathbf{P}_{108} = \left[1 - \mathbf{K}_{108} \mathbf{H} \right] \mathbf{P}_{108|100} \tag{105}$$

It is the end of the first step of the determination of the estimate of tracking target motion parameters.

The next step is more interesting because it demonstrates how one can use the measurements which were done by the direction finder. Accordingly:

For the first measurement we have:

$$\Delta t = 113.8 - 108 = 5.8 \tag{106}$$

$$\hat{\mathbf{x}}_{113.8|108} = \mathbf{F}(5.8)\hat{\mathbf{x}}_{108} \tag{107}$$

The covariance matrix of prediction is:

$$\mathbf{P}_{113.8|108} = \mathbf{F}(5.8)^T \mathbf{P}_{108} \mathbf{F}(5.8) + \mathbf{Q}(5.8) \tag{108}$$

The prediction of the detection at the global coordinate system XYZ

$$\hat{\mathbf{z}}_{113.8|108}^{XYZ} = \begin{bmatrix} 1571.7 \\ 56647 \\ 2000 \end{bmatrix} \tag{109}$$

We are taking into consideration the direction finder N2. The prediction of the detection at the direction finder local coordinate system:

$$\hat{\mathbf{z}}_{113.8|108}^{N2} = \begin{bmatrix} 18775 \\ -0.4687 \\ 0.1015 \end{bmatrix} \tag{110}$$

The direction finder measures only azimuth $\mathbf{z}_{N2} = [-0.3853]$. We treat the direction finder as a radar which measures all coordinates. The missing coordinates are obtained utilizing the prediction $\hat{\mathbf{z}}_{113.8|108}^{N2}$. We obtain:

$$\tilde{\mathbf{z}}_{113.8}^{N2} = \begin{bmatrix} 18775 \\ -0.3853 \\ 0.1015 \end{bmatrix} \tag{111}$$

The precision matrix variables calculated using the prediction have zero precision:

$$\mathbf{W}_{113.8}^{N2} = \begin{bmatrix} 0 & 0 & 0 \\ 0 & 1/0.05^2 & 0 \\ 0 & 0 & 0 \end{bmatrix} \tag{112}$$

After the transformation to global XYZ coordinate system we obtain:

$$\tilde{\mathbf{z}}_{113.8}^{XYZ} = \begin{bmatrix} 2987.1 \\ 57291 \\ 2000 \end{bmatrix} \tag{113}$$

As before, we determine the precision matrix $\mathbf{W}_{113.8}^{XYZ}$ at the global XYZ coordinate system and the Kalman gain $\mathbf{K}_{113.8}$ as well as new estimate $\hat{\mathbf{x}}_{113.8}$ together with the corresponding covariance matrix $\mathbf{P}_{113.8}$:

$$\mathbf{W}_{113.8}^{XYZ} = \mathbf{J}_{\hat{z}}^{-1T} \mathbf{W}_{113.8}^{N2} \mathbf{J}_{\hat{z}}^{-1} \tag{114}$$

$$\mathbf{K}_{113.8} = \mathbf{P}_{113.8|108} \mathbf{H}^T \mathbf{W}_{113.8} \left[\mathbf{W}_{113.8} + \left[\mathbf{H}\mathbf{P}_{113.8|108}\mathbf{H}^T \right]^{-1} \right]^{-1} \left[\mathbf{H}\mathbf{P}_{113.8|108}\mathbf{H}^T \right]^{-1} \tag{115}$$

$$\hat{\mathbf{x}}_{113.8} = \hat{\mathbf{x}}_{113.8|108} + \mathbf{K}_{113.8} \left(\tilde{\mathbf{z}}_{113.8}^{XYZ} - \hat{\mathbf{z}}_{113.8|108} \right) \tag{116}$$

$$\mathbf{P}_{113.8} = \left[\mathbf{1} - \mathbf{K}_{113.8}\mathbf{H} \right] \mathbf{P}_{113.8|108} \tag{117}$$

It is obvious that in the case of the second bearing we act similarly:

$$\Delta t = 114 - 113.8 = 0.2 \tag{118}$$

$$\hat{\mathbf{x}}_{114|113.8} = \mathbf{F}(0.2)\hat{\mathbf{x}}_{108} \tag{119}$$

The covariance matrix of prediction:

$$\mathbf{P}_{114|113.8} = \mathbf{F}(0.2)^T \mathbf{P}_{113.8}\mathbf{F}(0.2) + \mathbf{Q}(0.2) \tag{120}$$

The prediction of the detection in global XYZ coordinate system:

$$\tilde{\mathbf{z}}_{114|113.8}^{XYZ} = \begin{bmatrix} 2087 \\ 56625 \\ 2000 \end{bmatrix} \tag{121}$$

Now, we consider the local coordinate system of the direction finder N3. The prediction of the detection at the direction finder local coordinate system:

$$\hat{\mathbf{z}}_{114|113.8}^{N3} = \begin{bmatrix} 28750 \\ -1.338 \\ 0.065 \end{bmatrix} \tag{122}$$

The direction finder measures only azimuth $\mathbf{z}_{N3} = \begin{bmatrix} -1.2433 \end{bmatrix}$.

$$\tilde{\mathbf{z}}_{114}^{N3} = \begin{bmatrix} 28750 \\ -1.2433 \\ 0.065 \end{bmatrix} \tag{123}$$

The precision matrix variables defined using the prediction have zero precision:

$$\mathbf{W}_{114}^{N3} = \begin{bmatrix} 0 & 0 & 0 \\ 0 & 1/0.05^2 & 0 \\ 0 & 0 & 0 \end{bmatrix} \tag{124}$$

After transformation to global XYZ coordinate system we obtain:

$$\tilde{\mathbf{z}}_{114}^{XYZ} = \begin{bmatrix} 2560.1 \\ 59332 \\ 2000 \end{bmatrix} \tag{125}$$

As before, we determine the precision matrix $\mathbf{W}_{114}^{XYZ}$ at the global XYZ coordinate system and the Kalman gain $\mathbf{K}_{114}$ as well as a new estimate $\hat{\mathbf{x}}_{114}$ together with the corresponding covariance matrix $\mathbf{P}_{114}$:

$$\mathbf{W}_{114}^{XYZ} = \mathbf{J}_{\tilde{z}}^{-1T} \mathbf{W}_{114}^{N3} \mathbf{J}_{\tilde{z}}^{-1} \tag{126}$$

$$\mathbf{K}_{114} = \mathbf{P}_{114|113.8} \mathbf{H}^T \mathbf{W}_{114} \left[\mathbf{W}_{114} + \left[\mathbf{H} \mathbf{P}_{114|113.8} \mathbf{H}^T \right]^{-1} \right]^{-1} \left[\mathbf{H} \mathbf{P}_{114|113.8} \mathbf{H}^T \right]^{-1} \tag{127}$$

$$\hat{\mathbf{x}}_{114} = \hat{\mathbf{x}}_{114|113.8} + \mathbf{K}_{114} \left(\tilde{\mathbf{z}}_{114}^{XYZ} - \hat{\mathbf{z}}_{114|113.8} \right) \tag{128}$$

$$\mathbf{P}_{114} = \begin{bmatrix} 1 - \mathbf{K}_{114} \mathbf{H} \end{bmatrix} \mathbf{P}_{114|113.8} \tag{129}$$

The presented example demonstrates the power of the described technique. Even a single bearing can improve the estimation of motion parameters of the tracking target. If several bearings are utilized, they need not to be simultaneous.

6. Summary

In the proposed technique the measurements carried out by different devices are treated in a simple and uniform manner. The key idea comprises in expansion the vector of the measured values and insertion of missing variables based on the prediction. The precision matrix is used to describe their dispersion. In order to obtain the estimate of the tracked target state we have to present all measurements and corresponding precision matrixes in a common coordinate system. The formulas (16-17, 18-21) should be used for static systems or the modified Kalman filters equations (27-40, 46) should be used for dynamic systems.

For linear objects, the Kalman filters are the optimal estimators. In the presented example dynamics of the tracked target has been described by linear differential equation The presented methodology can be applied also in the case when the dynamics of tracked target is nonlinear. In this case, instead of formula (27) we use linearization. Let the dynamics of the tracked process be described by a nonlinear function:

$$\mathbf{x}(t + \Delta t) = \mathbf{F}(\mathbf{x}(t), \Delta t) \tag{130}$$

In order to determine the covariance of the state prognosis we use linearization $\tilde{\mathbf{F}}(\mathbf{x}(t), \Delta t)$ of the function $\mathbf{F}(\mathbf{x}(t), \Delta t)$

$$\mathbf{P}(t + \Delta t) = \tilde{\mathbf{F}}(\mathbf{x}(t), \Delta t) \mathbf{P}(t) \tilde{\mathbf{F}}(\mathbf{x}(t), \Delta t)^{T} + \mathbf{Q}(\Delta t) \tag{131}$$

where:

$$\tilde{\mathbf{F}}(t, \Delta t) = \left[\frac{\partial f_i}{\partial x_j} \right] \tag{132}$$

In order to increase accuracy, we can decrease time interval. Calculations (131) could be done several times. Let:

$$\delta t = \frac{\Delta t}{n} \tag{133}$$

$$\mathbf{x}(t + \delta t) = \mathbf{F}(\mathbf{x}(t), \delta t) \tag{134}$$

$$\mathbf{x}(t + \Delta t) = \mathbf{F}\left(\mathbf{F}\left(\mathbf{F}\left(\mathbf{F}\left(...\mathbf{F}(\mathbf{x}(t)), \delta t...\right), \delta t\right), \delta t\right), \delta t\right) \tag{135}$$

Analogically:

$$\mathbf{P}(t + \delta t) = \tilde{\mathbf{F}}(\mathbf{x}(t), \delta t) \mathbf{P}(t) \tilde{\mathbf{F}}(\mathbf{x}(t), \delta t)^{T} + \mathbf{Q}(\delta t) \tag{136}$$

$$\mathbf{P}(t + \Delta t) = \tilde{\mathbf{F}}(\mathbf{x}, \delta t) \mathbf{P}\left(\tilde{\mathbf{F}}(\mathbf{x}, \delta t) \mathbf{P}\left(...\tilde{\mathbf{F}}(\mathbf{x}, \delta t) \mathbf{P}(\mathbf{x}) \tilde{\mathbf{F}}(\mathbf{x}, \delta t)^{T} + \mathbf{Q}(\delta t)...\right) \tilde{\mathbf{F}}(\mathbf{x}, \delta t)^{T} + \mathbf{Q}(\delta t)\right) \tilde{\mathbf{F}}(\mathbf{x}, \delta t)^{T} + \mathbf{Q}(\delta t) \tag{137}$$

In the past, this approach was rejected due to computer high power requirements. Presently it creates no problem at all.

An interesting case of proposed methodology application is the use of measurements from Doppler radar. The Doppler radar measures frequency shifting of received signal permitting to detect approaching speed of tracking targets. More precisely, it permits to determine speed radial component v_R of the tracking target. The methodology is analogical to the example presented before. The velocity vector is expanded in order to include all components:

$$\tilde{\mathbf{v}} = \begin{bmatrix} v_R \\ v_\beta \\ v_\phi \end{bmatrix} \qquad (138)$$

The missing values are calculated utilizing the prediction. The corresponding precision matrix has the following form:

$$\mathbf{W} = \begin{bmatrix} \delta_{vR} & 0 & 0 \\ 0 & 0 & 0 \\ 0 & 0 & 0 \end{bmatrix} \qquad (139)$$

After transformation to the global coordinate system XYZ we can use measurements from the Doppler radar to improve estimation of the tracking target motion parameters.

In all so-far considered examples, we assumed that time error does not exist. In practice, the time is always measured with certain error. If for example the measuring time is presented in seconds, then during one second the tracked target can move several hundreds meters. Fortunately, in our case the error of time measuring can be easily taken into consideration. Time error δt is transformed into the position measurements error δx according to formula:

$$\delta x = vx * \delta t \qquad (140)$$

Assuming that the time measuring error is independent from the position measurement error, the variances of both errors are added:

$$C_x^{2\,'} = C_x^2 + \delta_x^2 = C_x^2 + vx^2 \delta_t^2 \qquad (141)$$

The error of the time measurement can be easily taken into consideration by the proper increase of the error covariance value C_x^2 of the measurement component X. Of course, the same apply to the remaining components Y and Z.

When tracking the state of dynamic system by means of the modified Kalman filter the assumption that the measurements are done in regular time intervals has been abandoned. Nevertheless, it has been assumed that the received measurements are sequential: $t_1 \le t_2 \le t_3 \le \dots \le t_{n-1} \le t_n$. Sometimes even this assumption is unfulfilled. It may happen that due to a transition delay the measurements carried out at earlier time are received later then the ones obtained in later time. In order to consider the measurements delay the very simple but useful trick can be used. We store in memory the series of last measurements together with their estimated parameters values and the corresponding covariance matrices which describe accuracy of estimate. When the delayed measurement reaches the system we retrieve from the memory an estimation of time t_0 which occurred earlier then the time t_d (attributed to the delayed measurement). All measurements later then t_0 and delayed measurements are rearranged to be sequential in time. Then, the prediction and estimation procedures are performed consecutively for each measurement. This way, the delayed measurement is taken into consideration. It must be noted that storing only the values of estimated parameters is not enough. In order to make the correct estimation, it is necessary to know also the covariance matrix of these parameter estimates.

The presented method is in 100% correspondence with the well-known least squares method (LSM). The careful analysis of the presented examples with taking into consideration the different weights of particular errors, existing correlations between them as well as the flow of time will provide the same results for the estimated values as the LSM method. Therefore the proposed methodology do note give better estimators then the classical theory. However the presented methodology avoids the very complicated and arduous computations and presents all measurements in a uniform way. Such approach is very useful for the automatic data processing by an automatic tracking system.

Since the presented technique is in 100% correspondence with the LSM, it also it discloses the weak points of LSM. In particular, it is not robust methodology. It is not robust for the values drastically odd (different) and for large errors. Even the single erroneous measurement not fitting the assumed model of errors distribution results in the drastic increase of the final error of the estimator. Because of this, it is recommended that the preliminary preselection of tracking measurements should be done. Drastically different values should be rejected.

The Kalman filter represents an optimal estimator for linear system only if the real model of the process corresponds to the one accepted during Kalman filter (27-34) design. It is particularly important when considering the error values. Consider for example the ladar (laser radar) which measures the position of an object from the distance of a few tens of kilometers with accuracy up to 5 m. Naturally, the error $\delta_R = 5$ should be taken into consideration. However, usually other sources of errors exist. For example, the position of ladar is known only with the limited precision, the time of measurement is provided with a limited accuracy. The Earth curvature can also be a source of errors during calculations. Based on the author's experience some various processes which normally are not taken into consideration may be the source of casual and systematic errors. For this reason, asserting the error value $\delta_R = 5$ is a wrong idea. According to author it is better to use larger values of errors then the small ones. If erroneous values are assumed in the filter design and are larger then in reality the resulting filter is not optimal, but it is more flexible and robust with respect to drastically different values. For this reason, moderation and caution should be used in choosing the parameters of the model utilized to design a filter.

7. References

Petersen, Ian R.; Savkin, Andrey V. (1999) *Robust Kalman Filtering For Signals and Systems with Large Uncertainties* A Birkhäuser book ISBN 978-0-8176-4089-7

Brookner, Eli (1998) *Tracking and Kalman Filtering Made Easy* John Wiley ISBN: 9780471184072

Farina, A.; Studer F.A. (1986). *Radar Data Processing Techniques, vol. 2: Advauced Topics and Applications* Research Studies Press ISBN: 978-0471909491

Blackman, Samuel S; Popoli Robert (1999) *Design and Analysis of Modern Tracking Systems* Artech House ISBN: 9781580530064

Haykin, S. (2001) *Adaptive Filter Theory* Prentice Hall ISBN: 0-13-090126-1

12

Radar Target Classification Technologies

Dr. Guy Kouemou
EADS Deutschland GmbH
Germany

1. Introduction

Apart from object detection, plot extraction and tracking, automatic classification is becoming one of the challenges of modern sensor systems.

In civil applications for example, different classification techniques are used for air traffic control (ATC) purposes. The primary purposes of air traffic control systems are collision prevention, the organization and control of the air traffic; furthermore it should provide information as well as other support for pilots. In most cases the general traffic situation is well known (aircraft, flight number, flight road, flight departure and destination, etc.), but the exact prediction of natural processes (weather, bird migration, wind mill rotation activities, insect migration, etc.) still remains rather difficult. For this reason the demanded expectations are stable feature extraction methods and classification technologies.

In meteorological applications, weather radars use backscattered echo signals to locate precipitation, calculate its motion, estimate its type (rain, snow, wind, hail, etc.) and forecast its future position and intensity. For this purpose modern weather radars are mostly pulse-Doppler radars. These are able to detect the motion of rain droplets as well as to estimate the intensity of precipitation. The structure of storms and severe weather can be derived from these data. In order to improve the efficiency of new weather radar product families, strong feature extraction and classification algorithms are required.

In the automotive industry, due to higher safety measure requirements in road traffic, the adaptive cruise control (ACC) is becoming unavoidable. These systems often use a radar setup to slow the vehicle down when approaching another vehicle or other obstacles and accelerate again to the present speed as soon as traffic allows it. ACC technology is widely considered as a key component for future generation's smart cars, as a form of artificial intelligence that can be used as a driving aid. For that reason robust signal feature extraction and classification methods are required.

For airborne applications, two kind of radars can be integrated in air platforms, in order to support the navigation and missions. The first category encompasses primary radars, the second category secondary radars. Civil airborne primary radar systems support the pilot by providing actual information like position or weather. Civil secondary surveillance radars (SSR) are radar systems supporting air traffic control (ATC). These systems do not only detect and measure the position of aircraft but also request additional information from the aircraft itself such as its identity or altitude. While primary radar systems measure only the range and bearing of targets based on reflected radio signals, SSRs rely on radar transponders aboard aircrafts. These transponders reply to interrogation signals by transmitting responding signals containing specific encoded information. The SSR technology is based on the military "identification, friend or foe" (IFF) technology, which was developed during World War II. These two systems are still compatible today. Nowadays monopulse secondary surveillance radars (MSSR) represent a improved version of SSR. For military applications radars are used in fighter aircraft for finding enemy aircraft and controlling air-to-air missiles, rockets, and guns. It is used in bombers to find surface targets, fixed or moving, and to navigate and avoid obstacles. It is used in large aircraft as an airborne warning and control system, searching the skies over great distances for enemy aircraft, tracking them, and controlling interceptors. It also is used to search the seas for surface vessels or surfaced submarines.

Furthermore special space technology is used in commercial as well as non-commercial spaceflight activities. For such purposes spaceborne radar systems are often used in spacecraft, in order to locate patterns of activity. Such critical applications need robust preprocessing, feature extraction, pattern classification, fusion and recognition methods.

The knowledge of the target class has significant influence on the identification, threat evaluation and weapon assignment process of large systems. Especially, considering new types of threats in Anti Asymmetric Warfare the knowledge of a target class is of significant importance. The target class can also be used to optimize track and resource management of today's agile sensor systems.

This chapter consists of the following sections (cf. Fig. 1): 1. the data acquisition section part, 2. introduction to methods of signal preprocessing, 3. introduction to methods of feature extraction, 4. basics of classification and sub-classification methods, 5. introduction to fusion methods, 6. recognition, typing or identification basics, 7. some experimental results, 8. some product examples of modern radar systems and finally 9. a brief conclusion.

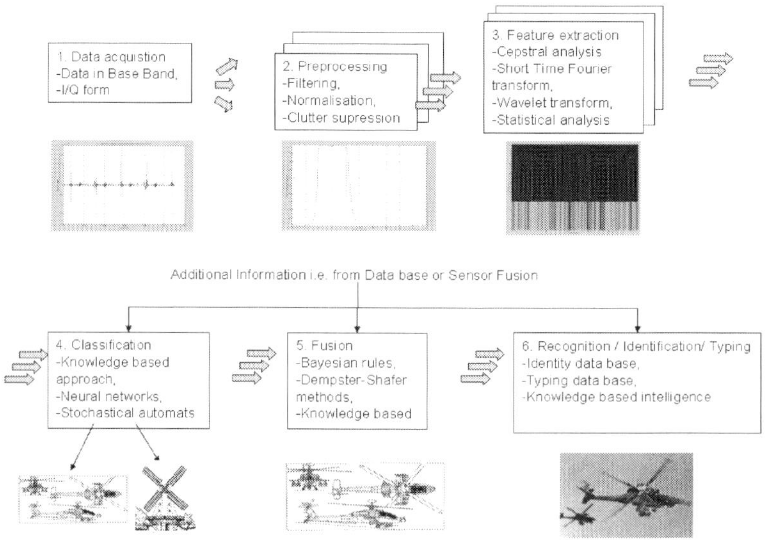

Fig. 1. Simplified structure of a modern radar classification function chain

2. Data acquisition

For the data acquisition part an active or passive radar frontend can be used. Either a primary or a secondary radar is usually taken into consideration. This radar uses self- or friendly generated waveform to reconstruct information from the environment with different objects or targets. The data acquisition part usually provides backscattered radar echo signal in I- and Q-form in the baseband.

For following chapter parts, we will assume that a backscattered radar echo signal is given. A simplified data acquisition chain can be resumed as following:

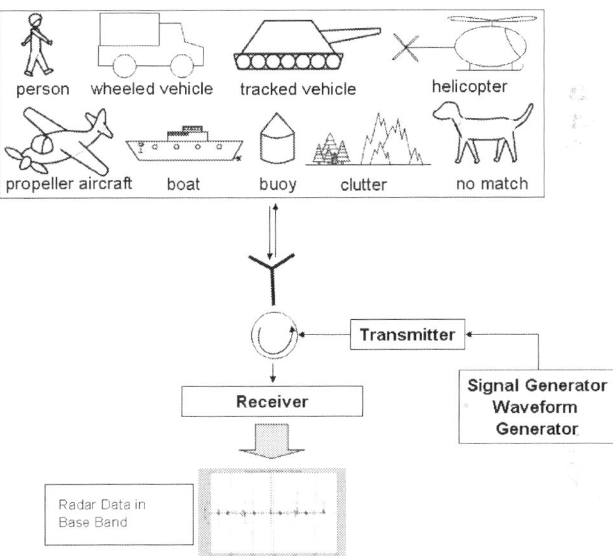

Fig. 2. Simplified diagram of a radar data acquisition process

3. Methods of signal preprocessing

The aim of signal preprocessing in modern radar classification function chains is to prepare and condition the acquired signal in order to simplify the feature extraction and later classification or recognition of the required patterns.

Generally the following three steps are indispensable:

1. Filtering and noise suppression; 2. Clutter suppression; 3. Normalisation

Filtering and noise suppression consists of all processes in the classification chain that are required to eliminate deterministical and well-known noise effects. For this purpose finite impulse response filters (FIR) as well as infinite impulse response filters (IIR) can be used. The filter design requires a good understanding of the defined radar, environment and target scenario. The following filter types are commonly used for suppressing noise in low, high, bandpass or stopband form (Stimson, 1998; Kouemou et al., 1996; Kouemou, 2000; Kammeyer & Kroschel, 2002):

Butterworth, Chebyshev filter, Elliptic (Cauer) filter, Bessel filter, Gaussian filter, Optimum "L" (Legendre) filter, Linkwitz-Riley filter, Gabor filter.
The following figure displays the normalized frequency responses of four common discrete-time filter in eighth-order (Butterworth, Chebyshev type 1, Chebyshev type 2, Elliptic)

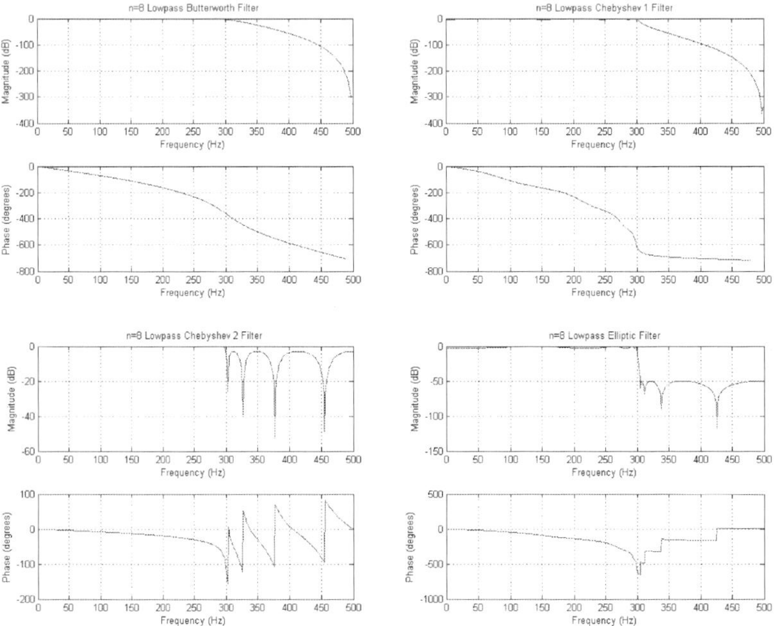

Fig. 3. Example of normalized frequency responses of typical simple discrete-time filters

The normalising process conditions the signal in a way, that, independent of the data acquisition situation, the samples, used to train a classifier, will be comparable to each other in the matter of characteristical features of a given object class, type or identity. It must be differentiated between statical and dynamical normalisation procedures. The problem of the statical normalisation procedure is that they often work very well in offline radar classification processes but fail in online classification situations. For this reason dynamical normalisation procedures are often required for online radar classification purpose. One of the simplest methods for designing dynamical normalisation procedure is the usage of moving average numeric (Kouemou et al., 1996; Kouemou, 2000).
A moving average is used to process a set of samples by creating a series of averages of different subsets by the given radar data set. A moving average is a vector of real numbers which are stored in a window function. The windows will be moved according to rules which will be designed by the radar classification experts depending on the given problem. Fig. 4 shows such an example for two simple weighted moving average windows.
The clutter suppression process eliminates land, sea, volume or space clutter information in order to reduce the complexity of the feature extraction process on one hand and to reduce the number of classes, types or identities of the pattern recognition procedure on the other

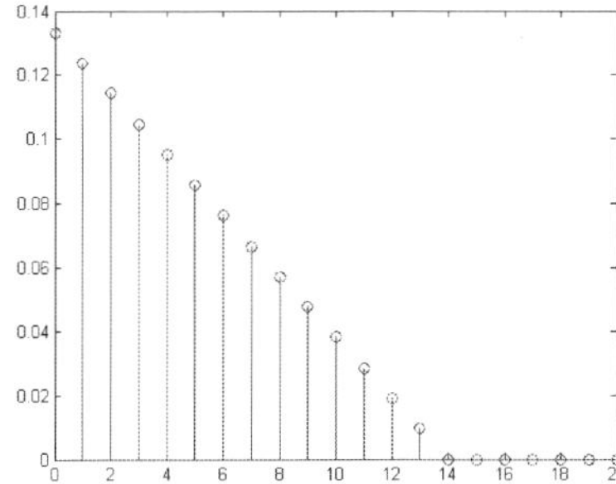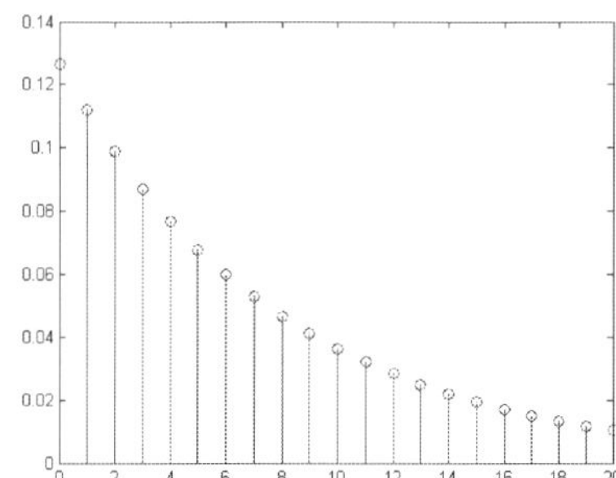

Fig. 4. Two examples of simple moving average weights, the left one of a weighted moving average, the right one of a exponential moving average

(Stimson, 1998; Skolnik, 1990; Barton, 1988). While designing modern radar systems one of the decisions of the designer experts is to decide where to integrate the clutter suppression procedure. Depending on the design philosophy this can be integrated in the parameter extraction and plot extraction function chain. Another design philosophy predicts advantages in integrating the clutter suppression process in a clutter learning map that can be updated live by an experienced operator. An advantage of such a philosophy is that the classifier learns how to recognise specific clutters offline or online. With this knowledge the radar clutter classifier could give a feedback to the parameter and track extractor in order to simplify the parameter extraction and plot procedure. This information could also be used to improve the results of the tracking system.

4. Feature extraction

For the feature extraction part several basic techniques can be used. One of the most successful philosophies while designing modern radar systems for classification purpose is the best handling of the feature extraction. This philosophy consists of best understanding of the physical behaviour of a radar system in its environment. Based on this understanding characteristical features must then be mathematically described depending on the given requirements. In the literature many methods are described for this purpose. In this chapter we will focus on the following basic methods:
1. Short-Time-Fourier transform; 2. Cepstral analysis; 3. Wavelet transform; 4. Fuzzy-logic.

4.1 Feature extraction using Short-Time-Fourier-analysis
The short-time Fourier transform (STFT) is a trade-off between time and frequency based views of a radar echo signal. It provides information about both when and what frequencies a signal event occurs. However, this information can only be obtained with limited precision, and that precision is determined by the size of the chosen window. A particular size for the time window is chosen. That window is the same for all frequencies. Many signals require a more flexible approach, one where we can vary the window size to determine more accurately either time or frequency.
The STFT is calculated by

$$X_h(t,\omega) = \int\limits_{-\infty}^{\infty} x(\tau) \cdot \overline{h}(\tau - t) \cdot e^{-i\omega\tau} d\tau \tag{1}$$

where h is a windowing function (Kouemou, 2000; Kammeyer & Kroschel, 2002; Kroschel, 2004). In Fig. 5 an example of a short-time Fourier transform of a time signal can be seen.

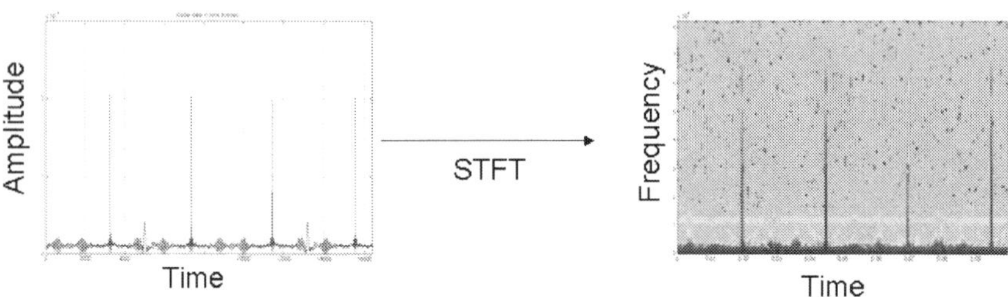

Fig. 5. Example of a short-time Fourier transform

4.2 Cepstral-analysis

The feature extraction process can also use a spectral-based technique similar to those which are used in speech processing, namely the Melscale Frequency Cepstral Coefficients (MFCC). It is well-known that moving targets create a modulated radar return signal whose characteristics in spectrum can be used to distinguish between the classes. This process is based directly on the complex I/Q radar Doppler signals. Due to several moving parts with different velocities of a target, the radar return signal may cover the whole frequency band, depending on the pulse repetition frequency (PRF), from $-\mathrm{PRF}/2$ to $\mathrm{PRF}/2$. Hence no linear filter is applied in order to retain any important frequencies. The common MFCC process is adapted to complex radar signals. The radar return signal of several hundred milliseconds is framed using a half-overlapping Hamming window in order to create signal segments representing the short quasi-stationary parts of the Doppler signal. The following feature extraction process is done for every frame and each frame results in a feature vector (Kouemou, 2000; Kouemou & Opitz, 2007a; Kouemou & Opitz, 2007b):

1. Apply the Fast Fourier Transform (FFT) to the signal resulting in the spectrum $\mathcal{F}\{s(n)\}, n = 1,...,T$, where T is the number of samples.

2. Calculate the power spectrum

$$P_f\{s(n)\} = \mathcal{F}\{s(n)\} \cdot \mathcal{F}_f\{s(n)\}^* \tag{2}$$

3. Mirror the power spectrum at zero frequency and add the negative frequencies $(n = 1,...,T/2)$ to the positive ones $(n = T/2+1,...,T)$ to get a positive spectrum $\tilde{P}_f\{s(n)\}$ of half the length as the sum of amplitudes of negative and positive frequencies, i.e. for $n = 1,...,T/2$:

$$\tilde{P}_f\{s(n)\} = P_f\{s(T/2+n)\} + P_f\{s(T/2-n)\} \tag{3}$$

4. Apply a k channel mel filter bank to $\tilde{P}_f\{s(n)\}$ of triangular shaped filters by multiplying with the transfer function $H_i(n)$ of the ith filter to get the filter bank results $Y(i)$. The number of channels is adapted to operator's capabilities.

5. Calculate the logarithm $\log(Y(i))$

6. Decorrelate the result with the Discrete Cosine Transform (DCT) to get the so called mel cepstrum

7. Take only the first m of the k coefficients of the mel cepstrum result

8. The feature vector can be extended adding dynamic features by using the first and second derivative of the coefficients with respect to time

9. The zero coefficient is replaced by a logarithmic loudness measure

$$c_0 = 10\log\left(\sum_{n=1}^{T/2}\tilde{P}_f\{s(n)\}\right) \tag{4}$$

The mel filter bank in step 4 is based on half-overlapping triangular filters placed on the whole frequency band of the signal. The lower edge of the first filter may be placed on a frequency greater than zero in order to filter out any dominant ground clutter.

As a result from the method above we achieve a sequence of MFCC feature vectors which represent the radar Doppler signal.

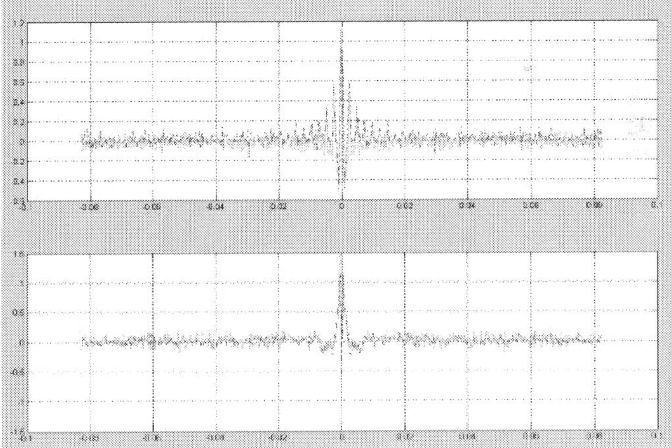

Fig. 6. Example of extracted cepstral based feature vectors of person (bottom), and tracked vehicle (top)

4.3 Wavelet-transform

The Wavelet transform $\mathcal{W}$ of a radar echo signal f (cf. Fig. 7) is basically defined by the following equations with normalisation factor a and time shift factor b:

$$\mathcal{W}\{f\}(a,b) = \int |a|^{-1/2} f(t)\psi\left(\frac{t-b}{a}\right)dt \tag{5}$$

$$W_{m,n}\{f\} = \int a_0^{-m/2} f(t) \psi \left(a_0^{-m} t - n b_0 \right) dt \tag{6}$$

where in both cases we assume that the "Mother Wavelet" ψ satisfies the condition

$$\int \psi \, dt = 0 \tag{7}$$

By restricting a and b to discrete values one can obtain formula (6) from (5): in this case $a = a_0^m$, $b = n b_0 a_0^m$ with $m, n \in \mathbb{Z}$, $a_0 > 1$, $b_0 > 0$ fixed.

The most popular Wavelets are: the Daubechies (Daubechies, 1992), Meyer, Mallat (Mallat, 1999), Coiflet, Symlet, Biorthogonal, Morlet and the Mexican Hat.

The Wavelet based feature extraction methodology that was developed for this study is decomposed in five main steps (Kouemou, & Opitz, 2005; Kouemou & Opitz, 2008b):

1. Design of the Wavelet type
2. Definition of the Wavelet observation window as subfunction of time on target
3. Definition of the Wavelet scaling function
4. Definition of the Wavelet dependency function to the radar operating pulse repetition frequency
5. Definition of the statistic adaptation model by combining the Wavelet extracted feature to the Discrete Hidden Markov Model

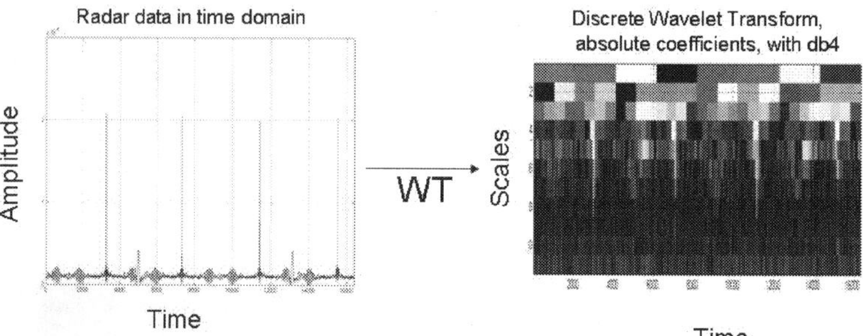

Fig. 7. Example of Wavelet extracted feature from a given air target using a Pulse-Doppler radar

4.4 Fuzzy-logic

The calculation by using the fuzzy logic module works slightly different. For this approach physical parameters, for example the velocity, the acceleration, the RCS, etc., of the target have to be measured. We need supporting points h_n to set up the fuzzy membership functions. Those membership functions have to reflect physical limits stored in knowledge data base (Kouemou et al., 2008; Kouemou et al., 2009; Kouemou & Opitz, 2008a).

For each class, membership functions are set up and membership values m depending on the supporting points are calculated as follows:

$$m(i,j) = \begin{cases} 1 & \text{if } w \leq h_1 \\ \frac{-1}{h_2 - h_1}(w - h_1) + 1 & \text{if } h_1 \leq w \leq h_2 \\ 0 & \text{else} \end{cases} \tag{8}$$

where w is the measured value for the considered physical attribute. With those membership values, the elements of a matrix $\tilde{P}$, containing probabilities for each parameter and each considered target class, are:

$$\tilde{P}_{i,j} = \frac{m(i,j)}{\sum_{k=1}^{N} m(i,k)} \tag{9}$$

An example membership function is illustrated in Fig. 8, where in the upper picture the membership function for a knowledge-based module, a special case of the fuzzy-logic approach, is shown, while in the lower picture the membership function for a trapezoidal fuzzy logic module can be seen.

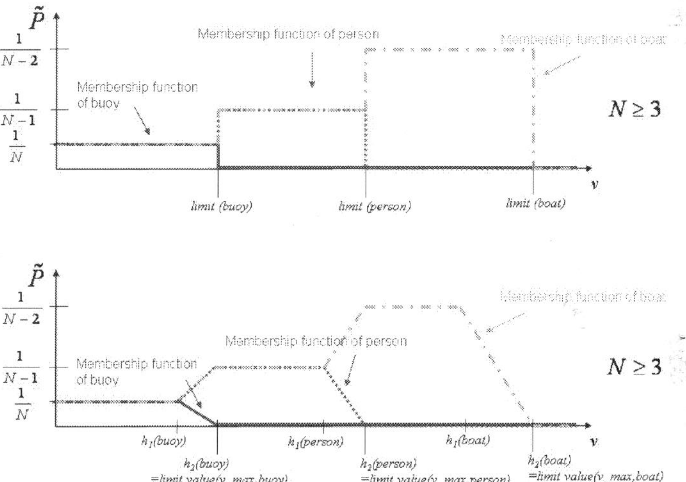

Fig. 8. Example of membership functions of target classes 'buoy', 'person' and 'boat' for knowledge-based and fuzzy logic module

The next step in order to receive the result vector is the introduction of a so called weighting matrix W. The weighting matrix contains elements $\omega_{i,j}$, with $i=1...M$ and $j=1...N$. The weighting matrix represents the influence of the single physical values i on the classification result for a given target class j.

The elements of W depend on several conditions, like available measurement data of radar, quality of knowledge database and terrain character, environmental or weather conditions.

Further feature extraction methods also considered in modern radar target classification are different types of autoregressive filters, auto- and crosscorrelation functional analysis as well as linear and non-linear prediciton-codeanalysis (Kammeyer & Kroschel, 2002; Kroschel, 2004).

Further applications use the properties of the Karhunen-Loewe transform as well as moments of higher order (Gardner, 1980; Gardner, 1987; Gardener, 1988; Gardner & Spooner, 1988; Fang, 1988; Fang, 1991).

5. Classification technologies

In this section two main philosophies for classification and subclassification will be presented. The first philosophy consists of a learning process. The second philosophy consists of knowledge based evidence. The different kind of classification and subclassification methods in most modern radar systems can be divided into deterministical methods, stochastical methods and neural methods. The deterministical methods in this section are essentially based on the handling of logical operators and knowledge based intelligence. The stochastical methods described in this section are based on finite stochastical automats. The finite stochastical automats which will be presented are usually based on different variants of learning Hidden Markov Models. Furthermore the neural methods illustrate the capability of solving pattern recognition problems in modern radar systems by using different kind of artificial neural networks. For specific classification or subclassification challenges in modern radar applications hybrid classifiers can also be recommended. These classifiers use – depending on the situation – learnable or non-learnable algorithms. The learnable algorithms can be designed using supervised or unsupervised learn concepts.

5.1 Classical knowledge based approach

For calculation with the classical knowledge-based approach, we need known limits for the considered physical values applying to the considered target classes. For example, the knowledge database needs to hold maximum velocities for persons as well as for tracked vehicles and all others. With the limit interval L from the knowledge database and the measured value v, we can calculate the elements of matrix $\tilde{P}$:

$$\tilde{P}_{i,j} = \begin{cases} \frac{1}{Q} & if\ v \in L \\ 0 & else \end{cases} \quad \forall\ i,j \tag{10}$$

Q represents the number of classes for which the measured value v is within the limits of the interval L. Therefore, the summing up of all $1/Q$ terms always has to yield 1.

Three technologies for radar target classification will be described in this section. One based on a classical knowledge based approach, one based on neural networks and the other based on stochastic automats.

5.2 Neural networks

Many different families of artificial neural networks are state-of-the-art in modern radar classification and identification issues. They can be divided into statical and dynamical networks (Rosenblatt, 1962) on one side, and into self-organising (Kohonen, 1982; Kohonen, 1984; Kohonen, 2001) and supervised learnable networks on the other side (Zell, 1994).

In this section the time delay neural networks (TDDN) as an example for dynamical feed-forward networks will be briefly introduced.

The time delay neural networks (TDNN) were developed by Waibel and Lang for classifying phonemes of speech in 1987. They belong to the class of forward networks and were firstly used in the speech recognition. Nowadays the classical architecture of TDNN was extended with special techniques for radar classification purposes. Classically, these networks consist of an input layer, an output layer and one or more hidden layers (cf. Fig.

9). The hidden layers lie between the input and the output layer. The number of hidden layers has to be determined depending on the application.

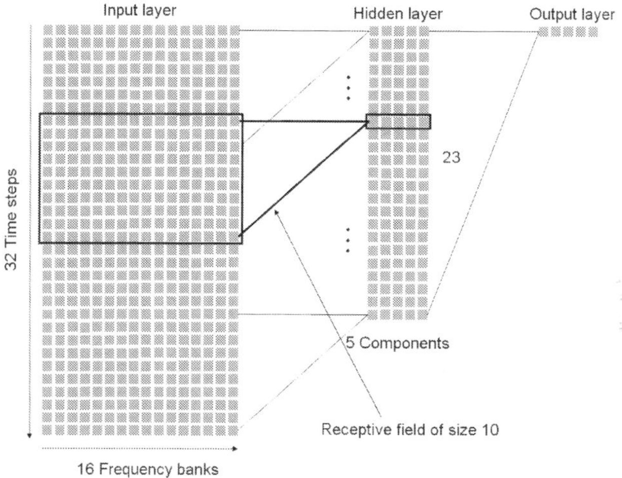

Fig. 9. Exemplary structure of a time delay neural network

Every layer is described by a matrix. The columns of the input layer represent the created frequency bank of the recorded Radar Doppler signals. The rows describe the time delay of the input pattern. A layer consisting of 32 rows denotes that an incoming information will be included in the calculation for 32 time units. The layers are divided into varying long time steps Δt according to the application.

The connection between the layers depends on the size of the so called receptive field. A receptive field means that a row of the subsequent layer is only connected to a defined number of rows of the preceding layer. For instance a receptive field of size 10 means that 10 rows of the first layer are each connected to a row of the subsequent layer. Thus a row of the subsequent layer can only see a short time period of the preceding layer.

Of course the neurons of a layer can be connected to every neuron of the preceding layer. Thus the recognition and training time of the networks would increase significantly. By the use of a receptive field of size r the number of connections can be reduced. Without this factor the combination of a layer of size 32x16 with a subsequent layer with 23x5 neurons would require 58,880 connections (32·16·23·5=58,880). For r=10 a row of the hidden layer sees only a temporal sequence over ten rows of the input layer. This reduces the number of connections significantly (instead of 58,880 only 10·16·23·5=18,400). Thereby the length of the subsequent layer is determined by the receptive size of the previous layer:

$$l = d - r + 1 \tag{11}$$

whereas the parameters are the following:
- l is the length of the subsequent layer
- d is the number of time pattern frames
- r is the size of the receptive field

The number of components (of columns) of the hidden layers is chosen according to the application. The more components exist, the more precise the recognition will be. However, the higher the number of components and the longer the layers are, the longer the recognition and training time will be.

The characteristic of this network lies in the recognition of features in time varying patterns at different positions and of different lengths. Their architecture permits to recognise the components of time varying patterns, in spite of time shift.

5.3 Stochastic automats

One of the most important family of classifiers nowadays in modern radar applications consists of finite stochastic automats. They are most the time derived from extended Bayesian networks. One of the most successfully integrated stochastical automats in technical radar systems applications are Markov-based learnable models and networks. In this section a brief introduction will made exemplary by taking the Hidden Markov Model (HMM) (Kouemou, 2000; Kouemou & Opitz, 2007a; Kouemou & Opitz, 2007b; Kouemou & Opitz, 2008a; Kouemou & Opitz, 2008b, Rabiner, 1989). Some experimental results of classification problems using HMMs will be shown in the experimental section.

A HMM consists mainly of five parts:

1. The N states $S = \{S_1, ..., S_N\}$;

2. M observation symbols per state $V = \{v_1, ..., v_M\}$;

3. State transition probability distribution $A = \{a_{ij}\}$, where a_{ij} is the probability that the state at time t+1 is S_j, given the state at time t was S_i;

4. Observation symbol probability distribution in each state $B = \{b_j(k)\}$, where $b_j(k)$ is the probability that symbol v_k is emitted in state S_j;

5. Initial state distribution $\pi = \{\pi_i\}$, where π_i is the probability that the model is in state S_i at time $t = 0$.

Mainly three different types of HMM exist in literature: the HMM with discrete outputs (DHMM), HMM with continuous output probability densities (CHMM) and the trade-off between both, the semi-continuous HMM (SCHMM).

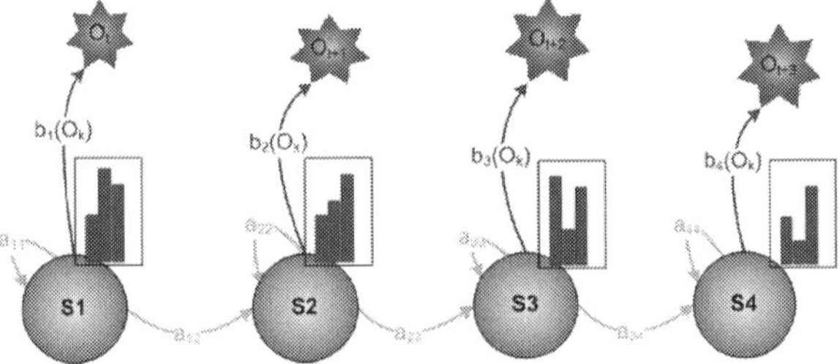

Fig. 10. Example of an HMM

Further methods for automatic target classification applied in modern radar systems are based on alternative methods such as classical learning Bayesian networks (Bayes, 1763; Kouemou, 2000; Pearl, 1986; Pearl & Russell, 2003)   as well as polynomial classifiers

(Schürmann, 1977), learning support vector machines (Ferris & Munson, 2002; Burges, 1998; Drucker, 1997) or evolutionary algorithm (Bäck, 1996; Bäck et al., 1997; Goldberg, 1989).

6. Classifier fusion technologies

Different techniques and strategies can be used in order to fuse information from different sensor systems. For example beside the Doppler information can also information of a radar tracking system be used for classification. From the information of a tracking system physical parameters such as velocity, acceleration, etc. of the target can be estimated (Kalman, 1960; Bar-Shalom, 1989; Blackman & Popoli, 1999). The introduced data fusion techniques can also be integrated in a stand-alone sensor system in order to produce a robust classification and recognition result. For this purpose three technologies will be presented in order to solve the given problems:
Bayesian networks based method, Dempster-Shafer rules based fusion methods and finally classical rule based methods.

6.1 Bayesian rules

For the application of Bayes' Theorem (Bayes, 1763) we assume that the considered physical parameters are stochastically independent. This requirement has to be verified. This can be done by considering the correlation between the physical values. If the correlation is nearby zero, the assumption can be held up.

The probabilities $P(j\,|\,I)$, i.e. the probabilities of the classes $j \in J$ under the condition of certain measured physical values, are searched.

From the classification one gets a matrix P whose items can be interpreted as conditional probabilities $P(i\,|\,j)$. The individual probabilities P_{ij} shall be combined for all j to one probability p_j of the class j.

The stochastically independence of the physical values is assumed, as mentioned before. If the classes are additionally independent, the following equation holds:

$$P(I\,|\,j) = P(i_1\,|\,j)\cdot P(i_2\,|\,j)\cdot .... \cdot P(i_M\,|\,j) \tag{12}$$

To simplify the calculation this assumption is also done.
Furthermore let all classes have the same a priori probability:

$$P(j) = \frac{1}{N} \quad \forall\, j \in J \tag{13}$$

The $j \in J$ generate a complete event system, i.e. $J = j_1 \cup j_2 \cup ... \cup j_N$ with $j_k \cap j_l = \varnothing$ for $k \neq l$. This yields:

$$P(j\,|\,I) = \frac{P(I\,|\,j)\cdot P(j)}{P(I)} = \frac{P(I\,|\,j)\cdot P(j)}{\displaystyle\sum_{k=1}^{N} P(I\,|\,k)P(k)} = \frac{P(I\,|\,j)}{\displaystyle\sum_{k=1}^{N} P(I\,|\,k)}$$

$$\approx \frac{P(i_1\,|\,j)\cdot P(i_2\,|\,j)\cdot .... \cdot P(i_M\,|\,j)}{\displaystyle\sum_{k=1}^{N} P(i_1\,|\,k)\cdot P(i_2\,|\,k)\cdot .... \cdot P(i_M\,|\,k)} = \frac{\displaystyle\prod_{\nu=1}^{M} P(i_\nu\,|\,j)}{\displaystyle\sum_{k=1}^{N} \prod_{l=1}^{M} P(i_l\,|\,k)} \tag{14}$$

This proceeding can be done inside a trackbased classification considering two or more physical parameters. But it can be also used to fuse different classifiers to get a combined classification (Kouemou et al., 2008; Kouemou & Opitz, 2008a).

6.2 Dempster-Shafer method

The Dempster-Shafer theory of evidence is a generalization of Bayes' theorem. It is based on the universal set $\mathcal{P}(J)$ of the set of all considered target classes J.

$$\begin{aligned}
\mathcal{P}(J) = \{ \ & \varnothing, \{person\}, \{tracked\ vehicle\}, \{helicopter\}, ..., \\
& \{person, tracked\ vehicle\}, \{helicopter, propeller\ aircraft\}, ... \\
& \{person, wheeled\ vehicle, tracked\ vehicle, helicopter, \\
& propeller\ aircraft, no\ match\} \ \}
\end{aligned} \tag{15}$$

For this approach 'No match' denotes the set of all considered target classes and all objects which can not be allocated to one of the defined classes. In the Bayesian approach 'No match' denotes only all objects not allocated to one of the classes is therefore strictly separated from the other classes.

Compared to the Bayesian approach the target class 'no match' is not strictly separated from the other classes when using Dempster-Shafer. Evidences are used instead of probabilities. So there is an opportunity to deal with uncertain information and illustrate ignorance explicitly (Dempster, 1968; Shafer, 1976; Shafer, 1990; Shafer & Pearl, 1990).

The task is to combine the results coming from Doppler sound classification and from trackbased classification. The elements of the vectors are considered to be evidences E.

The evidences are combined using Dempster's rule of combination:

$$E_l(j) \oplus E_{l+1}(j) = \frac{1}{1 - k_l} \cdot \sum_{Y \cap Z = j} E_l(Y) \cdot E_{l+1}(Z) \tag{16}$$

$$k_l = \sum_{Y \cap Z = \varnothing} E_l(Y) \cdot E_{l+1}(Z) \quad \text{with } Y, Z \in J \tag{17}$$

The factor k expresses the conflict between the propositions, $k=1$ stands for totally contradictory propositions. The result of this combination is the combined vector p, whose elements are also evidences. Using these evidences, degrees of belief and plausibility can be calculated:

$$bel(A) = \sum_{B:B \subseteq A} E(B) \tag{18}$$

$$pl(A) = \sum_{B:B \cap A \neq \varnothing} E(B) \tag{19}$$

In doing so, the degree of belief shows how well the evidences support the proposition. On the other side, the degree of plausibility shows how well the negation of one proposition is supported. With $bel(A)$ being a lower bound for the probability of proposition A and $pl(A)$ being an upper bound, one gets a limited interval for the probability of A.

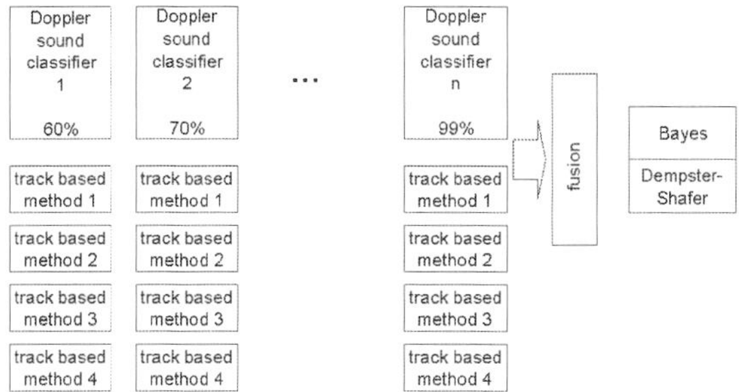

Fig. 11. Possible combinations of classifiers

As several HMMs are available for classification purposes as well as different methods to apply the track based classifier, a comparison of possible combinations is necessary. In Fig. 11 possible combinations are shown. In the track based classification, the evaluation of the target dynamics can either be performed in a classical knowledge-based way or by applying fuzzy logic, while the combination of the single membership values can be either done by applying Bayesian or Dempster-Shafer rules. These methods can be combined resulting in four different ways to use the track based classifier (Kouemou et al., 2008; Kouemou et. al, 2009).

6.3 Classical rule based approach

For this approach first two vectors p_1 and p_2, each containing probabilities for the several target classes, are considered. These vectors can be derived from each of the methods mentioned in the previous section. So the vectors are:

$$p_1 = \begin{bmatrix} p_{1,1} & p_{1,2} & \cdots & p_{1,N} \end{bmatrix} \tag{20}$$

$$p_2 = \begin{bmatrix} p_{2,2} & p_{2,2} & \cdots & p_{2,N} \end{bmatrix} \tag{21}$$

Both vectors shall be fused to one result vector using a rule based approach. This can be exemplary done with the following rule:

$$i = \arg\max(p_1)$$
$$p := \begin{cases} p_1, & \text{if } p_{2,i} \neq 0 \\ [0 \quad 0 \quad \cdots \quad 0 \quad 1], & \text{if } p_{2,i} = 0 \end{cases} \tag{22}$$

The second case is also called rejection, i.e. the classifier makes no decision due to lack of information or contradictory results of the stand-alone classifiers.

7. Object recognition

In modern radar systems, recorded data as well as recorded intelligence information can be used together with the classifier output or data fusion output information in order to exactly recognize, identify or type an object. This process is depicted in Fig. 12.
The recognition can be done for example with an identity data base, with a typing data base or with knowledge based intelligence (Kouemou & Opitz, 2005; Schürmann, 1996).

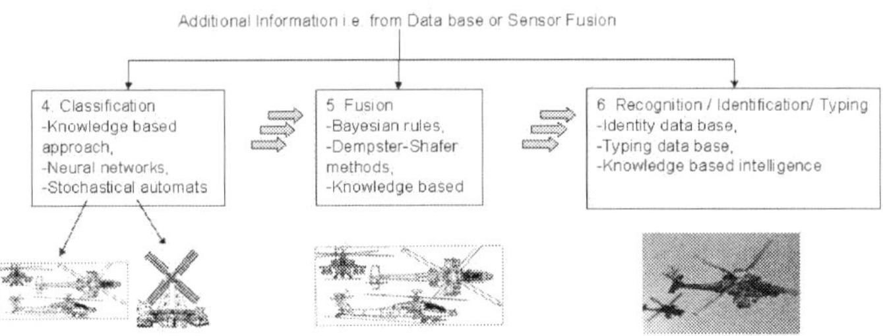

Fig. 12. Simplified illustration of an object recognition process

8. Some experimental results

In this section some classification results, where the introduced technologies were successfully tested, are presented. Due to company restrictions only results based on simulated data are presented.

8.1 Exemplary training algorithm

The following Fig. 13 illustrates a typical functional flow-chart of a supervised training procedure. This can be used for example to train a rapid backpropagation algorithm for an artificial network classification process. Similar structures can be used for the training of stochastical automats or support vector machines (Ferris & Munson, 2002; Burges, 1998; Drucker et al, 1997).

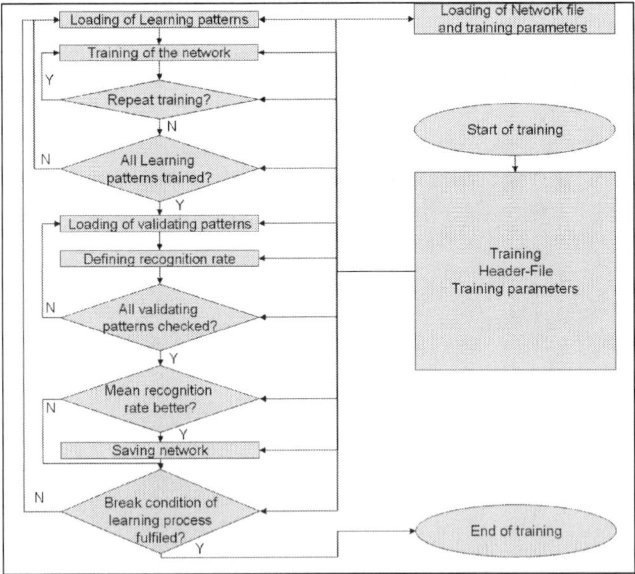

Fig. 13. Simplified illustration of an exemplary training proceeding

8.2 Exemplary identification results based on Hidden Markov Model and neural network technology

The following results were obtained by simulating a naval based radar target identification scenario using special neural network algorithms. It was presented at the International Radar Symposium in Berlin, Germany, 2005. Fig. 14 shows the two helicopters to be identified. A typical feature extraction process necessary as pre-classification step for the neural network is shown in Fig. 15. An exemplary confusion matrix of the classifier result is shown in Fig. 16.

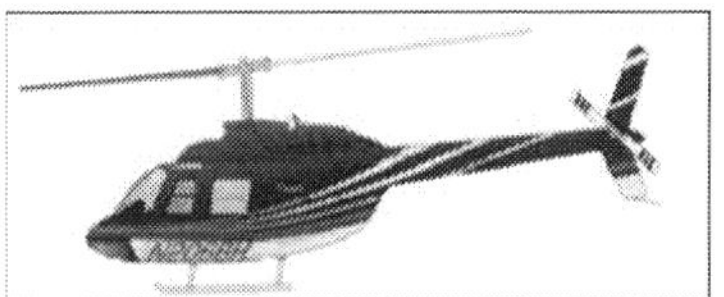

Fig. 14. Typical helicopter identification example, on the left side a Bell Jet Ranger 206 B and on the right side a Bell UH1D

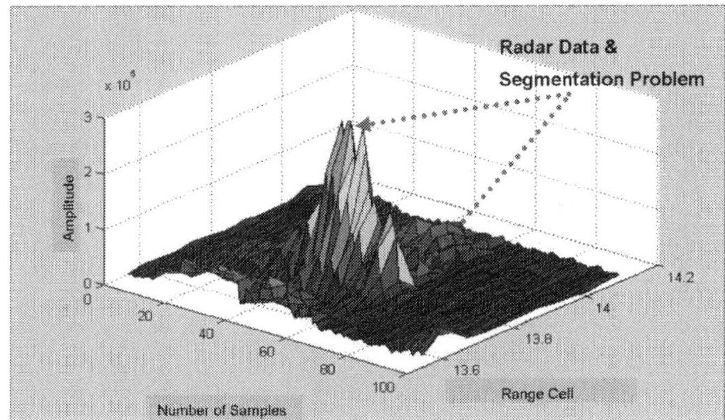

Fig. 15. Typical feature extraction process necessary as pre-classification step for the neural network of a simulated Bell jet helicopter.

	Bell UH 1D	Jet Ranger	No Match
Bell UH 1D	93,25	1,62	5,13
Jet Ranger	2,45	90,24	7,31

	Bell UH 1D	Jet Ranger	No Match
Bell UH 1D	94,75	0,93	4,32
Jet Ranger	1,75	91,21	7,04

Fig. 16. Two exemplary identification confusion matrix results using neural network on the left side and a Hidden Markov Model on the right side

8.3 Exemplary classifier results based on a hybrid system with a stochastical automat and Dempster-Shafer method

An Exemplary structure of the simplified hybrid classifier operating with a knowledge based fusion technique. It was presented at the IEEE Radar 2008 in Adelaide, Australia.

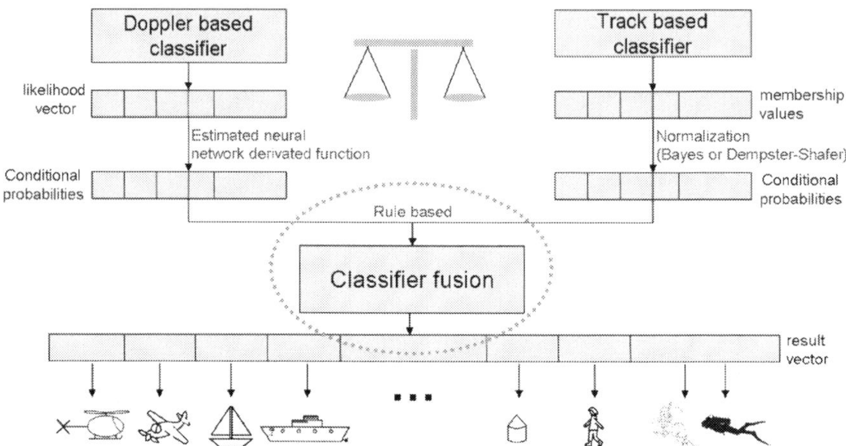

Fig. 17. Simplified structure of fusion process using Dempster-Shafer and knowledge based rules.

		person	wheeled vehicle	tracked vehicle	helicopter	propeller aircraft	buoy	boat	no match	rejection
labeled as	person	100	0	0	0	0	0	0	0	0
	wheeled vehicle	0	87	12	0	0	0	0	0	1
	tracked vehicle	0	26	74	0	0	0	0	0	0
	helicoper	0	0	0	89	0	0	0	0	11
	propeller aircraft	0	0	0	8	92	0	0	0	0
	buoy	0	0	0	0	0	84	16	0	0
	boat	0	0	0	0	0	0	100	0	0
	no match	31	0	0	11	0	0	0	58	0
						recognized as				

Fig. 18. Typical confusion matrix of a classifier fusion obtained after simulating a testing process as described in the scheme above (Fig. 17)

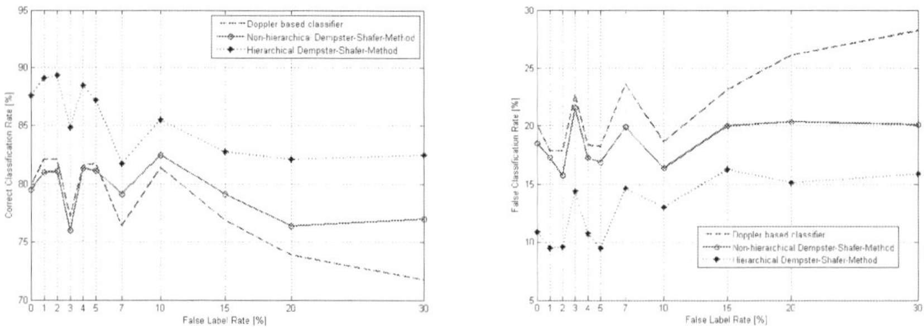

Fig. 19. Typical improvement obtained after fusing a stand-alone trained Doppler classifier with two Dempster-Shafer techniques.

9. Some application examples with modern radar systems

9.1 Example of an airport surveillance radar system (ASR-E)

ASR-E is the latest generation of a modern approach control radar for civil, military and dual use airports. ASR-E provides most advanced technologies. It uses a fully solid state S-band Primary Radar with outstanding, reliable detection performance and a monopulse Secondary Radar covering civil, military and Mode S interrogation modes. It is used for example by the German Air Force.

Fig. 20. Airport Surveillance System (Photo courtesy of EADS)

9.2 Example of a naval surveillance radar system (TRS-3D)

The TRS-3D is a multimode surface and air surveillance and target acquisition radar, designed for complex littoral environment with excellent clutter performance to detect small fast flying threats.

The TRS-3D is used for the automated detection, track initiation and tracking of all types of air and sea targets.

- Automatic detection and track initiation for air and sea targets
- Very low antenna weight
- Proven AAW weapon engagement support
- Gunfire control for sea targets (no FCR required)
- Helicopter automatic detection & classification
- Helicopter approach control
- Data correlation with integrated IFF interrogator
- Low risk integration with many CMS systems

The TRS-3D is used by various navies worldwide and has proven its operational performance from arctic to subtropical regions.

9.3 Example of a tactical radar ground surveillance system (TRGS)

The TRGS is a high performance ground surveillance radar system for the automatic detection, identification and classification of ground targets, sea targets and low flying air targets.

It is a vehicle integrated system with multi-sensor configuration. The electronically scanning antenna is one of the advanced technologies used in this system. One of the key features is the particularly high target location accuracy.

Fig. 21. TRS-3D on the Finnish guided missile boat "Hamina" (Photo courtesy of EADS)

Fig. 22. TRGS on the armoured vehicle "DINGO2" (Photo courtesy of EADS)

9.4 Example of an airborne radar integrated in a modern combat helicopter (Tiger)

The Tiger HAP is an air-to-air combat and fire support medium-weight (6 tonnes) helicopter fitted with 2 MTR 390 engines. It is daytime and night combat capable and is operable in NBC environments. Three basic parameters were taken into account right from the start of the development phase: low (visual, radar and infrared) detectability, which provides excellent survivability on the battlefield, maximum efficiency of the weapons and the associated fire control systems without heavier workload for the crew, and an optimized logistic concept offering minimum possession costs. The integrated airborne high-PRF radar fulfils all requirements needed in a critical combat environment.

9.5 Example of a spaceborne radar integrated in a satellite system (TerraSAR-X, TanDEM-X)

TanDEM-X ('TerraSAR-X add-on for Digital Elevation Measurement') is a radar observation satellite, which, together with the almost identical TerraSAR-X radar satellite, will form a high-precision radar interferometer system.

Fig. 23. Tiger HAP (Photo courtesy of EADS Eurocopter)

With this TerraSAR-X/TanDEM-X tandem formation, generating images similar to stereoscopic pictures, it will be possible to measure all the Earth's land surface (150 million square kilometres) within a period of less than three years. For a 12m grid (street width), surface height information can then be determined with an accuracy of under two meters. One goal is the production of a global Digital Elevation Model of unprecedented accuracy.

As for TerraSAR-X, the TanDEM-X project will be carried out under a Public–Private Partnership between Astrium and the German Space Agency DLR. TanDEM-X is due for launch in 2009, and is designed to operate for five years.

Use of the data for scientific applications will be the responsibility of the DLR's Microwaves and Radar Institute. Commercial marketing of the data will be managed by Infoterra GmbH, a wholly-owned subsidiary of Astrium.

Fig. 24. TanDEM-X in space (left illustration) and an Ariane 5 at launch (right photo) (Photo courtesy of EADS Astrium)

10. Conclusion

In this chapter basics of radar target classification technologies were introduced. A classification technology was presented, that decomposes a pattern recognition module of any modern radar system in the following components:
Data acquisition part, signal preprocessing and feature extraction part, classification and subclassification part, data and information fusion part and finally object recognition or identification or typing part.
For the data acquisition part an active or passive radar frontend can be used, that uses self- or friendly generated waveforms to reconstruct information from the environment with different objects or targets. The data acquisition part usually provides such backscattered radar echo signal in I- and Q-form in the baseband.
For the signal preprocessing part some basic techniques were described in order to filter and normalise the sampled signal. It was mentioned that some measures must be taken into consideration in order to respect the basics of information theory.
For the feature extraction part several basic techniques can be used. It was also mentioned that one of the most successful philosophies in designing modern radar systems for classification purpose is the best handling of the feature extraction. This philosophy consists of best understanding of the physical behaviour of a radar system in its environment. Based on this understanding characteristical feature must then been mathematically described depending on the given requirements. For this purpose the following basic methods were presented as central components of the feature extraction process: Short-Time-Fourier transform, cepstral analysis, wavelet transform and Fuzzy-logic.
For the classification and subclassification part two main philosophies were presented. The first philosophy consists of learning processes. The second philosophy consists of knowledge based evidence. The different kind of classification and subclassification methods in the most modern radar systems can be divided into deterministical methods, stochastical methods and neural methods. The deterministical methods introduced in this section were essentially based on the handling of logical operators and knowledge based intelligence. The stochastical methods described in this section were based on finite stochastical automats. The finite stochastical automats presented in this section were based on different variants of learning Hidden Markov Models. Furthermore the neural methods presented in this section illustrate the capability of solving pattern recognition problems in modern radar systems by using different kinds of artificial neural networks. It was also shown that for specific classification or subclassification challenges in modern radar applications hybrid classifiers can also be recommended. This classifier uses depending on the situation learnable or non-learnable algorithms. The learnable algorithms can be designed using supervised or unsupervised learn concepts.
For the data and information fusion part it was pointed out that different techniques and strategies can be used in order to fuse information from different sensor systems. It was also shown that the introduced data fusion techniques can also be integrated in a stand-alone sensor system in order to produce a robust classification and recognition result. For this purpose three technologies were presented in order to solve the given problems:
Bayesian networks based method, Dempster-Shafer rules based fusion methods and finally classical rule based methods.
For the object recognition, identification or typing part it was mentioned that in modern radar systems, recorded data as well as recorded intelligence information can additionally

be used together with the classifier output or data fusion output information in order to exactly recognize, identify or type an object.

11. References

Bäck, T. (1996). *Evolutionary Algorithms in Theory and Practice: Evolution Strategies, Evolutionary Programming, Genetic Algorithms*, Oxford Univ., ISBN 0195099710, USA

Bäck, T.; Fogel, D. & Michalewicz, Z. (1997). *Handbook of Evolutionary Computation*, Oxford Univ. Press

Bar-Shalom, Y. (1989). *Multitarget – Multisensor Tracking: Applications and Advances*, Vol. 1, Artech House, Inc

Barton, D.K. (1988). *Modern Radar System Analysis*, Artech House, Inc

Bayes, T. (1763). An Essay towards solving a Problem in the Doctrine of Chances, *Philosophical Transactions of the Royal Society of London*, 53, pp. 370–418

Blackman, S. & Popoli, R. (1999). *Design and Analysis of Modern Tracking Systems*, Artech House, Inc

Burges, C. (1998). A Tutorial on Support Vector Machines for Pattern Recognition, *Data Mining and Knowledge Discovery 2*, pp. 121–167

Daubechies, I. (1992). I. *Ten lectures on Wavelets*, SIAM Philadelphia

Dempster, A. (1968). A Generalization of Bayesian Inference, *Journal of the Royal Statistical Society Series B 30*, pp. 205–247

Drucker, H. et. al (1997). Support Vector Regression Machines, *Advances in Neural Information Processing Systems 9, NIPS 1996*, MIT Press, pp. 155-161

Fang, T.T. (1988). I- and Q-Decomposition of Self-Noise in Square-Law Clock Regeneration, *IEEE Trans. Commun. 36*, pp. 1044–1051

Fang, T.T. (1991). Analysis of Self-Noise in Fourth-Power Clock Regenerator, *IEEE Trans. Commun. 39*, pp. 133–140

Ferris, M. & Munson, T. (2002). Interior-point methods for massive support vector machines,. *SIAM Journal on Optimization 13*, pp. 783–804

Gardener, W.A. (1980). A Unifying View of Second-Order Measures of Quality for Signal Classification, *IEEE Trans. Commun. 28*, pp. 807–816

Gardener, W.A. (1987). Spectral Correlation of Modulated Signals: Part I – Analog Modulation, Part II – Digital Modulation, *IEEE Trans. Commun. 35*, pp. 584–601

Gardener, W.A. (1988). Signal Interception: A Unifying Theoretical Framework for Feature Detection, *IEEE Trans. Commun. 36*, pp. 897–906

Gardener, W.A. & Spooner, C.M. (1988). Cyclic Spectral Analysis for Signal Detection and Modulation Recognition, *IEEE MILCOM Conference Proceedings*, pp. 24.2.1–24.2.6

Goldberg, D.E. (1989). *Genetic Algorithms in Search, Optimization and Machine Learning*, Kluwer Academic Publishers, Boston, MA

Jahangir, M. et al .(2003). Robust Doppler classification Technique based on Hidden Markov models, IEE Proc., *Radar Sonar Navigation*, 150, (1) , pp. 33-36

Kalman, R. (1960). A New Approach to Linear Filtering and Prediction Problems, *Transactions of the ASME-Journal of Basic Engineering*, 82, pp. 35-45

Kammeyer, K. & Kroschel, K. (2002). *Digitale Signalverarbeitung (5th ed.)*, Teubner, Stuttgart

Kohonen, T.(1982). Self-organized formation of topologically correct feature maps, *Biological Cybernetics*, 43, pp. 59-69

Kohonen, T. (1984). *Self-Organization and Associative Memory*, Springer, Berlin

Kohonen, T. (2001). *Self-Organizing Maps (3rd ed.)*, Springer, Berlin

Kouemou, G. (2000). *Atemgeräuscherkennung mit Markov-Modellen und Neuronalen Netzen beim Patientenmonitoring*, Dissertation an der Fakultät für Elektrotechnik und Informationstechnik der Universität Karlsruhe

Kouemou, G.; Köhle, R.; Stork, W.; Müller-Glaser, K.-D. & Lutter, N. (1996). Einsatz von Methoden der elektronischen Spracherkennung für die Analyse von Atemgeräuschen im Rahmen des Patientenmonitoring, *BMT Kongress*, Zürich

Kouemou, G.; Neumann, C. & Opitz, F. (2008). Sound and Dynamics of Targets – Fusion Technologies in Radar Target Classification, *International Conference on Information Fusion*, pp. 561-567, Cologne Germany

Kouemou, G.; Neumann, C. & Opitz, F. (2009). Exploitation of Track Accuracy Information in fusion Technologies for Radar Target Classification using Dempster-Shafer Rules, *International Conference on Information Fusion*, Seattle USA

Kouemou, G. & Opitz, F. (2005). Wavelet-based Radar Signal Processing in Target Classification, *International Radar Symposium*, Berlin Germany

Kouemou, G. & Opitz, F. (2007a). Hidden Markov Models in Radar Target Classification, *International Conference on Radar Systems*, Edinburgh UK

Kouemou, G. & Opitz, F. (2007b). Automatic Radar Target Classification using Hidden Markov Models, *International Radar Symposium*, Cologne Germany

Kouemou, G. et al. (2008a). Radar Target Classification in Littoral Environment with HMMs Combined with a Track Based classifier, *RADAR Conference*, Adelaide, Australia

Kouemou, G. & Opitz, F. (2008b). Impact of Wavelet based signal processing methods in radar classification systems using Hidden Markov Models, *International Radar Symposium*, Warsaw, Poland

Kroschel, K. (2004). *Statistische Informationstechnik: Signal- und Mustererkennung, Parameter- und Signalschätzung (4th ed.)*, Springer, ISBN-10: 3540402373, Berlin

Mallat, S. (1999). *A Wavelet Tour of Signal Processing (2nd ed.)*, Academic Press, San Diego

Pearl, J. (1986). Fusion, propagation, and structuring in belief networks, *Artificial Intelligence 29(3)*, pp. 241–288

Pearl, J. & Russell, S. (2003). Bayesian Networks, In: *Handbook of Brain Theory and Neural Networks*, Arbib, M.A. (editor), pp. 157–160, MIT Press, ISBN 0-262-01197-2, Cambridge, MA

Rabiner, L.R. (1989): A Tutorial on Hidden Markov Models and Selected Applications in Speech Recognition, *Proceedings of the IEEE*, 77, Nr. 2, pp. 257–286

Rosenblatt, F. (1962). *Principles of Neurodynamics*, Spartan Books

Schukat-Talamazzini, E.-G. (1995). *Automatische Spracherkennung*, Vieweg, 1.Auflage

Schürmann, J. (1977). *Polynomklassifikatoren für die Zeichenerkennung*, R. Oldenburg Verlag, ISBN 3-486-21371-7, München

Schürmann, J. (1996). *Pattern Classification: A Unified View of Statistical and Neural Approaches*, Wiley&Sons, ISBN 0-471-13534-8, New York

Shafer, G. (1976). *A Mathematical Theory of Evidence*, Princeton University Press

Shafer, G. (1990). Perspectives on the Theory and Practice of Belief Functions, *International Journal of Approximate Reasoning*, pp. 323-362

Shafer, G. & Pearl, J. (1990). *Readings in Uncertain Reasoning*, Morgan Kaufmann Publishers Inc., San Mateo

Stimson, G.W. (1998). *Introduction to Airborne Radar (2nd ed.)*, SciTech Publishing, Inc.

Skolnik, M.I. (1990). *Radar Handbook (2nd ed.)*, McGraw-Hill, 0-07-057913-X, New York

Zell, A. (1994). *Simulation neuronaler Netze*, Addison Wesley Longman Verlag GmbH

13

Use of Resonance Parameters of Air-intakes for the Identification of Aircrafts

Janic Chauveau, Nicole de Beaucoudrey and Joseph Saillard
IREENA Laboratory, University of Nantes
France

1. Introduction

Since the development of radar systems, extensive researches have been conducted on the detection and the identification of radar targets. From the scattered electromagnetic field, the aim is to detect and characterize radar targets. In a military context, these targets try to get stealthy to assure their security. This can be provided by using composite materials which absorb electromagnetic waves in usual radar frequency bands. Consequently, studies are involved in lower frequencies. These lower frequency bands correspond to the resonance region for object dimensions of the same order as electromagnetic wavelengths. Therefore, the energy scattered by the target significantly fluctuates and resonance phenomena clearly appear in this region.

Extensive studies have been performed to extract these resonances, mainly the Singularity Expansion Method (SEM) introduced by Baum (Baum, 1976), (Baum, 1991). For years, the SEM is used to characterize the electromagnetic response of structures in both the time and the frequency domains. The SEM was inspired by observing that typical transient temporal responses of various scatterers (e.g. aircrafts, antennas ...) behave as a combination of exponentially damped sinusoids. Such temporal damped sinusoids correspond, in the complex frequency domain, to complex conjugate poles called "natural" poles. The knowledge of these singularities is an useful information for the discrimination of radar targets and it can be used for different purposes of recognition and identification. For example, the E-pulse technique consists in synthesizing "extinction-pulses" signals from natural poles of an expected target, then in convolving them with the measured late-time transient response of a target under test, what leads to zero responses if both targets match (Chen et al., 1986), (Rothwell et al., 1987), (Toribio et al., 2003). In fact, the mapping of these natural poles in Cartesian complex plane behaves as an identity card enabling to recognize the detected target by comparison with a data base of mapping of poles, created before experiments for a set of possible targets. Moreover, the information contained in natural poles can give some indications on the general shape, the nature and the constitution of the illuminated target.

Among radar targets identification problems, the scattering characterization of aircrafts is a relevant topic. Indeed, jet-engine inlets give a very significant contribution to the overall Radar Cross Section (RCS) of an airplane structure. More generally, apertures such as inlets, open ducts and air-intakes can be used for aircraft identification. Thus, numerous studies on the scattering from open cavities are found in the literature. The reader is invited to read the paper of Anastassiu (Anastassiu, 2003), which presents a quite extensive review of methods used to calculate the RCS of such open structures (more than 150 references).

Particularly, when illuminated in a suitable frequency band, an aperture in a radar target body can give access to high-Q internal resonances. Hence the use of natural frequencies of resonance, mainly internal ones, is a relevant basis for open targets identification.

In section 2, we first explain what are the poles corresponding to resonance phenomena and we present the extraction of these natural poles from the simulated transfer function of a target. Using examples of canonical targets (a sphere and a dipole), we consider resonance parameters in the usual Cartesian representation in the complex plane, with real part σ and imaginary part ω, but also in a new representation with quality factor Q and natural pulsation of resonance ω_0. We then introduce the use of these resonance parameters to characterize objects. Afterwards, we extend our study to resonances of targets with apertures, which can give access to internal resonances. In section 3, we first choose the simple example of a perfectly conducting (PC) rectangular open cavity to show how resonance parameters depend on object dimensions and permit to distinguish internal and external resonances. Finally, in section 4, we present a more realistic example in the case of a simulated aircraft with air-intakes, in order to show how to take advantage of internal resonances phenomena of air-intakes for the identification of aircrafts.

2. Resonance parameters of a radar target

A radar target is illuminated in far field by an incident broadband plane wave including resonant frequencies of the target. Consequently, induced resonances occur at these particular frequencies. In the frequency domain, the scattered-field transfer function H is given by the ratio of the scattered field to the incident field, for each frequency. In the time domain, the scattered transient response is composed of two successive parts. First, the impulsive part, $h_E(t)$, corresponding to the early time, comes from the direct reflection of the incident wave on the object surface. In general, for a monostatic configuration, in free space, this forced part is of duration $0 < t \le T_L = 2D/c$, where D is the greatest dimension of the target and c the celerity of light (Kennaugh & Moffatt, 1965). Next, during the late time $(t \ge T_L)$, the oscillating part, $h_L(t)$, is due to resonance phenomena of the target. These resonances have two origins (Chen, 1998): resonances occurring outside the object are called "external resonances" and correspond to surface creeping waves. Conversely, resonances occurring inside the object are called "internal resonances" and correspond to potential cavity waves. In the case of a perfectly conducting closed target, only external resonances occur. The resonant behaviour of the late time is characteristic of the studied target and can be used to define a method of identification.

2.1 Extraction of natural poles

The Singularity Expansion Method (SEM) (Baum, 1976), (Baum, 1991) provides a convenient methodology, describing the late time response of various scatterers as a finite sum of exponentially damped sinusoids

$$h_L(t) \approx \sum_{m=1}^{M} 2|R_m| \exp(\sigma_m t)\cos(\omega_m t + \varphi_m) \tag{1}$$

Conversely, the Laplace transform of equation (1) gives the scattered-field transfer function H(s) corresponding to the sum of pairs of complex conjugate poles in the complex frequency plane

$$H(s) \approx \sum_{m=1}^{M} \left(\frac{R_m}{s - s_m} + \frac{R_m^*}{s - s_m^*} \right)$$ (2)

where $s = \sigma + j\omega$ is the complex variable in the Laplace plane. M is the total number of modes of the series. For the mth singularity, R_m is the residue associated to each natural pole $s_m = \sigma_m + j\omega_m$ (R_m^* and s_m^* are complex conjugate of R_m and s_m). The imaginary part, ω_m, is the resonance pulsation. The real part, σ_m, is negative, indeed corresponding to a damping due to radiation losses and losses on the surface and, eventually, inside dielectric targets. In the following, $|\sigma_m|$ is named the damping coefficient.

In order to choose a technique of extraction of such singularities, we present here a brief state of the art of existing methods. In time domain, Prony proposed as soon as 1795 (Prony, 1795) to model transient responses as a sum of damped sinusoids (equation (1)). Concerning the characterization of targets, this led to the development of algorithms for finding natural poles and their associated residues, by studying either the impulse response of targets in time domain or their transfer function in frequency domain (Baum, 1991), (Sarkar & Pereira, 1995), (Kumaresan, 1990).

In time domain, the most popular algorithms of poles extraction are based on Prony's methods: The original method of Prony (Prony, 1795), very sensitive to noise in measured data, has been later improved in Least Square-Prony (LS-Prony) (Householder, 1950) and Total Least Square-Prony (TLS-Prony) (Rahman & Yu, 1987) methods, by using Singular Value Decomposition (SVD) (Hua & Sarkar, 1991). However, in the case of a low signal-to-noise environment, only a few modes can be reliably extracted by this inherently ill-conditioned algorithm. More recently developed, the Matrix Pencil Method (MPM) (Sarkar & Pereira, 1995) is more robust to noise in the sampled data, it has a lower variation of estimates of parameters than Prony's methods and it is also computationally more efficient. Moreover, the use of the Half Fourier Transform better separates the early and the late time responses (Jang et al., 2004), through a precise determination of the beginning of the late time T_L.

An alternative approach to time domain estimators, such as Prony's and MPM methods, consists in extracting poles and residues directly from the frequency data set. Frequency domain methods are advantageous when measurements are performed in the frequency domain. Thus one does not need to perform an inverse Fourier transform to obtain time domain response. Cauchy showed in 1821 that it is possible to approximate a function by a rational function approximation (Cauchy, 1821). The original frequency domain method of poles extraction is based on this approximation for the transfer function (Adve et al., 1997), (Kottapalli et al., 1991). Many papers present and compare these various techniques of poles extraction, e.g. in signal processing (Kumaresan, 1990), (Kumaresan, 1990) as well as in object scattering (Tesche, 1973), (Moffatt & Shubert, 1977), (Licul, 2004), (Li & Liang, 2004). Whereas temporal methods solve polynomial equations of order equal to the number of existing poles, frequency domain methods require to solve higher order polynomial equations than the number of poles. Consequently, frequency domain methods are not suitable when this number of poles is very large. However, in scattering problems, these methods are advantageous in comparison with temporal methods because they avoid the choice of the beginning of the late time response.

For all the examples presented in this paper, we can use any of these existing methods to extract poles. We choose to use numerical data obtained with an electromagnetic simulation software, which is based on the Method of Moments (FEKO) and gives the scattered field at different frequencies. Consequently, we choose a frequency domain method (Cauchy, 1821), (Moffatt & Shubert, 1977) and (Kumaresan, 1990).

As an example, we present in figure 1 the modulus of the scattered-field transfer function $H(\omega)$ of two perfectly conducting canonical targets: a dipole of length L = 0.15 m and aspect-ratio L/D = 150 where D is the diameter, and a sphere of diameter D = 0.15 m. Both targets are studied in free-space, in the frequency range [50 MHz – 12.8 GHz] which contains the resonance region of the studied targets (the pulsation range is [3.10^8 rad/sec – 8.10^{10} rad/sec]). We note that for a very resonant object, such as the dipole, resonance peaks are narrow and clearly appear in the response $|H(\omega)|$, what is not the case for a less resonant object as the sphere. For efficient target characterization, it is important to define a frequency range adapted to scatterer dimensions, such as it really contains not only the fundamental pulsation of resonance but also further harmonic pulsations (Baum, 1991).

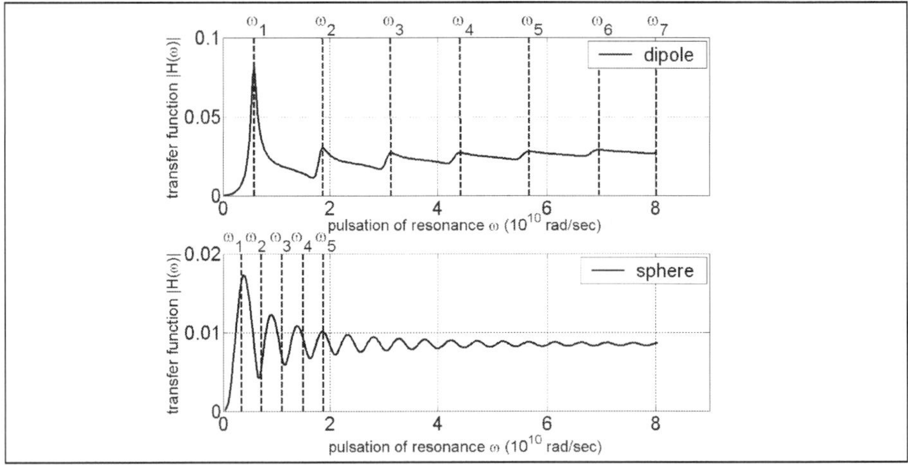

Fig. 1. Modulus of the transfer function $H(\omega)$ of the dipole (upper) and the sphere (lower). 256 samples, monostatic configuration.

In order to extract poles of resonance, the Cauchy's method (Cauchy, 1821) is applied for approximating the transfer function, H, by a ratio of two complex polynomials (Tesche, 1973), (Moffatt & Shubert, 1977), (Licul, 2004), (Li & Liang, 2004). The zeros of the denominator polynomial, B(s), are the poles of H(s). We obtain

$$H(s) \approx \frac{A(s)}{B(s)} \approx \sum_{n=1}^{N} \frac{R_n}{s - s_n} \qquad (3)$$

where N is the total number of singularities of the development and R_n is the residue associated to each pole s_n.

Among these N poles, we expect to get the M pairs of poles corresponding to resonances, called "natural" poles. As stated in (2), they are complex conjugate, with negative real part σ_m, and should be independent of the form of the fitting function and the order N of the rational function. It is assumed that all natural poles are simple (Baum, 1976). In fact, we find not only these natural poles but also "parasitical" poles, which are not complex conjugate by pair and/or have a positive real part. Some of theses parasitical poles correspond to the early time effects, which cannot be represented by exponentially damped sinusoids. Moreover, these parasitical poles depend of the order N in (3). To be sure to get the whole set of natural poles, we choose a high value of the order N (N = 50 for instance). In order to separate the 2M natural poles and the N-2M parasitical poles, we vary the value of N of some units (N = 50 ± 2). Consequently, extracted poles are natural ones not only if they are complex conjugate by pair and have a negative real part, but also if their value is stable when N varies. In our examples, we find 2M = 14 (7 resonances) for the studied dipole and 2M = 10 (5 resonances) for the sphere.

In figure 2, we plot the mapping of natural poles for the studied targets, in one quarter of the complex plane ($\sigma < 0$ and $\omega > 0$), because poles have a negative real part and are complex conjugate. Indeed, for canonical targets (sphere, cylinder, dipole ...), poles are distributed over branches joining the fundamental pulsation of resonance, ω_1, and harmonic pulsations (Chen, 1998) (Chauveau, 2007-a). For a very resonant target as a dipole, we can notice that resonance peaks of $|H(\omega)|$ occur at pulsations of resonance ω_m (figure 1 - upper). However, in the case of a weakly resonant target as the sphere, peaks of resonance, corresponding to pulsation of resonance, ω_m, overlay and cannot be distinguished in the modulus of the transfer function $|H(\omega)|$ (Figure 1 - lower). Indeed, low resonant targets have natural poles with high value of damping factor, $|\sigma_m|$, corresponding to wide peaks. Moreover, because of this low resonant behaviour, natural poles of high order ($s_m > s_5$ for the studied sphere) become difficult to obtain. Only the fundamental pole and some harmonic poles can be obtained. On the contrary, for the dipole, we succeed in extracting all poles existing in the studied pulsation range $[\omega_{min} ; \omega_{max}]$.

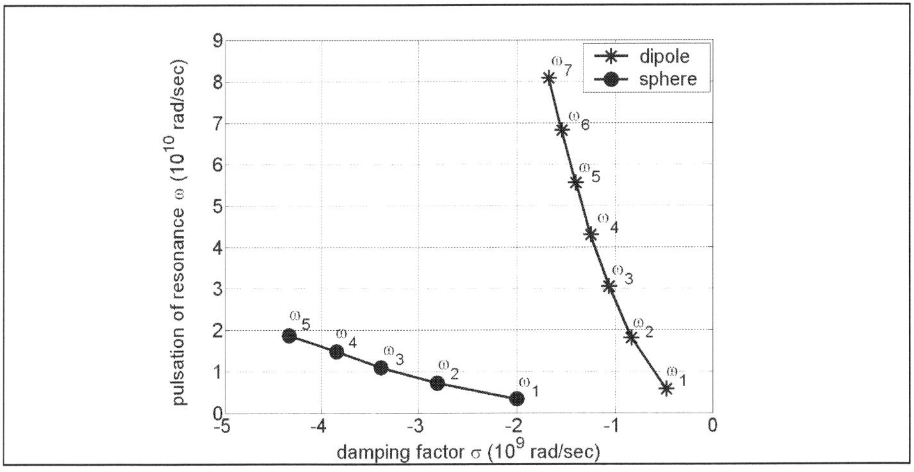

Fig. 2. Mapping of natural poles extracted from H(ω) of figure 1 in the complex plane {σ,ω}, (M = 7 for the dipole, M = 5 for the sphere).

The main advantage of using such natural poles is that only 3 parameters $\{\omega_m\,;\,\sigma_m\,;\,R_m\}$ are required to define each resonance mode. Moreover, in a homogeneous medium, the mapping of natural poles $\{\sigma_m\,;\,\omega_m\}$ is independent of the target orientation relatively to the excitation (Berni, 1975) and can be used as a discriminatory of targets. Furthermore, among the whole set of these possible poles, only a few of them appreciably contribute to the target response and are thus sufficient to characterize a radar target (Chauveau, 2007-a). In general, the selected poles are those which are close to the vertical axis.

In order to show that the representation of resonances with natural poles is an efficient way to characterize the resonance behaviour of targets, we compare the time domain responses, h(t), calculated by inverse Fourier transform of H(ω) and $h_{rec}(t)$, reconstructed using (1) from M pairs of natural poles. In figure 3, for the dipole, both temporal responses are almost identical, in the late time domain corresponding to resonance phenomena. This comparison is performed in the time domain $(t > T_L)$ and not in the frequency domain, because the early time effect is spread all over the frequency band of the initial response, forbidding any comparison between initial and reconstructed frequency responses.

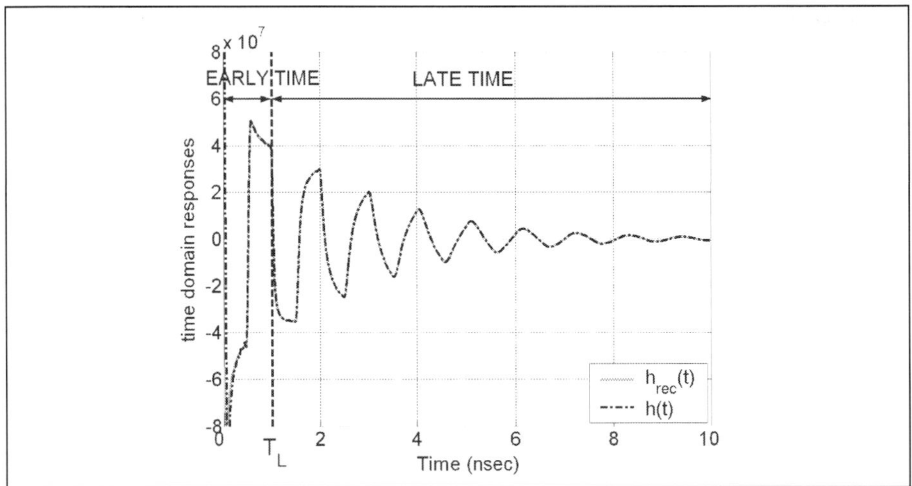

Fig. 3. Time domain response h(t) compared with $h_{rec}(t)$ reconstructed with M = 7 natural pairs of poles of Fig. 2 (dipole).

To quantitatively measure the difference between h(t) and $h_{rec}(t)$, we used the normalised mean square error

$$MSE = \sqrt{\sum_t \left| h(t) - h_{rec}(t) \right|^2} \Bigg/ \sum_t \left| h(t) \right| \qquad (4)$$

both sums being calculated for $t > 2T_L$, in order to be sure that the resonances are well established. In fact, T_L can be precisely determined using the Half Fourier Transform (Jang et al., 2004). For the dipole response plotted in figure 3, we get a MSE of 10^{-4}, with M = 7 pairs of poles. For the studied sphere, we get a MSE of 5.10^{-3}, with M = 5 pairs of poles.

2.2 Representation of natural poles with the quality factor and the natural pulsation of resonance

In (Chauveau, 2007-b), we propose to represent a natural pole ($s_m = \sigma_m + j\omega_m$) not only in Cartesian coordinates, $\{\sigma_m; \omega_m\}$, but also in the form $\{\omega_{0,m}; Q_m\}$ with the natural pulsation of resonance, $\omega_{0,m}$, and the quality factor, Q_m.

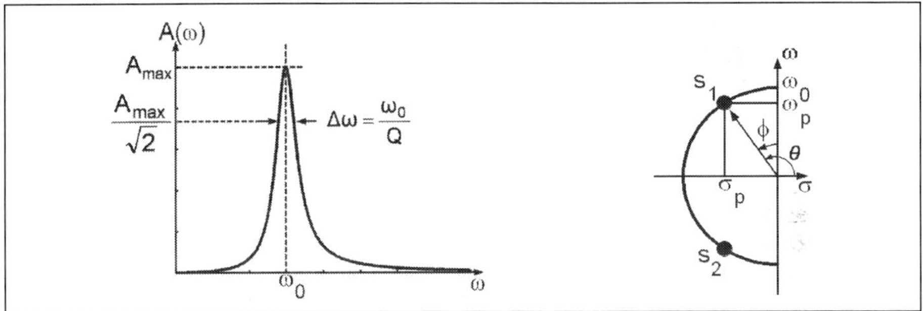

Fig. 4. Resonator: transfer function (a) and representation of poles in complex plane (b).

The transfer function of a resonator, mechanical as well as electrical, with a natural pulsation of resonance ω_0 and a quality factor Q, is given by (figure 4-a)

$$A(\omega) = \frac{A_{max}}{1 + jQ\left(\dfrac{\omega}{\omega_0} - \dfrac{\omega_0}{\omega}\right)} \tag{5}$$

In order to determine the poles of such resonator, we replace, in equation (5), $j\omega$ by s, complex variable in the Laplace plane. Thus, the s-plane transfer function is

$$A(s) = A_{max} \frac{s(\omega_0/Q)}{s^2 + s\omega_0/Q + \omega_0^2}$$

$$= A_{max} \frac{\omega_0}{Q} \frac{s}{(s-s_1)(s-s_2)} = \frac{r_1}{(s-s_1)} + \frac{r_2}{(s-s_2)} \tag{6}$$

with s_1 and s_2, the two roots of the denominator of $A(s)$

$$s_{1,2} = -\frac{\omega_0}{2Q} \pm j\omega_0\sqrt{1 - \left(\frac{1}{2Q}\right)^2} \tag{7}$$

For $Q > 1/2$, the two poles $s_{1,2}$ are complex conjugate, with respective residue $r_{1,2}$

$$r_1 = r_2^* = A_{max} \frac{\omega_0}{Q} \frac{s_1}{(s_1 - s_2)} \tag{8}$$

In the complex plane (figure 4-b), the poles, $s_{1,2} = \sigma_p \pm j\omega_p$, are located on the half-circle of radius equal to the natural pulsation of resonance, ω_0. The real part of poles (damping factor) σ_p and the imaginary part (damped pulsation) ω_p are given by

$$\sigma_p = -\frac{\omega_0}{2Q} = -\frac{\Delta\omega}{2} \qquad \omega_p = \omega_0\sqrt{1 - \left(\frac{1}{2Q}\right)^2} \qquad (9)$$

When Q value is high, ω_p is very close to ω_0. Consequently, the damped pulsation ω_P is often improperly used instead of the natural pulsation ω_0.

In polar coordinates, we get: $s_{1,2} = \omega_0 \exp \pm i\theta$, with ω_0 the modulus and θ the angle of $s_{1,2}$. We prefer to use a modified polar representation in the half complex plane ($\sigma < 0$): $\{\omega_0 ; \Phi\}$ where $\Phi = \theta - \pi/2$ is the angle between the pole direction and the imaginary axis ω. We have

$$2\sin\Phi = -\frac{2\sigma_p}{\omega_0} = -\frac{\Delta\omega}{\omega_0} = \frac{1}{Q} \qquad (10)$$

Indeed, Φ is related to the selectivity (i.e. the width $\Delta\omega$ of the peak of resonance (figure 4-a) divided by the pulsation of resonance ω_0) which is equal to $1/Q$. Thus, a high Q corresponds to a low Φ and the associated pole is close to the vertical axis.

Finally, instead of using the Cartesian representation of natural poles in $\{\sigma_p ; \omega_p\}$, it is interesting to use this new representation in $\{\omega_0 ; Q\}$.

We apply now this resonator point of view to the scattering transfer function H(s) of a radar target, which can be expressed as a sum of transfer functions $A_m(s)$ (equation (5)) of elementary resonators $\{\omega_{0,m} ; Q_m\}$.

For the mth singularity ($s_m = \sigma_m + j\omega_m$), the natural pulsation of resonance, $\omega_{0,m}$, and the quality factor, Q_m, are respectively given by

$$\omega_{0,m} = |s_m| \qquad Q_m = -\frac{\omega_{0,m}}{2\sigma_m} \qquad (11)$$

As an example, natural poles of the sphere and the dipole, plotted in Cartesian coordinates $\{\sigma_m ; \omega_m\}$ in figure 2, are now plotted in $\{\omega_{0,m} ; Q_m\}$ representation in figure 5.

2.3 Use of resonance parameters to characterize objects

Resonance parameters can be used to characterize objects. Indeed, the representation of natural poles in $\{\omega_{0,m} ; Q_m\}$ provides an efficient means for that because it better separates information than the usual Cartesian mapping in the complex plane: $\omega_{0,m}$ gives some indications on dimensions of the target and Q_m brings out the resonance behaviour of targets. Moreover, Q_m is a discriminatory of the aspect ratio of targets and consequently gives some indications on the general shape of targets. We now investigate separately each resonance parameter, $\omega_{0,m}$ and Q_m, for both canonical examples, the sphere and the dipole (figure 5).

2.3.1 Natural pulsation of resonance ω_0

First, we compare the natural pulsation of resonance of the fundamental pole of a target, $\omega_{0,1}$, to the natural pulsation of resonance, $(\omega_0)_P$, of a creeping wave travelling on the surface of this object along a given perimeter P with a wavelength equal to this perimeter (Chauveau, 2007-b)

$$(\omega_0)_P \approx \frac{2\pi c}{P} \qquad (12)$$

with c, the speed of light in vacuum, and P, the path travelled by the wave over the surface.

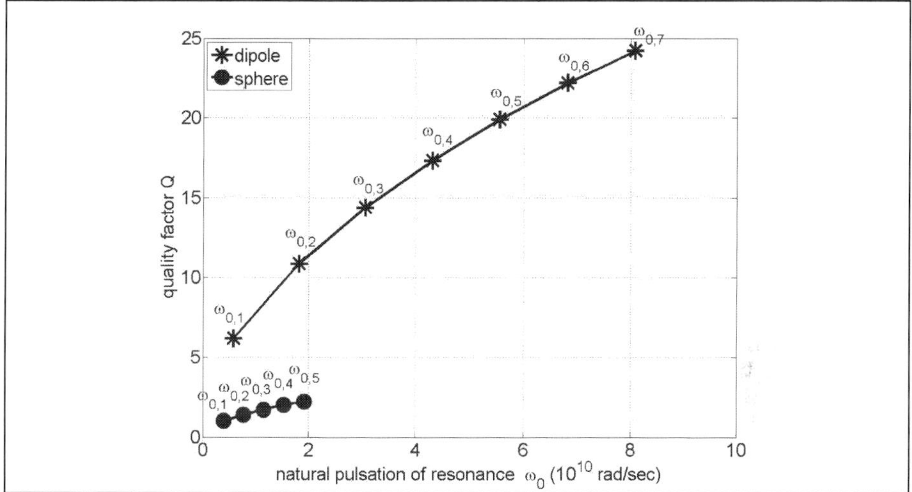

Fig. 5. Natural poles of the sphere and the dipole in the $\{\omega_{0,m} ; Q_m\}$ representation.

For the sphere, the perimeter is $P = \pi D$. With a diameter $D = 0.15$ m, we get $P = 0.471$ m and, from equation (12), $(\omega_0)_P = 4.002 \, 10^9$ rad/sec, to be compared to $\omega_{0,1} = 4.001 \, 10^9$ rad/sec, while $\omega_1 = 3.465 \, 10^9$ rad/sec. We can notice that $(\omega_0)_P$ nearly equals $\omega_{0,1}$, the modulus of s_1, and differs from ω_1, the imaginary part of s_1. Indeed, the representation in $\{\omega_{0,m} ; Q_m\}$ permits to show that this external resonance belongs to the creeping wave.

For the dipole, with $L = 0.15$ m and $D = 0.001$ m, we get $P = 2(L+D) \approx 2L = 0.30$ m. From (Harrington, 1961), (Kinnon, 1999), the natural pulsation of resonance of the dipole, $(\omega_0)_d$, is given by $(\omega_0)_d = 0.95 \, (\omega_0)_P$, with $(\omega_0)_P$ given by equation (12), thus $(\omega_0)_d = 5.969 \, 10^9$ rad/sec, to be compared to $\omega_{0,1} = 5.826 \, 10^9$ rad/sec, while $\omega_1 = 5.807 \, 10^9$ rad/sec. We notice here that the natural pulsation $\omega_{0,1}$ and the damped pulsation ω_1 are nearly equal as can be seen from equation (9) because the dipole is a very resonant object with high Q-factor ($Q_1 \approx 6.19$ for the fundamental pole s_1) .

Harmonic pulsations of the fundamental pulsation of resonance follow the same behaviour. For the sphere, $\omega_{0,m} = m \, \omega_{0,1}$ ($m = 1,2,\ldots$), while for the dipole only odd harmonic pulsations are present, that is $\omega_{0,m} = (2m-1) \, \omega_{0,1}$ ($m = 1,2,\ldots$) (Harrington, 1961), (Kinnon, 1999).

2.3.2 Quality factor Q

Secondly, we examine the Q parameter: we can see that the conducting sphere is a weakly resonant target ($Q_1 \approx 1$ for the fundamental pole s_1). Indeed, it has been shown (Long, 1994) (Moser & Überall, 1983), (Moffatt & Mains, 1975} that more a conducting object is voluminous as the sphere, less it is resonant (low Q), because more the object surface is large relatively to its dimension, more losses on the surface are important. Moreover, this low resonant behaviour can also come from the degeneracy phenomenon of the external poles, due to geometrical symmetries of the sphere (Long, 1994), (Rothwell & Cloud, 1999). On the contrary, the dipole is a very resonant object, with a high quality factor ($Q_1 \approx 6.19$ for the fundamental pole s_1). Consequently, the quality factor can give information on the aspect ratio of the target (Chauveau, 2007-b).

In this general presentation of resonance parameters, we only considered closed objects so as to simplify our purpose. Thus, in this case, we only get external resonances. But, we now intend to focus on the resonance behaviour of targets with apertures, which can give access to internal resonances. First, we choose the simple example of an open rectangular cavity (section 3). Next, we present a more realistic example in the case of an aircraft with air-intakes (section 4).

3. Study of a PC rectangular open cavity

When illuminated in a suitable frequency band, an aperture in a radar target body can give access to high-Q internal resonances. Hence the use of natural frequencies of resonance, mainly internal ones, is a relevant basis for open targets identification. In this way, Rothwell and Cloud (Rothwell & Cloud, 1999) calculate analytically natural frequencies of a canonical target: a hollow perfectly conducting sphere with a circular aperture. Their study presents an interesting behaviour of poles in the complex plane depending on whether the poles originate from internal or external sphere resonances. However, they conclude that the sphere is probably not a good candidate for target identification studies, because of its modal degeneracy. Our approach is different: we intend to show, from the RCS of a target, how resonance parameters depend on its dimensions. For this purpose, we choose to study a perfectly conducting rectangular cavity with a rectangular aperture. On one hand, this example is a more realistic model of air-intake than a spherical cavity; moreover its resonance pulsations are well-known. On the other hand, the search of poles no longer uses an analytical method but a numerical one, based on SEM and therefore applicable to any target and not only to canonical ones. However, the SEM method extracts the whole set of poles, without separating internal and external poles. Consequently, our main objective is to show how we can discriminate the two origins of these poles: external poles corresponding to creeping waves on the surface of the target and internal poles corresponding to internal cavity waves. For this purpose, we first compare poles of the rectangular cavity with those of a closed rectangular box of the same size (section 3.2). Next, we study the variation of these poles with dimensions of the cavity (section 3.3).

3.1 Parameters of the PC rectangular cavity
The studied rectangular cavity is open on one side with a centered slot (figure 6). Its characteristic dimensions are given in table 1: height h, width w, depth d, and slot height s. The configuration of the excitation is

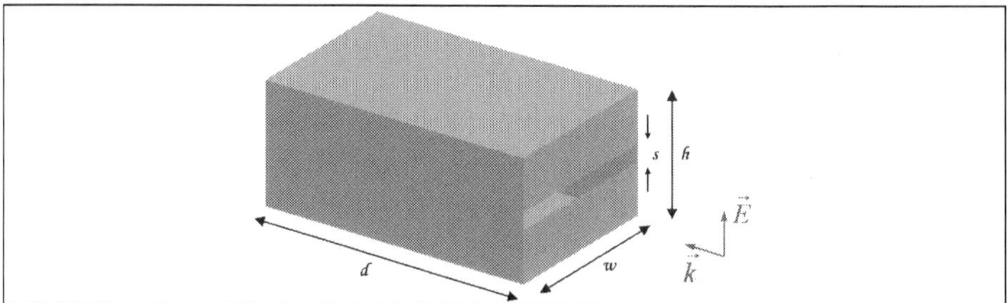

Fig. 6. Geometry of the rectangular cavity (slot centered in the front wall).

- frequency band of investigation: [50MHz ; 685MHz]
- excitation: electric field $\vec{E}$ parallel to the vertical direction h and direction of propagation $\vec{k}$ perpendicular to the aperture plane (w,h)
- monostatic study
- target (inside and outside) in vacuum

dimensions (m)	theoretical resonance pulsations (10^9 rad/sec)
h = 0.35	internal cavity modes
w = 0.50	$(\omega_0)_{011} = 2.22$
d = 0.80	$(\omega_0)_{012} = 3.02$
s = 0.05	$(\omega_0)_{013} = 4.00$
	external modes
$P_{sw} = 2(w + s) = 1.10$	$(\omega_0)_{sw} = 1.71$
$P_{hw} = 2(h + w) = 1.70$	$(\omega_0)_{hw} = 1.11$
$P_{hd} = 2(h + d) = 2.30$	$(\omega_0)_{hd} = 0.82$
$P_{wd} = 2(w + d) = 2.60$	$(\omega_0)_{wd} = 0.72$

Table 1. Characteristic dimensions of the cavity and theoretical natural pulsations of resonance.

The internal modes of a closed rectangular cavity have natural pulsations of resonance given by equation (13) (Harrington, 1961)

$$(\omega_0)_{mnp} = \frac{\pi}{(\varepsilon\mu)^{1/2}}\left[\left(\frac{m}{h}\right)^2 + \left(\frac{n}{w}\right)^2 + \left(\frac{p}{d}\right)^2\right] \tag{13}$$

where ε and μ are respectively the permittivity and the permeability of the medium inside the cavity; and with m = 0,1,2,... ; n = 0,1,2,... ; p = 1,2,3,... ; m = n = 0 being excepted. For an open cavity, the presence of the slot slightly perturbs the resonant pulsation ω_0 and makes the Q-factor finite, corresponding to a non-zero damping factor, σ_m.

In the frequency band of investigation, with such dimensions of the target and such orientations of the electric field $\vec{E}$ and the slot, there are only three possible cavity modes, $(\omega_0)_{011}$, $(\omega_0)_{012}$ and $(\omega_0)_{013}$, satisfying equation (13). Their values are given in table 1 for corresponding dimensions and $\varepsilon = \varepsilon_0$ and $\mu = \mu_0$ for vacuum.

The external modes of resonance are waves creeping on the outside surface of the PC cavity. There are four fundamental natural pulsations, given by equation (12), which correspond to each perimeter, P_{sw}, P_{hw}, P_{hd} and P_{wd}. Values of these perimeters and their corresponding fundamental natural pulsations are given in table 1. Moreover, for each characteristic dimension, it is possible to find further harmonic natural pulsations.

3.2 Comparison with a PC closed rectangular box

In order to distinguish internal and external resonances, we propose to compare the PC open rectangular cavity with the PC closed rectangular box of same dimensions without the slot, both objects being studied with the same excitation configuration. The box being closed, internal resonances cannot be excited from an outside illumination, hence only external resonances are present.

Figure 7 compares the modulus of the scattered-field transfer function $|H(\omega)|$ for the open cavity (solid line) and the PC closed box (dashed line). The cavity response presents narrow peaks of resonance occurring at expected resonance pulsations of each cavity mode, $(\omega_0)_{011}$, $(\omega_0)_{012}$ and $(\omega_0)_{013}$. We can see another peak of resonance occurring near the natural pulsation corresponding to the resonance of the slot $(\omega_0)_{sw}$. Wider peaks are also present corresponding to lower quality of resonance. Indeed, these wider peaks also exist for the box response.

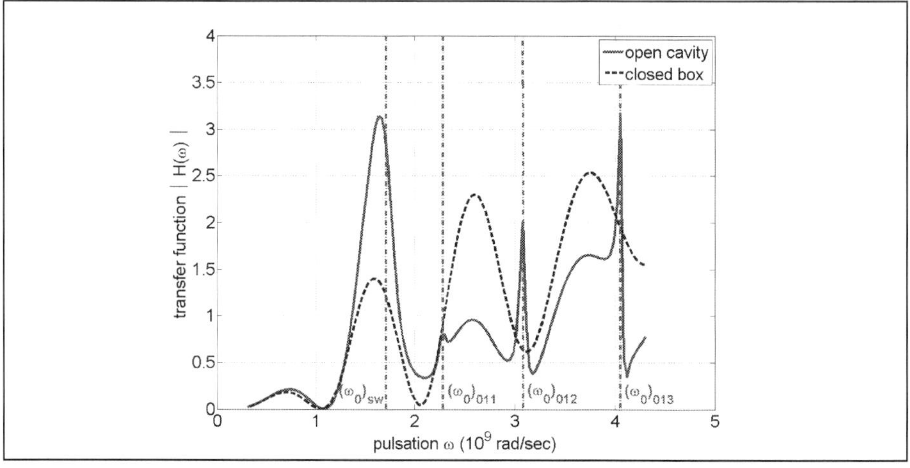

Fig. 7. Comparison between $|H(\omega)|$ of the open cavity and the closed box.

Figures 8 and 9 compare the cavity and the box using both representations of resonance parameters (see section 2), respectively the Cartesian mapping of natural poles in the complex plane $\{\sigma_m\,;\,\omega_m\}$ and the quality factor as a function of the natural pulsation of resonance $\{\omega_{0,m}\,;\,Q_m\}$. First, we examine poles existing only for the open cavity, *i.e.* those numbered '1', '2', '3' and '4'. We can see that resonance pulsations of poles of the open cavity numbered '1', '2' and '3' almost correspond to theoretical pulsations of the closed cavity modes, $(\omega_0)_{011}$, $(\omega_0)_{012}$ and $(\omega_0)_{013}$ (see table 1 and figure 9). Indeed, these three pulsations of resonance coincide with the narrowest resonance peaks of the open cavity response (figure 7). Accordingly, these three poles have a very low damping coefficient $|\sigma_m|$ (figure 8) and correspond to a high quality factor Q_m (figure 9). About the pole '4', its resonance pulsation corresponds to the resonance pulsation $(\omega_0)_{sw}$ of the slot with P_{sw} in equation (12). This pole has a higher damping coefficient $|\sigma_m|$ and a lower quality factor Q_m than poles '1', '2' and '3'. Indeed, the peak of resonance occurring at the resonance pulsation of the slot $(\omega_0)_{sw}$ is wider than previous peaks corresponding to internal resonances '1', '2' and '3'.

Concerning natural poles '5', '6', '7' and '8' of the open cavity, we can see that they are very close to natural poles of the box. Consequently, we can state that these four poles correspond to creeping waves on the outside surface of the perfectly conducting cavity. Following equation (12), these natural poles depend on various perimeters of the target given in table 1. For example, the pole numbered '5' can correspond to both characteristic perimeters P_{hd} or P_{hw}. The three other poles appear to correspond to harmonic poles which mainly depend on dimensions h and d.

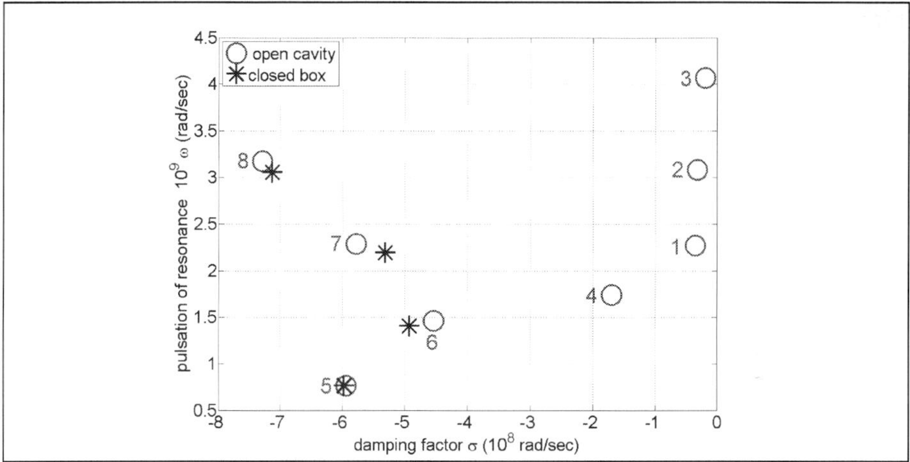

Fig. 8. Comparison of the mapping of poles $\{\sigma_m ; \omega_m\}$ of the open cavity and the closed box.

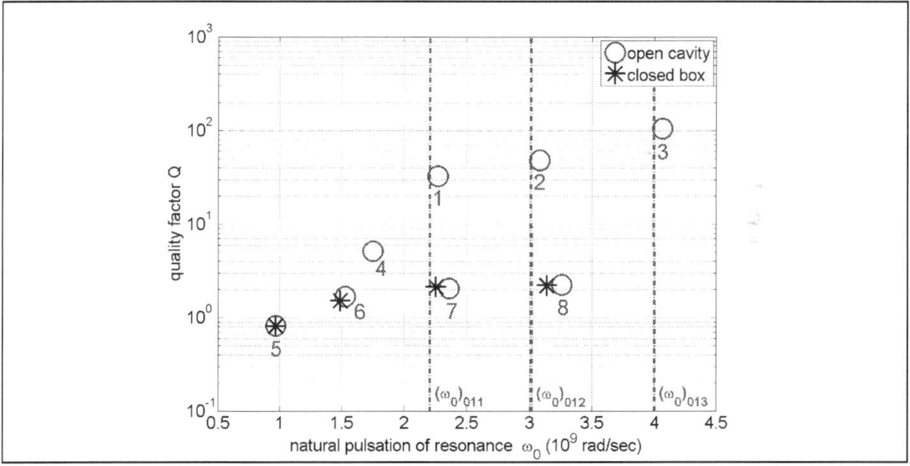

Fig. 9. Comparison of resonance parameters in $\{\omega_{0,m} ; Q_m\}$ representation of the open cavity and the closed box.

3.3 Effect of dimensions of the rectangular cavity

In (Chauveau, 2009), we vary each characteristic dimensions of the open cavity and we examine how natural poles move. To quantify the dependence of the natural pulsation of resonance ω_0 on each dimension, we compute the variation of ω_0 with each dimension of the cavity ($y = h$, w, d or s), $i.e.$ $d\omega_0/dy$, relatively to ω_0. Results obtained in (Chauveau, 2009) are given in table 2: the mean of the absolute value of this variation: $<|d\omega_0/dy|/\omega_0>$ for each dimension of the cavity and each pole. We consider that the natural pulsation of resonance of a pole appreciably depends on a dimension of the cavity when this relative variation is significant (noted in bold in table 2).

Poles numbered '1', '2' and '3' are internal resonances corresponding to the first modes of the cavity. Indeed, their pulsations of resonance only depend on dimensions w and d because of the orientation of the incident electric field $\vec{E}$ which points to the direction of h (figure 6) involving m = 0 in equation (13).

The pole numbered '4' corresponds to the resonance of the slot and consequently is an external resonance. ω_0 depends strongly on the slot width w and to a smaller degree on the slot height s.

In the same way, poles numbered '6', '7' and '8' are external resonances with a similar behavior with the variation of dimensions of the cavity: their pulsations of resonance are strongly affected by the variation of dimensions h and d. Thus, we deduce that these poles depend on the perimeter P_{hd} of the cavity. In fact, these three poles are harmonic poles corresponding to the perimeter P_{hd}. Moreover the pole '7' is affected by the slot, its pulsation of resonance slightly varying as a function of s.

The pole numbered '5' is an external resonance, with ω_0 mainly depending on the dimension h, but also on w and d. Consequently, we cannot determine which perimeter is associated to this pole.

Anyway, the four poles numbered '5', '6', '7' and '8', due to external creeping waves on the surface of the cavity, are very close to those of the perfectly conducting box with same dimensions, even if they are modified by the slot.

Concerning the quality of resonance, the Q-factor of internal cavity modes, '1', '2' and '3' decreases when the slot height, s, increases, anyway, Q always remains higher than Q of external poles, '4', '5', '6', '7' and '8' (Chauveau, 2009). Indeed, radiating losses are much stronger for external resonances than for internal resonances, consequently, internal resonances are predominant.

pole \ dimension	h	w	d	S
1	0.05	**1.14**	**0.35**	0.15
2	0.05	**0.62**	**0.70**	0.29
3	0.06	**0.35**	**0.89**	0.25
4	0.02	**1.72**	0.04	**0.60**
5	**1.29**	0.51	**0.44**	0.06
6	**1.10**	0.06	**0.73**	0.08
7	**0.74**	0.15	**0.75**	**0.46**
8	**0.79**	0.11	**1.03**	0.19

Table 2. Relative variation of ω_0 with dimensions of the cavity $<|d\omega_0/dy|/\omega_0>$ in m^{-1}).

4. Application to air-intakes of a simulated aircraft

We have shown in section 3 that resonances of a PC open object (a rectangular cavity) have two origins: external resonances corresponding to external surface creeping waves and internal resonances corresponding to internal cavity waves. Because of their high Q-factor, internal resonances are predominant, and consequently, they are more easily extracted. In this section, we propose to use this property in order to characterize a simulated aircraft with air-intakes.

A complex shape target as an aircraft can often be modelled as a combination of canonical objects (figure 10-a), with resonances corresponding as well to canonical objects (here: cylinder, cone, three triangular wings and two cylindrical cavities) as structures created by the assembly of these canonical objects (for instance: dihedral created by the junction between each wing and the cylinder) (Chauveau, 2007-b). Indeed, air-intakes of an aircraft are cavities which can exhibit strong resonances for suitable frequency bands. Thus, we propose to use their internal resonance parameters to characterize aircrafts.

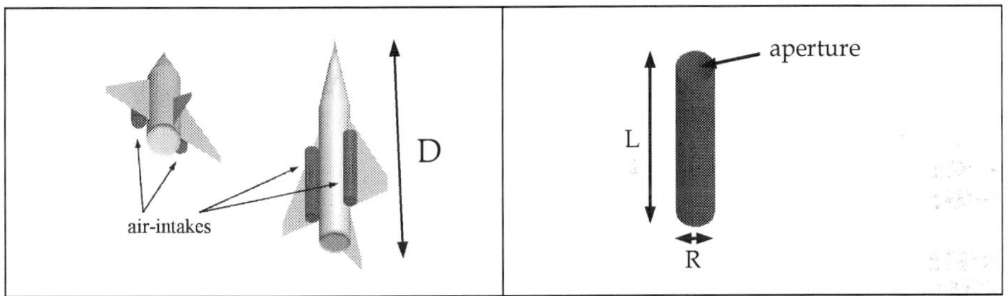

Fig. 10. Simulated aircraft with air-intakes (a) and open cylindrical cavity (b).

We first study the PC cylindrical open cavity alone (figure 10-b). This cavity has a length $L = 0.5$m and a radius $R = 0.04$m, it is open on one side and it is studied in free space in the pulsation range [1.2 10^{10} rad/sec ; 1.8 10^{10} rad/sec]. We choose this pulsation range in order to include resonance pulsations corresponding to the first modes of the cavity. Figure 11 presents extracted resonance parameters of the cylindrical cavity, in both representations. We can see that we have a branch of 5 poles with low damping factor σ and high Q-value ($Q_1 \approx 250$).

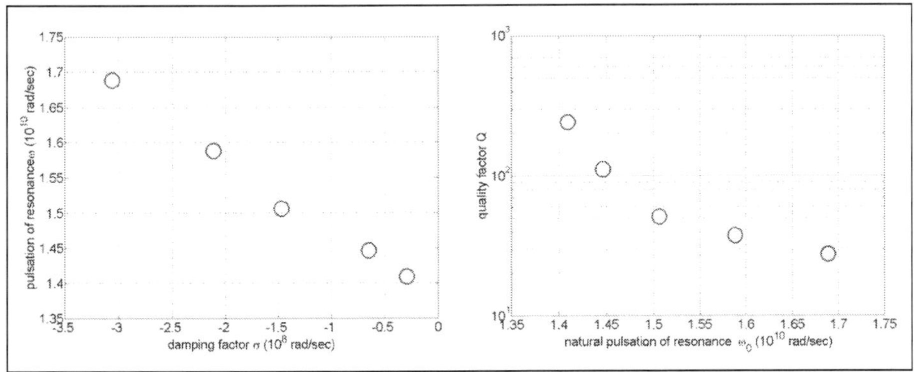

Fig. 11. Poles of the open cylindrical cavity: $\{\sigma_m ; \omega_m\}$ representation (left) and $\{\omega_{0,m} ; Q_m\}$ representation (right).

In order to verify that these poles correspond to internal modes of the cavity, we calculate the natural pulsation of resonance of a closed cylindrical cavity of same dimensions, L and R, with equation (14) (Harrington, 1961)

$$\left(\omega_r\right)_{mnp} = \frac{c}{R\sqrt{\varepsilon}} \sqrt{x_{mn}^2 + \left(\frac{p\pi R}{L}\right)^2} \qquad (14)$$

where ε is the permittivity of the medium inside the cavity; x_{mn} are ordered zeros of Bessel functions $J_m(x)$ given in (Harrington, 1961).

In the frequency band of investigation and with such dimensions of the target, only 5 cavity modes satisfy equation (14): m = 1; n = 1; p = [1:5]. In table 3, their values are compared with natural pulsations of resonance $\omega_{0,m}$ (noted ω_{open_cavity}) extracted from the transfer function $H(\omega)$ of the cylindrical open cavity. We can see that resonance pulsations ω_{open_cavity} of extracted poles of the open cavity are nearly equal to theoretical resonance pulsations ω_{closed_cavity} of the closed cavity. Thus, these extracted poles actually correspond to internal resonances.

p	ω_{closed_cavity} (10^{10} rad/sec)	ω_{open_cavity} (10^{10} rad/sec)	$\omega_{aircraft}$ (10^{10} rad/sec)
1	1.39	1.41	1.42
2	1.43	1.45	1.43
3	1.49	1.51	1.52
4	1.57	1.59	/
5	1.67	1.69	/

Table 3. Comparison of calculated resonance pulsations of the closed cavity ω_{closed_cavity}, extracted resonance pulsations of the open cavity ω_{open_cavity} and extracted resonance pulsations of the aircraft $\omega_{aircraft}$.

We now compare the previous open cavity alone (figure 10-b) with the simulated aircraft (figure 10-a), in the same configuration (free space, pulsation range, excitation …). The aircraft has a characteristic dimension D = 1.5 m and both air-intakes have the same dimensions as the open cavity.

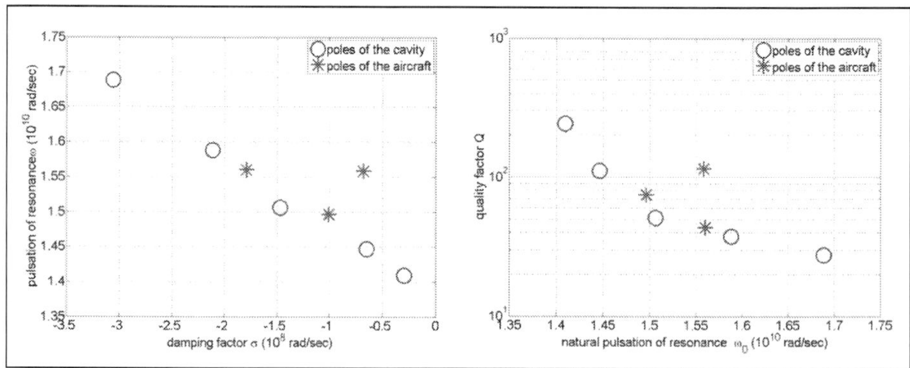

Fig. 12. Comparison of poles of the open cylindrical cavity and the simulated aircraft in wideband response: $\{\sigma_m ; \omega_m\}$ representation (left) and $\{\omega_{0,m} ; Q_m\}$ representation (right).

We are interested only in internal resonances of air-intakes, that is why we choose the pulsation range [1.2 10^{10} rad/sec ; 1.8 10^{10} rad/sec], corresponding to the 5 first modes of the cavity. In figure 12, we can see that only 3 poles are extracted for the aircraft and these poles cannot be associated with poles of the cavity alone.

In order to intend to retrieve more precisely poles of the open cavity, we calculate poles of the aircraft in narrow frequency bands. Indeed, in (Chauveau, 2007-c), we show the possibility to extract natural poles one-by-one in tuning narrow frequency bands to frequencies of wanted poles. Thus, we apply this method of poles extraction to the simulated aircraft in order to find the three first poles of the open cavity. We respectively search the first pole in the pulsation range [1.35 10^{10} rad/sec; 1.44 10^{10} rad/sec], the second pole in [1.40 10^{10} rad/sec; 1.46 10^{10} rad/sec] and the third pole in [1.47 10^{10} rad/sec; 1.55 10^{10} rad/sec]. We can see in figure 13 and in table 3 that the pulsations of these poles are now correctly extracted. This shows the advantage of the narrow band extraction of poles.

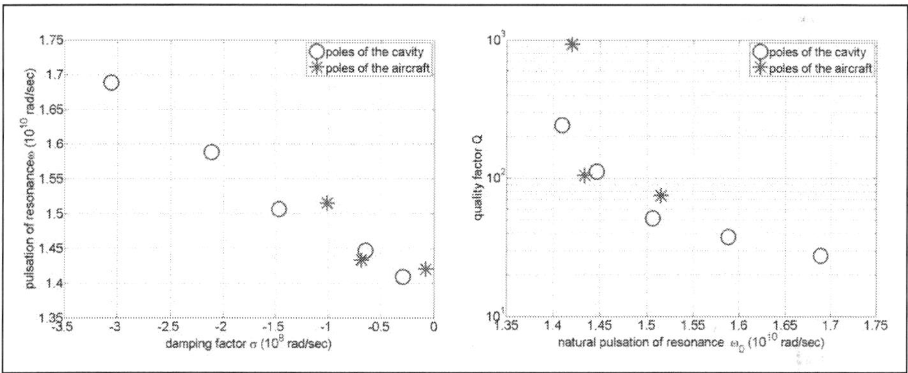

Fig. 13. Comparison of poles of the open cylindrical cavity and the simulated aircraft in narrow band response: $\{\sigma_m ; \omega_m\}$ representation (left) and $\{\omega_{0,m} ; Q_m\}$ representation (right).

5. Conclusion and perspectives

In this chapter, we have first presented how to extract natural poles from the simulated transfer function of a target and how to use them for identification purposes. We have introduced a new representation of poles with quality factor Q and natural pulsation of resonance ω_0 in order to better separate information: the Q parameter permits to bring out clearly the resonance behaviour of targets (Q is a discriminatory of the aspect ratio of targets), and the natural pulsation of resonance ω_0 depends on dimensions of targets. Next, we have extended our study to resonances of targets with apertures. The simple example of a PC rectangular open cavity has permitted to show how resonance parameters depend on object dimensions and how internal and external resonances can be distinguished, by comparison of poles of the open cavity and the closed box. Internal resonances having a lower damping coefficient $|\sigma_m|$ than external resonances, they have a higher quality of resonance Q_m, and can therefore be more easily extracted. That is why, we have used this interesting property of internal resonances in order to identify an aircraft with air-intakes. Thus, we have shown that the use of selective narrow frequency bands permits a better extraction of poles in the case of complex objects, as aircrafts with air-intakes.

6. References

Adve, R.S.; Sarkar, T.K.; Rao, S.M.; Miller, E.K. & Pflug, D.R. (1997). Application of the Cauchy Method for Extrapolating/Interpolating Narrow-Band System Responses. *IEEE Transaction on Microwave Theory and Techniques*, Vol. 45, No. 5, May 1997, pp. 837-845

Anastassiu, H.T. (2003). A review of electromagnetic scattering analysis for inlets, cavities and open ducts. *IEEE Antennas and Propagation Magazine*, vol. 45, No. 6, 2003, pp. 27-40

Baum, C.E. (1976). The Singularity Expansion Method, In: *Transient Electromagnetic Field*, L.B. Felsen (Ed. New York: Springer – Verlag) pp. 129-179

Baum, C.E.; Rothwell, E.J.; Chen, K.M. & Nyquist, D.P. (1991). The Singularity Expansion Method and Its Application to Target Identification. *Proceedings of the IEEE*, Vol. 79, No. 10, October 1991, pp. 1481-1492

Berni, A.J. (1975). Target Identification by Natural Resonance Estimation. *IEEE Transaction on Aerospace Electronic Systems*, Vol. 11, No. 2, 1975, pp. 147-154

Cauchy, A.L. (1821). Sur la formule de Lagrange relative à l'interpolation. *Analyse algébrique*, Paris, 1821.

Chauveau, J.; de Beaucoudrey, N.; & Saillard, J. (2007-a). Selection of Contributing Natural Poles for the Characterization of Perfectly Conducting Targets in Resonance Region. *IEEE Transaction on Antennas and Propagation*, Vol. 55, No. 9, September 2007, pp. 2610-2617

Chauveau, J.; de Beaucoudrey, N.; & Saillard, J. (2007-b). Characterization of Perfectly Conducting Targets in Resonance Domain with their Quality of Resonance. *Progress In Electromagnetic Research*, Vol. 74, 2007, pp. 69-84

Chauveau, J.; de Beaucoudrey, N.; & Saillard, J. (2007-c). Determination of Resonance Poles of Radar Targets in Narrow Frequency Bands. *EuMW/Eurad*, Germany, Munich, October 2007.

Chauveau, J.; de Beaucoudrey, N.; & Saillard, J. (2009). Resonance behaviour of radar targets with aperture. Example of an open rectangular cavity. Submitted in *IEEE Transaction on Antennas and Propagation*

Chen, C-C. (1998). Electromagnetic resonances of Immersed Dielectric Spheres. *IEEE Transaction on Antennas and Propagation*, Vol. 46, No. 7, July 1998, pp. 1074-1083

Chen, K.M.; Nyquist, D.P.; Rothwell, E.J.; Webb, L.L. & Drachman, B. (1986). Radar Target Discrimination by Convolution of Radar Return with Extinction-Pulses and Single-Mode Extraction Signals. *IEEE Transaction on Antennas and Propagation*, Vol. 34, No. 7, 1986, pp. 896-904

Harrington, R.F. (1961). *Time-harmonic electromagnetic fields*, McGraw-Hill book company Inc., New York, 1961

Householder, A.S. (1950). On Prony's method of fitting exponential decay curves and multiple-hit survival curves. Oak RidgeNat. Lab., Oak Ridge, TN, Tech. Rep. ORNL-455, February 1950.

Hua, Y. & Sarkar, T.K. (1991). On SVD for Estimating Generalized Eigenvalues of Singular Matrix Pencil in Noise. *IEEE Transaction on Signal Processing*, Vol. 39, No. 4, April 1991, pp. 892-900

Jang, S.; Choi, W.; Sarkar, T.K.; Salazar-Palma, M.; Kim, K. & Baum, C.E. (2004). Exploiting Early Time Response Using the Fractional Fourier Transform for Analyzing

Transient Radar Returns. *IEEE Transaction on Antennas and Propagation*, Vol. 52, No. 11, November 2004, pp. 3109-3121

Kennaugh, E.M. & Moffatt, D.L. (1965). Transient and Impulse Response Approximations. *Proceedings of the IEEE*, Vol. 53, August 1965, pp. 893-901

Kinnon, D. (1999). Identification of Landmines Using Microwave Resonance Signatures. *The University of Queensland*, Thesis Report, October 1999.

Kottapalli, K.; Sarkar, T.K.; Hua Y.; Miller, E.K. & Burke, G.J. (1991). Accurate Computation of Wide-Band Response of Electromagnetic Systems Utilizing Narrow-Band Information. *IEEE Transaction on Microwave Theory and Techniques*, Vol. 39, No. 4, April 1991, pp. 682-687

Kumaresan, R. (1990-a). On a frequency domain analog of Prony's method. *IEEE Transaction on Acoustics, Speech, and Signal Processing*, Vol. 38, January 1990, pp. 168-170

Kumaresan, R. (1990-b). Identification of Rational Transfer Function from Frequency Response Samples. *IEEE Transaction on Aerospace Electronic Systems*, Vol. 26, No. 6, November 1990, pp. 925-934

Li, L. & Liang, C.H. (2004). Analysis of resonance and quality factor of antenna and scattering systems using complex frequency method combined with model-based parameter estimation. *Progress In Electromagnetic Research*, Vol. 46, 2004, pp. 165-188

Licul, S. (2004). Ultra-wideband antenna characterization and measurements. *Ph.D. dissertation*, Blacksburg, VA: Virginia Tech., 2004

Long, Y. (1994). Determination of the Natural Frequencies for Conducting Rectangular Boxes. *IEEE Transaction on Antennas and Propagation*, Vol. 42, No. 7, 1994, pp. 1016-1021.

Moffatt, D.L. & Mains, R.K. (1975). Detection and Discrimination of radar targets. *IEEE Transaction on Antennas and Propagation*, Vol. 23, 1975, pp. 358-367

Moffatt, D.L. & Shubert, K.A. (1977). Natural Resonances via Rational Approximants. *IEEE Transaction on Antennas and Propagation*, Vol. 25, No. 5, September 1977, pp. 657-660

Moser, P.J. & Überall, H. (1983). Complex eigenfrequencies of axisymmetic perfectly conducting bodies: Radar spectroscopy. *Proceedings of the IEEE*, Vol. 71, 1983, pp. 171-172

Prony, R. (1795). Essai expérimental et analytique sur les lois de la dilatabilité de fluides élastiques et sur celles de la force expansive de la vapeur de l'alcool à différentes températures. *Journal de l'École Polytechnique*, vol. 1, cahier 2, 1795, p. 24-76

Rahman, M.D. A. & Yu, K.B. (1987). Total least squares approach for frequency estimation using linear prediction. *IEEE Transaction on Acoustic, Speech Signal Proceeding*, Vol. 35, October 1987, pp. 1440-1454

Rothwell, E.J.; Chen, K.M. & Nyquist, D.P. (1987). Extraction of the Natural Frequencies of a Radar Target from a Measured Response Using E-Pulse Techniques. *IEEE Transaction on Antennas and Propagation*, Vol. 35, No. 6, June 1987, pp. 715-720

Rothwell, E.J. & Cloud, M.J. (1999). Natural frequencies of a conducting sphere with a circular aperture. *Journal of Electromagnetic Waves and Applications*, Vol. 13, 1999, pp. 729-755

Sarkar, T.K. & Pereira, O. (1995). Using the Matrix Pencil Method to Estimate the Parameters of a Sum of Complex Exponentials. *IEEE Antennas and Propagation Magazine*, Vol. 37, No. 1, February 1995, pp. 48-55

Tesche, F. M. (1973). On the Analysis of Scattering and Antenna Problems Using the Singularity Expansion Technique. *IEEE Transaction on Antennas and Propagation*, Vol. 21, No. 1, January 1973, pp. 53-62

Toribio, R.; Pouliguen, P. & Saillard, J. (2003). Identification of radar targets in resonance zone: E-Pulse techniques. *Progress In Electromagnetic Research*, Vol. 43, 2003, pp. 39-58.

14

Bistatic Synthetic Aperture Radar Synchronization Processing

Wen-Qin Wang
School of Communication and Information Engineering,
University of Electronic Science and Technology of China,
Chengdu,
P. R. China

1. Introduction

Bistatic synthetic aperture radar (BiSAR), operates with separate transmitter and receiver that are mounted on different platforms (Cherniakov & Nezlin, 2007), will play a great role in future radar applications (Krieger & Moreira, 2006). BiSAR configuration can bring many benefits in comparison with monostatic systems, such as the exploitation of additional information contained in the bistatic reflectivity of targets (Eigel et al., 2000; Burkholder et al., 2003), improved flexibility (Loffeld et al., 2004), reduced vulnerability (Wang & Cai, 2007), forward looking SAR imaging (Ceraldi et al., 2005). These advantages could be worthwhile, e.g., for topographic features, surficial deposits, and drainage, to show the relationships that occur between forest, vegetation, and soils. Even for objects that show a low radar cross section (RCS) in monostatic SAR images, one can find distinct bistatic angle to increase their RCS to make these objects visible in BiSAR images. Furthermore, a BiSAR configuration allows a passive receiver, operating at a close range, to receive the data reflected from potentially hostile areas. This passive receiver may be teamed with a transmitter at a safe place, or make use of opportunistic illuminators such as television and radio transmitters or even unmanned vehicles [Wang, 2007a].

However, BiSAR is subject to the problems and special requirements that are neither not encountered or encountered in less serious form for monostatic SAR (Willis, 1991). The biggest technological challenge lies in synchronization of the two independent radars: time synchronization, the receiver must precisely know when the transmitter fires (in the order of nanoseconds); spatial synchronization, the receiving and transmitting antennas must simultaneously illuminate the same spot on the ground; phase synchronization, the receiver and transmitter must be coherent over extremely long periods of time. The most difficult synchronization problem is the phase synchronization. To obtain focused BiSAR image, phase information of the transmitted pulse has to be preserved. In a monostatic SAR, the co-located transmitter and receiver use the same stable local oscillator (STALO), the phase can only decorrelate over very short periods of time (about 1×10^{-3} sec.). In contrast, for a BiSAR system, the transmitter and receiver fly on different platforms and use independent master oscillators, which results that there is no phase noise cancellation. This superimposed phase noise corrupts the received signal over the whole synthetic aperture time. Moreover, any phase noise (instability) in the master oscillator is magnified by frequency multiplication. As a consequence, the low phase noise requirements imposed on the oscillators of BiSAR are much more higher than the monostatic cases. In the case of indirect phase synchronization using identical STALOs in the transmitter and receiver, phase stability is required over the coherent integration time. Even the toleration of low frequency or quadratic phase synchronization errors can be relaxed to 90°, the requirement of phase stability is only achievable with ultra-high-quality oscillators (Wei β, 2004). Moreover, aggravating circumstances are accompanied for airborne platforms because of different platform motions, the performance of phase stability will be further degraded.

Although multiple BiSAR image formation algorithms have been developed (Wang et al., 2006). BiSAR synchronization aspects have seen much less development, at least in open literature. The requirement of phase stability in BiSAR was first discussed in (Auterman, 1984), and further investigated in (Krieger et al., 2006; Krieger & Younis, 2006), which conclude that uncompensated phase noise may cause a time variant shift, spurious sidelobes and a deterioration of the impulse response, as well as a low-frequency phase modulation of the focused SAR signal. The impact of frequency synchronization error in spaceborne parasitic interferometry SAR is analyzed in (Zhang et al., 2006) and an estimation of oscillator's phase offset in bistatic interferometry SAR is invstigated in (Ubolkosold et al., 2006). In an alike manner, linear and random time synchronization errors are discussed in (Zhang et al., 2005).

As a consequence of these difficulties, there is a lack of practical synchronization technique for BiSAR. But its application is of great scientific and technological interest, several authors have proposed some potential synchronization techniques or algorithms, such as ultra-high-quality oscillators (Gierull, 2006), a direct exchange of radar pulses (Moreira et al., 2004), a ping-pong interferometric mode in case of full-active systems (Evans, 2002) and an appropriate bidirectional link (Younis et al., 2006a; Younis et al., 2006b; Eineder, 2003). The practical work is to develop a practicable synchronization technique without too much alteration to existing radars.

This chapter concentrates on general BiSAR synchronization, aims at the development of a practical solution for time and phase synchronization aspects without too much alteration to existing radars. The remaining sections are organized as follows. In Section 2, the impact of synchronization errors on BiSAR systems are analysed by using an analytical models. A conclusion is made that some synchronization compensation techniques must be applied to focus BiSAR raw data. Then, possible time synchronization and phase synchronization approaches are investigated in Section 3 and Section 4, respectively. Finally, Section 5 concludes the chapter with some possible future work.

2. Impact of synchronization errors on BiSAR systems

2.1 Fundamental of phase noise

The instantaneous output voltage of a signal generator or oscillator $V(t)$ is (Lance et al., 1984)

$$V(t) = [V_o + \delta\varepsilon(t)]\sin\left[2\pi v_o t + \phi_o + \delta\phi(t)\right] \tag{1}$$

where V_o and v_o are the nominal amplitude and frequency, respectively, ϕ_o is a start phase, $\delta\varepsilon(t)$ and $\delta\phi(t)$ are the fluctuations of signal amplitude and phase, respectively. Notice that, here, we have assumed that (Wang et al., 2006)

$$\frac{\delta\varepsilon(t)}{V_o} \ll 1 \text{, and } \frac{\partial\left[\delta\phi(t)\right]}{\partial t} \ll 1 \tag{2}$$

It is well known that $S_\phi(f)$ defined as the spectral density of phase fluctuations on a 'per-Hz' is the term most widely used to describe the random characteristics of frequency stability, which is a measure of the instantaneous time shifts, or time jitter, that are inherent in signals produced by signal generators or added to signals as it passes through a system (Wall & Vig, 1995). Although an oscillator's phase noise is a complex interaction of variables, ranging from its atomic composition to the physical environment of the oscillator, a piecewise polynomial representation of an oscillator's phase noise exists and is expressed as (Rutman, 1978)

$$S_\phi(f) = \sum_{\alpha=2}^{2} h_{\alpha-2} f^{\alpha-2} \tag{3}$$

where the coefficients $h_{\alpha-2}$ describe the different contributions of phase noise, and f represents the phase fluctuation frequency. As modeled in the Eq. (3), they can be represented by several physical mechanisms which include random walk frequency noise, flicker frequency noise. Random walk frequency noise (Vannicola & Varshney, 1983) is because of the oscillator's physical environment (temperature, vibration, and shocks etc.). This phase noise contribution can be significant for a moving platform, and presents design difficulties since laboratory measurements are necessary when the synthesizer is under vibration. White frequency noise originates from additive white thermal noise sources inside the oscillator's feedback loop. Flicker phase noise generally is produced by amplifiers, and white phase noise is caused by additive white noise sources outside the oscillator's feedback loop (Donald, 2002).

In engineering, for the condition that the phase fluctuations occurring at a rate of f and are small compared with 1 rad, a good approximation is

$$L(f) = \frac{S_\phi(f)}{2} \tag{4}$$

where $L(f)$ is defined as the ratio of the power in one sideband referred to the input carrier frequency on a per-Hertz of bandwidth spectral density basis to the total signal power at Fourier frequency f from the carrier per device.

2.2 Model of phase noise

One cannot foresee to simulate the phase noise if one does not have a model for the phase noise. In (Hanzo et al., 2000), a white phase noise model is discussed, but it cannot describe the statistical process of phase noise. In (Foschini & Vannucci, 1988), a Wiener phase noise model is discussed, but it cannot describe the low-frequency phase noise, since this part of phase noise is an unstationary process. As different phase noise will bring different effects on BiSAR (see Fig. 1), the practical problem is that how to develop an useful and comprehensive model of frequency instability that can be understood and applied in BiSAR processing. Unfortunately, Eq. (3) is a frequency-domain expression and not convenient in analyzing its impact on BiSAR. As such, we have proposed an analytical model of phase noise, as shown in Fig. 2. This model uses Gaussian noise as the input of a hypothetical low-pass filter and its output is then considered as phase noise, that is this model may represent the output of a hypothetical filter with impulse response $h(t)$ receiving an input signal $x(t)$.

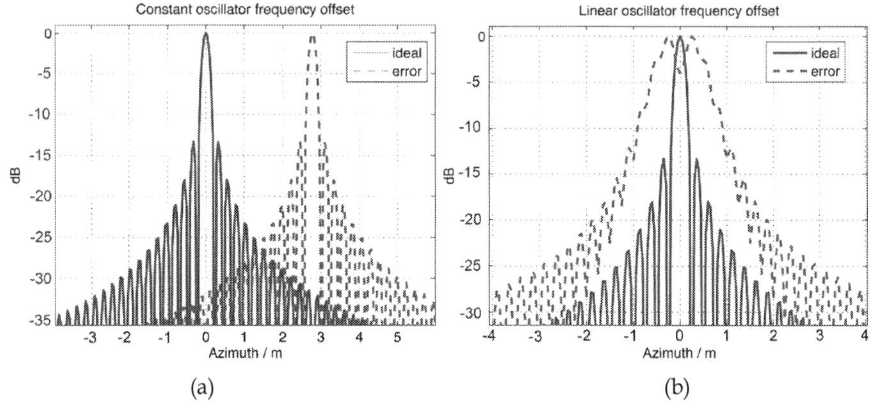

Fig. 1. Impacts of various oscillator frequency offsets: (a) constant offset, (b) linear offset, (c) Sinewave offset, (d) random offset.

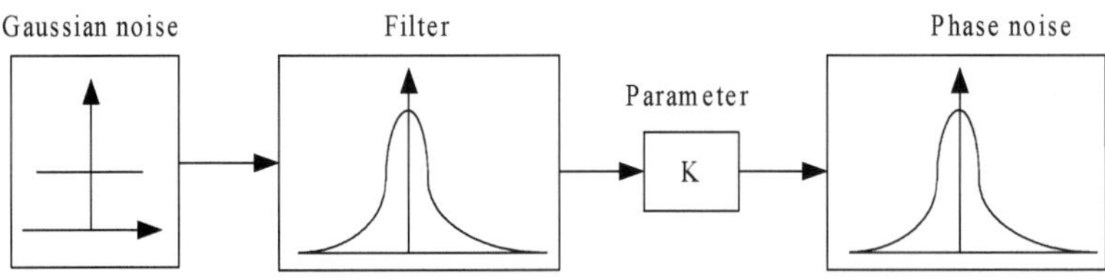

Fig. 2. Analytical model of phase noise.

It is well known that the power spectral density (PSD) of the output signal is given by the product $S_x(f)|H(f)|^2$, where the filter transfer function $H(f)$ is the Fourier transform of $h(t)$. Notice that, here, $|H(f)|^2$ must be satisfied with

$$|H(f)|^2 = \begin{cases} S_\varphi(f), & f_l \leq |f| \leq f_h \\ S_\varphi(f_l), & |f| \leq f_l \\ 0, & \text{else} \end{cases} \tag{5}$$

where a sharp up cutoff frequency f_h and a sharp down cutoff frequency f_l are introduced. Notice that time domain stability measures sometimes depend on f_h and f_l which must then be given with any numerical result, although no recommendation has been made for this value. $f_h = 3kHz$ and $f_l = 0.01Hz$ are adopted. Thereby, the PSD of phase noise in Fig. 2 can be analytical expressed as

$$S_\varphi(f) = KS_x(f)|H(f)|^2 \tag{6}$$

where $S_x(f)$ is the PSD of Gaussian noise and also the input of filter, and K is a constant. An inverse Fourier transform yields

$$\varphi(t) = \sqrt{K}x(t) \otimes h(t) \tag{7}$$

where $\varphi(t)$ and $\otimes$ denote the phase noise in time domain and a convolution, respectively.

2.3 Impact of phase synchronization errors

Since only STALO phase noise is of interest, the modulation waveform used for range resolution can be ignored and the radar can be simplified into an azimuth only system (Auterman, 1984). Suppose the transmitted signal is sinusoid whose phase argument is

$$\phi_T(t) = 2\pi f_T t + M\varphi_T(t) \tag{8}$$

The first term is the carrier frequency and the second term is the phase, and M is the ratio of the carrier frequency to STALO frequency. After reflection from a target, the received signal phase is that of the transmitted signal delayed by the round-trip time τ. The receiver output signal phase $\phi(t)$ results from demodulating the received signal with the receiver STALO which has the same form as the transmitter STALO

$$\phi_R(t) = 2\pi f_R t + M\varphi_R(t) \tag{9}$$

Hence we have

$$\phi(t) = 2\pi(f_R - f_T)t + 2\pi f_T \tau + M(\varphi_R(t) - \varphi_T(t-\tau)) \tag{10}$$

The first term is a frequency offset arising from non-identical STLO frequencies, which will result focused image with a drift. Because this drift can easily be corrected using ground calibrator, it can be ignored here. The second term forms the usual Doppler term as round-trip time to the target varies, it should be preserved. The last term represents the effect of STALO frequency instability which is of interest. As a typical example, assuming a X-band airborne SAR with a swath of 6km. Generally, a typical STALO used in current SAR has a frequency accuracy (δf) of $10^{-9}/1s$ or better (Wei β, 2004). As a typical example, assuming a BiSAR system with the following parameters: radar carrier frequency is $1 \times 10^{10} Hz$, the speed of light is $3 \times 10^{8} m/s$, the round-trip from radar to target is $12000m$, and then the phase error in fast-time is found to be

$$M\left[\phi_T(t) - \phi_T(t - \tau)\right] = 2\pi \cdot \delta f \cdot \tau$$
$$\leq 2\pi \times 10^{10} \times 10^{-9} \times \frac{12000}{3^8} = \frac{8\pi}{3} \times 10^{-4} \ (rad) \tag{11}$$

which has negligible effects on the synchronization phase. Hence, we have an approximative expression

$$\phi_T(t) \doteq \phi_T(t - \tau) \tag{12}$$

That is to say, the phase noise of oscillator in fast-time is negligible, we can consider only the phase noise in slow-time.
Accordingly, the phase error in BiSAR can be modelled as

$$\phi_B(t) = M\left[\varphi_T(t) - \varphi_R(t)\right] \tag{13}$$

It is assumed that $\varphi_T(t)$ and $\varphi_R(t)$ are independent random variables having identical PSD $S_\varphi(f)$. Then, the PSD of phase noise in BiSAR is

$$S_{\varphi_B}(f) = 2M^2 S_\varphi(f) \tag{14}$$

Where the factor 2 arises from the addition of two uncorrelated but identical PSD. This is true in those cases where the effective division ratio in frequency synthesizer is equal to the small integer fractions exactly. In other instances, an experimental formula is (Kroupa, 1996)

$$S_{\varphi_B}(f) = 2\left[M^2 S_\varphi(f) + \frac{10^{-8}}{f} + 10^{-14}\right] \tag{15}$$

Take one $10MHz$ STALO as an example, whose phase noise parameters are listed in Table 1. This STALO can be regarded as a representative example of the ultra stable oscillator for current airborne SAR systems. Predicted phase errors are shown in Fig. 3 for a time interval of $10s$. Moreover, the impacts of phase noise on BiSAR compared with the ideal compression results in azimuth can be founded in Fig. 4(a). We can draw a conclusion that oscillator phase instabilities in BiSAR manifest themselves as a deterioration of the impulse response function. It is also evident that oscillator phase noise may not only defocus the SAR image, but also introduce significant positioning errors along the scene extension.
Furthermore, it is known that high-frequency phase noise will cause spurious sidelobes in the impulse function. This deterioration can be characterized by the integrated sidelobe ratio (ISLR) which measures the transfer of signal energy from the mainlobe to the sidelobes. For

an azimuth integration time, T_s, the ISLR contribution because of phase errors can be computed in dB as

$$ISLR = 10\log\left[\int_{1/T_s}^{\infty} 2M^2 S_\varphi(f)\,df\right] \tag{16}$$

Frequency, Hz	1	10	100	1k	10
$S_\varphi(f), dBc/Hz$	-80	-100	-145	-145	-160

Table 1. Phase noise parameters of one typical STALO

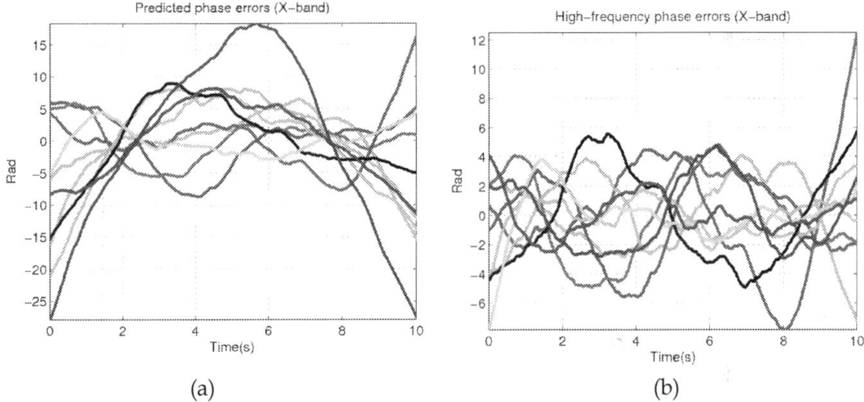

Fig. 3. Simulation results of oscillator phase instabilities with ten realisations: (a) predicted phase noise in $10s$ in X-band (linear phase ramp corresponding to a frequency offset has been removed). (b) predicted high-frequency including cubic and more phase errors.

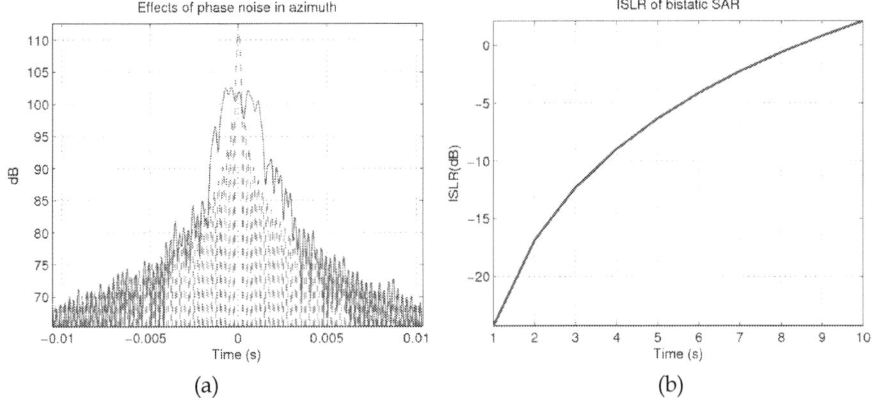

Fig. 4. Impacts of phase noise on BiSAR systems: (a) impact of predicted phase noise in azimuth. (b) impact of integrated sidelobe ratio in X-band.

A typical requirement for the maximum tolerable ISLR is $-20dB$, which enables a maximum coherent integration time T_s of $2s$ in this example as shown in Fig. 4(b). This result is coincident with that of (Krieger & Younis, 2006).

Generally, for $f \leq 10Hz$, the region of interest for SAR operation, $L(f)$ can be modelled as (Willis, 1991)

$$L(f) = L_1 10^{-3\log f} \tag{17}$$

Note that L_1 is the value of $L(f)$ at $f = 1Hz$ for a specific oscillator. As the slope of Eq. (17) is so high, there is

$$\log\left[\int_{1/T_s}^{\infty} L(f)\,df\right] \approx L\left(\frac{1}{T_s}\right) \tag{18}$$

Hence, the deterioration of ISLR may be approximated as

$$ISLR \approx 10\log\left(4M^2 L_1 10^{3\log T_s}\right) = 10\log\left(4M^2 L_1\right) + 30\log T_s \tag{19}$$

It was concluded in (Willis, 1991) that the error in this approximation is less than $1dB$ for $T_s > 0.6s$.

2.4 Impact of time synchronization errors

Timing jitter is the term most widely used to describe an undesired perturbation or uncertainty in the timing of events, which is a measurement of the variations in the time domain, and essentially describes how far the signal period has wandered from its ideal value. For BiSAR applications, timing jitter becomes more important and can significantly degrade the performance of image quality. Thus a special attenuation should be given to study the effects of timing jitter in order to predict possible degradation on the behavior of BiSAR systems. Generally speaking, we can model jitter in a signal by starting with a noise-free signal and displacing time with a stochastic process. Figure 5 shows a square wave with jitter compared to an ideal signal. The instabilities can eventually cause slips or missed signals that result in loss of radar echoes.

Because bistatic SAR is a coherent system, to complete the coherent accumulation in azimuth, the signals of same range but different azimuths should have the same phase after between the echo window and the PRF (pulse repetition frequency) of the receiver system would be a fixed value to preserve a stable phase relationship. But once there is clock timing jitter, the start time of the echo sampling window changes with certain time difference between the echo sampling window (or PRI, pulse repetition interval) and the real echo signal, as shown in Fig. 5. Consequently, the phase relation of the sampled data would be destroyed.

To find an analytical expression for the impact of time synchronization error on BiSAR images, we suppose the transmitted radar is

$$S_T\left(\hat{t}\right) = rect\left[\frac{\hat{t}}{T_r}\right] \exp\left(j\omega_o \hat{t} + j\pi\gamma\hat{t}^2\right) \tag{20}$$

where $rect[\bullet]$ is the window function, T_r is the pulse duration, ω_o is the carrier angular frequency and is the chirp rate, respectively. Let $e(t)$ denote the time synchronization errors of BiSAR, the radar echo from a scatterer is given by

$$S_r(t) = rect\left[\frac{t - R_{ref}/c - e(t)}{T_w}\right] rect\left[\frac{t-\tau}{T_r}\right] \exp\left[j\omega_o(t-\tau) + j\pi\gamma(t-\tau)^2\right] \tag{21}$$

where the first term is the range sampling window centered at R_{ref} , having a length of T_w , c is the speed of light, and τ is the delay corresponding to the time it takes the signal to travel the distance transmitter-target-receiver distance, R_B .

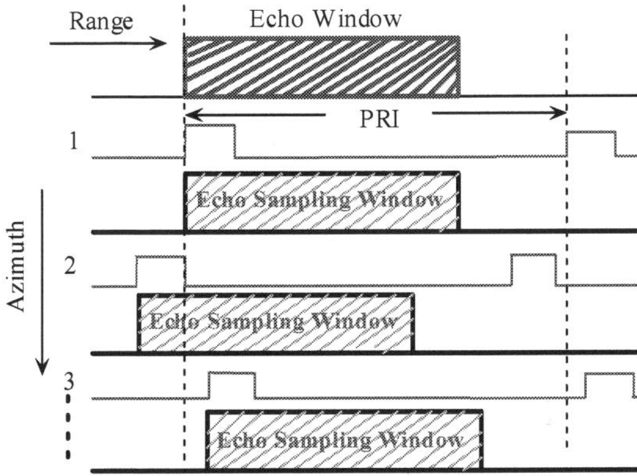

Fig. 5. Impacts of time synchronization error on BiSAR data.

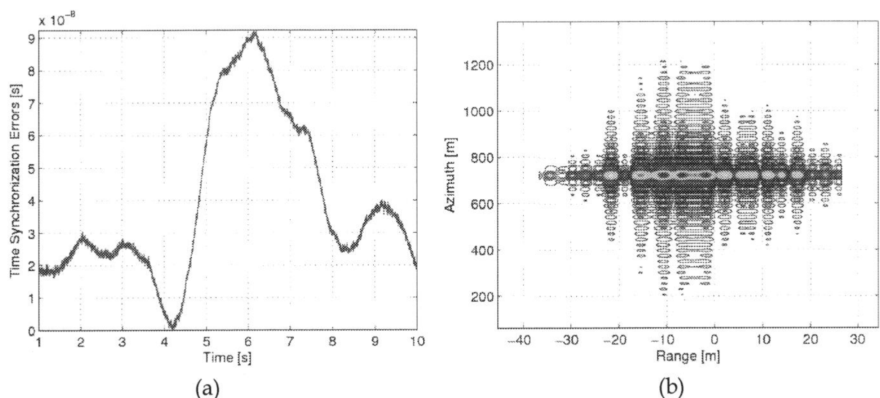

(a)   (b)

Fig. 6. Impact of time synchronization errors: (a) predicted time synchronization errors in $10s$. (b) impact on BiSAR image for one point target.

Considering only time synchronization error, that is to say, phase synchronization is ignored here, we can obtain the demodulated signal as

$$S_r^{'}(t) = rect\left[\frac{t - R_{ref}/c - e(t)}{T_w}\right] rect\left[\frac{t - \tau}{T_r}\right] \exp\left[j\omega_o\tau + j\pi\gamma(t - \tau)^2\right] \qquad (22)$$

Suppose the range reference signal is

$$S_{ref}(r) = \exp\left[-j\pi\gamma t^2\right] \qquad (23)$$

The signal, after range compression, can be expressed as

$$S_{ref}(r) = rect\left[\frac{r - R_{ref}}{cT_w}\right] sinc\left[\frac{B(r - R_B + c \cdot e(t))}{c}\right] \exp\left[-j\frac{R_B \cdot \omega_o}{c}\right] \qquad (24)$$

where B is the radar signal bandwidth and $\Delta R = c \cdot e(t)$ is the range drift because of time synchronization errors.

From Eq. (24) we can notice that if the two clocks deviate a lot, the radar echoes will be lost due to the drift of echo sampling window. Fortunately, such case hardly occurs for current radars. Hence we considered only the case that each echo can be successfully received but be drifted because of clock timing jitter. In other words, the collected data with the same range but different azimuths are not on the same range any more. As an example, Fig. 6(a) illustrates one typical prediction of time synchronization error. From Fig. 6(b) we can conclude that, time synchronization errors will result unfocused images, drift of radar echoes and displacement of targets. To focus BiSAR raw data, some time synchronization compensation techniques must be applied.

Notice that the requirement of frequency stability may vary with applications. Image generation with BiSAR requires a frequency coherence for at least the coherent integration time. For interferometric SAR (InSAR) (Muellerschoen et al., 2006), however this coherence has to be expanded over the whole processing time (Eineder, 2003).

3. Direct-path Signal-based synchronization approach

A time and phase synchronization approach via direct-path signal was proposed in (Wang et al., 2008). In this approach, the direct-path signal of transmitter is received with one appropriative antenna and divided into two channels, one is passed though an envelope detector and used to synchronize the sampling clock, and the other is downconverted and used to compensate the phase synchronization error. Finally, the residual time synchronization error is compensated with range alignment, and the residual phase synchronization error is compensated with GPS (global positioning systems)/INS (intertial navigation system)/IMU (intertial measurement units) information, then the focusing of BiSAR image may be achieved.

3.1 Time synchronization

As concluded previously, if time synchronizes strictly, intervals between the echo window and the PRF (pulse repetition frequency) of the receiver would be a fixed value to preserve a stable phase relationship. But once there is time synchronization error, the start time of the echo sampling window changes with certain time difference between the echo sampling window (or PRI, pulse repetition interval) and the real echo signal. As a consequence, the phase relation of the sampled data would be destroyed.

It is well known that, for monostatic SAR, the azimuth processing operates upon the echoes which come from target points at equal range. Because time synchronization errors (without considering phase synchronization which are compensated separately in subsequent phase synchronization processing) have no effect on the initial phase of each echo, time synchronization errors can be compensated separately with range alignment. Here the spatial domain realignment (Chen & Andrews, 1980) is used. That is, let $f_{t_1}(r)$ and $f_{t_2}(r)$ denote the recorded complex echo from adjacent pulses where $t_2 - t_1 = \Delta t$ is the PRI and r is the range assumed within one PRI. If we consider only the magnitude of the echoes, then $m_{t_1}(r + \Delta r) \approx m_{t_2}(r)$, where $m_{t_1}(r) \triangleq |f_{t_1}(r)|$. The Δr is the amount of misalignment, which we would like to estimate. Define a correlation function between the two waveforms $m_{t_1}(r)$ and $m_{t_2}(r)$ as

$$R(s) \triangleq \frac{\displaystyle\int_{-\infty}^{\infty} m_{t_1}(r) m_{t_2}(r-s)\, dr}{\left[\displaystyle\int_{-\infty}^{\infty} m_{t_1}^{2}(r)\, dr \int_{-\infty}^{\infty} m_{t_2}^{2}(r-s)\, dr\right]^{1/2}} \tag{25}$$

From Schwartz inequality we have that $R(s)$ will be maximal at $s = \Delta r$ and the amount of misalignment can thus be determined. Note that some other range alignment methods may also be adopted, such as frequency domain realignment, recursive alignment (Delisle & Wu, 1994), and minimum entropy alignment. Another note is that, sensor motion error will also result the drift of echo envelope, which can be corrected with motion compensation algorithms. When the transmitter and receiver are moving in non-parallel trajectories, the range change of normal channel and synchronization channel must be compensated separately. This compensation can be achieved with motion sensors combined with effective image formation algorithms.

3.2 Phase synchronization

After time synchronization compensation, the primary causes of phase errors include uncompensated target or sensor motion and residual phase synchronization errors. Practically, the receiver of direct-path can be regarded as a strong scatterer in the process of phase compensation. To the degree that motion sensor is able to measure the relative motion between the targets and SAR sensor, the image formation processor can eliminate undesired motion effects from the collected signal history with GPS/INS/IMU and autofocus algorithms. This procedure is motion compensation that is ignored here since it is beyond the scope of this paper. Thereafter, the focusing of BiSAR image can be achieved with autofocus image formation algorithms, e.g., (Wahl et al., 1994).

Suppose the n th transmitted pulse with carrier frequency f_{T_n} is

$$x_n(t) = s(t)\exp\left[j\left(2\pi f_{T_n} t + \varphi_{d(n)} \right) \right] \tag{26}$$

where $\varphi_{d(n)}$ is the original phase, and $s(t)$ is the radar signal in baseband

$$s(t) = rect\left[\frac{t}{T_r}\right] \exp\left(j\pi\gamma t^2\right) \tag{27}$$

Let t_{dn} denote the delay time of direct-path signal, the received direct-path signal is

$$s'_{dn}(t) = s(t - t_{dn}) \exp\left[j2\pi\left(f_{Tn} + f_{dn}\right)\left(t - t_{dn}\right)\right] \exp\left(j\varphi_{d(n)}\right) \tag{28}$$

where f_{dn} is Doppler frequency for the n th transmitted pulse. Suppose the demodulating signal in receiver is

$$s_f(t) = \exp\left(-j2\pi f_{Rn} t\right) \tag{29}$$

Hence, the received signal in baseband is

$$S_{dn}(t) = s(t - t_{dn}) \exp\left(-j2\pi\left(f_{Tn} + f_{dn}\right)t_{dn}\right) \cdot \exp\left(j2\pi\Delta f_n t\right) \cdot \exp\left(j\varphi_{d(n)}\right) \tag{30}$$

with $\Delta f_n = f_{Tn} - f_{Rn}$, where $\varphi_{d(n)}$ is the term to be extracted to compensate the phase synchronization errors in reflected signal. A Fourier transform applied to Eq. (30) yields

$$\begin{aligned}
S_{dn}(f) &= rect\left[\frac{f - \Delta f_n}{\gamma T_r}\right] \cdot \exp\left[-j2\pi\left(f - \Delta f_n\right)t_{dn}\right] \cdot \exp\left[\frac{-j\pi\left(f - \Delta f_n\right)^2}{\gamma}\right] \\
&\times \exp\left[-j2\pi\left(f_{Tn} + f_{dn}\right)t_{dn} + j\varphi_{dn}\right]
\end{aligned} \tag{31}$$

Suppose the range reference function is

$$S_{ref}(t) = rect\left[\frac{t}{T_r}\right] \exp\left(-j\pi\gamma t^2\right) \tag{32}$$

Range compression yields

$$\begin{aligned}
y_{dn}(t) &= \left(\gamma T_o - \Delta f_n\right) \operatorname{sin} c\left[\left(\gamma T_o - \Delta f_n\right)\left(t - t_{dn} + \Delta f_n/\gamma\right)\right] \\
&\times \exp\left[j\pi\Delta f_n\left(t - t_{dn} + \Delta f_n/\gamma\right)\right] \cdot \exp\left\{-j\left[2\pi\left(f_{dn} + f_{Rn}\right)t_{dn} - \frac{\pi\Delta f_n^2}{\gamma} + \varphi_{d(n)}\right]\right\}
\end{aligned} \tag{33}$$

We can notice that the maxima will be at $t = t_{dn} - \Delta f_n/\gamma$, where we have

$$\exp\left[j\pi\Delta f_n\left(t - t_{dn} + \Delta f_n/\gamma\right)\right]\Big|_{t = t_{dn} - \Delta f_n/\gamma} = 1 \tag{34}$$

Hence, the residual phase term in Eq. (33) is

$$\psi(n) = -2\pi\left(f_{dn} + f_{Rn}\right)t_{dn} - \frac{\pi\Delta f_n^2}{\gamma} + \varphi_{d(n)} \tag{35}$$

As Δf_n and γ are typical on the orders of $1kHz$ and $1\times10^{13}\,Hz/s$, respectively. $\pi\Delta f_n^2/\gamma$ has negligiable effects. Eq. (35) can be simplified into

$$\psi(n) = -2\pi\left(f_{dn} + f_{Rn}\right)t_{dn} + \varphi_{d(n)} \tag{36}$$

In a like manner, we have

$$\psi(n+1) = -2\pi\left(f_{d(n+1)} + f_{R(n+1)}\right)t_{d(n+1)} + \varphi_{d(n+1)} \tag{37}$$

Let

$$f_{d(n+1)} = f_{d0} + \delta f_{d(n+1)}, \qquad f_{R(n+1)} = f_{R0} + \delta f_{R(n+1)} \tag{38}$$

where f_{d0} and f_{R0} are the original Doppler frequency and error-free demodulating frequency in receiver, respectively.

Accordingly, $\delta f_{d(n+1)}$ and $\delta f_{R(n+1)}$ are the frequency errors for the $(n+1)$ th pulse. Hence, we have

$$\begin{aligned}
\varphi_{d(n+1)} - \varphi_{dn} &= \left[\psi(n+1) - \psi(n)\right] - 2\pi\left(f_{R0} + f_{d0}\right)\left(t_{d(n+1)} - t_{dn}\right) \\
&\quad - 2\pi\left(\delta f_{d(n+1)} + \delta f_{R(n+1)}\right)\left(t_{d(n+1)} - t_{dn}\right)
\end{aligned} \tag{39}$$

Generally, $\delta f_{d(n+1)} + \delta f_{R(n+1)}$ and $t_{d(n+1)} - t_{dn}$ are typical on the orders of $10Hz$ and $10^{-9}s$, respectively, then $2\pi\left(\delta f_{d(n+1)} + \delta f_{R(n+1)}\right)\left(t_{d(n+1)} - t_{dn}\right)$ is founded to be smaller than $2\pi \times 10^{-8}$ rad, which has negligiable effects. Furthermore, since $t_{d(n+1)}$ and t_{dn} can be obtained from GPS/INS/IMU, Eq. (39) can be simplified into

$$\varphi_{d(n+1)} - \varphi_{dn} = \psi_e(n) \tag{40}$$

With $\psi_e(t) = \left[\psi(n+1) - \psi(n)\right] - 2\pi\left(f_{R0} + f_{d0}\right)\left(t_{d(n+1)} - t_{dn}\right)$. We then have

$$\begin{aligned}
\varphi_{d(2)} - \varphi_{d(1)} &= \psi_e(1) \\
\varphi_{d(3)} - \varphi_{d(2)} &= \psi_e(2) \\
&\cdots\cdots\cdots\cdots\cdots \\
\varphi_{d(n+1)} - \varphi_{d(n)} &= \psi_e(n)
\end{aligned} \tag{41}$$

From Eq. (41) we can get $\varphi_{d(n)}$, then the phase synchronization compensation for reflected channel can be achieved with this method. Notice that the remaining motion compensation errors are usually low frequency phase errors, which can be compensated with autofocus image formation algorithms.

In summary, the time and phase synchronization compensation process may include the following steps:

Step 1, extract one pulse from the direct-path channel as the range reference function;
Step 2, direct-path channel range compression;
Step 3, estimate time synchronization errors with range alignment;
Step 4, direct-path channel motion compensation;

Step 5, estimate phase synchronization errors from direct-path channel;
Step 6, reflected channel time synchronization compensation;
Step 7, reflected channel phase synchronization compensation;
Step 8, reflected channel motion compensation;
Step 9, BiSAR image formation.

4. GPS signal disciplined synchronization approach

For the direct-path signal-based synchronization approach, the receiver must fly with a sufficient altitude and position to maintain a line-of-sight contact with the transmitter. To get around this disadvantage, a GPS signal disciplined synchronization approach is investigated in (Wang, 2009).

4.1 System architecture

Because of their excellent long-term frequency accuracy, GPS-disciplined rubidium oscillators are widely used as standards of time and frequency. Here, selection of a crystal oscillator instead of rubidium is based on the superior short-term accuracy of the crystal. As such, high quality space-qualified 10MHz quartz crystal oscillators are chosen here, which have a typical short-term stability of $\sigma_{Allan}(\Delta t = 1s) = 10^{-12}$ and an accuracy of $\sigma_{rms}(\Delta t = 1s) = 10^{-11}$. In addition to good timekeeping ability, these oscillators show a low phase noise.

As shown in Fig. 7, the transmitter/receiver contains the high-performance quartz crystal oscillator, direct digital synthesizer (DDS), and GPS receiver. The antenna collects the GPS L1 (1575.42MHz) signals and, if dual frequency capable, L2 (1227.60MHz) signals. The radio frequency (RF) signals are filtered though a preamplifier, then down-converted to

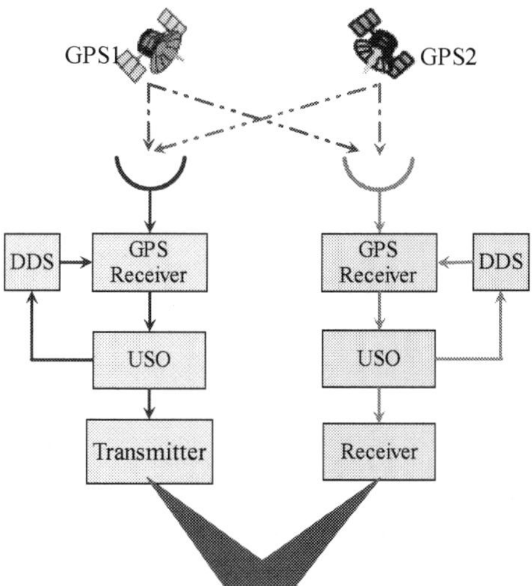

Fig. 7. Functional block diagram of time and phase synchronization for BiSAR using GPS disciplined USOs.

intermediate frequency (IF). The IF section provides additional filtering and amplification of the signal to levels more amenable to signal processing. The GPS signal processing component features most of the core functions of the receiver, including signal acquisition, code and carrier tracking, demodulation, and extraction of the pseudo-range and carrier phase measurements. The details can be found in many textbooks on GPS (Parkinson & Spilker, 1996).

The USO is disciplined by the output pulse-per-second (PPS), and frequency trimmed by varactor-diode tuning, which allows a small amount of frequency control on either side of the nominal value. Next, a narrow-band high-resolution DDS is applied, which allows the generation of various frequencies with extremely small step size and high spectral purity. This technique combines the advantages of the good short-term stability of high quality USO with the advantages of GPS signals over the long term. When GPS signals are lost, because of deliberate interference or malfunctioning GPS equipment, the oscillator is held at the best control value and free-runs until the return of GPS allows new corrections to be calculated.

4.2 Frequency synthesis

Since DDS is far from being an ideal source, its noise floor and spurs will be transferred to the output and amplified by 2 (denotes the frequency multiplication factor) in power. To overcome this limit, we mixed it with the USO output instead of using the DDS as a reference directly. Figure 8 shows the architecture of a DDS-driven PLL synthesizer. The frequency of the sinewave output of the USO is 10MHz plus a drift Δf, which is fed into a double-balanced mixer. The other input port of the mixer receives the filtered sinewave output of the DDS adjusted to the frequency Δf. The mixer outputs an upper and a lower sideband carrier. The desired lower sideband is selected by a 10MHz crystal filter; the upper sideband and any remaining carriers are rejected. This is the simplest method of simple sideband frequency generation.

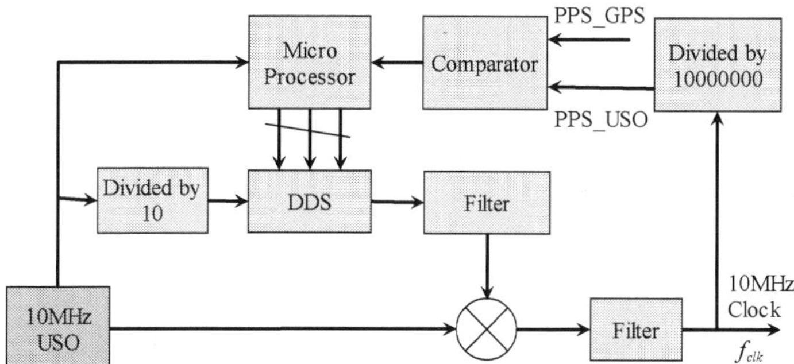

Fig. 8. Functional block diagram of GPS disciplined oscillator.

The DDS output frequency is determined by its clock frequency f_{clk} and an M-bit number $2^{j} (j \in [1, M])$ written to its registers, where M is the length of register. The value 2^{j} is added to an accumulator at each clock uprate, and the resulting ramp feeds a sinusoidal look-up table followed by a DAC (digital-to-analog convertor) that generates discrete steps at each update, following the sinewave form. Then, the DDS output frequency is (Vankka, 2005)

$$f = \frac{f_{clk} \cdot 2^j}{2^M}, j \in [1,2,3,....,M-1]$$ (42)

Clearly, for the smallest frequency step we need to use a low clock frequency, but the lower the clock frequency, the harder it becomes to filter the clock components in the DDS output. As a good compromise, we use a clock at about 1MHz, obtained by dividing the nominal 10MHz USO output by 10. Then, the approximate resolution of the frequency output of the DDS is $df = 1MHz/2^{48} = 3.55 \cdot 10^{-9} Hz$. Here, $M = 48$ is assumed. This frequency is subtracted from the output frequency of the USO. The minimum frequency step of the frequency corrector is therefore $3.55 \cdot 10^{-9} Hz / 10^6$, which is $3.55 \cdot 10^{-16}$. Thereafter, the DDS may be controlled over a much larger frequency range with the same resolution while removing the USO calibration errors. Thus, we can find an exact value of the 48-bit DDS value M to correct the exact drift to zero by measuring our PPS, divided from the 10MHz output, against the PPS from the GPS receiver.

However, we face the technical challenge of measuring the time error between the GPS and USO pulse per second signals. To overcome this difficulty, we apply a high-precision time interval measurement method. This technique is illustrated in Fig. 9, where the two PPS signals are used to trigger an ADC (analog-to-digital convertor) to sample the sinusoid that is directly generated by the USO. Denoting the frequency of PPS_GPS as f_o, we have

$$T_1 = \frac{\phi_B - \phi_A}{2\pi f_o}$$ (43)

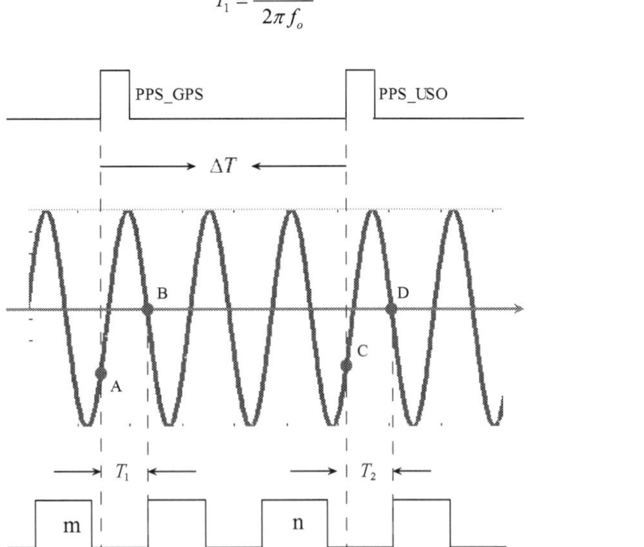

Fig. 9. Measuring time errors between two 1PPS with interpolated sampling technique.

Similarly, for PPS_USO, there is

$$T_2 = \frac{\phi_D - \phi_C}{2\pi f_o} \tag{44}$$

Hence, we can get

$$\Delta T = (n-m)T_0 + T_1 - T_2 \tag{45}$$

Where n and m denote the calculated clock periods. Since there is $\phi_B = \phi_D$, we have

$$\Delta T = \left(nT_0 + \frac{\phi_C}{2\pi f_o} \right) - \left(mT_0 + \frac{\phi_A}{2\pi f_o} \right) \tag{46}$$

To find a general mathematical model, suppose the collected sinewave signal with original phase $\phi_i (i \in (A,C))$ is

$$x(n) = \cos(2\pi f_o n + \phi_i) \tag{47}$$

Parting $x(n)$ into two nonoverlapping subsets, $x_1(n)$ and $x_2(n)$, we have

$$S_1(k) = FFT\left[x_1(n) \right], \quad S_2(k) = FFT\left[x_2(n) \right] \tag{48}$$

Thereby we have

$$\left| S_1(k_1) \right| = \left| S_1(k) \right|_{\max}, \quad \left| S_2(k_2) \right| = \left| S_2(k) \right|_{\max} \tag{49}$$

Thus, $\phi_i (i \in (A,C))$ can be calculated by

$$\phi_i = 1.5 \arg\left[S_1(k_1) \right] - 0.5 \arg\left[S_2(k_2) \right] \tag{50}$$

Since the parameters m, n, ϕ_C, ϕ_A and f_o are all measurable, the time error between PPS_GPS and PPS_USO can be obtained from (50). As an example, assuming the signal-to-noise ratio (SNR) is $50dB$ and $f_o = 10MHz$, simulations suggest that the RMS (root mean square) measurement accuracy is about $0.1ps$. We have assumed that some parts of the measurement system are ideal; hence, there may be some variation in actual systems. The performance of single frequency estimators has been detailed in (Kay, 1989).

Finally, time and phase synchronization can be achieved by generating all needed frequencies by dividing, multiplying or phase-locking to the GPS-disciplined USO at the transmitter and receiver.

4.3 Residual synchronization errors compensation

Because GPS-disciplined USOs are adjusted to agree with GPS signals, they are self-calibrating standards. Even so, differences in the PPS fluctuations will be observed because of uncertainties in the satellite signals and the measurement process in the receiver (Cheng et al., 2005). With modern commercial GPS units, which use the L1-signal at 1575.42MHz, a standard deviation of 15ns may be observed. Using differential GPS (DGPS) or GPS common-view, one can expect a standard deviation of less than 10ns. When GPS signals are lost, the control parameters will stay fixed, and the USO enters a so-called free-running mode, which further degrades synchronization performance. Thus, the residual synchronization errors must be further compensated for BiSAR image formation.

Differences in the PPS fluctuations will result in linear phase synchronization errors, $\varphi_0 + 2\pi\Delta f \cdot t = a_0 + a_1 t$, in one synchronization period, i.e., one second. Even though the USO used in this paper has a good short-term timekeeping ability, frequency drift may be observed in one second. These errors can be modeled as quadratic phases. We model the residual phase errors in the i-th second as

$$\varphi_i(t) = a_{i0} + a_{i1}t + a_{i2}t^2, \quad 0 \le t \le 1 \tag{51}$$

Motion compensation is ignored here because it can be addressed with motion sensors. Thus, after time synchronization compensation, the next step is residual phase error compensation, i.e., autofocus processing.

We use the Mapdrift autofocus algorithm described in (Mancill & Swiger, 1981). Here, the Mapdrift technique divides the i-th second data into two nonoverlapping subapertures with a duration of 0.5 seconds. This concept uses the fact that a quadratic phase error across one second (in one synchronization period) has a different functional form across two half-length subapertures, as shown in Fig. 10 (Carrara et al., 1995). The phase error across each subapertures consists of a quadratic component, a linear component, and an inconsequential constant component of $\Omega/4$ radians. The quadratic phase components of the two subapertures are identical, with a center-to-edge magnitude of $\Omega/4$ radians. The linear phase components of the two subapertures have identical magnitudes, but opposite slopes. Partition the i-th second azimuthal data into two nonoverlapping subapertures. There is an approximately linear phase throughout the subaperture.

$$\varphi_{ei}(t+t_j) = b_{0j} + a_{1j}t, \quad |t| < \frac{T_a}{4} \tag{52}$$

with $((2j-1)/2-1)/2, j \in [1,2]$. Then the model for the first subaperture $g_1(t)$ is the product of the error-free signal history $s_1(t)$ and a complex exponential with linear phase

$$g_1(t) = s_1(t)\exp(b_{01} + b_{11}t) \tag{53}$$

Similarly, for the second subaperture we have

$$g_2(t) = s_s(t)\exp(b_{02} + b_{12}t) \tag{54}$$

Let

$$g_{12}(t) = g_1(t)g_1^*(t) = s_1(t)s_2^*(t)\exp\left[j(b_{01} - b_{02}) + j(b_{11} - b_{12})t\right] \tag{55}$$

After applying a Fourier transform, we get

$$G_{12}(\omega) = \int_{-1/4}^{1/4} g_{12}(t)e^{-j\omega t}dt = \exp\left(j(b_{01} - b_{02})\right)S_{12}(\omega - b_{11} + b_{12}) \tag{56}$$

where $S_{12}(\omega)$ denotes the error-free cross-correlation spectrum. The relative shift between the two apertures is $\Delta_\omega = b_{11} - b_{12}$, which is directly proportional to the coefficient a_{i2} in Eq. (51).

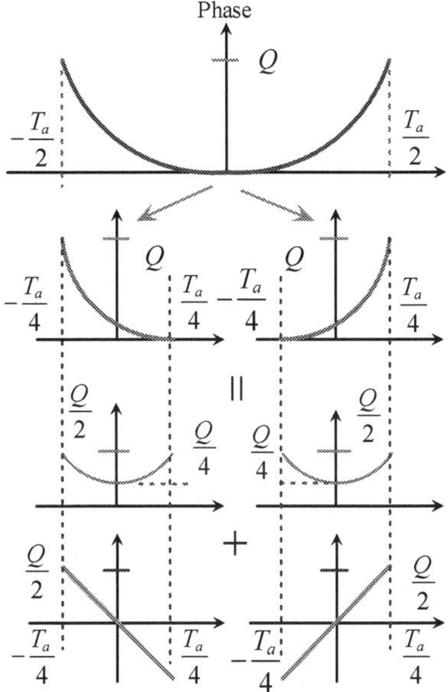

Fig. 10. Visualization of quadratic phase error.

$$a_{i2} = \Delta_\omega = b_{11} - b_{12} \tag{57}$$

Next, various methods are available to estimate this shift. The most common method is to measure the peak location of the cross-correlation of the two subapterture images.

After compensating for the quadratic phase errors a_{i2} in each second, Eq. (51) can be changed into

$$\varphi_i^c(t) = a_{i0} + a_{i1}t, \quad 0 \le t \le 1 \tag{58}$$

Applying again the Mapdrift described above to the i-th and (i+1)-th second data, the coefficients in (58) can be derived. Define a mean value operator $\langle \varphi \rangle_2$ as

$$\langle \varphi \rangle_2 \equiv \int_{-1/2}^{1/2} \varphi \, dt \tag{59}$$

Hence, we can get

$$a_{1i} = \frac{\left\langle \left(t - \bar{t}\right)\left(\varphi_{ei} - \overline{\varphi_{ei}}\right)\right\rangle_2}{\left\langle \left(t - \bar{t}\right)^2 \right\rangle_2}, \quad a_{0i} = \overline{\varphi}_{ei} - b_{1i}\langle t \rangle_2 \tag{60}$$

where $\overline{\varphi}_{ei} \equiv \langle \varphi_{ei} \rangle_2$. Then, the coefficients in (51) can be derived, i.e., the residual phase errors can then be successfully compensated. This process is shown in Fig. 11.

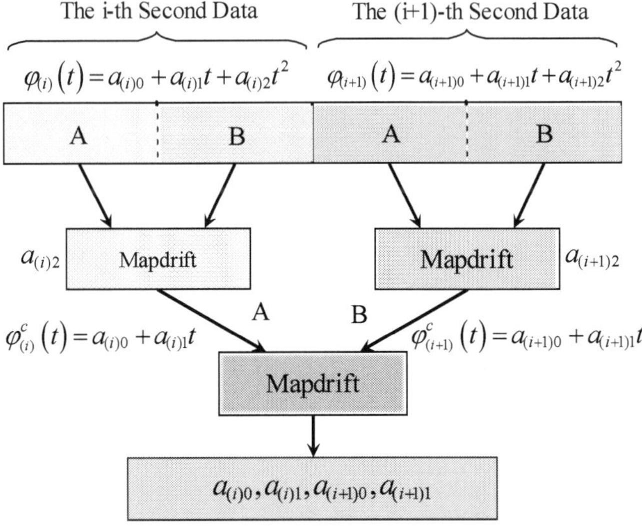

Fig. 11. Estimator of residual phase synchronization errors

Notice that a typical implementation applies the algorithm to only a small subset of available range bins, based on peak energy. An average of the individual estimates of the error coefficient from each of these range bins provides a final estimate. This procedure naturally reduces the computational burden of this algorithm. The range bins with the most energy are likely to contain strong, dominant scatterers with high signal energy relative to clutter energy. The signatures from such scatterers typically show high correlation between the two subaperture images, while the clutter is poorly correlated between the two images.

It is common practice to apply this algorithm iteratively. On each iteration, the algorithm forms an estimate and applies this estimate to the input signal data. Typically, two to six iterations are sufficient to yield an accurate error estimate that does not change significantly on subsequent iterations. Iteration of the procedure greatly improves the accuracy of the final error estimate for two reasons First, iteration enhances the algorithm's ability to identify and discard those range bins that, for one reason or another, provide anomalous estimates for the current iteration. Second, the improved focus of the image data after each iteration results in a narrower cross-correlation peak, which leads to a more accurate determination of its location. Notice that the Mapdrift algorithm can be extended to estimate high-order phase error by dividing the azimuthal signal history in one second into more than two subapertures. Generally speaking, N subapertures are adequate to estimate the coefficients of an *Nth*-order polynomial error. However, decreased subaperture length will degrade both the resolution and the signal-to-noise ratio of the targets in the images, which results in degraded estimation performance.

5. Conclusion

Although the feasibility of airborne BiSAR has been demonstrated by experimental investigations using rather steep incidence angles, resulting in relatively short synthetic aperture times of only a few seconds, the time and phase synchronization of the transmitter and receiver remain technical challenges. In this chapter, with an analytical model of phase noise, impacts of time and phase synchronization errors on BiSAR imaging are derived. Two synchronization approaches, direct-path signal-based and GPS signal disciplined, are investigated, along with the corresponding residual synchronization errors.

One remaining factor needed for the realization and implementation of BiSAR is spatial synchronization. Digital beamforming by the receiver is a promising solution. Combining the recorded subaperture signals in many different ways introduces high flexibility in the BiSAR configuration, and makes effective use of the total signal energy in the large illuminated footprint.

6. Acknowledgements

This work was supported in part by the Specialized Fund for the Doctoral Program of Higher Education for New Teachers under contract number *200806141101*, the Open Fund of the Key Laboratory of Ocean Circulation and Waves, Chinese Academy of Sciences under contract number *KLOCAW0809,* and the Open Fund of the Institute of Plateau Meteorology, China Meteorological Administration under contract number *LPM2008015*.

7. References

Cherniakov, M. & Nezlin, D. V. (2007). *Bistatic Radar: Principles and Practice,* John Wiley, ISBN, New York, ISBN: 978-0-470-02630-4.

Krieger, G. & Moreira, A. (2006). Spaceborne bi- and multistatic SAR: potential and challenge. *IEE Proc. Radar Sonar Navig.,* Vol. 153, No.3, pp. 184−198.

Eigel, R., Collins, P., Terzuoli, A., Nesti, G., & Fortuny, J. (2000). Bistatic scattering characterization of complex objects. IEEE Trans. Geosci. Remote Sens., Vol. 38, No. 5, pp. 2078−2092.

Burkholder, R. J., Gupta, I. J. & Johnson, J. T. (2003). Comparison of monostatic and bistatic radar images. IEEE Antennas Propag. Mag., Vol. 45, No. 3, pp. 41−50.

Loffeld, O., Nies, H., Peters, V. & Knedlik, S. (2004). Models and useful relations for bistatic SAR processing. IEEE Trans. Geosci. Remote Sens., Vol. 42, No. 10, pp. 2031−2038.

Wang, W. Q. & Cai, J. Y. (2007). A technique for jamming bi- and multistatic SAR systems. IEEE Geosci. Remote Sens. Lett., Vol. 4, No. 1, pp. 80−82.

Ceraldi, E., Franceschetti, G., Iodice, A. & Riccio, D. (2005). Estimating the soil dielectric constant via scattering measurements along the specular direction. IEEE Trans. Geosci. Remote Sens., Vol. 43, No. 2, pp. 295−305.

Wang, W. Q. (2007a). Application of near-space passive radar for homeland security. Sens. Imag. An Int. J., Vol. 8, No. 1, pp. 39−52.

Willis, N. J. (1991). *Bistatic Radar,* Artech House, ISBN, Boston, MA, ISBN: 978-0-890-06427-6.

Wei β, M. (2004). Time and phase synchronization aspects for bistatic SAR systems. Proceedings of Europe Synthetic Aperture Radar. Symp., pp. 395−398, Ulm, Germany.

Wang, W. Q., Liang, X. D. & Ding, C. B. (2006). An Omega-K algorithm with inte grated synchronization compensation. Proceedings of Int. Radar. Symp., pp. 395−398, Shanghai, China.

Krieger, G., Cassola, M. R., Younis, M. & Metzig, R. (2006). Impact of oscillator noise in bistatic and multistatic SAR. Proceedings of IEEE Geosci. Remote Sens. Symp., pp. 1043−1046, Seoul, Korea.

Krieger, G. & Younis, M. (2006). Impact of oscillator noise in bistatic and multistatic SAR. IEEE Geosci. Remote Sens. Lett., Vol. 3, No. 3, pp. 424−429.

Zhang, Y. S., Liang, D. N. & Wang, J. G. (2006). Analysis of frequency synchronization error in spaceborne parasitic interferometric SAR system. Proceedings of Europe Synthetic Aperture Radar. Symp., Dresden, Germany.

Unolkosold, P., Knedlik, S. & Loffeld, O. (2006). Estimation of oscillator's phase offset, frequency offset and rate of change for bistatic interferometric SAR. Proceedings of Europe Synthetic Aperture Radar. Symp., Dresden, Germany.

Zhang, X. L., Li, H. B. & Wang, J. G. (2005). The analysis of time synchronization error in bistatic SAR system. Proceedings of IEEE Geosci. Remote Sens. Symp., pp. 4615−4618, Seoul, Korea.

Gierull, C. (2006). Mitigation of phase noise in bistatic SAR systems with extremely large synthetic apertures. Proceedings of Europe Synthetic Aperture Radar. Symp., Dresden, Germany.

Moreira, A., Krieger, I., Hajnsek, M., Werner, M., Hounam, D., Riegger, E. & Settelmeyer, E. (2004). TanDEM-X: a TerraSAR-X add-on satellite for single-pass SAR interferometry. Proceedings of IEEE Geosci. Remote Sens. Symp., pp. 1000−1003, Anchorage, USA.

Evans, N. B., Lee, P. & Girard, R. (2002). The Radarsat-2/3 topgraphic mission. Proceedings of Europe Synthetic Aperture Radar. Symp., Cologne, Germany.

Younis, M., Metzig, R. & Krieger, G. (2006a). Performance prediction and verfication for bistatic SAR synchronization link. Proceedings of Europe Synthetic Aperture Radar. Symp., Dresden, Germany.

Younis, M., Metzig, R. & Krieger, G. (2006b). Performance prediction of a phase synchronization link for bistatic SAR. IEEE Geosci. Remote Sens. Lett., Vol. 3, No. 3, pp. 429−433.

Eineder, M. (2003). Oscillator clock shift compensation in bistatic interferometric SAR. Proceedings of IEEE Geosci. Remote Sens. Symp., pp. 1449−1451, Toulouse, France.

Lance, A. L. Seal, W. D. & Labaar, F. (1984). Phase-noise and AM noise measurements in the frequency domain. Infrared Millim. Wave, Vol. 11, No. 3, pp. 239−289.

Wang, W. Q., Cai, J. Y. & Yang, Y. W. (2006). Extracting phase noise of microwave and millimeter-wave signals by deconvolution. IEE Proc. Sci. Meas. Technol., Vol. 153, No. 1, pp. 7−12.

Walls, F. L. & Vig. J. R. (1995). Fundamental limits on the frequency stabilities of crystal oscillators. IEEE Trans. Ultra. Ferroelectr. Freq. Control., Vol. 42, No. 4, pp. 576−589.

Rutman, J. (1978). Characterization of phase and frequency instabilities in precision frequency sources: fifteen years of progress. Proceeding of IEEE, Vol. 66, No. 9, pp. 1048−1073.

Vannicola, V. C. & Varshney, P. K. (1983). Spectral dispersion of modulated signals due to oscillator phase instability: white and random walk phase model. IEEE Trans. Communications, Vol. 31, No. 7, pp. 886−895.

Donald, R. S. (2002). *Phase-locked loops for wireless communications: digital, Analog and Optical Implementations,* Kluwer Academic Publishers, New York, ISBN: 978-0792376026.

Hanzo, L., Webb, W. & Keller, T. (2000). Signal- and multi-carrier quadrature amplitude modulation−principles and applications for personal communications, WLANs and broadcasting. John Wiley & Sons Ltd, ISBN: 978-0471492399.

Foschini, G. J. & Vannucci, G. (1988). Characterizing filtered light waves corrupted by phase noise. IEEE Trans. Info. Theory, Vol. 34, No. 6, pp. 1437−1488.

Auterman, J. L. (1984). Phase stability requirements for a bistatic SAR. Proceedings of IEEE Nat. Radar Conf, pp. 48−52, Atlanta, Georgia.

Wei β, M. (2004). Synchronization of bistatic radar systems. Proceedings of IEEE Geosci. Remote Sens. Symp., pp. 1750−1753, Anchorage.

Kroupa, V. F. (1996). Close-to-the carrier noise in DDFS. Proceedings of Int. Freq. Conrol. Symp., pp. 934−941, Honolulu.

Muellerschoen, R. J., Chen, C. W., Hensley, S. & Rodriguez, E. (2006). Error analysis for high resolution topography with bistatic single-pass SAR interferometry. Proceedings of IEEE Int. Radar Conf., pp. 626−633, Verona, USA.

Eineder, M. (2003). Oscillator clock drift compensation in bistatic interferometric SAR. Proceedings of IEEE Geosci. Remote Sens. Symp., pp. 1449−1451, Toulouse, France.

Wang, W. Q., Ding, C. B. & Liang, X. D. (2008). Time and phase synchronization via direct-path signal for bistatic SAR systems. IET Radar, Sonar Navig., Vol. 2, No. 1, pp. 1−11.

Chen, C. C. & Andrews, H. C. (1980). Target-motion-induced radar imaging. IEEE Trans. Aerosp. Electron. Syst., Vol. 16, No. 1, pp. 2−14.

Delisle, G. Y. & Wu, H. Q. (1994). Moving target imaging and trajectory computation using ISAR. IEEE Trans. Aerosp. Electron. Syst., Vol. 30, No. 3, pp. 887−899.

Wahl, D. E., Eichel, P. H., Ghilia, D. C. & Jakowatz, C. V. Jr. (1994). Phase gradient autofocus − a robust tool for high resolution SAR phase correction. IEEE Trans. Aerosp. Electron. Syst., Vol. 30, No. 3, pp. 827−835.

Wang, W. Q. (2009). GPS based time and phase synchronization processing for distributed SAR. IEEE Trans. Aerosp. Electron. Syst., in press.

Parkinson, B. W., & Spilker, J. J. (1996). *Global Position System: Theory and Applications.* American Institute of Aeronautics and Asronautics, Washington, D.C.

Vankka, J. (2005). *Digital Synthesizers and Transmitters for Software Radio.* Springer, Netherlands, ISBN: 978-1402031946.

Kay, S. (1989). A fast and accurate single frequency estimator. IEEE Trans. Acoustics, Speech, Sig. Process., Vol. 27, pp. 1987−1990.

Cheng, C. L., Chang, F. R. & Tu, K. Y. (2005). Highly accurate real-time GPS carrier phase-disciplined oscillator. IEEE Trans. Instrum. Meas. , Vol. 54, pp. 819−824.

Carrara, W. G., Goodman, R. S. & Majewski, R. M. (1995). Spotlight Synthetic Aperture Radar: Signal Processing Algorithms. Artech House, London, ISBN: 978-0890067284.

TOPIC AREA 4: Radar Subsystems and Components

15

Planar Antenna Technology for mm-Wave Automotive Radar, Sensing, and Communications

Joerg Schoebel and Pablo Herrero
Braunschweig University of Technology
Germany

1. Introduction

Planar antennas are common components in sensing applications due to their low cost, low profile and simple integration with systems. They can be used commonly at frequencies as high as 77 GHz and above, for example in automotive sensing. Also, new ultra-wideband communication systems operating around 60 GHz will heavily rely on planar antennas due to their unique properties. One special advantage of planar elements is that they can easily form array structures combining very simple elements, like microstrip patches. Also phased and conformal arrays can be built. Due to all these characteristics, planar antennas are very good candidates for building front ends for mm-wave applications in radar, sensing or communications.

For some applications, which are either established today or which will become commercial in a few years, general requirements can be given. In addition to automotive radar, this mainly applies to 60 GHz ultra-broadband wireless indoor communications. Microwave imaging at 94 GHz and above is still much under research, as well as other sensing applications in the millimeter-wave region above 100 GHz. Prominent frequency bands are in the 122 and 140 GHz range[1].

Typical antenna requirements are multi-beam or scanning capability, high gain up to 30..36 dBi and moderate to low sidelobes. In monopulse radar systems, the sidelobes may be included in the angle determination scheme, then the sidelobe level requirements are rather relaxed.

Loss is generally an important issue. Comparing a planar approach to a commercially available dielectric lens, the planar antenna exhibits significantly higher losses, especially if the beamforming device is included. The antenna efficiency of 77 GHz planar array columns is roughly 50%. These losses are caused by dielectric losses and conductor losses. To attain reliable predictions of the losses in full-wave simulations, care has to be taken in the

---

[1] The FCC frequency allocations contain an amateur/mobile band at 122.25-123 GHz and an ISM (industrial, scientific, medical) band at 122-123 GHz. The bands 136-141 GHz and 141-148.5 GHz are allocated (among others such as radio astronomy) for amateur plus radiolocation and mobile plus radiolocation services, respectively by FCC and ECC. Therefore, a lot of bandwidth is within technological reach.

modeling of the conductors. Even if the conductors mainly are copper, a typical surface finishing consists of 3 to 5 µm nickel and a few hundred nanometers of gold. Comparing these numbers with the skin depth, it is found that a significant part of the current flows in the nickel. A layered impedance model for the metallization therefore improves the loss prediction significantly.

On the other hand, fabrication aspects have to be taken into account. Nickel acts as a diffusion barrier between gold and copper. Although practically not much interdiffusion will occur at room temperature, the nickel layer may still prove necessary as soon as wirebonding or other mounting technologies at elevated temperature are employed.

Generally planar antennas are well suitable for mass fabrication with low-cost processing. The required minimum feature sizes are realizable with common circuit board processes and the reproducibility of an industrial circuit board process is sufficiently good to achieve consistent antenna characteristics from lot to lot. For a low cost approach the RF functions should be limited to a single layer, which may be complemented with additional layers of low-frequency material (e.g. FR-4) for mechanical stabilization and accommodation of power supply and baseband circuitry.

2. Planar antennas for automotive radar applications

Due to the large number of groups working on planar antennas for automotive radar and the multitude of respective publications, an extensive review of the literature is not provided here. However, some examples for planar antenna implementations will be given below, which illustrate specific and interesting particular solutions.

Requirements for a multitude of different automotive-radar-based comfort and safety functions are compiled in (Mende & Rohling, 2002). While for the long range ACC applications an angular range of ±4° to ±8° is usually sufficient, short or medium range applications have different requirements in distance and angular range. The angular range is covered with a number of mutually overlapping beams intersecting approximately at their 3 dB points.

Conventionally, the beams are generated for example with a dielectric lens, which is fed from different positions in the focal plane. If the feed elements are realized as patches, a pre-focusing may be necessary for optimum illumination of the lens. This can be achieved with dielectric rods placed on top of the patches (Kühnle et. al., 2003). Fig. 2-1 a) shows the

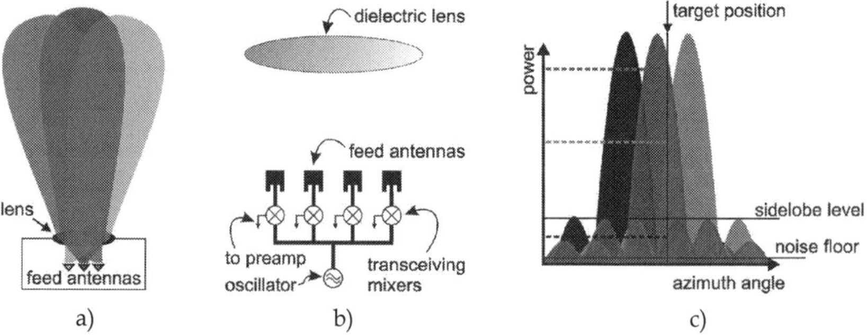

Fig. 2-1. a) Beamforming in conventional lens-based radar systems; b) Block diagram of a lens-based automotive radar, c) Target angle determination with monopulse technique

concept of beam generation with a dielectric lens. A simple and low-cost circuit concept for an FMCW radar multi-beam front end is given in Fig. 2-1 b). All beams are illuminated simultaneously. The transmit signal also drives the transceiving mixers, which down convert the individual receive signals to baseband (Kühnle et. al., 2003).

The angular position of a radar target is determined by the amplitude ratio of the received signals in adjacent radar beams, typically referred to as monopulse technique, see Fig. 2-1 c). To achieve a reasonable angular accuracy, a number of 2 to 5 mutually overlapping beams is required to cover the azimuthal angular range of typical long or medium range applications. The beam width depends on the aperture size of the antenna, which is limited by practical reasons such as available mounting space at the bumper region.

Whereas the monopulse technique has shown excellent performance in automotive radar sensors for a long time, there are extensions and alternative algorithms, such as complex monopulse for multi-target situations (Schoebel, 2004b; Schoebel & Schneider, 2007) and high-resolution algorithms (Schneider et. al., 2002). Alternatively, a faster but less accurate determination of the target angular position can be realized using a larger number of beams and determining the target angle in a maximum search of the receive signals of the beams. In (Kawakubo et. al., 2004) a system providing pre-crash safety and ACC functions is described, which employs 41 digitally generated beams covering an angular range of ±10°.

A crucial point is the required radar range, which often exceeds 100 to 200 m. Concomitant to the narrow beam angles this also calls for a large antenna gain, which may exceed 30 dBi. As a result the size of a planar array may become quite large and special attention has to be paid to antenna and feed losses. Polarization is not very critical. Some systems use 45° polarization, so that radar signals from cars travelling in opposite directions and facing each other will be polarization decoupled to reduce interference. Alternatively, vertical polarization is less prone to multipath propagation between a car's bottom and the road surface (Schneider & Wenger, 1999), so that signal processing is facilitated by a complexity reduction of the radar scenario.

An early example of a Rotman lens employed for azimuthal beam forming in automotive radar is given in (Russell et. al., 1997). In elevation, series-fed columns are employed. Such planar antenna columns are elaborated in detail in (Freese et. al., 2000), where element widths are tapered to achieve a better sidelobe suppression (see examples in Fig. 3-2 a), b). 45° polarization is added to this concept in (Iizuka et. al., 2002) by employing inclined patches, which are connected to the feed line on a corner (see Fig. 3-2 c). A rather complex beam forming system based on several Rotman lenses was developed by (Metz et. al., 2001).

3. Design of patch elements, patch columns and arrays

3.1 Single patches

The design of single patch elements has been treated extensively in the literature (Balanis, 2005) and is not repeated here in detail. There are many methods for analyzing and designing patches, divided into empirical models, semi empirical models and full-wave analysis (Bhartia et. al., 1990). In this work, a rectangular patch is designed with the Transmission Line Model with the aid of a Matlab program to obtain a starting point for a parametric design with a full-wave simulator. With the substrate chosen (5 mil Ro3003), parameters of the simulation are patch width and length.

A single cell is here designed and simulated, in order to obtain the optimum parameters when connected in the array. The model consists of the rectangular patch with its initial dimensions. The patch is connected from both sides with the high impedance microstrip lines. The impedance of this line is designed to be 100 Ohm, which will make feed arrangements easier. This impedance is equivalent for this substrate to a width of 85 µm, which can be processed with conventional "fine-feature" circuit board technology.

The patch model is tuned with the patch length to resonate at the desired operation frequency (e.g. around 140 GHz). The patch width is used for balancing the maxima of the E-field at the radiating edges of the patch (see Fig. 3-1 a). With the aid of the E-field diagrams and by deembedding the input port, the optimum line length is found for a 360° insertion phase between input and output of the cell. The respective input and output positions are located at the points where the field is minimum on the line (see arrows in Fig. 3-1 a). These calculations make the design of the final array arrangement and also the calculations of the number of elements very easy. A good design with very stable and processable dimensions is found. This model resonates at a frequency of 138 GHz (compare Fig. 3-1 b), so this model is chosen for designing the whole array.

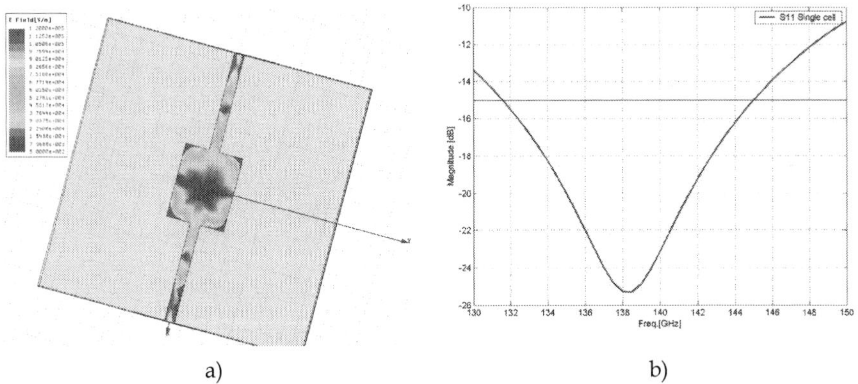

a)   b)

Fig. 3-1. a) Single patch cell simulation for column design; b) Return loss of the patch cell

3.2 Patch array pattern modeling

Useful array patterns can be derived most straightforwardly by operating the patch in the ground (TM01) mode. Therefore, the radiation model needs to cover this mode only. (Carver & Mink, 1981) give analytical radiation patterns, which can easily be implemented in a numerical model. We implemented this model in a Matlab code, which allows to specify positions, excitation voltages, and phases of all individual patches in a planar array. Array patterns can easily be optimized or analyzed with this tool. Patch and feed losses are empirically taken into account in the excitation. Fig 3-2 e) gives antenna gain measurement and simulation of a single patch column of 20 elements in E and H planes. The correspondence is very good, only at very large angles some discrepancies can be seen, which are partly caused by the feed structure.

The designs presented in this work are all realized on 5 mil Ro3003 substrate. In the V and W bands this substrate is electrically sufficiently thin, so that surface waves play no significant role. Consequently, the pattern modeling may neglect surface waves and still reproduces the measured patterns very well (cf. Fig. 3-2 e). As patch antennas mainly excite surface waves on their radiating edges in the E plane, the surface waves will travel along the series-fed column (see below) but their effect on the amplitude and phase of the excitation is negligible. The H plane azimuthal patterns are not much affected by surface waves, because their excitation from patch elements in this plane is low. Overall, the effect of surface waves is more pronounced at higher frequencies, in our case above 100 GHz. As the examples shown below are optimized in 3D or 2.5D simulations, partly for finite substrates, the effects of surface waves are implicitly taken into account.

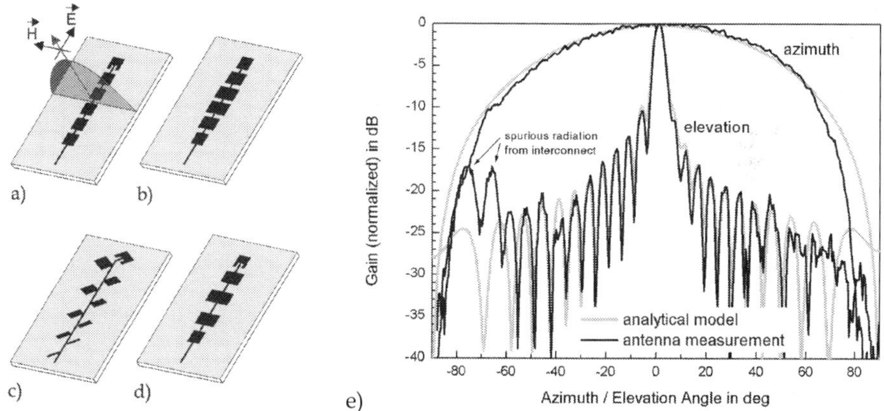

Fig. 3-2. a) Patch column with uniform series feed; b) Amplitude-tapered patch column; c) Amplitude-tapered inclined patch column; d) Phase and amplitude optimized column; e) Gain simulation and measurement of a 20-element patch column for 77 GHz

3.3 Series-fed columns

The mutual distance of the antenna columns in azimuth is restricted to not much more than half a free-space wavelength to suppress grating lobes, so the space for a feed network within each column is very limited. As the column consists of more than a few antenna elements, series feeding is the only possible approach and the series feed also has to provide the beam forming.

Once that the single patch cell is designed adjusting the complete column is easy. The most critical parameter affecting the matching of the whole array is the length of the high impedance line, because it ensures perfect impedance transformation at the edge of the patch. This length also contributes to the control of the phase distribution across the array and therefore influences the radiation profile.

Rectangular patches are used as antenna elements. They are directly coupled to the feed line at their edges and they are equidistantly placed on the column. As the input impedance of the feed point at the patch edge is rather large, the interconnecting lines are also designed for this impedance. The main beam should point in the broadside direction, so the interconnecting lines' lengths are adjusted for 360° phase shift between the input edges of consecutive patches (see cell design in section 3.1).

The feed point of the last element is shifted towards the inside of the element to provide a termination of the column (see Fig. 3-2 a). The inset length controls the global input reactance, so this parameter was used for improving the matching without much affecting the center frequency. Finally, the length of the patch is slightly increased to decrease the resonant frequency.

3.4 Series-fed columns with uniform elements for 77 GHz

For the 77 GHz array, all elements on the column have the same size and emit approximately 10% of the respective input power except for the last element. Therefore, the power distribution on the column decreases exponentially except for the terminating element, which emits all the remaining power.

The phase distribution on the column (except the last element) is almost uniform with a small phase shift of approximately 3.3° between the elements. Far-field antenna pattern calculations for a 20-element column with the Matlab model described above are shown in Fig 3-2 e). As long as the phase distribution is uniform, the far-field pattern is symmetrical despite the unsymmetrical excitation amplitude pattern. The sidelobe level is between -9 to -11 dB for 10 to 20 elements. Due to the shifted feed point of the terminating element, a phase shift of this element with respect to the previous element of -50° occurs. The power emitted from the terminating element is a function of the number of elements in the column. This variation is sufficient to result in a visible squint of the main lobe and also in an unsymmetry of the pattern, as can be seen in Fig. 3-2 e).

This effect was studied experimentally. Fig. 3-3 a) shows antenna gain and squint angle as a function of the number of elements in the column. Due to the exponential power distribution and the element and inter-element losses, the gain tends to saturate for larger numbers of elements. The total loss of the column is around 2.5 dB for 20 elements. The squint is also reproduced satisfactorily taking into account an angular measurement uncertainty (antenna mounting and angular calibration) of approximately ±0.5° to ±1°. These results are confirmed in 3D electromagnetic simulations using HFSS.

3.5 Optimization of column patterns

An optimization of the elevation characteristics with respect to lowering the sidelobe level is beneficial for reducing the ground clutter, especially the near-range clutter from the road surface. To a lesser extent multi-path propagation due to reflections on the road surface will be mitigated. A straightforward solution is the implementation of a low-sidelobe excitation pattern on the series-fed column, as can be found for example in (Freese et. al., 2000). This can easily be achieved by variation of the patch width (cf. Fig. 3-2 b) and c). Due to the dependence of the available power of an element on the previous elements, antenna elements with rather large emission efficiencies are required. However, the maximum patch width is restricted by the excitation of the first transversal mode in the patch and also by a minimum distance between the columns of an array in order to suppress coupling between the columns.

For small patch widths, a reliable determination of the emission efficiency is difficult, because the radiated power is small and comparatively large measurement or numerical errors can easily occur in the development process. In general, the maximum achievable sidelobe suppression is a function of the maximum and minimum available patch widths.

An additional complication results from the matching of the patch elements, which requires a length adjustment as a function of patch width. Hence, for equidistant positions of the patches, phase errors are introduced, which deteriorate a low-sidelobe pattern. Adjusting the positions of the elements to ensure a uniform phase distribution will also affect the antenna pattern as the spacing becomes non-uniform. Therefore, we propose an optimization procedure, which uses patch width and patch position as adjustable parameters to attain a low-sidelobe pattern with a series-fed antenna column. The excitation phase, which results from the position, is taken into account. Generally, this procedure will result in a non-uniform phase distribution and non-equidistant patch positions. It has been shown, that by controlling the phases of the elements of a linear array, an unsymmetrical pattern with low sidelobes on one side of the main beam can be achieved, even if the amplitude distribution is uniform (Trastoy & Ares, 1998). In an automotive application, only the sidelobes pointing to the road need to be reduced, therefore an unsymmetrical antenna pattern can be employed.

Fig. 3-3 b) shows antenna patterns calculated for Taylor 1-parameter and Chebyshev amplitude distributions for 8 elements implemented with a loss-free series feed and certain restrictions of maximum and minimum available patch widths. The excitation phase of all elements is assumed to be uniform. Due to the limitation of the range of available patch widths, the maximum sidelobe suppression is limited.

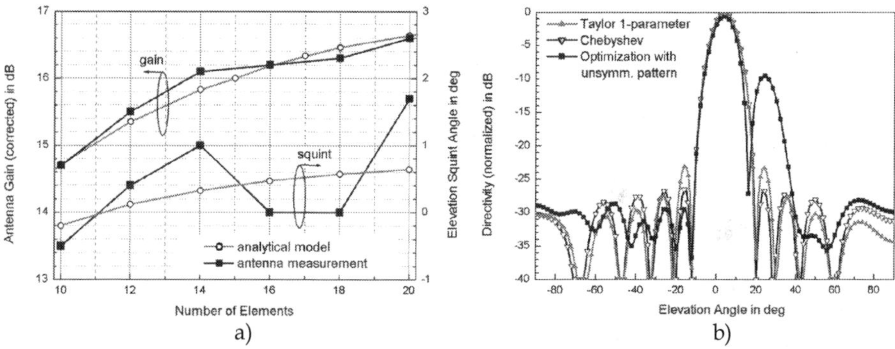

Fig. 3-3. a) Antenna gain and elevation squint angle as a function of antenna elements in the column. The gain is corrected with respect to the loss of the feed line and interconnect;
b) Antenna patterns of series-fed columns restricted by the same maximum and minimum patch widths

Under the same conditions the patch widths and positions (and implicitly the excitation phases) are optimized for a low sidelobe level on one side of the main lobe. The optimization was implemented in Matlab using multidimensional nonlinear function minimization with Monte-Carlo and direct search algorithms. As a result, the first sidelobe level improves from -22.9 dB (Taylor one-parameter) or -26.4 dB (Chebyshev) to -28.8 dB. Correspondingly, the directivity reduces from 0 dB (Taylor, reference) or -0.2 dB (Chebyshev) to -0.7 dB.

It should be noted that to attain a low sidelobe level on a series feed, a similar optimization procedure has to be executed on a conventional amplitude distribution (such as Taylor or Chebyshev) in order to account for the changes in electrical and mechanical patch lengths, which result from the different patch emission efficiencies. Therefore, the improvement of the sidelobe level for the unsymmetrical pattern comes at virtually no extra effort in the design.

3.6 Patch array

To achieve a low sidelobe pattern in azimuth, the power divider feeding a corporate array has to generate a certain power distribution, here we choose a 20 dB Chebyshev pattern. In order to mitigate pattern distortions, which may be caused by reflections on the antenna input, the power divider is realized as a 3 stage Wilkinson setup employing uneven dividing ratios. As the size and parasitics of available thin-film resistors do not allow to mount them directly between the outputs, transformers are introduced between the resistor and the outputs of the dividers. This also allows to use other than the ideal resistor values. The complete divider is validated in an HFSS simulation including appropriate models of the resistors. Fig. 3-4 shows the HFSS model, the complete 8 column / 12 row array and a detail of the Wilkinson dividers with mounted resistors.

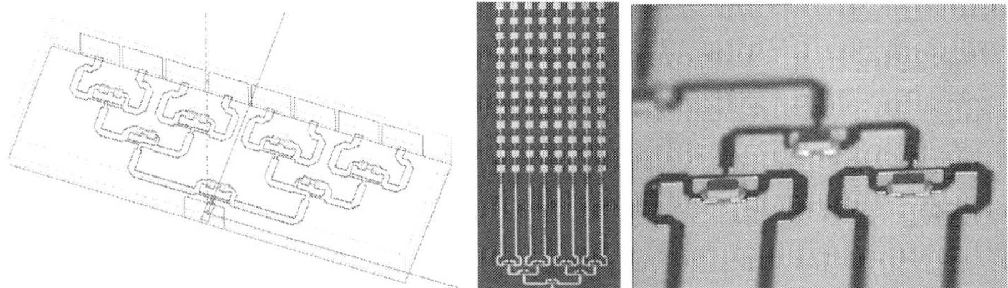

Fig. 3-4. HFSS model and photographs of circuit board layout and detail of Wilkinson divider with mounted resistors for an 8-output 20dB sidelobe level pattern

The dividers were analyzed on a wafer prober and excellent agreement of the power and phase distribution with the design was obtained. Results are given in Fig. 3-5. Amplitude tolerances amount to a maximum of ±7% and phase tolerances of ±6°, this is to be compared with probe placement tolerances of ±4°. The total loss is approximately 4.7 dB, where 1.5 dB are caused by radiation. The latter number was determined from HFSS simulations. The azimuthal pattern is given in Fig. 4-5 a).

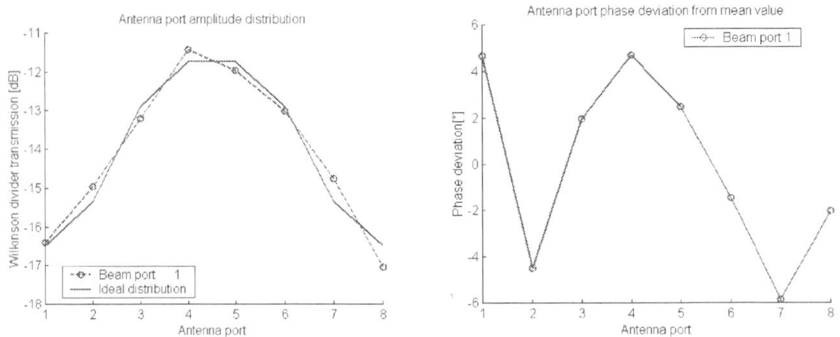

Fig. 3-5. Characterization results of 3-stage 8-output Wilkinson divider

4. Beam forming with planar Rotman lenses

In this chapter we present exemplary designs of a beam forming subsystem for automotive radar. These subsystems were originally intended to be used with a RF-MEMS based multi-throw switch, so that a single beam is active at a time (Schoebel et al., 2004 and 2005). But the same beam forming system can also be employed in an approach, in which all beams are active simultaneously (Hansen et al., 2006) similarly to the current automotive radar sensors as depicted in Fig. 2-1 b).

The planar Rotman lens consists of a parallel-plate waveguide with the beam ports on one side and antenna ports on the other side. The antenna ports are connected to the antenna elements via delay lines, which introduce an additional, individual phase shift. The lens has an ideal focus on its middle symmetry axis (shown red in Fig. 4-1 a) and two additional symmetrical ideal focal points. One is shown blue in Fig. 4-1 a) and the other one lies on a symmetrical position in the lower half. The focal lengths corresponding to the focal points can be different from each other.

Compared to an optical lens, the delay line network comprises the actual lens (position dependent phase shift), while the parallel plate waveguide corresponds to the free space region between the focal plane and the lens surface.

If additional beam ports are needed, focal spots are found on an approximately circular arc between and beyond the exact foci. In principle, the usable beam and antenna port contours extend to the point of their intersection (dashed in Fig. 4-1 a), but only the inner parts of these edge contours are used, as this improves the field distribution on the beam and antenna ports. The edges between the end points of the port contours are terminated (grey in Fig. 4-1 a), e.g. with microstrip ports, which are terminated at their ends. Rubber absorber glued on top of 1 to 2 cm of microstrip line is a practical termination for millimeter waves.

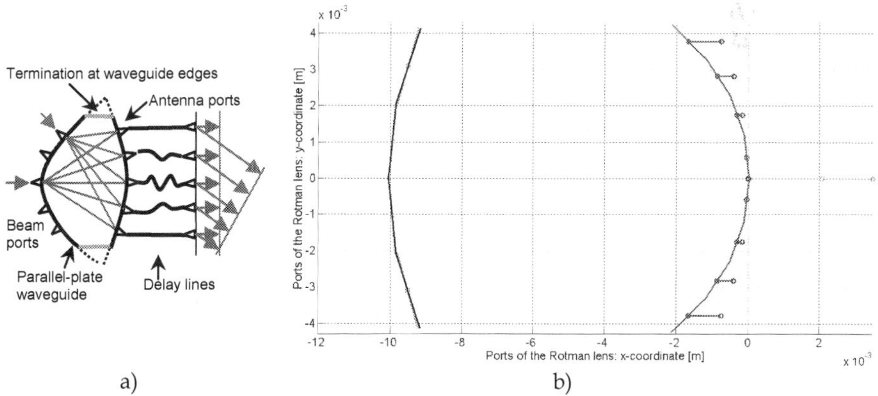

Fig. 4-1. a) Operation principle of the Rotman lens, b) Analytical design of a Rotman lens

4.1 Design principle

The design of the Rotman lens is conducted using the classical analytical framework of (Rotman & Turner, 1963), which has been rewritten in terms of phase shift between the antenna elements. From the input parameters, such as desired phase shift, focal lengths and geometry and transmission line parameters the positions of the antenna ports and the lengths of the delay lines are determined. In case that more beam ports than the 3 ideal foci are required, a numerical optimization of the positions of the additional beam ports is conducted. The result of the solution of the analytical design equations is shown in Fig. 4-1 b). The beam ports are located on the left and the antenna ports are on the right (circles). The delay lines are indicated by the horizontal lines at the antenna ports. The feed structures are microstrip tapers ("planar horn antennas") radiating into the parallel plate line. The front faces of the tapers are also shown in Fig. 4-1 b).

We use the total phase error (sum of absolute values of the phase deviation) versus position to determine the location of the non-ideal focusing points (beam ports) numerically. The same procedure is also employed at the ideal beam port foci and antenna ports[2]. Results are given in Fig. 4-2 a). Here, the results are restricted to two ports on one side of the lens; for the other two ports, the outcome is symmetrical. The dark spots give the location of the beam and antenna port foci. In our example, good phase accuracy is found for location tolerances of a few hundred micrometers. Generally, the light (yellow) regions indicate total phase deviations of a few degrees, which would generally be acceptable.

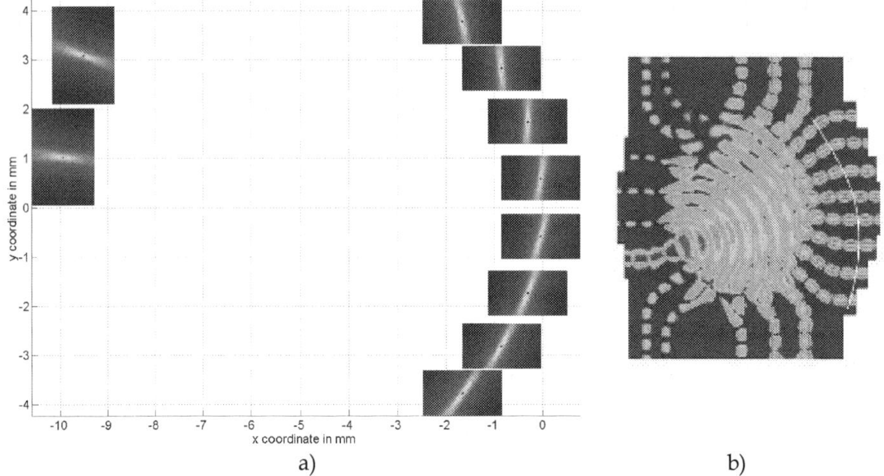

a)   b)

Fig. 4-2. a) Phase errors versus port positions; b) 3D simulation of the electric field distribution in a Rotman lens. The white line indicates the phase shift of the antenna signals

It is interesting to note that the low phase error regions at the beam ports have the form of an elongated valley. As long as the beam port is on the bottom of this valley, low phase errors are obtained. From this observation, we can postulate a new design paradigm: the longitudinal orientation of the beam port feeding tapers should be along these valleys. In this case, uncertainties in the position of the phase centers of these tapers will have little influence on the lens performance, as the phase center of the taper will in any case be on its symmetry axis. Classically, the beam port faces are oriented tangentially on a circle around

---

[2] At the antenna ports the phase error is given by the deviation of electrical length between beam port and antenna port location. The latter is varied. In Fig. 4-2 a) the phase errors for all beam ports are added at each antenna port location.

the center of the antenna port contour (position (0,0) in Fig. 4-1b). The intersections of linear extrapolations of the valleys with the x axis are given in Fig 4-1 b) as circles to the right of the antenna port contour, also the beam port tapers' front faces are given for both cases[3] (black for the classical circular approach and green for orientation along the valleys).

Comparing the electrical lengths of the lens and delay line structure using the ideal focus points with the desired phase distributions on the antenna ports yields a very good agreement, as shown in Fig. 4-3 a).

The design of the feed tapers has to be conducted with some care. 3D or 2.5D simulations have to be employed to determine the phase centers of the tapers for different taper widths. The length of the tapers should be fixed to a value which ensures good matching, e.g. a few wavelengths. As the design is optimized empirically, the width of the tapers may change during the design process. As shown in Fig. 4-1 b) the width of the tapers is typically chosen so that negligible space between their edges remains.

Once the sizes of the feed tapers are know, the transmission in the parallel-plate waveguide can be modeled with a 2D radiation model as described in (Smith & Fong, 1983). The radiation characteristics of the tapers are calculated assuming a uniform E-field distribution at the front face. Taking into account the propagation path length and the overlap between beam and antenna port radiation patterns, the transmission properties of the lens and beam pointing losses are calculated. Results are given in Fig. 4-3 b).

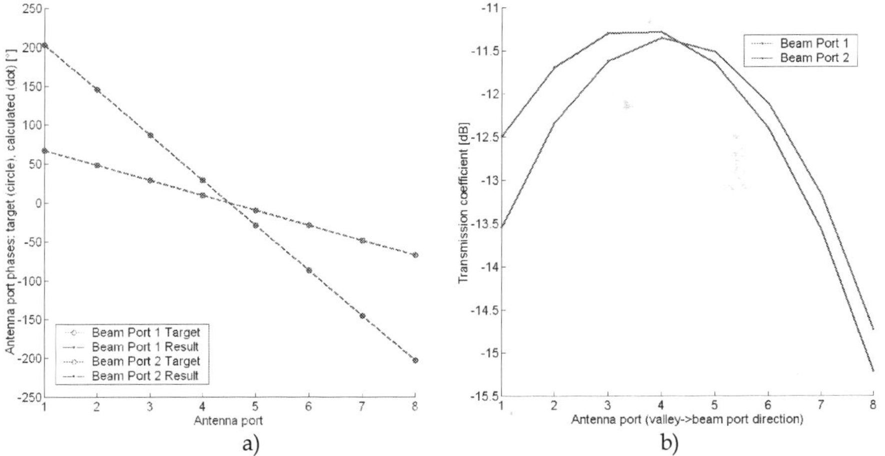

Fig 4-3. a) Antenna port phase distribution; b) Antenna port amplitude distribution

With these results, the lens is empirically optimized for optimum transmission. Typically we find that the focal lengths of on- and off-axis focal points should be equal. In addition, lateral extensions of beam and antenna port contours should also roughly be equal (compare Fig. 4-1). With the overall size of the lens the focusing of the port tapers can be controlled, as

---

[3] Here, the actual phase centers of the ports are not taken into account. In the final layout, the front faces of the tapers are somewhat shifted so that the phase centers of the tapers lie exactly on the focus points.

a larger feed will focus the radiation more strongly towards the middle of the opposite face of the parallel-plate waveguide. This may be used to control the sidelobe level.

Finally, the complete lens design can be validated in a 2.5D or 3D simulation, an example is given in Fig. 4-2 b).

4.2 Design examples

The design in Fig. 4-1 b) is conducted for 5 mil Ro3003 substrate ($\varepsilon = 3$) and 50 Ω delay lines ($\varepsilon_{eff} = 2.428$), focal lengths F=G=10 mm and 18° inclination between F and G. The figure is drawn to scale. The resulting beam pointing angles are ±6° and ±18° for an 8 element array in azimuth, which correspond to a prototypical short-to-medium range automotive radar application. The lens combined with 12-element antenna columns is depicted in Fig. 4-4 a) and the respective antenna pattern is presented in Fig. 4-5 a). This lens employs a conventional delay line network, which consists of a first, mostly linear section in order to change the lateral/azimuthal distance of the lines, so that their mutual distance is that of the final array ("Azimuthal distance control" in Fig. 4-4 b). Then, sinus-shaped sections are added, with which the specific delay is realized. The loss of the lens with delay line network amounts to approx. 4 dB.

Additional designs were conducted for long-range applications, for which a large antenna gain and beam angles of ±4° and ±8° are envisaged. Different focal lengths were tested, as can be seen in Fig. 4-4 a). If the size of the parallel-plate waveguide is made larger, the delay line network becomes shorter. We employ a novel routing scheme, which consists of a number of lines and circular arcs. This network is generally smaller than the conventional one and induces less radiation loss due to the larger radii of the arcs (compare Fig. 4-4 b).

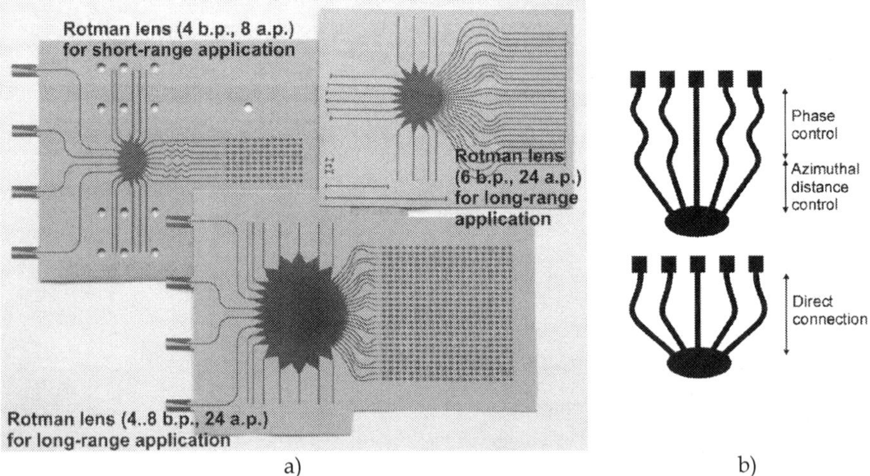

a)   b)

Fig 4-4. a) Rotman lens examples for different automotive radar applications; b) Delay line network layout principles

Additionally, multiples of a guided wavelength are added towards the outer antenna ports, so that an additional amplitude taper comes in effect. This normally reduces the sidelobe level. Fig. 4-5 b) shows the antenna pattern of the largest lens in comparison to a commercial dielectric-lens based radar sensor (Kühnle et. al., 2003). As can be seen, the pattern of the latter can be reproduced quite well by the inner four beams of the Rotman lens, but the overall gain of the Rotman-lens based antenna is smaller. The Rotman lens with delay line network induces losses of 6.5 dB.

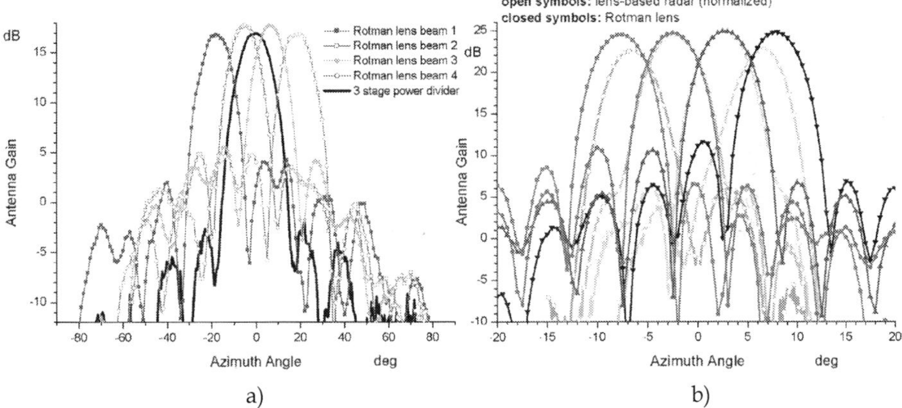

Fig 4-5. a) Antenna patterns for Wilkinson divider and first Rotman lens; b) Antenna pattern for 8-beam-port Rotman lens (beams for the 4 inner beam ports are shown) compared to dielectric-lens based automotive radar system (gain maximum normalized to Rotman lens)

The design with 8 beam ports resulted from the empirical optimization of the lens performance. The desired 4 beam ports would occupy only a small part of the beam port contour, so additional ports were added. If the beam ports were increased in size, their radiation into the parallel-plate waveguide would be too much focused and would not illuminate the antenna ports at the edges sufficiently.

5. Patch antenna arrays for future automotive radar

5.1 Rationale

Even from the early beginning in automotive radar the key driver of all these investigations has been the idea of collision avoidance. Currently these systems are moving from autonomous cruise control and crash warning to real pre-crash reaction. Control systems of the vehicle (throttle, brakes) act when the radar system detects an unavoidable collision. The future concept of these systems is to create a virtual safety belt around the vehicle with multiple sensors, enhancing the current scenario towards the goal of autonomous driving. To achieve this goal, higher resolution radars and better accuracy is required.

A frequency of about 140 GHz is proposed here because radar antennas with better resolution and narrower beams than the actual 77 GHz sensors can be developed. Also, the very high frequency allows an easy integration, because the antenna size is very small. In addition, this band has been investigated as a serious candidate for future automotive radar systems (Schneider, 2007).

5.2 A case study for 140 GHz

The proposed antenna consists of a number of parallel patch columns, which are driven by a corporate feed network printed on one side of a grounded substrate as shown in Fig. 5.1 a) (Herrero et. al., 2008). Patches are connected by high impedance lines, while the main feeder is a 50 Ohm microstrip line. This feeding concept enables beam steering in an easy way and the area consumption is kept low. In this arrangement, the main lobe points to the broadside direction, which is usually required in an automotive radar antenna sensor. Moreover, as the main feeder is a corporate line, this direction is independent of the frequency. Sidelobe level in elevation should be low, as it determines the amount of clutter that the final system may have. The performance in this aspect was improved by using a tapered power distribution across the patch column, so more power is concentrated in the broadside direction. This is done by changing the width of the patch along the column.

For the single element, a patch antenna was chosen, because of its flexibility and easiness of design and fabrication. For designing the whole array, first a single cell was optimized for operation at a frequency of 140 GHz (see 3.1 above). The array calculator of HFSS was then used for finding the right number of cells for a 12 dBi gain array. The last patch was designed with an inset to allow an easy control of the whole antenna matching.

Results of matching show very good agreement between the simulated design and the measured one. The matching bandwidth (Fig. 5-1 b) is about 3 GHz, which is enough for a radar of this kind. Best matching is moved in frequency with respect to the original design due to inaccuracies of the lab-scale fabrication. Nevertheless, the fabrication process can be optimized because the behavior of the photolithography and etching process is very predictable. Also, a proper oversizing of layouts may be employed or dry etching techniques may be used. The radiation pattern measured also shows good agreement with the simulated one, showing a gain of about 11 dBi. A main lobe of about 20° width at -3 dB is observed, which makes the array suitable for short-to-medium range radar or other sensing applications.

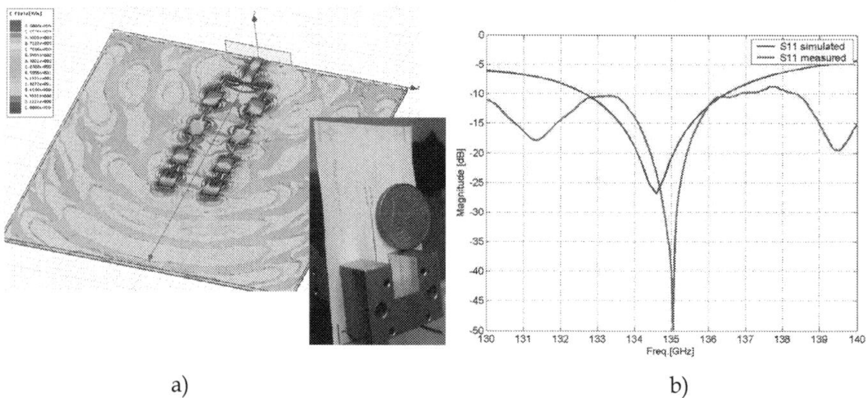

<table>
<tr><td align="center">a)</td><td align="center">b)</td></tr>
</table>

Fig. 5-1. a) Prototypical 140 GHz array with E-field simulation and photograph compared to a 1-Eurocent coin; b) Return loss of the array

A setup for planar antenna measurement in the 110-170 GHz range was designed and realized in-house. It was used for all antenna measurements. The system is designed to measure return loss and radiation pattern in E and H planes in an accurate and repeatable way.

In consists of a vectorial network analyzer with external WR6 test heads, a rectangular waveguide-to-microstrip transition and some accessories, like a horn antenna and a rotary station controlled by a software programmed in python in conjunction with the network analyzer software.

The transition is proposed here as a low-cost solution for waveguide and microstrip interconnection and as a feed structure for planar antennas. Fig. 5-2 a) shows the concept of the proposed transition. It consists of a microstrip line printed on a dielectric substrate terminated with a radial stub. The ground plane of this microwave circuit has an H-shaped slot aligned with the waveguide aperture. Proper placement is achieved using alignment holes on the waveguide flange and a housing, which also contains a rectangular cavity designed to avoid power loss by back radiation. An H-shaped slot is used here to increase the bandwidth of the transition. It also contributes to improve the front-to-back radiation ratio. Moreover, the transition is easier to scale to other frequency ranges, contrary to a classical rectangular slot approach.

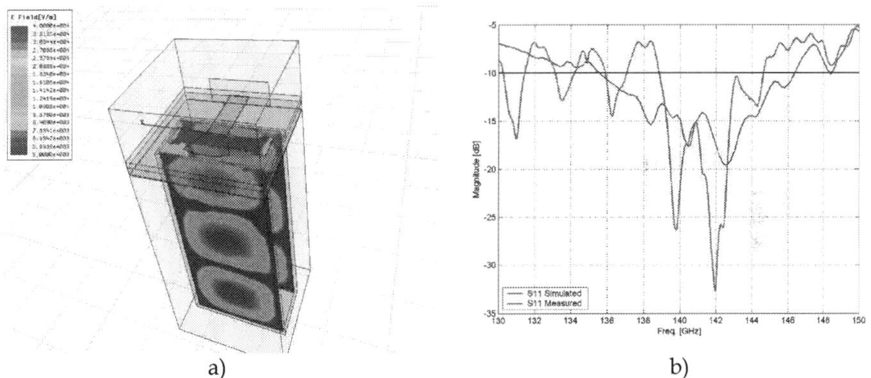

a)   b)

Fig. 5-2. a) Concept of the waveguide to microstrip transition with E-field simulation; b) Return loss of the waveguide port in the back-to-back double configuration

A value of 0.5 dB loss per interconnect can be obtained within the band of interest. Measured return loss yields a bandwidth around 10 GHz, typically enough for the requirements here. Matching also shows consistency between simulation and measurement. Performance of an exemplary transition centred at the frequency of 140 GHz is shown in Fig. 5-2 b). The fringes in the return loss are caused by standing waves in the characterized back to back configuration consisting of two transitions connected with a microstrip line.

6. Antennas for next generation applications in communications and sensing

In addition to the automotive radar, many other applications are expected to arise in the higher mm-wave range. In fact, an ISM (Industrial, Scientific and Medic) free to use band is allocated at the frequency of 122 GHz. This will partially solve the problem of the overcrowded spectrum in the low frequency ISM bands, like 2.45 GHz, where the current most popular microwave systems work. The 122 GHz band will have at least 1 GHz of bandwidth, enough for various future sensing applications.

Probably the most appealing feature of going to such a high frequency would be the integration of systems. Sensors working at a frequency of 122 GHz can be even as tiny as the size of a coin, and therefore their integration possibilities increase substantially. Examples of these could be sensing devices for detecting gas particles, or microwave imaging systems. As an example, a very small sensor built in the safety belt, airbag cap, or dashboard of a regular car can improve the functionality of the airbag system by sensing distance and position of the persons in the front seats.

6.1 Patch antenna arrays

As demonstrated in the chapter above, planar antenna systems can be used at such high frequencies as 140 GHz. In this part we describe more patch arrays working at the frequency of 122 GHz for future sensing applications (Herrero et. al., 2009). Two arrays are presented, designed keeping in mind the typical requirements of a mm-wave radar sensor: low sidelobes for one antenna and scan capability for the other antenna.

Fig. 6-1 a) shows the concept of the proposed antennas. The first one shown is a serially fed patch array. The second one is a corporate fed patch array. The rectangular patch used as a single element in both arrays was chosen due to its simple design and manufacturing, and because is very flexible when building phased and conformal arrays.

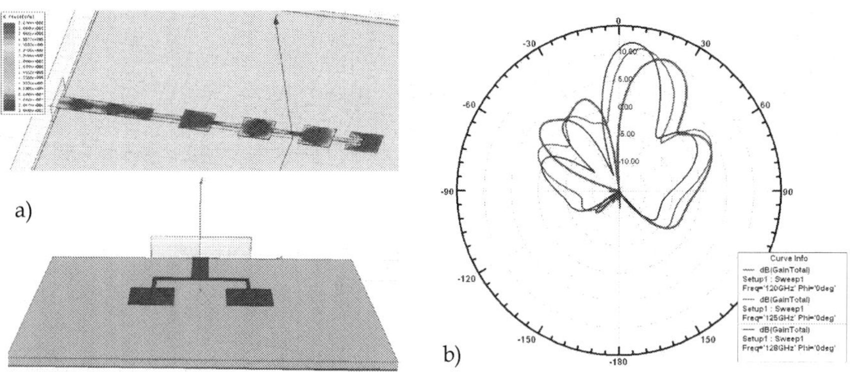

Fig. 6-1. a) Concepts of 122 GHz series and corporate feed arrays; b) Frequency-dependent E-plane antenna patterns of the series-fed array

The array column consists of a traveling wave transmission line, where fractions of power are coupled into the single elements. Last patch in the column is a matching element, avoiding the array to be a resonant array. This is done by introducing an inset in the patch, canceling the reactance at the microstrip main feed input. The whole configuration consumes very little area of substrate and it does not have impedance transformations, which keeps the loss of the feeding line low, which is important at such frequencies. This type of configuration allows beam steering in a simple way by changing the operating frequency of the array. The number of elements determine the scan range of the array

(Bhartia et. al., 1990) being 4 in this configuration. Simulated radiation patterns in Fig. 6-1 b) show the scan capability of the antenna.

The corporate line fed patch array consists in this case of one input and two outputs terminated by radiating patches. The power distribution chosen in this configuration is uniform. This is done by using a symmetrical structure, which also yields a uniform phase at the patch inputs. This makes the main beam independent of frequency, and points to the broadside direction. The width of the radiating elements was tuned so the matching to the high impedance line is optimum, improving the array's bandwidth.

For the measurements of the array column, we will focus on the elevation plane of the radiation pattern. The antenna has scan capability of the main lobe in this plane when changing the frequency of operation. Fig. 6-2 a) shows gain vs. angle and frequency in the E-plane. The scanning capability and range of the device can be clearly seen. Scan range is approx. between 25° and 5° with only four elements. Gain of the main lobe in this plane is about 6 dBi. Matching of the antenna is shown in Fig. 6-3 a) showing about 5 GHz bandwidth under -10 dB.

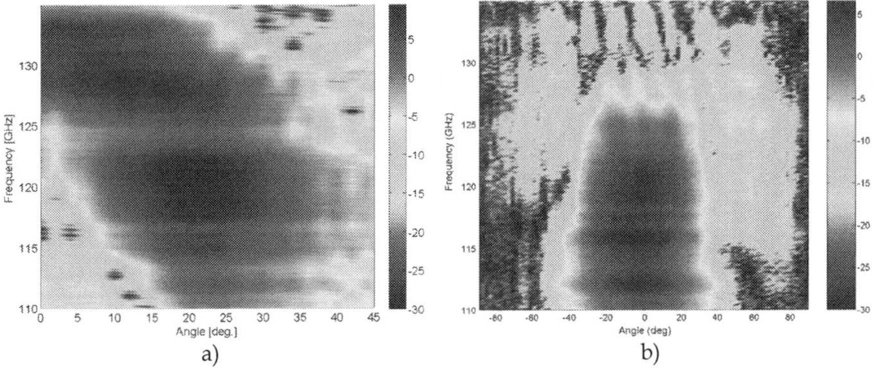

a)   b)

Fig. 6-2. a) Frequency-dependent E-plane radiation pattern of the series-fed array; b) Frequency-dependent H-plane radiation pattern of the corporate array

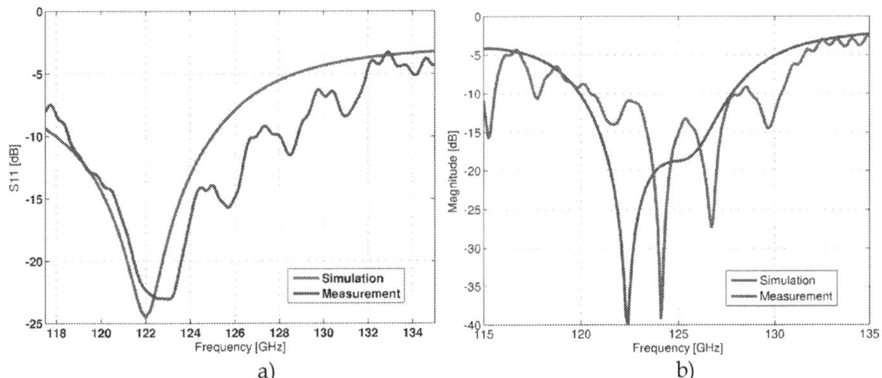

a)   b)

Fig. 6-3. a) Return loss of the series-fed array; b) Return loss of the corporate array

For the corporate feed array configuration, return loss and both radiation patterns on E and H planes were measured. Matching of the antenna is depicted in Fig. 6-3 b) showing an excellent agreement between measurement and simulation. Bandwidth under -10 dB is about 7 GHz. A complete radiation profile including all the frequencies is shown in Fig. 6-2 b) for the in azimuth (H-) plane. It clearly shows two sidelobes more than -10 dB below the main beam. The radiation pattern is very stable for all frequencies. Subtracting losses of the transition and the microstrip line, which are known, the gain is found to be about 5 dBi.

6.2 Monopoles and dipoles

Another typical sensor antenna is the well known monopole. Planar monopoles yield a wide bandwidth, an omnidirectional radiation pattern, a simple low-profile structure and are easy to fabricate. Because of these reasons modern communication systems like UWB or WLAN use such antennas. In our case a metal strip of 100 µm width, extended above the ground plane and fed by a 50 Ohm microstrip line serves as radiating element (see Fig. 6-4 a). Different designs for 122 GHz were developed, fabricated and measured. Fig. 6-4 b) and Fig. 6-4 c) show the matching and the radiation patterns (H (yz) plane) of an exemplary antenna. The measured bandwidth accounts for 11 GHz between 121 and 132 GHz limited by the waveguide transition of the measurement setup. The simulation of the return loss in Fig. 6-4 b) yields a very large bandwidth of approx. ~30 GHz without the waveguide transition, so in the final application a similarly large bandwidth would be usable.

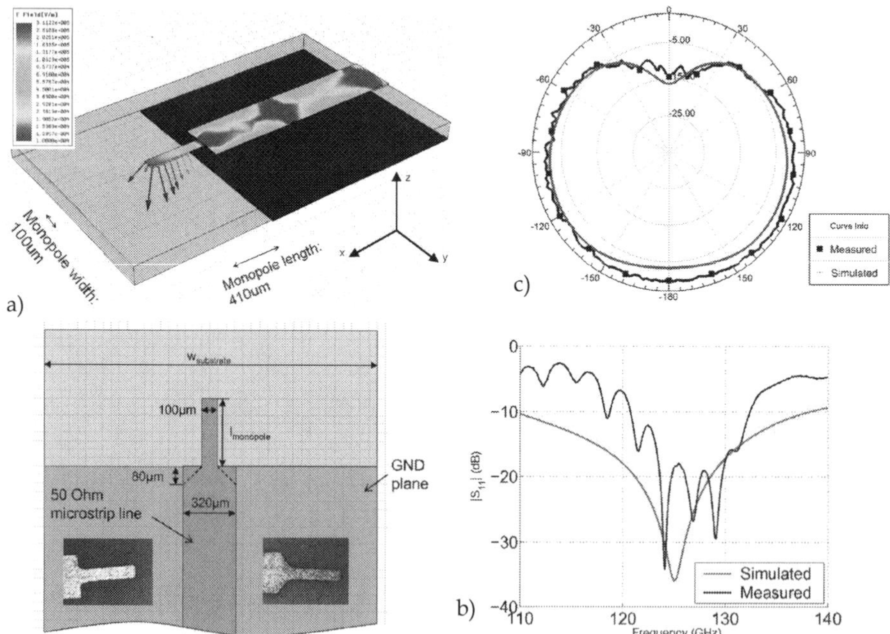

Fig. 6-4. a) Concept of the monopole with E-field simulation, dimensions and photographs; b) Return loss of the monopole; c) H (yz) plane radiation pattern

The microstrip dipole is a very good choice for integrating antennas and arrays in a future generation of high frequency applications. It is quite popular in WLAN systems because is intrinsically a quite broadband element, mandatory in bandwidth demanding applications. Also is a very compact and flexible structure when building arrays. In these, the level of cross coupling between elements is typically very low. Moreover, the structure is easily fabricated using conventional photolithography techniques.

Fig. 6-5 a) shows the concept of one of the proposed antennas. The double sided printed dipole consists of two arms formed by microstrip lines printed on different faces of a dielectric substrate and linear matching stubs (Tefiku & Grimes, 2000). The structure is fed by a 50 Ohm microstrip line. Its ground plane is truncated where the antenna element is connected to the line. The microstrip line is fed by a rectangular waveguide using a transition to microstrip line, allowing accurate and repeatable measurements.

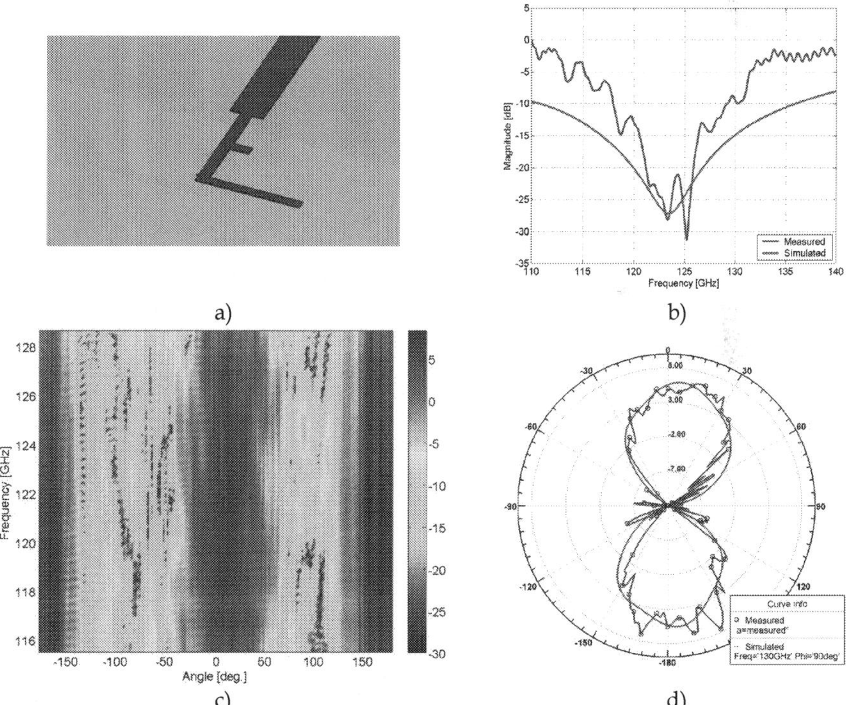

Fig. 6-5. a) Concept of the dipole; b) Return loss of the dipole; c) E plane radiation pattern versus frequency; d) E plane radiation pattern at 122 GHz

Key points in the design are the length of the dipole, which controls the main resonance. The length and width of the stub and the distance between the dipole and the stub are used to transform the "single ended" 50 Ohm microstrip line to the impedance of the dipole's balanced structure, thus acting as a balun. Dimensions of the whole structure are chosen keeping in mind also repeatable fabrication. The structure so designed is simple and does not need a complicated balun to match the microstrip line. This balun usually consists of a tapered transition, which would be very complicated to implement at this frequency. The target frequency is around the 122 GHz ISM band. The bandwidth envisaged is at least 10 GHz for enabling future multimedia and communication applications (cf. matching in Fig. 6-5 b). The antenna is built on a commercial Rogers Ro3003 (12 μm metallization, 130 μm dielectric, ε = 3) substrate. Common photolithography tools and techniques are used in fabrication, yielding a very low-cost approach.

The radiation pattern observed is dipole like with two beams in the azimuth plane and a fan-like beam in the endfire directions (Fig. 6-5 d). The radiation characteristics are very uniform over the operation frequency range (Fig. 6-5 c). This characteristic, combined with the very integrable shape of the antenna makes the structure very suitable for future sensing systems using very wide bandwidths.

7. Conclusions

We have presented a selection of different planar antenna designs with different properties suitable for a multitude of applications in the higher mm-wave range.

Beamforming with power dividers or Rotman lenses was discussed in detail. We investigated the focusing properties of the Rotman lens and concluded with a new design concept for the positioning and orientation of the beam ports.

Smaller arrays and monopole and dipole elements were demonstrated in the 122 and 140 GHz ranges, which are interesting candidates for future applications in radar and sensing.

It was shown that a low-cost approach relying on commercial circuit board processes is feasible. The fabricated antennas generally exhibit properties very close to the design values, so detrimental effects such as fabrication inaccuracies and surface waves can be well under control, even for frequencies as high as 140 GHz.

The results shown in this work are generally also applicable to indoor ultra-wideband communications in the 60 GHz band. Here, antenna gains on the order of 15 dBi for transmitter and receiver are required to achieve a sufficiently high signal-to-noise ratio supporting data rates of several Gb/s under today's technological constraints (mainly transmitter power and receiver noise figure). Due to the high gain, the transmission is very directed and may be blocked, e.g. by moving people. This calls for beamsteering solutions such as the Rotman lens, which can be integrated in an essentially flat device setup. Ultimately, such a flat antenna front ends could be easily mounted on a laptop cover lid or in a PCMCIA card.

8. References

Balanis, C. (2005). *Antenna Theory: Analysis and Design,* 3rd ed., Wiley and Sons, 047166782X, Hoboken, New Jersey, USA

P. Bhartia, P.; Rao, K.V.S.; Tomar R.S. (1990). *Millimeter wave microstrip and printed circuit antennas,* Artech House, 0890063338, Boston, Massachusetts, USA

Carver, K.R.; Mink, J.W. (1981). Microstrip Antenna Technology, *IEEE Trans. Antennas Propagat.,* vol. AP-29, pp. 2-24

Freese, J.; Blöcher, H.-L.; Wenger, J.; Jakoby, R. (2000). Microstrip Patch Arrays for a Millimeter-Wave Near Range Radar Sensor, *Proc. German Radar Symposium GRS 2000,* pp. 149-153, Berlin, Germany, Oct. 2000

Hansen, T.; Schneider, M.; Schoebel, J.; Brueggemann, O. (2006). *Monostatic Planar Multibeam Radar Sensor*, WO2006/029926 A1.

Herrero, P.; Schoebel, J. (2008). Planar Antenna Array at D-Band Fed By Rectangular Waveguide for Future Automotive Radar Systems, *European Microwave Conference 2008*, Amsterdam, The Netherlands, October 2008.

Herrero, P.; Schoebel, J. (2009). Microstrip Patch Array Antenna Technology for 122 GHz Sensing Applications, *German Microwave Conference 2009*, Munich, Germany, March 2009.

Iizuka, H.; Sakakibara, K.; Watanabe, T.; Sato, K.; Nishikawa, K. (2002). Millimeter-Wave Microstrip Array Antenna with High Efficiency for Automotive Radar Systems, *R&D Review of Toyota CRDL*, Vol. 37, No. 2, pp. 7-12.

Kawakubo, A.; Tokoro, S.; Yamada, Y.; Kuroda, K.; Kawasaki, T. (2004). *Electronically Scanning Millimeter-Wave Radar for Forward Objects Detection*, SAE Technical Paper 2004-01-1122.

Kühnle, G.; Mayer, H.; Olbrich, H.; Steffens, W.; Swoboda, H.-C. (2003). Low-Cost Long-Range Radar for Future Driver Assistance Systems, *Auto Technology*, Issue 4/2003, pp. 2-5, ISSN 1616-8216

Mende, R.; Rohling, H. (2002). New Automotive Applications for Smart Radar Systems, *Proc. German Radar Symposium GRS 2002*, pp. 35-40, Bonn, Germany, Sept. 2002.

Metz, C.; Lissel, E.; Jacob, A.F. (2001). Planar multiresolutional antenna for automotive radar, *Proc. 31st European Microwave Conf.*, London, UK, October 2001

Rotman, W.; Turner, R.F. (1963). Wide-Angle Microwave Lens for Line Source Applications, *IEEE Trans. Antennas Propagat.*, vol. AP-11, pp. 623-632

Russell, M.E.; Crain, A.; Curran, A.; Campbell, R.A.; Drubin, C.A.; Miccioli, W.F. (1997). Millimeter-Wave Radar Sensor for Automotive Intelligent Cruise Control (ICC), *IEEE Trans. Microwave Theory Tech.*, Vol. 45, pp. 2444-2453

Schneider, M.; Groß, V.; Focke, T.; Hansen, T.; Brüggemann, O.; Schöberl, T.; Schoebel, J. (2002). Automotive Short Range Radar (SRR) Sensors at 24 GHz with Smart Antennas, *Proc. German Radar Symposium 2002*, pp. 175-179, Bonn, Germany, September 2002

Schneider, R.; Wenger, J. (1999). System aspects for future automotive radar, *IEEE MTT-S International Microwave Symposium 1999*, pp.293-296, Anaheim, California, USA, June 1999

Schneider, M. (2007). Advances in Microsystems Technology and RF Microelectronics for Highly Integrated 77 GHz Automotive Radar Sensors, *Workshop on Low-Cost, Integrated Automotive and Industrial Radar Sensors, International Microwave Symposium*, Honolulu, Hawaii, USA, June 2007

Schoebel, J.; Buck, T.; Reimann, M.; Ulm, M. (2004a). W-Band RF-MEMS Subsystems for Smart Antennas in Automotive Radar Sensors, *Proc. 34th European Microwave Conf.*, pp. 1305-1308, Amsterdam, The Netherlands, October 2004

Schoebel, J. (2004b). *Method of direction determination for multiple target situations for multi-lobe monopulse radar*, EP 1 500 952 B1.

Schoebel, J.; Buck, T.; Reimann, M.; Ulm, M.; Schneider, M.; Jourdain, A.; Carchon, G. and Tilmans, H. A. C. (2005). Design Considerations and Technology Assessment of RF-MEMS Systems in W-Band Smart Antenna Frontends for Automotive Radar Applications, *IEEE Trans. Microwave Theory Tech.*, Vol. 53, no. 6, pp. 1968-1975

Schoebel, J.; Schneider, M. (2007). Smart antennas for automotive radar front-ends: from beam steering to high-resolution algorithms, *Workshop on Smart Antennas, European Microwave Conference 2007*, Munich, Germany, October 2007.

Smith, M.S.; Fong, A.K.S. (1983). Amplitude Performance of Ruze and Rotman Lenses, *The Radio and Electronic Engineer*, vol. 53, No. 9, pp. 329-336

Tefiku, F. and Grimes, C. (2000). Design of broad band and dual band antennas comprised of series fed printed strip dipole pairs, *IEEE Trans. Antennas and Propagat.*, vol. 48, pp. 895–899

Trastoy, A.; Ares, F. (1998). Phase-only synthesis of continuous linear aperture distribution patterns with asymmetric side lobes, *Electron. Lett.*, vol. 34, pp. 1916-1917

16

High-gain Millimeter-wave Planar Array Antennas with Traveling-wave Excitation

Kunio Sakakibara
Nagoya Institute of Technology
Japan

1. Introduction

High-gain and large-aperture antennas with fixed beams are required to achieve high S/N ratio for point-to-point high-speed data-communication systems in the millimeter-wave band. Furthermore, beam-scanning antennas are attractive to cover wide angle with high gain for applications of high-speed data-communication systems and high-resolution sensing systems. High-gain pencil-beam antennas are used for mechanical beam-scanning antennas. Although high antenna efficiency can be obtained by using dielectric lens antennas or reflector antennas (Kitamori et al., 2000, Menzel et al., 2002), it is difficult to realize very thin planar structure because they essentially need focal spatial length. By using printed antennas such as microstrip antennas, the RF module with integrated antennas can be quite low profile and low cost. Array antennas possess a high design flexibility of radiation pattern. However, microstrip array antennas are not suitable for high-gain applications because large feeding-loss of microstrip line is a significant problem when the size of the antenna aperture is large. They are applied to digital beam forming (DBF) systems since they consist of several sub-arrays, each of which has small aperture and requires relatively lower gain (Tokoro, 1996, Asano, 2000, Iizuka et al., 2003).
Slotted waveguide planar array antennas are free from feeding loss and can be applied to both high-gain antennas and relatively lower-gain antennas for sub-arrays in beam-scanning antennas. Waveguide antennas are more effective especially in high-gain applications than low-gain since a waveguide has the advantage of both low feeding loss and compact size in the millimeter-wave band even though the size of the aperture is large (Sakakibara et al., 1996). However, the production cost of waveguide antennas is generally very high because they usually consist of metal block with complicated three-dimensional structures. In order to reduce the production cost without losing a high efficiency capability, we propose a novel simple structure for slotted waveguide planar antennas, which is suitable to be manufactured by metal injection molding (Sakakibara et al., 2001).
We have developed two types of planar antenna; microstrip antenna and waveguide antenna. It is difficult to apply either of them to all the millimeter-wave applications with different specifications since advantages of the antennas are completely different. However, most applications can be covered by both microstrip antennas and waveguide antennas. Microstrip antennas are widely used for relatively lower-gain applications of short-range wireless-systems and sub-arrays in DBF systems, not for high-gain applications. Waveguide antennas are suitable for high-gain applications over 30 dBi.

With regard to the microstrip antennas, comb-line antennas are developed in the millimeter-wave band. In the comb-line antenna, since radiating array-element is directly attached to the feeding line, feeding loss could be quite small in comparison with other ordinary patch array antennas connected via microstrip branch from feeding lines. The branch of the comb-line antenna is no longer feeding circuit but radiating element itself. Radiation from the discontinuity at the connection of the radiating element joins to the main radiation from the element. Consequently, equivalent circuit model can not be used in the design any more. Electromagnetic simulator must be used to estimate the amplitude and phase of radiation from the elements accurately. Traveling-wave excitation is assumed in the design of comb-line antennas. Reflection waves from the elements in the feeding line degrade the performance of the antenna. When all the radiating elements are excited in phase for broadside beam, reflection waves are also in-phase and return loss grows at the feeding point. Furthermore, reflection waves from elements re-radiate from other elements. Radiation pattern also degrades since it is not taken into account in the traveling-wave design. Therefore, reflection-canceling slit structure is proposed to reduce reflection from the radiating element. Feasibility of the proposed structure is confirmed in the experiment (Hayashi et al, 2008).

On the other hand in the design of conventional shunt and series slotted waveguide array antennas for vertical and horizontal polarization, radiation slots are spaced by approximately a half guided wavelength for in-phase excitation. Interleave offset and orientation from waveguide center axis are necessary to direct the main beam toward the broadside direction (Volakis, 2007). Since the spacing is less than a wavelength in free space, grating lobes do not appear in any directions. For bidirectional communication systems in general, two orthogonal polarizations are used to avoid interference between the two signals. In the case of automotive radar systems, 45-degrees diagonal polarization is used so that the antenna does not receive the signals transmitted from the cars running toward the opposite direction (Fujimura 1995). However, in the design of the slotted waveguide array antenna with arbitrarily linear polarization such as 45-degrees diagonal polarization for the automotive radar systems, slot spacing is one guided wavelength which is larger than a wavelength in free space. All the slots are located at the waveguide center with an identical orientation in parallel unlike conventional shunt and series slotted waveguide array. Consequently, grating lobes appear in the radiation pattern. Antenna gain is degraded significantly and ghost image could be detected in the radar system toward the grating-lobe direction. In order to suppress the grating lobes, dielectric material is usually filled in the waveguide (Sakakibara et al. 1994, Park et al. 2005). However, it would cause higher cost and gain degradation due to dielectric loss in the millimeter-wave band.

We have proposed a narrow-wall slotted hollow waveguide planar antenna for arbitrarily linear polarization (Yamamoto et al., 2004). Here, we developed two different slotted waveguide array antennas with 45-degrees diagonal linear polarization. One is quite high gain (over 30 dBi) two-dimensional planar array antenna (Mizutani et al. 2007) and the other one is a relatively lower gain (around 20 dBi) antenna which can be used for a sub-array in beam-scanning antennas (Sakakibara et al. 2008). Microstrip comb-line antenna is also developed for lower-gain applications of the sub-array. Both waveguide antennas consist of the same waveguides with radiating slots designed by traveling-wave excitation. The number of the radiating waveguide and the structures of the feeding circuits are different in the two antennas. Traveling-wave excitation is common technique in the designs of the slotted waveguide array antennas and the microstrip comb-line antenna. Array fed by traveling-wave excitation suffer beam shift from frequency deviation, which causes narrow frequency bandwidth. However, in the case of narrow band application, traveling-wave excitation is quite effective to achieve high antenna efficiency.

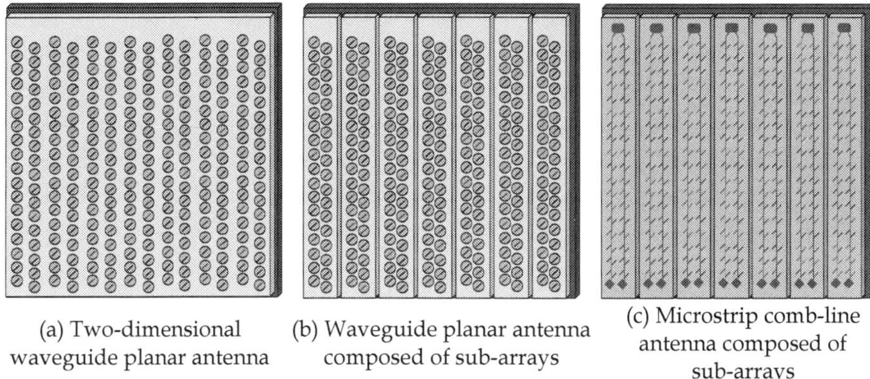

(a) Two-dimensional waveguide planar antenna

(b) Waveguide planar antenna composed of sub-arrays

(c) Microstrip comb-line antenna composed of sub-arrays

Fig. 1. Configurations of three planar antennas

2. Antenna configurations

Three different planar antennas are developed in the millimeter-wave band. Configurations of the antenna systems are shown in Fig. 1. Figure 1(a) shows a high-gain slotted waveguide antenna which has only one feeding port. Feeding network is included in the antenna. Second one in Fig. 1(b) is also a slotted waveguide antenna. However, the antenna system consists of some sub-arrays, each one of which has its own feeding port. Feeding network could be DBF systems or RF power divider with phase shifters for beam scanning. The waveguide antennas can be replaced by microstrip comb-line antenna for sub-arrays as shown in Fig. 1(c).

2.1 Slotted waveguide array antenna

We developed a design technology of slotted waveguide array antennas for arbitrarily linear polarization without growing grating lobes of two-dimensional array. Here, we designed an antenna with 45-degrees diagonal polarization to apply to automotive radar systems. Novel ideas to suppress grating lobes are supplied in the proposed structure of the two slot antennas in Fig. 1(a) and (b), which is still simple in order to reduce production cost. The configurations of the proposed antennas are shown in Fig. 2(a) and (b). All the radiating slots are cut at the center of the narrow wall of the radiating waveguides and are inclined by identically 45 degrees from the guide axis (x-axis) for the polarization requirement. The slotted waveguide planar antenna is composed of one feeding waveguide and two or 24 radiating waveguides. Spacing of slots in x-direction is one guided wavelength of the radiating waveguide. It is larger than a wavelength in free space. Grating lobes appear in zx-plane for one-dimensional array. Therefore, the radiating waveguides are fed in alternating 180 degrees out of phase since adjacent waveguides are spaced in a half guided wavelength $1/2\ \lambda_{gf}$ of the feeding waveguide. Slots are arranged with a half guided wavelength shift alternately in x-direction on each waveguide in order to compensate the phase difference between the adjacent waveguides. Consequently, the grating lobes do not appear in zx-plane because the slot spacing becomes about a half guided wavelength in x-direction. However, the grating lobes still appear in the plane including z-axis and diagonal kk'-direction due to the triangular lattice arrangement since the slot spacing in kk'-direction becomes the maximum as is shown in Fig. 2(a). In order to suppress the grating lobes, we propose the post-loaded waveguide-slot with open-ended cavity shown in Fig. 3.

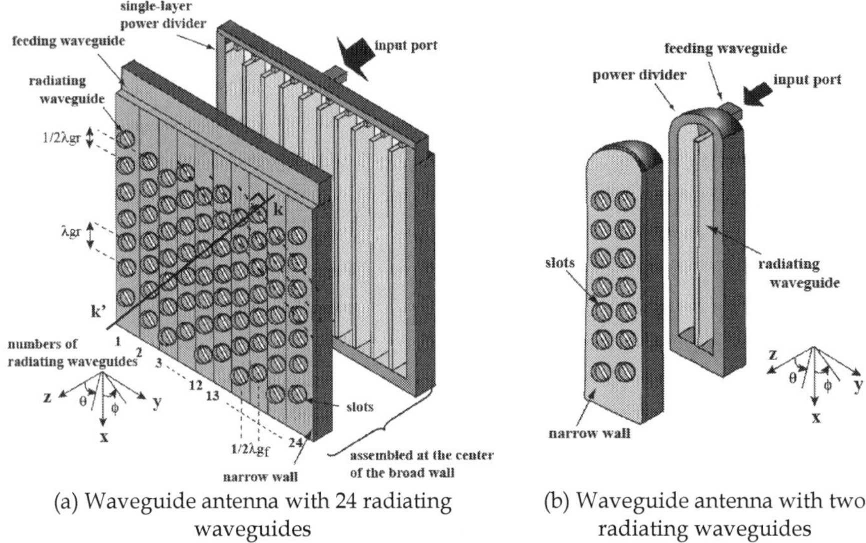

(a) Waveguide antenna with 24 radiating waveguides

(b) Waveguide antenna with two radiating waveguides

Fig. 2. Configuration of slotted waveguide planar antennas

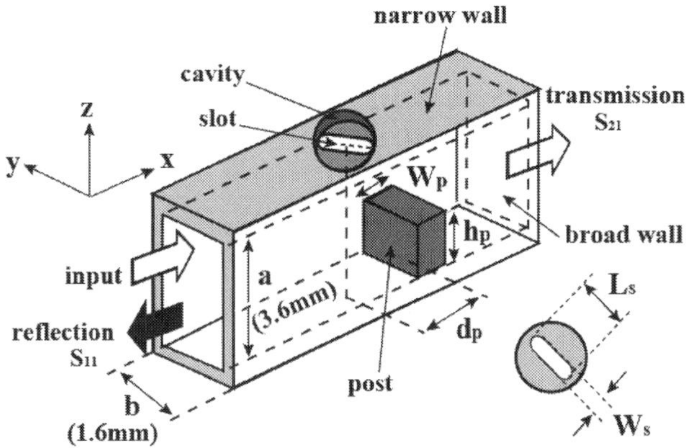

Fig. 3. Configuration of a post-loaded waveguide slot with open-ended cavity

A slot is cut on the narrow wall of the radiating waveguide. The spacing of radiating waveguides in y-direction can be short because the narrow-wall width is much smaller than the broad-wall width which is designed as a large value to reduce guided wavelength and slot spacing in x-direction. Broad-wall width can be designed independently from slot spacing in y-direction. Slot length is designed for required radiation. Resonant slot is not used in this design. Radiation is not enough because slot length is limited by narrow-wall width. Furthermore, since broad-wall width is designed to be large in order to reduce guided wavelength, power density is small around the slot. So, in order to increase radiation, we propose to locate a post on the opposite side of the slot at the bottom of the radiating waveguide as is shown in Fig. 3. In the case that post is located in the waveguide, as the size of the cross section of the waveguide is small above the post, power density increases around the slot. Thus, radiation from the slot increases depending on the height of the post.

In order to improve the return loss characteristic of the array, previously mentioned conventional slotted waveguide arrays are often designed to have some degrees beam-tilting. However, in this case, it becomes the cause to generate grating lobes or to enhance their levels because the visible region of array factor changes. In terms of the proposed antenna, the post in the waveguide is designed to cancel the reflections from the slot and from the post by optimization of their spacing. Therefore, it is not necessary to use the beam-tilting technique because the reflection from each element has already been small due to the effect of the post. Furthermore, we set an open ended cavity around each slot. Since the cavity shades the radiation toward the low elevation angle, the grating lobe level is reduced effectively. Thus, we can suppress the grating lobes in the diagonal direction.

The proposed structure has a following additional advantage for low loss. The antenna is assembled from two parts, upper and lower plates to compose a waveguide structure. Since radiating slots are cut on the narrow wall of the waveguides, cut plane of the waveguide is xy-plane at the center of the broad wall, where the current flowing toward z-direction is almost zero. Therefore, transmission loss of the waveguide could be small. High efficiency is expected even without close contact between the two parts of the waveguide. The electric current distribution would be perturbed by existence of slots. However, since the current in z-direction is still small at the cut plane, it is expected that the proposed antenna structure is effective to reduce transmission loss in the antenna feed.

2.2 Microstrip comb-line antenna

A microstrip comb-line antenna is composed of several rectangular radiating elements that are directly attached to a straight feeding line printed on a dielectric substrate (Teflon-compatible Fluorocarbon resin film, thickness $t = 0.127$ mm, relative dielectric constant $\varepsilon_r = 2.2$ and loss tangent tan $\delta = 0.001$) with a backed ground plane, as shown in Fig. 4. The width of the feeding microstrip line is 0.30 mm. The characteristic impedance of this line is 60 Ω. The radiating elements are inclined 45 degrees from the feeding microstrip line for the polarization requirement of automotive radar systems. The radiating elements with length L_{en} and width W_{en} are arranged on the both sides of the feeding line, which forms an interleaved arrangement in a one-dimensional array. The resonant length L_{en} is identical to a half guided wavelength. Element spacing d_{en} is approximately a half guided wavelength so that all the elements on the both sides of the microstrip line are excited in phase. A matching element is designed to radiate in phase all the residual power at the termination of the feeding line. Coupling power of radiating element is controlled by width W_{en} of the radiating element. Wide element radiates large power.

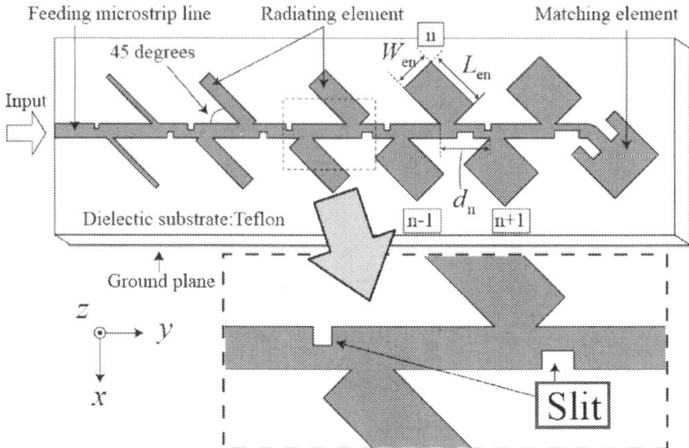

Fig. 4. Configuration of microstrip comb-line antenna with reflection-canceling slit structure

A radiation pattern with broadside beam is often used in many applications. However, when all the radiating elements are designed to excite in phase, all the reflections are also in phase at the feeding point, thus significantly degrading the overall reflection characteristic of the array. In the conventional design with beam tilting by a few degrees, reflections are canceled at the feed point due to the distributed reflection phases of the radiating elements. This means that the design flexibility of beam direction is limited by the reflection characteristics.

To solve this problem, we propose a reflection-canceling slit structure as shown in Fig. 4. A rectangular slit is cut on the feeding line near the radiating element. A reflection from each radiating element is canceled with the reflection from the slit. As the reflection from a pair of radiating element and slit is suppressed in each element, a zero-degree broadside array can be designed without increasing the return loss of the array. Because the sizes of all the radiating elements are different for the required aperture distribution, the slit dimensions and spacing of slit from the radiating element are optimized for each radiating element. Simple design procedure is required in the array design.

3. Design of linear array with traveling-wave excitation

Both waveguide antenna and microstrip antenna are designed in common procedure based on the traveling-wave excitation. Reflection wave is neglected in the design since reflection-canceling post and slit are used for the waveguide antenna and the comb-line antenna, respectively. Simple and straight-forward design procedure is expected in traveling-wave excitation. Here, design procedure based on traveling-wave excitation of the waveguide antenna is presented in this section.

A configuration of a post-loaded waveguide slot with open ended cavity is shown in Fig. 3. A slot element with post is designed at 76.5 GHz. The slot is cut on the waveguide narrow wall and is inclined by 45 degrees from the guide axis. The slot spacing becomes one guided wavelength which is larger than a wavelength in free space. The guided wavelength of the TE_{10} mode in the hollow waveguide is given by

$$\lambda_g = \frac{\lambda_0}{\sqrt{1-\left(\dfrac{\lambda_0}{2a}\right)^2}} \tag{1}$$

where λ_0 is a wavelength in free space and a is the broad-wall width of the waveguide. In order to become the guided wavelength short, the broad-wall width is determined to be large within the limit that only TE_{10} mode propagates. Broad-wall width is 3.92 mm in which cutoff frequency of TE_{10} mode is 76.5 GHz. The width is designed to be 3.6 mm taking production error and required frequency bandwidth into account. The guided wavelength of the radiating waveguide (3.6 mm X 1.6 mm) is 4.67 mm which is shorter than 5.06 mm of standard waveguide (3.1 mm X 1.55 mm). Slot spacing in x-direction can be short by increasing the broad-wall width a of the waveguide. The spacing in y–direction is about 2.6 mm because the wall thickness is about 1.0 mm. Consequently, the spacing in kk'-direction becomes about 3.48 mm (0.89 λ_0) and the grating lobe would be suppressed.

Radiation is controlled by both the slot length L_s and the post height h_p. The slot width W_s is 0.4 mm. Edge of slot forms semicircle of radius 0.2 mm for ease in manufacturing. Narrow-wall width b of waveguide is determined as 1.6 mm to cut the 45-degrees inclined slot of maximum slot length 2.0 mm. The post width W_p is 0.5 mm. Each slot element with post is designed to obtain the desired radiation and reflection lower than −30 dB at the design frequency. The reflection characteristic is controlled by changing the post height h_p and the post offset d_p from the center of the slot for several slot lengths. They are calculated by using the electromagnetic simulator of finite element method. Coupling C is defined as radiating from the waveguide to the air through the slot and is given by

$$C = \{\, 1-|S_{11}|^2-|S_{21}|^2 \,\} \times 100 \;[\%] \tag{2}$$

where S_{11} is reflection and S_{21} is transmission of the waveguide with slot in the analysis model shown in Fig. 3. Figure 5 shows simulated frequency dependency of reflection S_{11}, transmission S_{21} and coupling C in the case of maximum slot length 2.0 mm. The coupling was approximately 56.9%, where h_p and d_p are 1.6 and 0.6 mm, respectively. It is more than three times as large as coupling 18% from the slot without post. It is confirmed that large coupling is achieved due to the post even though the broad-wall width is large. Figure 6 shows the slot length L_s, the post height h_p and the post offset d_p from the center of the slot depending on coupling. Large slot length L_s is required for large coupling. In order to cancel the reflections from slot and post, amplitude of the reflections should be equal and phase difference should be 180 degrees out of phase. The post height h_p is also large for large coupling to satisfy the amplitude condition. The post offset d_p from the center of the slot gradually increases for the phase condition because perturbation of reflection phase grows for increasing slot length.

An open ended cavity is set on each slot. Since the cavity shades the radiation from the slots to the low elevation angle, the grating lobe level is reduced. Figure 7 shows calculated element radiation pattern and array factors in kk'-z plane. Element radiation pattern without cavity is ideally identical to isotropic radiation pattern because it is the radiation pattern of slot element in the perpendicular plane to the slot axis (E-plane). Difference of the levels in the element radiation patterns with and without cavity is approximately 14 dB at 90 degrees from broadside direction. Sidelobes of total radiation pattern are suppressed in large angle

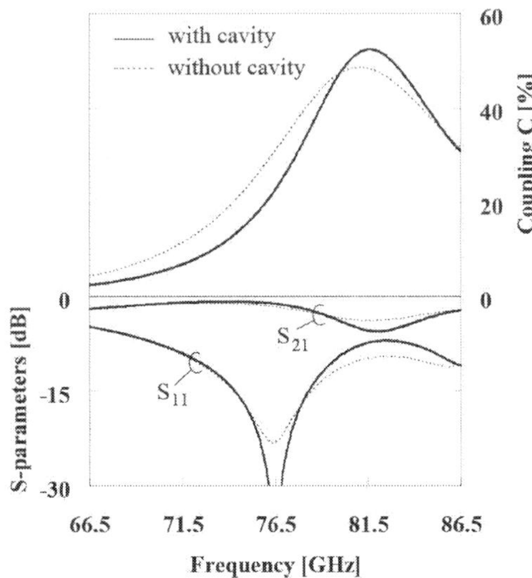

Fig. 5. Simulated frequency dependency of S_{11}, S_{21} and coupling C

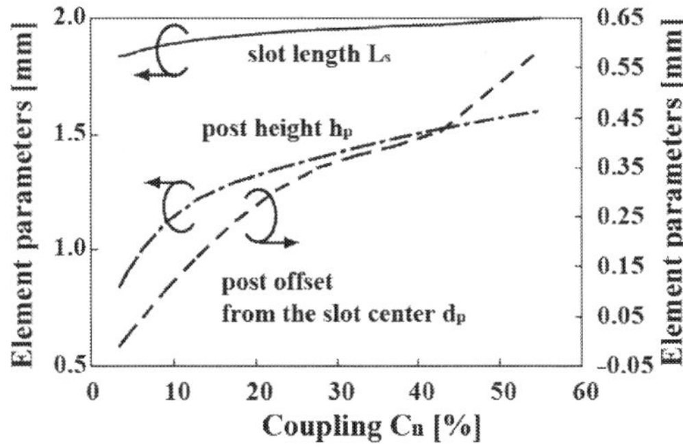

Fig. 6. Slot length, post height and offset from the center of the slot versus coupling

from the broadside. It is observed that the grating lobe level of antenna without cavities is
–22 dB, which is suppressed to –36 dB by using cavity. However, radiation pattern near
broadside is almost the same and independent on the cavity. No mutual coupling between
slots are taken into account in the design because the mutual coupling is very small due to
the element radiation pattern of cavity. Simple design procedure can be applied. The effect
of the circular cavity to the slot impedance and coupling is shown in Fig. 5. High-Q
resonance characteristic of cavity structure is observed, that is, maximum coupling power is
larger when cavity is installed, on the other hand, coupling power from slot with cavity is
smaller than without cavity in lower frequency than resonance. Since optimum parameters
for minimum S_{11} are slightly different between with and without cavity, level of S_{11} without
cavity does not fall down at the design frequency.

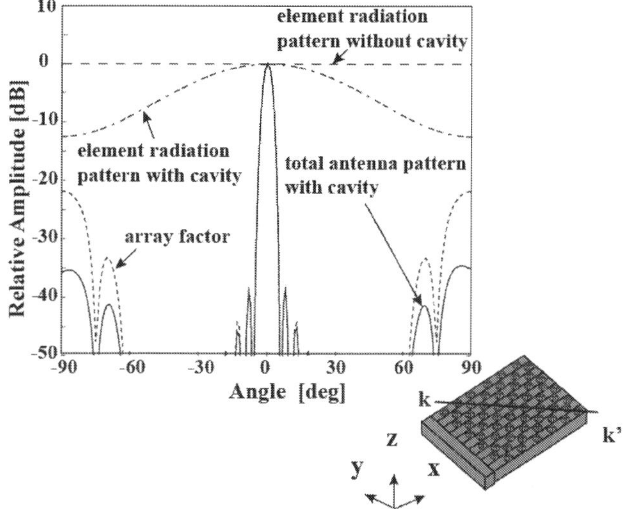

Fig. 7. Element radiation pattern of slot with cylindrical cavity and total radiation pattern which is product of array factor and element radiation pattern of slot with cylindrical cavity in diagonal plane (kk'-z plane)

A 13-element array is designed at 76.5 GHz. Thirteen radiating slots are arranged on one radiating waveguide, which corresponds to a linear array antenna. A terminated element composed of a post-loaded slot and a short circuit is used at the termination of each radiating waveguide. All the remaining power at the termination radiates from the element and also contributes antenna performance. Reflections from all the elements are suppressed by the function of post-loaded slot. So, design procedure for traveling-wave excitation is implemented (Sakakibara, 1994). Thirteen slot elements are arrayed and numbered from the feed point to the termination. Required coupling from slots are assigned for Taylor distribution on the aperture to be a sidelobe level lower than -20 dB. Incidence $P_w(n)$, transmission $P_w(n+1)$ and radiation $P_r(n)$ of nth slot shown in Fig. 8 are related by

$$P_w\,(n+1) = P_w(n) - P_r(n) \tag{3}$$

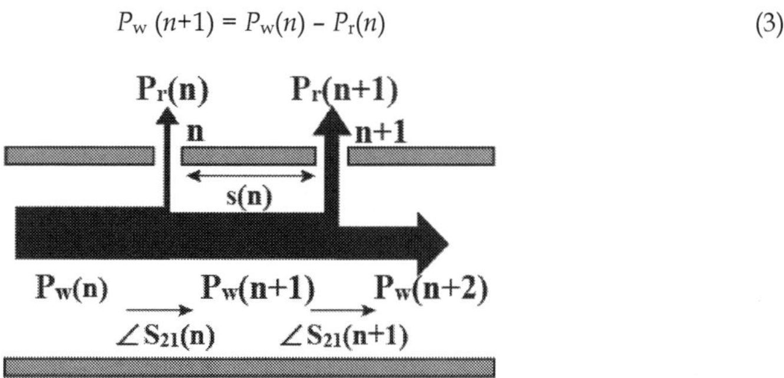

Fig. 8. Relation of radiation and transmission versus input. Slot spacing is related to transmission phase.

Coupling C_n is defined by

$$C_n = \frac{P_r(n)}{P_w(n)} \times 100(\%) \qquad (n = 1,2,3,\cdots,12) \tag{4}$$

The previously mentioned parameters L_s, h_p and d_p are optimized in each slot element shown in Fig. 9. A required variety of coupling is 3.5% ~ 54.9%. Element spacings $s(n)$ are determined to realize uniform phase distribution. This condition imposes

$$k_g s(n) = 2\pi + \angle S_{21}(n) \tag{5}$$

$$s(n) = \lambda_g + \frac{\angle S_{21}(n)}{2\pi} \lambda_g \tag{6}$$

where k_g ($= 2\pi/\lambda_g$) is the wave number of the waveguide and $\angle S_{21}(n)$ is phase advance perturbed by the slot element as is shown in Fig. 8. As $\angle S_{21}(n)$ is positive value, element spacing becomes slightly larger than a guided wavelength. This phase perturbation is simulated accurately by using electromagnetic simulator. So, the above design procedure dispenses with iteration.

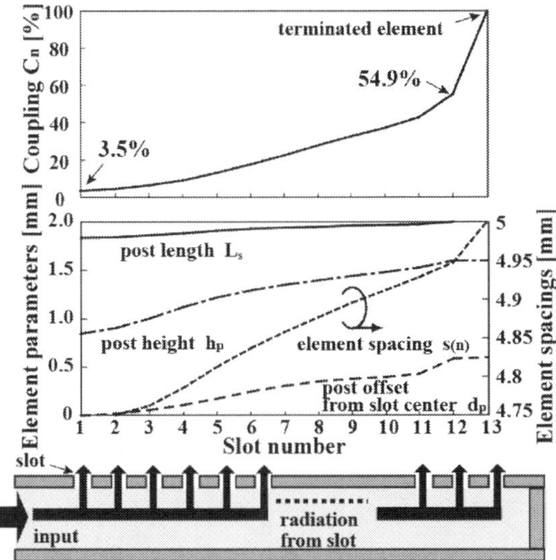

Fig. 9. Assigned coupling and dimensions of array elements for Taylor distribution on the aperture to be a sidelobe level lower than −20 dB.

4. Design of feeding circuits

The developed linear arrays are arranged to compose two-dimensional planar array. Required feeding circuits depend on the transmission lines and the number of the linear arrays. Waveguide 24-way and two-way power dividers are developed to feed the waveguide antennas. Microstrip-to-waveguide transition is also developed to feed the microstrip comb-line antenna from waveguide.

4.1 Waveguide feeding 24-way power divider

In the development of the two-dimensional planar waveguide antennas, a single-layer 24-way power divider composed of E-plane T-junctions feeding narrow-wall slotted waveguide planar array are designed as is shown in Fig. 2(a). It is composed of one feeding waveguide and 24 radiating waveguides slotted on the narrow walls. The antenna input port is located at the center of the feeding waveguide. All the radiating waveguides are fed from the feeding waveguide. The radiating waveguides are connected on the broad wall of the feeding waveguide, which forms a series of E-plane T-junctions shown in Fig. 10(a) (Mizutani et al. 2005). The broad-wall width of the feeding waveguide is determined so that the guided wavelength of feeding waveguide corresponds just twice the narrow-wall width of the radiating waveguide including the wall thickness between the radiating waveguides since adjacent waveguides are fed in an alternating 180 degrees out of phase. A coupling window is opened at each junction. Coupling to the radiating waveguide is controlled by the window width W_f in the H-plane. A post is set at the opposite side of the coupling window to obtain large coupling into the radiating waveguide and to cancel the reflections from the window and the post out of phase. Reflection level and phase of the post are adjusted to cancel both reflections by changing the post length L_f and the post offset d_f from the center of the window, respectively. Two edge radiating waveguides are fed from

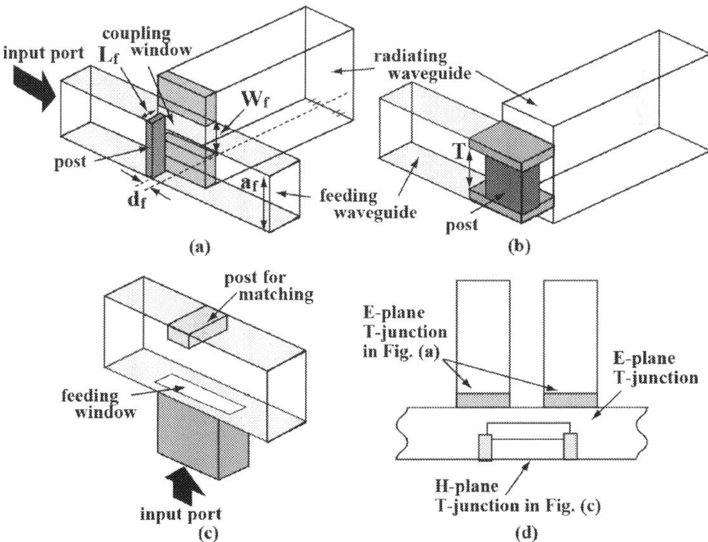

Fig. 10. Configuration of several parts of feeding circuit. (a) T-junction. (b) Terminated E-bend. (c) H-plane T-junction for input. (d) Top view and analysis model of the input H-plane T-junction on bottom plate around the antenna feed port with adjacent radiating waveguide.

terminated E-bends, shown in Fig. 10(b), in order to make all the remaining power contribute to the antenna performance. The size of the post is designed for matching and the width T of the waveguide is a parameter for phase adjustment of wave into the radiating waveguide. The feeding waveguide is fed through the H-plane T-junction at the input port shown in Fig. 10(c). Since the two adjacent radiating waveguides are very close to the input port, the H-plane T-junction is designed taking the effect of the two radiating waveguides into consideration as the analysis model is shown in Fig. 10(d). Phase perturbation of each E-plane T-junction is evaluated by electromagnetic simulator. The phase delay is compensated by adjustment of the spacing between radiating waveguides.

In order to feed radiating waveguides in alternating 180 degrees out of phase, we designed the E-plane T-junctions. The broad-wall width a_f of the feeding waveguide is designed to be 2.45 mm. So, the guided wavelength of the feeding waveguide is 6.6 mm calculated by equation (1). Required coupling to each radiating waveguide is assigned for Taylor distribution on the aperture as is shown in Fig. 11 to be a sidelobe level lower than −20 dB as well as the design of slotted linear array mentioned in the previous section. Geometrical parameters of each junction are also shown in this figure. Input port is at the center of the feeding waveguide and aperture distribution is designed to be symmetrical. So, only one half from port 13 to 24 is shown here. A required variety of coupling is 13.8% ~ 51.6%. The previously mentioned parameters L_f, W_f and d_f are optimized for each T-junction by using the electromagnetic simulator of finite element method. The window width W_f gradually increase with port number because required coupling increases. Relatively large windows are used at the center of the port numbers 12 and 13 to compensate the effect of the mutual coupling from the closely-located input port. The post length L_f also increases with port number to cancel the reflection from the window because the reflection coefficient of the large window is large. Furthermore, the post offset d_f from the center of the window gradually increases for the phase condition to cancel reflections because perturbation of reflection phase grows for increasing window width W_f and post length L_f.

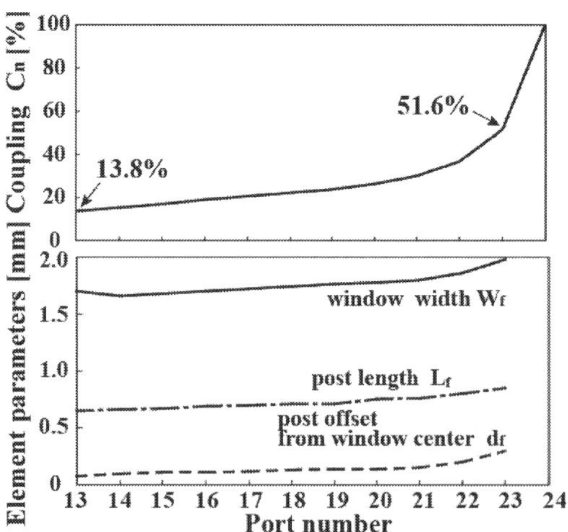

Fig. 11. Coupling of T-junctions for Taylor distribution and geometrical parameters of T-junction to be a sidelobe level lower than −20 dB.

A 24-way power divider is designed at 76.5 GHz. Field amplitude and phase distributions of the twenty four output ports are shown in Fig. 12. The simulated and measured results almost agree well with the design having the error smaller than 1 dB in amplitude and 5 degrees in phase. Simulated frequency dependency of reflection of the input T-junction with and without all the twenty four input ports is shown in Fig. 13. Resonant frequency is observed at the design frequency 76.5 GHz. Bandwidth of the reflection below −20 dB is approximately 8 GHz.

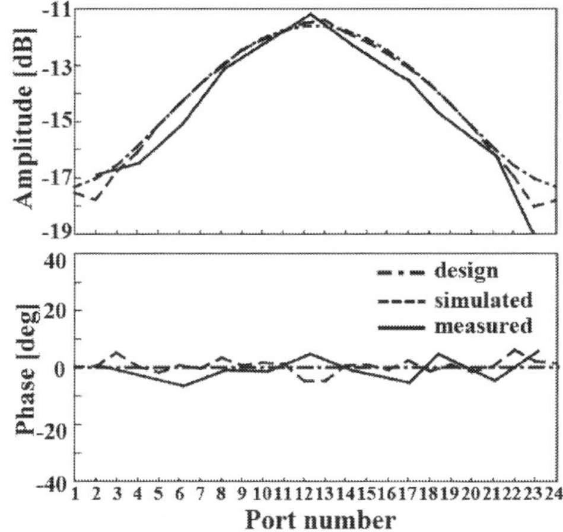

Fig. 12. Output amplitude and phase distributions of the single layer power divider

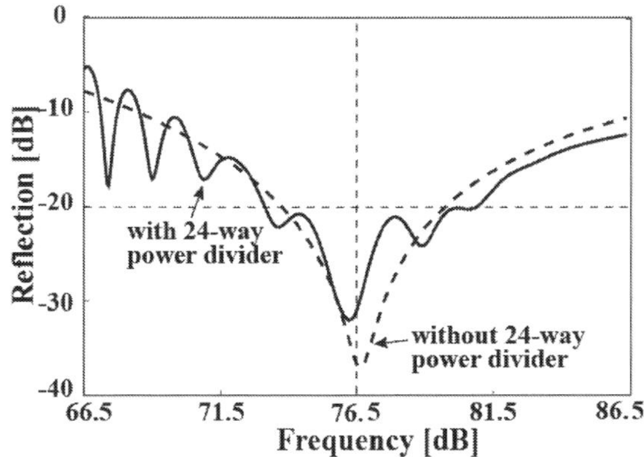

Fig. 13. Simulated frequency dependency of reflection of the input T-junction with and without 24-way power divider

4.2 Waveguide feeding two-way power divider

In order to excite all the slots in phase with a triangular lattice arrangement in the two-waveguide antenna, two radiating waveguides are fed in 180 degrees out of phase each other. We propose the compact power divider for the feeding circuit of the sub-array as shown in Fig. 14. The feeding waveguide is connected at the junction of the two radiating waveguides from the opposite side of the slots. There is a feeding window at the boundary between the radiating waveguide and the feeding waveguide for matching. A matching post is installed at the opposite side of the feeding window. The reflection characteristic is controlled by changing the size of the feeding window W_a, W_b and the height h_p of the matching post. The size of the feeding waveguide is W_{a0}, W_{b0} (3.10 X 1.55 mm) and the broad wall width of the radiating waveguide is h_{p0} (3.6mm). Figures 15(a), (b) and (c) shows the reflection characteristic depending on the height of the matching post h_p, the broad width W_a and the narrow width W_b of the feeding waveguide, respectively. Minimum reflection is obtained when the height h_p of the matching post is 0.40 h_{p0} and the broad width W_a and narrow width W_b of the feeding window are 1.0 W_{a0} and 0.65 W_{b0}, respectively.

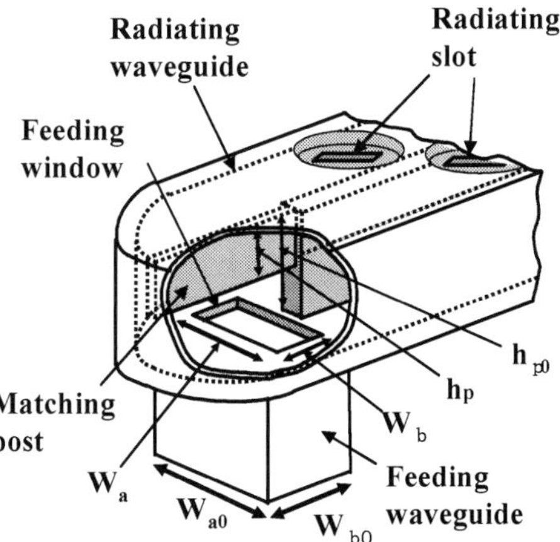

Fig. 14. Configuration of waveguide two-way power divider.

4.3 Design of microstrip-to-waveguide transition for feeding microstrip antenna

For feeding circuit of microstrip comb-line antenna from waveguide, microstrip-to-waveguide transition is developed. Ordinary microstrip-to-waveguide transitions require back-short waveguide on the substrate. In order to reduce number of parts and assembling error of the back-short waveguide, transition with planar structure is developed (Iizuka et al. 2002). Figure 16 shows a configuration of the planar microstrip-to-waveguide transition. A microstrip substrate with metal pattern is on the open-ended waveguide. Microstrip line is inserted into the ground pattern of waveguide short on the upper plane of the substrate. Electric current on the microstrip line is electromagnetically coupled to the current on the patch in the aperture at the lower plane of the substrate. Via holes surround the waveguide in the substrate to prevent leakage. Figure 17 shows S_{11} and S_{21}. Resonant frequency of S_{11} is observed at the design frequency 76.5 GHz. Insertion loss of the transition is approximately 0.3 dB at 76.5 GHz.

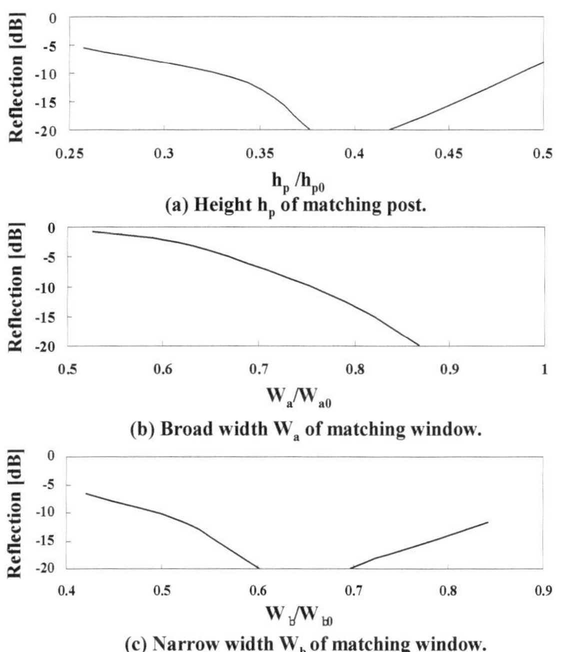

(a) Height h_p of matching post.

(b) Broad width W_a of matching window.

(c) Narrow width W_b of matching window.

Fig. 15. Reflection characteristics of the waveguide two-way power divider

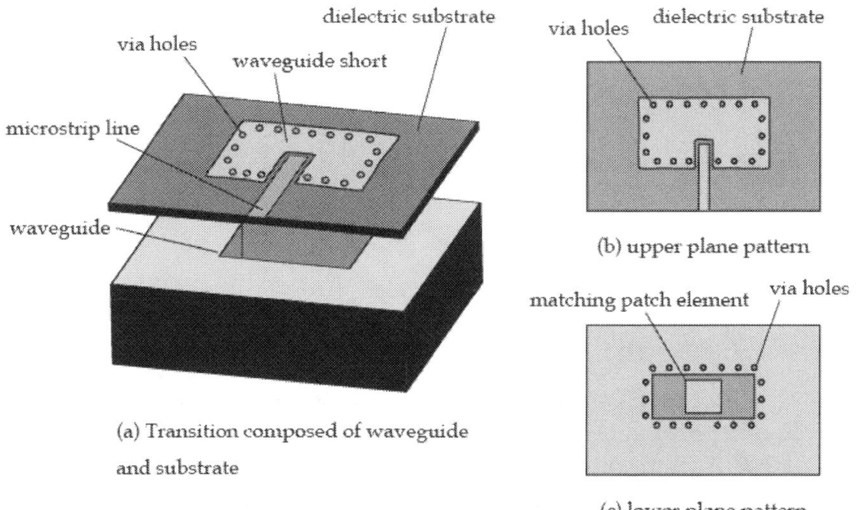

(a) Transition composed of waveguide and substrate

(b) upper plane pattern

(c) lower plane pattern

Fig. 16. Microstrip-to-waveguide transition with planar structure

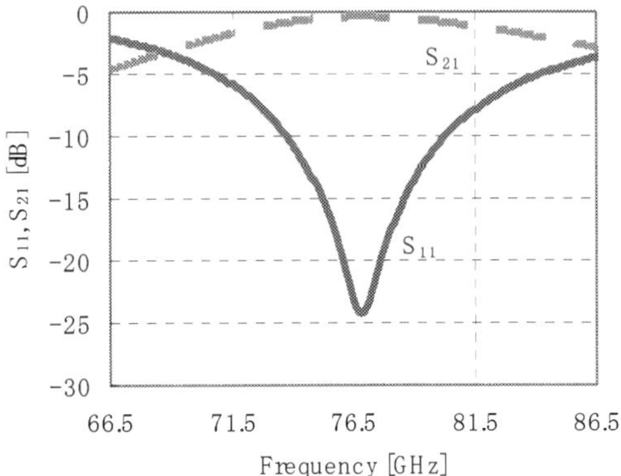

Fig. 17. Simulated S-parameters of the transition

5. Experiments

5.1 24-waveguide antenna

A 24-waveguide planar antenna is fabricated to evaluate the antenna performance. The photograph of the antenna is shown in Fig. 18. The fabricated antenna is assembled from two parts, upper and bottom aluminum plates with groove structures to compose waveguides. Cut plane is at the center of the broad wall of the waveguide and are fixed by screws. Twenty-four waveguides with 13 slots are arranged in parallel. Consequently, aperture size of antenna is 71.5 mm (in x-direction) × 64.7 mm (in y-direction).

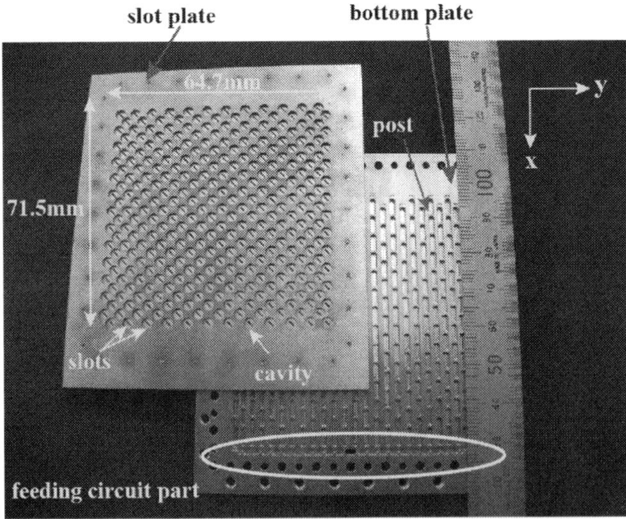

Fig. 18. Photograph of the fabricated antenna composed of the two plates

We discuss performance of the fabricated planar antenna in this section. Figure 19(a) shows the measured and designed radiation patterns in zx-plane at 76.5 GHz. The measured main beam direction results in −0.5 degrees from z-axis. This beam squint is due to error in the estimation of phase perturbation for transmission through the radiating elements of slots.

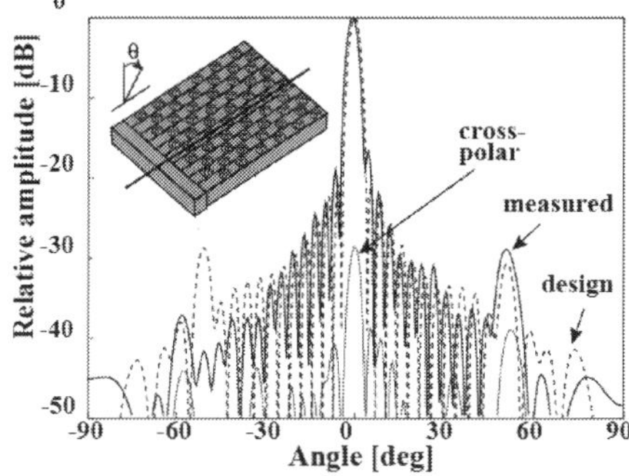

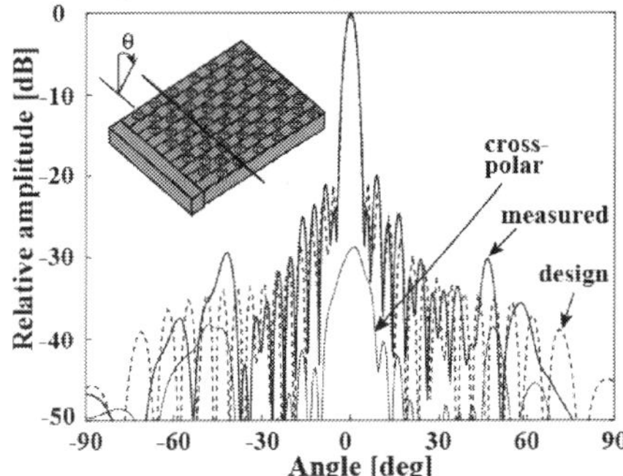

Fig. 19(a) Radiation patterns in zx-plane   Fig. 19(b) Radiation patterns in yz-plane

The sidelobe level is −16.8 dB which is 3.2 dB higher than design. Figure 19(b) shows radiation patterns in yz-plane. The main beam directs to the broadside as the same with design. Measured sidelobe level is −20.0 dB, which also almost agrees well with the design. Measured cross-polar patterns are also shown in Figs. 19(a) and (b). XPD (cross polarization discrimination) is 28.7 dB. Figure 20 shows the measured two-dimensional radiation patterns. In contrast to the general slotted waveguide arrays, maximum grating lobe level of the proposed antenna is suppressed to −28.6 dB. Figure 21 shows the measured gain and antenna efficiency at the frequency from 74 to 78 GHz. The center frequency shifts in 500 MHz from the design frequency 76.5 to 76.0 GHz. The gain is 33.2 dBi and the antenna efficiency is 56% at 76.0 GHz. Total loss 44% is estimated to consist of mismatch 3%, directivity of Taylor distribution 1% and cross-polarization 1%. Rest of them 39% is considered to be a feeding loss due to the imperfect waveguide in the millimeter-wave band. Bandwidth of gain over 30 dBi is approximately 1.9 GHz. High gain and high antenna

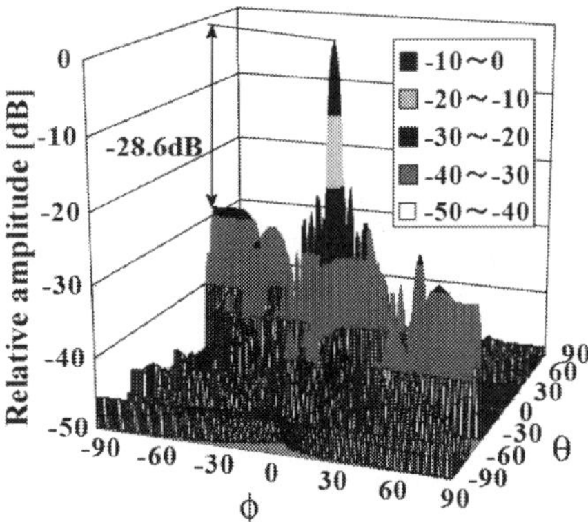

Fig. 20. Two-dimensional radiation pattern

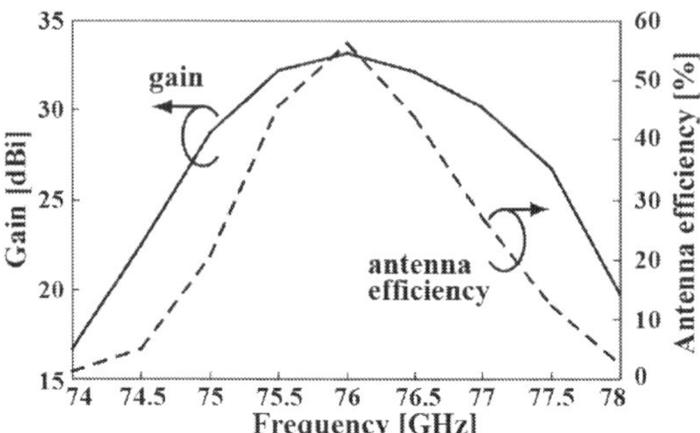

Fig. 21. Measured gain and antenna efficiency

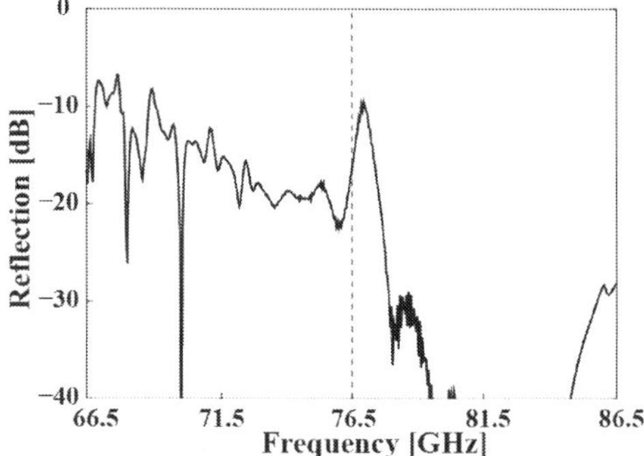

Fig. 22. Measured frequency dependency of reflection at the input port

efficiency are achieved. The measured reflection characteristics are indicated in Fig. 22. The measured reflection level is −22.0 and −16.5 dB at 76.0 and 76.5 GHz, respectively. On the other hand, large reflection is observed at 77.0 GHz, whose level is −10 dB. It is one of the causes of gain degradation. The cause of reflection increasing at 77.0 GHz would be that the proposed slot element is narrow frequency band width as is shown in Fig. 5. All the reflections from antenna elements due to frequency shift of fabrication error would be summed up in phase at 77.0 GHz.

5.2 Two-line waveguide antenna

The designed antenna was fabricated and feasibility was confirmed by experiments. Photograph of the developed antenna is shown in Fig. 23. Two metal plates of aluminium alloy were screwed together. Cut plane is at the center of the waveguide broad wall as well as the 24-waveguide antenna shown in the previous section. Posts were located in the waveguide to increase radiation from slots and to improve reflection characteristics. The cavity was set on each slot.

Figure 24(a) shows measured and simulated radiation patterns in the plane parallel to the waveguide axis at the design frequency 76.5 GHz. Beam direction was approximately 0 degree as was the same with the broadside beam design. Sidelobe level was around −20 dB as was almost the same level with the design of Taylor distribution for −20 dB sidelobe level.

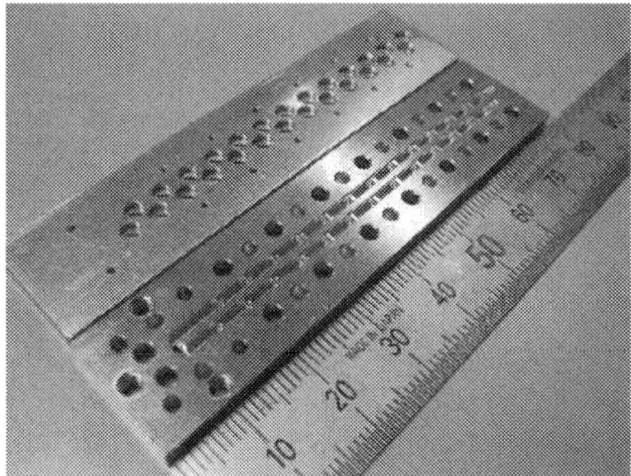

Fig. 23. Photograph of the two-line waveguide antenna

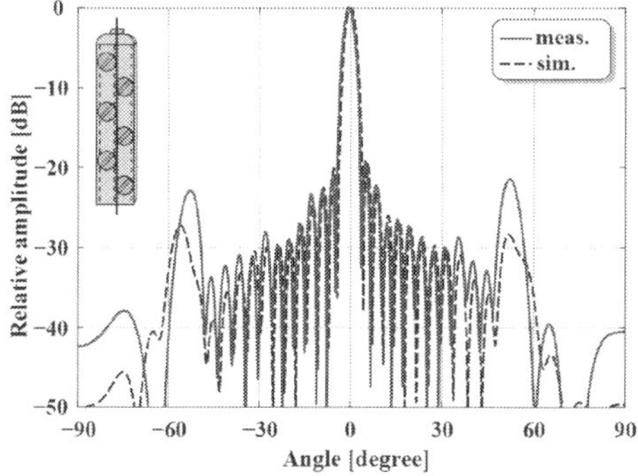

Fig. 24(a) Radiation pattern in the plane parallel to the waveguide

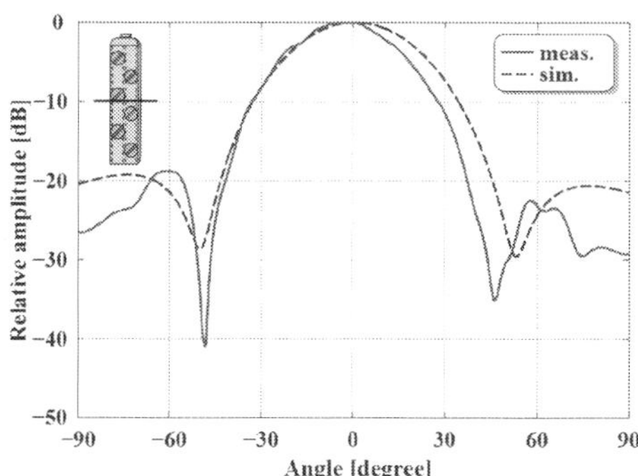

Fig. 24(b) Radiation patterns in the plane perpendicular to the waveguide

Some portion of the grating lobes were observed in 50 degrees which were about 7 dB higher than the simulation and still lower than −20 dB. Figure 24(b) shows measured and simulated radiation patterns at 76.5 GHz in the plane perpendicular to the waveguide. Almost symmetrical radiation pattern was obtained in the experiment. Sidelobe level was around −20 dB as was the same with the simulation. Figure 25 shows reflection characteristics. Since the resonant frequency corresponded to the design frequency 76.5 GHz, reflection level was lower than −20 dB at the frequency. Although the bandwidth was wider than 3 GHz for reflection lower than −10 dB, the center frequency of the bandwidth shifted by a few GHz lower than the design frequency. Figure 26 shows gain and antenna efficiency. Gain and antenna efficiency were 21.1 dBi and 51 %, respectively. They were degraded in the lower frequency band due to the return loss mentioned in Fig. 25. However, the efficiency was still relatively high compared with other millimeter-wave planar antennas.

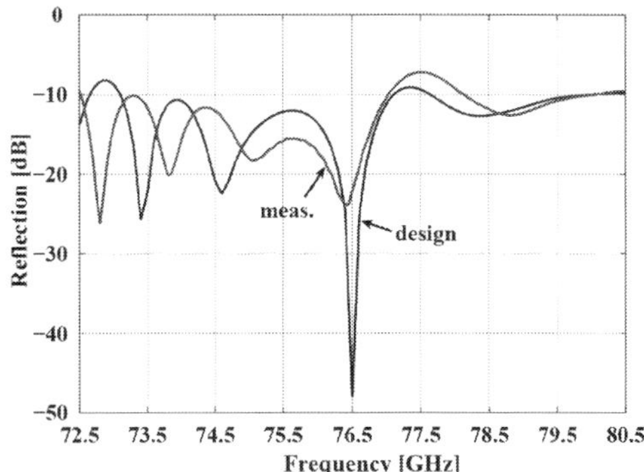

Fig. 25. Reflection characteristics

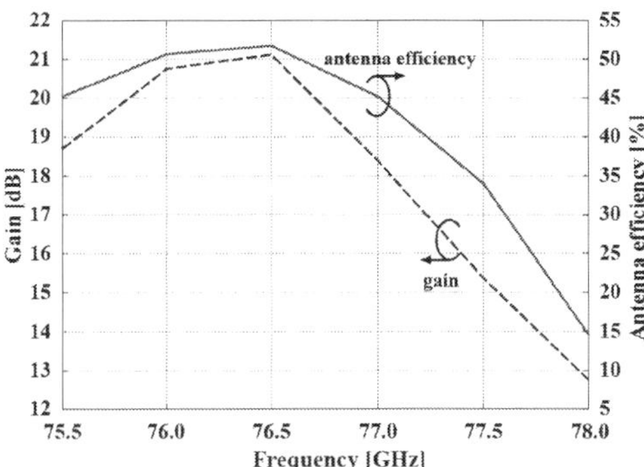

Fig. 26. Gain and antenna efficiency

5.3 Microstrip comb-line antenna

Microstrip comb-line antenna with two lines of 27 elements and with broadside beam is fabricated for experiments as is shown in Fig. 27. Reflection level of the fabricated antenna is −12.9dB at the design frequency 76.5 GHz as shown in Fig. 28. Measured beam direction in the plane parallel to the feeding line is −1.0 degree, and sidelobe level is −17.9 dB shown in Fig. 29(a). Symmetrical radiation pattern is obtained in the plane perpendicular to the feeding line as shown in Fig. 29(b). The measured radiation pattern almost agrees well with the array factor. High antenna efficiency 55 % with antenna gain 20.3 dBi is obtained at the design frequency 76.5 GHz. The efficiency is almost the same level with the two-waveguide antenna.

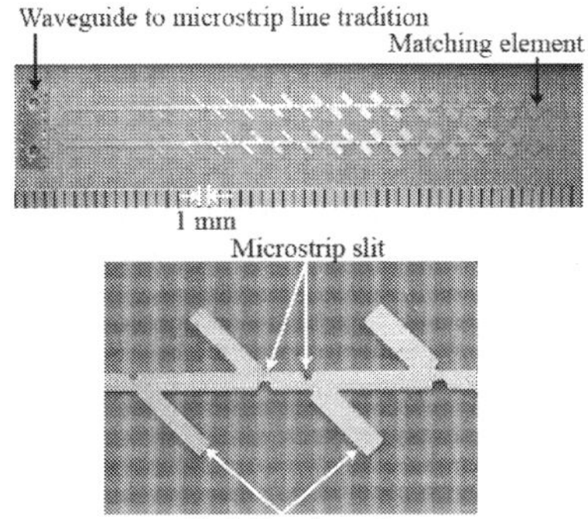

Fig. 27. Photographs of the developed microstrip comb-line antenna

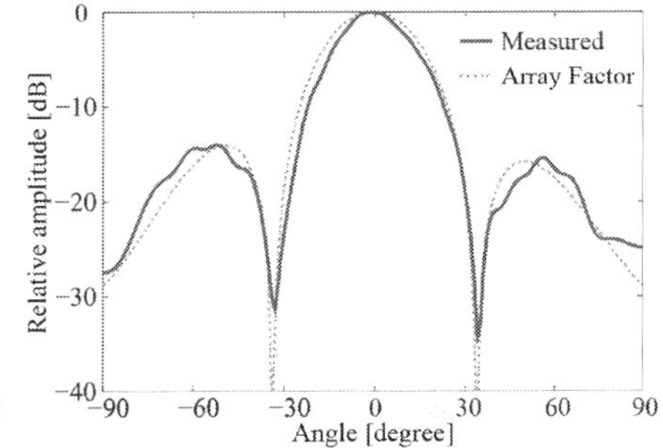

Fig. 28. Measured reflection characteristics

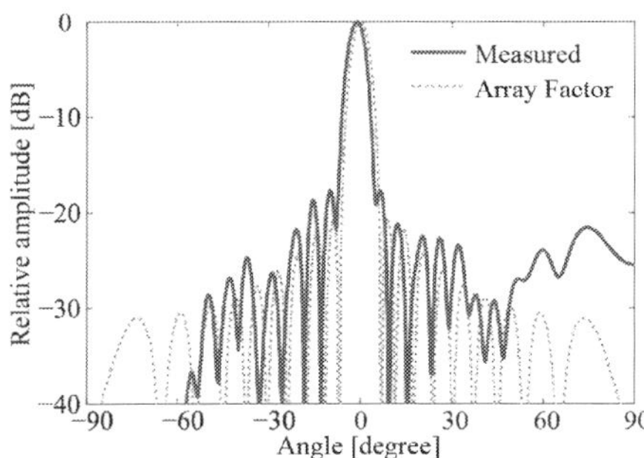

Fig. 29(a) Radiation pattern in the plane parallel to the feeding line

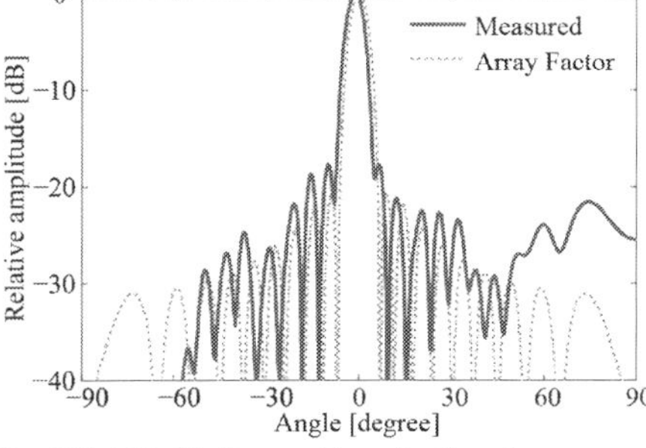

Fig. 29(b) Radiation pattern in the plane perpendicular to the feeding line

6. Conclusion

We have developed three types of millimeter-wave low-profile planar antennas; high-gain two-dimensional planar waveguide array antenna with 24 waveguides, two-line waveguide antenna and microstrip comb-line antenna which can be applied to the sub-arrays for beam-scanning antennas. Microstrip comb-line antenna is advantageous at the points of low cost and lower feeding loss compared with other microstrip array antennas. High efficiency is achieved, which is almost the same level with the waveguide when the aperture size and gain is relatively small. Waveguide antenna possesses much higher performance capability due to the low loss characteristic of waveguide feeding when the aperture size and gain is large. However, cost reduction is one of the most serious problems for mass production. Metal injection molding could be a solution for the waveguide antenna.

7. References

Asano, Y. (2000). Millimeter-wave Holographic Radar for Automotive Applications, *2000 Microwave workshops and exhibition digest, MWE 2000*, pp. 157-162, Yokohama, Japan, Dec. 2000

Fujimura, K. (1995). Current Status and Trend of Millimeter-Wave Automotive Radar, *1995 Microwave workshops and exhibition digest, MWE'95*, pp. 225-230, Yokohama, Japan, Dec. 1995

Hayashi, Y.; Sakakibara, K.; Kikuma, N. & Hirayama, H., (2008). Beam-Tilting Design of Microstrip Comb-Line Antenna Array in Perpendicular Plane of Feeding Line for Three-Beam Switching, *Proc. 2008 IEEE AP-S International Symposium and USNC/URSI National Radio Science Meeting*, 108.5, ISBN 978-1-4244-2042-1, San Diego, CA, July 2008

Iizuka, H.; Watanabe, T.; Sato, K.; Nishikawa, K. (2002). Millimeter-Wave Microstrip Line to Waveguide Transition Fabricated on a Single Layer Dielectric Substrate," IEICE Trans. Commun., vol. E85-B, NO.6, June 2002, pp.1169-1177, ISSN 0916-8516

Iizuka, H.; Watanabe, T.; Sato, K. & Nishikawa, K. (2003). Millimeter-Wave Microstrip Array Antenna for Automotive Radars," *IEICE Trans. Commun.*, vol. E86-B, NO. 9, Sept. 2003, pp. 2728-2738, ISSN 0916-8516

Kitamori, N.; Nakamura, F.; Hiratsuka, T.; Sakamoto, K. & Ishikawa, Y. (2000). High-ε Ceramic Lens Antenna with Novel Beam Scanning Mechanism, *Proc. 2000 Int. Symp. Antennas and propagat., ISAP 2000*, vol. 3, pp.983-986, Fukuoka, Japan, Aug. 2000

Menzel, W.; Al-Tikriti, M. & Leberer, R. (2002). A 76 GHz Multiple-Beam Planar Reflector Antenna, *European Microw. Conf.*, pp. 977-980, Milano, Italy, Sept. 2002

Mizutani, A.; Yamamoto, Y.; Sakakibara, K.; Kikuma N. & Hirayama H., (2005). Design of Single-Layer Power Divider Composed of E-plane T-juctions Feeding Waveguide Antenna, *Proc. 2005 Int. Symp. Antennas and propagat., ISAP 2005*, vol. 3, pp. 925-928, Seoul, Korea, Aug. 2005.

Mizutani, A.; Sakakibara, K.; Kikuma, N. & Hirayama, H., (2007). Grating Lobe Suppression of Narrow-Wall Slotted Hollow Waveguide Millimeter-Wave Planar Antenna for Arbitrarily Linear Polarization, *IEEE Trans. Antennas and Propag.*, Vol. 55, No. 2, Feb. 2007

Park, S.; Okajima, Y.; Hirokawa, J. & Ando, M., (2005). A Slotted Post-Wall Waveguide Array With Interdigital Structure for 45° Linear and Dual Polarization, *IEEE Trans. Antennas Propag.*, vol. 53, no. 9, Sept. 2005, pp.2865-2871, ISSN 0018-926X

Sakakibara, K.; Hirokawa, J.; Ando, M. & Goto, N., (1994). A Linearly-Polarized Slotted Waveguide Array Using Reflection-Cancelling Slot Pairs," *IEICE Trans. Commun.*, vol. E77-B, NO. 4, Apr. 1994, pp. 511-518, ISSN 0916-8516

Sakakibara, K.; Hirokawa, J.; Ando, M. & Goto, N. (1996). Single-layer slotted waveguide arrays for millimeter wave application," *IEICE Trans. Commun.*,vol. E79-B, NO. 12, Dec. 1996, pp. 1765-1772, ISSN 0916-8516

Sakakibara, K.; Watanabe, T.; Sato, K.; Nishikawa, K. & Seo, K., (2001). Millimeter-Wave Slotted Waveguide Array Antenna Manufactured by Metal Injection Molding for Automotive Radar Systems, *IEICE Trans. Commun.*, vol. E84-B, NO. 9, Sept. 2001, pp. 2369-2376, ISSN 0916-8516

Sakakibara, K.; Kawasaki, A.; Kikuma, N. & Hirayama, H., (2008). Design of Millimeter-wave Slotted-waveguide Planar Antenna for Sub-array of Beam-scanning Antenna, Proc. 2008 Int. Symp. Antennas and Propagation, ISAP 2008, pp. 730-733, Taipei, Taiwan, Oct. 2008

Tokoro, S. (1996). Automotive Application Systems Using a Millimeter-wave Radar," *TOYOTA Technical Review*, vol.46 No.1, May 1996, pp. 50-55

Volakis, J. L. (2007). *Antenna Engineering Handbook*, Chap. 9, McGraw-Hill, ISBN 0-07-147574-5, New York

Yamamoto, Y.; Sakakibara, K.; Kikuma, N. & Hirayama, H., (2004). Grating Lobe Suppression of Narrow Wall Slotted Waveguide Array Antenna Using Post, *Proc. 2004 Int. Symp. Antennas and propagat., ISAP'04*, vol. 4, pp. 1233-1236, Sendai, Japan, Aug. 2004

17

Wideband Antennas
for Modern Radar Systems

Yu-Jiun Ren and Chieh-Ping Lai
Advanced A&M Technologies, Inc.
U.S.A.

1. Introduction

Wideband antennas have received much attention in wireless communications and broadcasting systems. With a wide bandwidth, the signals can be transmitted by ultra-short impulses or by using multi-band groups. More information and applications can be carried through the radio frequency channels with a high data rate and accuracy. Also, there are numerous military and commercial sensing systems that require wide bandwidth to increase the range accuracy, imaging resolution, jamming/anti-jamming, and hardware flexibility. For various modern radars, such as air defense, air traffic control, astronomy, Doppler navigation, terrain avoidance, and weather mapping, the design of wideband antennas would have different considerations and specifications in order to meet the requirements of specified signals and waveforms.

The key issues of emerging radar antennas are low-profile, low cost, conformal, and compact design. Currently many researchers focus on omni-directional wideband antennas because their widely-covered beamwidth enables them to detect and communicate with targets anywhere. However, this may cause the radar to confuse where the target is located. Therefore, in certain aspects, directional antennas present more interest. The effective isotropic radiated power of a directional antenna is directed to reduce power consumption at the transmitter. This is also beneficial for the receiver's link budget. According to the radar range equation, the antennas can significantly determine the radar performance owing to its power handling, aperture dimension, beamwidth resolution, and potentials of beamforming and scanning. It may spend large amount of the budget for the antenna system cost depending on required functions.

In this chapter, we investigate state of the art wideband radar antennas and focus on applications for airborne radars, ground penetrating radars, noise radars, and UWB radars. The design concepts and considerations of these antennas will be addressed. Following the performance overview of these antennas, we introduce state of the art wideband antennas recently developed for new radar applications. Design details, examples, and profiles of the wideband antennas will be reviewd and discussed.

2. Antenna design and performance requirements

In this section, we introduce important parameters to describe the antenna performance. Instead of theoretically deriving them, we just simply pose these parameters and connect them with performance. Here, only basic antenna knowledge for understanding the characteristics of antenna and array antenna will be presented.

2.1 Antenna basics

Before knowing the antenna basic parameters, we should know in what range the antenna works in its normal function. The radiation filed close to the antenna includes both radiation energy and reactive energy. For most radar applications, the target is far from the antenna that satisfies the far-field region condition. In this region, only the radiating energy is present and the power variation is related to direction, independent of distance, which is at a radius R given by

$$R = 2L^2/\lambda \qquad (1)$$

where L is the diameter of the antenna and λ is the free space wavelength of the operation frequency. Within that radius is the near-field region (Fresnel region) in which the radiation field changes rapidly. In the far-field region (Fraunhofer region), the wavefronts are actually spherical, but a small section of the wavefronts can be approximately considered as a plane wave. This way, only the power radiated to a particular direction is concerned, instead of the shape of the antenna. The above condition is determined based on the error due to distance phase ($< \pi/8$) from the observing point to the antenna. If requiring more accurate phase, the far-field radius has to be increased (Balanis, 1982).

The radiation pattern is a plot of the power radiated from the antenna per unit solid angle, which is usually plotted by normalizing the radiation intensity to its maximum value. The characteristics of the radiation pattern are determined by the aperture shape of the antenna and the current distribution on the aperture. Important parameters to present the radiation patterns are briefly described. The mainlobe expresses the maximum radiation direction that is opposite to the backlobe, and the other minor lobes are called sidelobes, which are separated by nulls where no radiations. The sidelobe level (SLL) means the amplitude of the peak sidelobe; the sidelobe just beside the mainlobe is called the first sidelobe. The front-to-back ratio (FBR) is the amplitude ratio of the peak mainlobe and the backlobe. The beamwidth, or more specifically, half power beamwidth (HPBW), is the angle subtended by the half-power points of the mainlobe.

The directivity D of an antenna is a measure of radiation intensity as a function of direction, i.e., the ratio of the radiation intensity in direction (θ, ϕ) to the mean radiation intensity in all directions, or to the radiation intensity of an isotropic antenna radiating the same total power. An approximate value of the D can be calculated in terms of the beam solid angle Ω, or the HPBW in the two orthogonal plane, i.e.,

$$D = \frac{4\pi}{\Omega} = \frac{4\pi}{\theta_{HPBW}\phi_{HPBW}} \cong \frac{41253}{\theta_{HPBW}\phi_{HPBW}} (\text{deg.}) \qquad (2)$$

In general, the directivity without specifying a direction means the maximum value of the D. Most radars usually adopt the antenna with large aperture and high directivity. The relationship between the directivity and the power gain G are

$$G = D \cdot \eta_{eff} = \frac{4\pi A}{\lambda^2} \cdot \eta_{eff} = \frac{4\pi A_{eff}}{\lambda^2} \qquad (3)$$

where A is the antenna aperture area, η_{eff} is the radiation efficiency of the aperture, and A_{eff} = $\eta_{eff}A$ is the effective aperture area. The power gain G is the product of directivity and radiation efficiency, which is the ratio of power radiated to power fed into the antenna. This is due to the antenna mismatch and transmission loss, which can be observed from the transmission line theory shown in Figure 1. The source impedance $Z_s = R_s + jX_s$ and the antenna impedance $Z_a = R_r + R_l + jX_a$ are complex conjugates so that the transmitter is matched to the antenna and the maximum power can be delivered to the antenna and radiated. Then the radiation efficiency can be expresses by

$$\eta_{eff} = \frac{R_r}{R_r + R_l} \tag{4}$$

Since the power source is connected to the antenna, as the load, through the transmission line (feed-line), the antenna has to be matched to the feed-line to effectively radiate the power into the space. If mismatched, the part of power will be reflected and form standing wave, which can be measured using the reflection coefficient Γ or the voltage standing wave ratio (VSWR):

$$\Gamma = \frac{Z_a - Z_s}{Z_a + Z_s}; \ VSWR = \frac{1 + |\Gamma|}{1 - |\Gamma|} \tag{5}$$

The power radiated out is equivalent to the source power (P_s) minus the reflected power (P_{ref}) at the terminal of the antenna and feed-line, and the power dissipated at the antenna (P_l), i.e., $P_t = P_s - P_{ref} - P_l = P_s(1 - |\Gamma|^2) - P_l$. In fact, the efficiency consists of reflection efficiency, conductor efficiency, and dielectric efficiency, so $\eta_{eff} = \eta_{ref} \times \eta_{cond} \times \eta_{di}$. For a well designed antenna, the conductor and dielectric efficiencies are close to 1 and the reflection efficiency is dependent on the impedance matching.

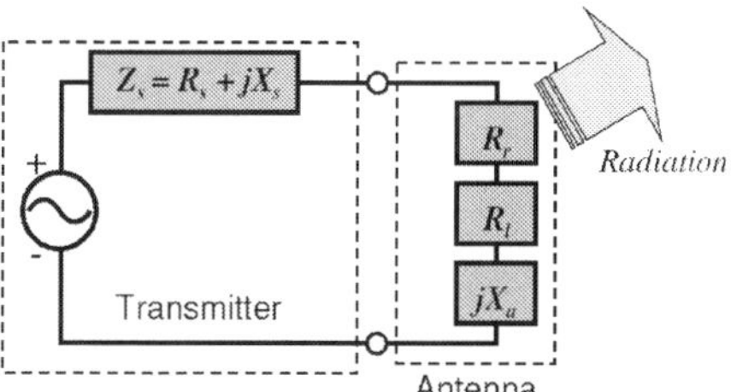

Fig. 1. Equivalent circuit of the transmitting antenna.

The polarization of an antenna can be determined by the electric field in the far-field region consisted of E_θ and E_ϕ. These two components will have a phase difference. When the phase difference is 0º or 180º, in the plane orthogonal to the propagation direction, these vectors oscillate in a straight line and this situation is called linearly polarized. When the phase difference is ±90º, it is called circularly polarized. For any other phase differences, it is generally called elliptically polarized. In practice, perfect circularly polarization does not exist, which means certain amount of ellipticity exists. This amount is expressed by the axial ratio (AR). Perfect circular polarization has AR = 1 (0 dB), and generally a circularly polarized antenna requires AR ≤ 3 dB.

2.2 Antenna array

Array antennas are often used in communication and radar systems to improve sensitivity and range extension without increasing power levels. Here, important design parameters for pattern synthesis and array scanning will be discussed. The configuration of the array antenna for different radar systems will be dressed and design considerations will be highlighted later.

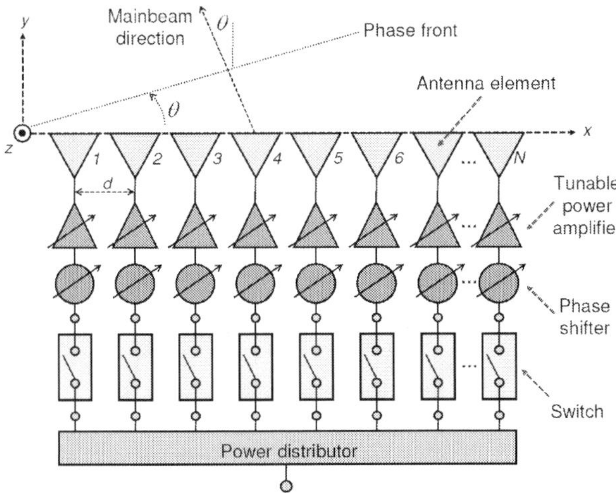

Fig. 2. Configuration of a linear array antenna.

A) Array pattern and grating lobe

The radiation pattern $F(\theta)$ of an array antenna is the product of the antenna element factor $F_e(\theta)$ and the array factor (AF) $F_a(\theta)$, i.e., $F(\theta) = F_e(\theta) \cdot F_a(\theta)$. The antenna element factor is the radiation pattern of a single antenna. To obtain the AF, we can start from the simplest array configuration, the linear array of N elements, as shown in Figure 2. The elements are separated in a distance d and it is assumed the phase reference is at the first element. The radiated field in far field from the array is given by

$$AF = \sum_{n=1}^{N} A_{n-1} exp(j(n-1)\psi) \tag{6}$$

with

$$\psi = kd \cos\theta + \beta \tag{7}$$

where $k = (2\pi/\lambda_0)$ is the wave number, θ is the arrival angle of the far-field, and β is the excitation phase for each element. The amplitude term A_{n-1} is used for the array amplitude taper that can be controlled by the tunable power amplifier; the phase term ψ is used for the phase taper that can be controlled by the phase shifter. If the phase reference is set at the center of the array that excited with a uniform amplitude equal to one, i.e., $A_n = 1$, the radiated field can be normalized as

$$AF_N = \frac{1}{N} \cdot \frac{1 - e^{jN\psi}}{1 - e^{j\psi}} = \frac{1}{N} \cdot \frac{\sin(N\psi/2)}{\sin(\psi/2)} \tag{8}$$

It can be observed that the pattern has peak vaule of 1 every 2π. The mainlobe at $\psi = 0$ is the mainbeam and the mainlobe at $\psi = \pm 2n\pi$ are the grating lobes. The grating lobe should be avoided becasue of receiving from unwanted directions. The condition to have the element spacing d satisfies

$$\frac{d}{\lambda} < \frac{1}{1 + |\sin\theta_{0,max}|} \tag{9}$$

where $\theta_{0,max}$ is the maximum directional angle that the mainbeam points to. The theory of the array scanning will be shortly described later. Here, let's consider the broadside array first which means the mainbeam is directed to boresight ($\theta_0 = 0^\circ$).

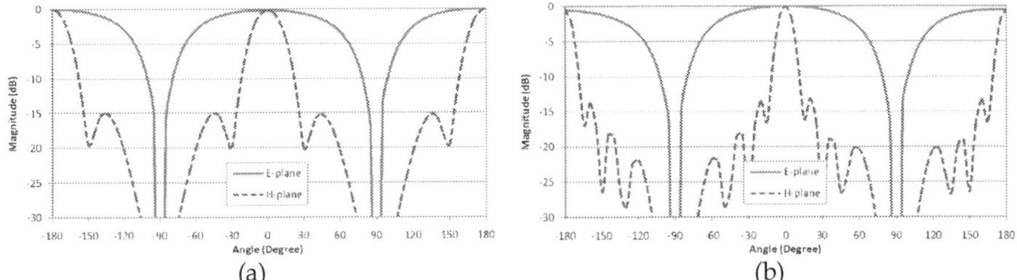

(a)   (b)

Fig. 3. Radiation patterns of a broadside uniform linear array with 8 elements: (a) $d = \lambda_0/4$ and (b) $d = \lambda_0/2$.

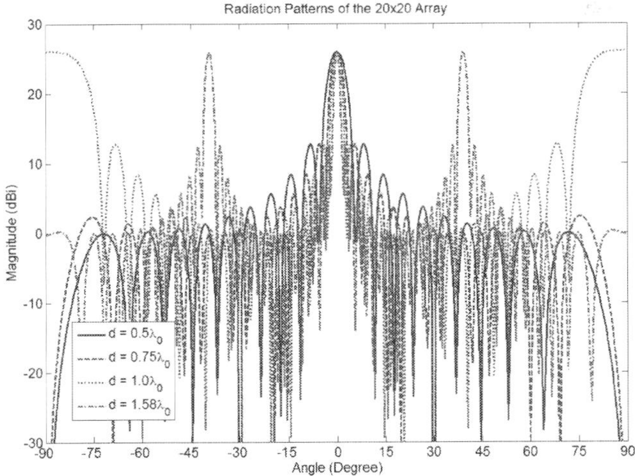

Fig. 4. Radiation patterns of a 20-element array with different element distances, while the mainbeam is at broadside direction.

A broadside array of N elements has a radiation pattern perpendicular to the array plane while the amplitude and the phase of each element are equal. In this case, Equation (7) can be reduced to

$$\psi = kd \cos \theta \qquad (10)$$

Then the maximum field occurs at $\theta = \pi/2$ and $\theta = 3\pi/2$. Figure 3 shows the simulated radiation patterns of an 8-element broadside linear array with different element spacings, where the antenna element is an elliptical patch antenna. Comparing these two results, the patterns on the E-plane are almost the same. However, the H-plane pattern with $d = \lambda_0/2$ has a narrower HPBW compared to that with $d = \lambda_0/4$. As the element distance increases, the mainlobe become sharper but the grating lobe may appear which is not desired.

Figure 4 shows the radiation patterns with different element distances ($0.5\lambda_0$, $0.75\lambda_0$, $1\lambda_0$, and $1.58\lambda_0$), when the antenna elements is a microstrip patch antenna. When the mainbeam is at the broadside direction, the array with $d = 0.5\lambda_0$ does not have any grating lobe and the magnitude of grating lobes of the array with $d = 0.75\lambda_0$ are very small. However, when the spacing is larger than one wavelength, there are two obvious grating lobes. In the case of $d = 1.58\lambda_0$, the angle range between the mainbeam and the grating lobe is only 40º, which significantly interference the desired signals.

B) Amplitude tapering and sidelobes

Array antennas can be employed in advanced radar systems to achieve the required power, high-resolution beam pattern, electronic beam steering, and interference suppression through nulling sidelobes. An array antenna can be excited by uniform or non-uniform amplitude taper, which can be implemented by using the tunable power amplifiers. As shown in Figure 2, all antenna elements can be turned on/off by the switch, which can decide how many elements to be used. Various combinations may happen, for example, eight or more elements could be used simultaneously as a sub-array. All of these elements are then connected to a power distributor, and then the summed signals can be processed and analyzed.

To implement a non-uniform amplitude taper for the array, several amplifiers are needed, as shown in Figure 2. In this case, assume all switches are turned on. To avoid the result of amplitude taper being obscured by grating lobe effects, the element spacing has to be equal to or less than half-wavelength. The easiest way to apply the amplitude taper is to linearly decrease the contribution of the array elements from the center to the edge of the array, which is called triangular amplitude taper. For an 8-element linear array, a triangular amplitude taper are given by: $A_1 = A_8 = 0.25$; $A_2 = A_7 = 0.5$; $A_3 = A_6 = 0.75$; $A_4 = A_5 = 1$. The uniform and triangular tapered radiation patterns of this broadside array with $d = \lambda_0/2$ are shown in Figure 5. It is observed that the triangular amplitude taper leads a lower SLL where the highest sidelobe is now −28 dB, compared to −13 dB of the uniform amplitude taper array. However, its mainbeam is broadened a little and increases from 12.8° to 16.4°.

It is possible to remove all sidelobes at the same time by using a proper amplitude taper such as the coefficients of a binomial series. The binominal amplitude taper for an 8-element is given by: $A_1 = A_8 = 0.03$; $A_2 = A_7 = 0.2$; $A_3 = A_6 = 0.6$; $A_4 = A_5 = 1$. Similar to the previous two cases, the resulting patterns are shown in Figure 5. While the sidelobes are suppressed, it can be seen that a more broadened mainbeam is the price of removing the sidelobes. In this case, the HPBW is about 22.2°.

In this example, it seems that using a linear method is the simplest approach for implementation in an array with only eight elements. However, as the element number increases, different non-uniform amplitude tapers should be chosen to narrow the mainlobe beamwidth or to suppress the sidelobe levels. Besides the binomial and trianular distributions, popular techniques for the amplitude tapering and sidelobe level control include circularly symmetric distribution, Bayliss Distribution, Hamming distribution, Hanning distribution, Chebyshev distribution, Taylor distribution, and Elliott distribution *et al* (Milligan, 2005).

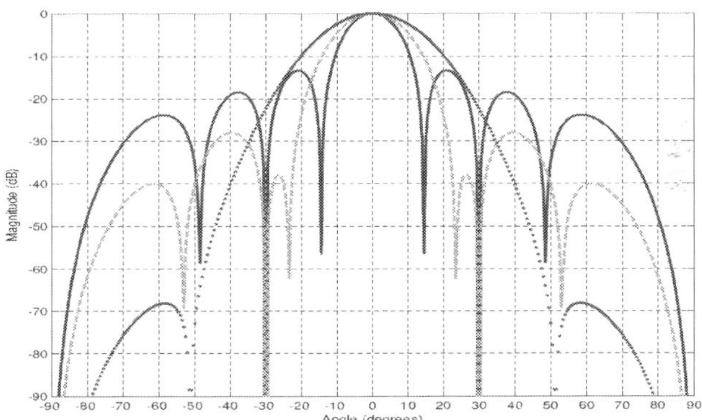

Fig. 5. Radiation patterns of an 8-element broadside linear array with different amplitude taperings. Solid line: uniform amplitude taper; dash line: triangular amplitude taper; dot line: binomial amplitude taper.

C) Phased array antenna

Based on the equation of the AF, if we want to direct the mainbeam of the array antenna to a desired direction θ_0, the AF will have the maximum value while ψ = 0 and hence β = -kdcos θ_0. Then the AF can be rewritten as:

$$AF = \sum_{n=1}^{N} A_{n-1}exp(j(n-1)kd(\sin\theta - \sin\theta_0))$$ (11)

The concept of the linear array can be easily extended to the planar array such as rectangular array and circular array (Rudge *et al.*, 1983). Same as the linear array, the mainbeam of a planar array can scan toward any interested point. It is straight forward to have the AF of a x-y rectangular array directed to (θ_0, ϕ_0):

$$AF_{rec} = \sum_{m=1}^{M} A_m exp(j(m-1)kd_x(\sin\theta\cos\phi - \sin\theta_0\cos\phi_0)) \cdot$$

$$\sum_{n=1}^{N} A_n exp(j(n-1)kd_y(\sin\theta\sin\phi - \sin\theta_0\sin\phi_0))$$ (12)

Where A_m is the amplitude taper of the element in x-axis, A_n is the amplitude taper of the element in y-axis, d_x is the element spacing in x-axis, and d_y is the element spacing in y-axis. Figure 6 shows the radiation patterns of a 20×20 array with d = 0.5λ_0. It can be seen there is no grating lobe even if the mainbeam is directed at 60º. As the scanning angle increases, the sidelobes close to ±90º becomes larger. However, they are still about 17 dB lower than the mainlobe. Therefore, from the view point of beamsteering, the element spacing, theoretically, has to be less than half wavelength if the mainbeam need scan to ±90º. In practical, we can have only ±60º scanning range at most, due to the degrading of the radiation patterns and the limitation of hardware.

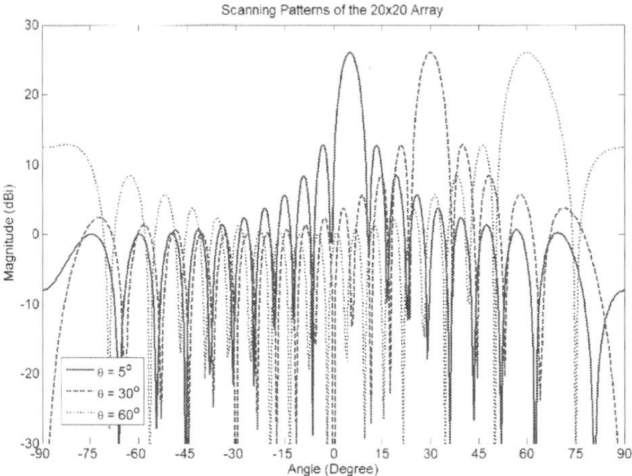

Fig. 6. Radiation patterns of a 20×20 array with different scanning angles, when $d = 0.5\lambda_0$.

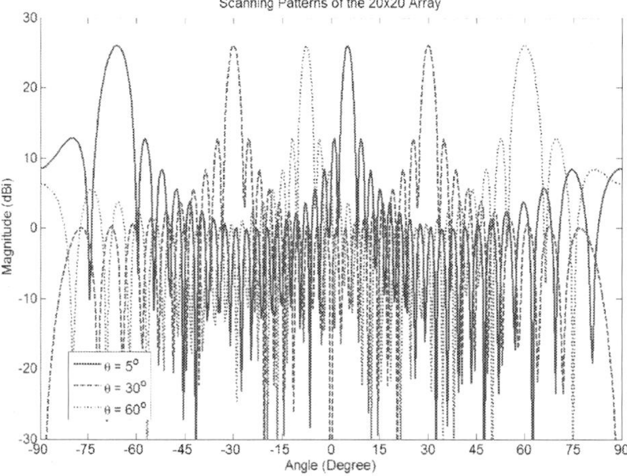

Fig. 7. Radiation patterns of a 20×20 array with different scanning angles, when $d = 1\lambda_0$.

Figure 7 shows the radiation patterns of $d = 1.05\lambda_0$, while the mainbeam is directed at different angles. Here radiation patterns of three scanning angles ($\theta = 5$º, 30º, and 60º) are shown. No matter how large angle the mainbeam scans, besides the mainlobe, there is at least one grating lobe directed at the unwanted direction. In Figure 4, it can be observed, when $d = 1.58\lambda_0$ there are two grating lobe when the mainbeam is at the broadside. The angle range between mainlobe and grating lobe decrease when the element spacing increases. From these results, it can be observed how the raidaiton pattern is affected by the element spacing.

There are many design considerations for the phased array antenna including mutual coupling and correlation, beam broadening, phase shifting, nulling undesired lobes, and feed network. Interested readers can review these topics in many published resources (Skolnik, 1990; Mailloux, 2005; Volakis *et al*, 2007). It should keep in mind that, from above discussion, to obtain a desired array pattern, we can vary: (1) array geometry, (2) array element spacing, (3) pattern of individual element, (4) element excitation amplitude, and (5) element excitation phase. After giving required array antenna characteristics, preliminary aperture and feed network designs will be conducted. Aperture distribution (amplitude taper) can be determined by aperture systhesis techniques, considering gain, beamwidth, sidelobe level, and etc. After optimization of the array antenna design, the fabrication and the testing can be implemented to verify the performance and readjust the design. The design process of an array antenna is shown in Figure 8. Based on these important concepts just reviewed, now we can look forward to see some antennas designed for different radar systems. In next section, we will look state-of-the-art radar antennas where their design concept and performance will be presented.

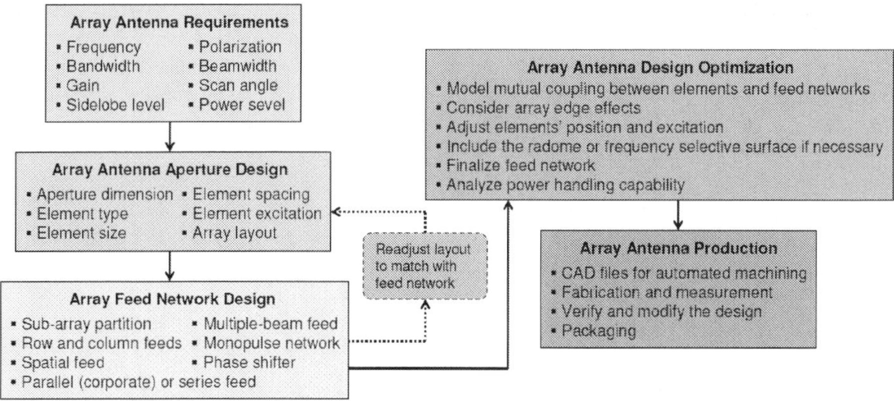

Fig. 8. Array antenna design process. One or more design iterations will be needed to achieve the design goal.

3. Radar antennas

3.1 Antennas for the airborne radar

Popular antennas for the ground-based and airborne radar systems include horn antennas, reflector antennas, and slotted waveguide antennas (Skolnik, 2001). These antennas can provide narrow beamwidth and high power handling. The airborne survillance radars and the air fighter radars usually need phased array antennas to conduct the beam steering for the prupose of detection and tracking, which requires narrow mainbeam and very low sidelobes. The slotted waveguide antennas are better candidates for the large array antenna design because of the waveguide structure, which can couple energy in a very precise manner (low loss) and can be used for the feed network of the array antenna to simplify the phased array antenna design. The slotted waveguide array antennas (multiple slots on a single waveguide) have potential of low cost, low weight, ease of fabrication, high gain, and easier amplitude tapering for extremely low sidelobe. Today, they have been widely used in the airborne phased array radar systems.

The radiation magnitude and phase of the slotted waveguide antenna can be determined by the size and orientation angle of the slot. The apperance of grating lobes and the elemetn spacing govern the locations of slots. To optimze the bandwidth, the slots can resonate for either resonant array (standing wave array) or non-resonant array (traveling wave array). Basically, the radiation power is proportional to the slot resistance or conductance. Normalized series-slot resistance (r) and shunt-slot conductance (g) are given by

$$r = 2P/I^2; g = 2P/V^2 \tag{13}$$

where P is the required power; V and I are the voltage and current across the slot, respectively, which can be determined by the transmission line or scattering matrix theories, and hence r and g. Then, the slot array can be designed using on known, or measured, slot characteristics or design curves based on the slot geometry. Watson and Stevenson pioneered the slot characteristics for different slot geometries in a rectangular waveguide (Waston, 1949; Stevenson, 1948). Most of the closed form solutions can correctively predict the performance of the single slot. However, since the mutual coupling effect is neglected, the slot array based on these analyses will have higher VSWR and the aperture illumincation may not be the desired distribution. Several techniques have been proposed to overcome this difficulty (Volakis *et al*, 2007).

A) Resonant array (standing wave array)

The slotted waveguide array antenna cab be fed at edge or center with shorted port at the waveguide end. When the element spacing d_s is $\lambda_g/2$, the electrical fields inside the waveguide form standing waves, and hence the array antenna is called a standing wave array or resonant array. Figure 9 shows two types of the resonant arrays. Since the standing wave array is a narrow band antenna, to have the desired bandwidth, the slot number on each waveguide has to be analytically designed. Then, several subarrays, which consisted of several waveguides with different slot numbers, are shunt connected to form a planar array

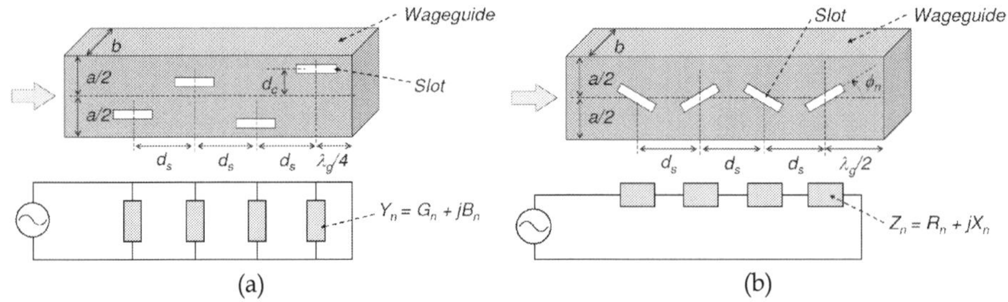

Fig. 9. The resonant slotted waveguide array antennas and their equivalent circuits using (a) shunt slots and (b) series slots.

that excited by another waveguide coupling the energy to the subarrays. Generally, the more slots on a waveguide, the narrower bandwidth. When the operation frequency offsets from the designed center frequency, the gian decreases and the SLL increases, due to the change of the electric field (standing wave) distributions in the waveguide that deteriorate the amplitude and phase excited. Large planar arrays of uniformly spaced longitudinal-shunt-slots are commonly used for many airborne radar applications.

B) Non-resonant array (traveling wave array)

The traveling wave array is fed from one port of the waveguide and the other port connects a matched load to terminate the waveguide that absorbing energy not radiated. The non-resonant array has more slots on each waveguide than the resonant array. The element spacing of this array is not eqaul to $\lambda_g/2$ (may be larger or smaller) to avoid reflection, so the reflected wave from each slot do not generate large standing wave due to constructively field accumulation. The reflected power from the termination has to keep sufficiently small so the spuriois beam is below the required SLL. Also, becasuse of unequal element spacing, there are phase differences between slot elements, which results in the mainbeam deviates the broadside direction and changes with the operation frequencies.

C) Frequency scanning array

One of the important applications of the slotted waveguide array antenna is to design the frequency scaning array (FSA) antenna, which is a kind of electrically scanning array by controlling the operation frequencies. Frequency scanning array have two feeding meethods: series-fed and cooperated-fed. It usually uses slow wave to feed the slotted waveguide. The feed line network of a linear FSA can be analyzed in equivalent slow wave structure, whose phase speed (v_p) is

$$v_p = \frac{d}{L} \cdot \frac{\lambda_g}{\lambda_0} \cdot c \tag{14}$$

Where L is the length difference of the feed-lines for neighboring elements, c is the light speed, and λ_g is the wavelength in the waveguide, which is given by

$$\lambda_g = \frac{\lambda_0}{\sqrt{1-(\lambda_0/\lambda_{cf})^2}} = \frac{\lambda_0}{\sqrt{1-(\lambda_0/2a)^2}} \tag{15}$$

λ_{cf} is the cutoff wavelength of the waveguide, which is equal to twice of the waveguide width ($2a$). Assume the frequency scanning occurs relative to a desired center frequency (f_s). From the far-field pattern of the slotted waveguide array, it can be shown that the maximum pattern exists while

$$\frac{2\pi}{\lambda_0} d \sin(\psi_m) + \pi - \delta = 2m\pi, m = 0,\pm1,\pm2,..., \tag{16}$$

$\delta = 2\pi d/\lambda_g$ is the phase difference between slot elements. Using the condition of broadside ($\psi_m = 0$) at the center frequency and Equation (15), we can derivate

$$\sin(\psi_m) = \sqrt{1-(\frac{f_{cf}}{f})^2} - \frac{\lambda_0}{\lambda_g} \cdot \frac{\lambda_0}{\lambda_s} \tag{17}$$

Now, let $f_{cf} = \kappa \cdot f_s$, then this equation can be re-written as

$$\sin(\psi_m) = \sqrt{1 - \kappa^2 (\frac{f_s}{f})^2} - \sqrt{1 - \kappa^2} \cdot \frac{f_s}{f} \tag{18}$$

This equation gives the relationship between these frequencies and how to scan the mainbeam by tuning the operation frequency for the slotted waveguide array. In fact, the scanning angle with frequency is rather confined, due to the fixed path length limited by the waveguide configuration. If a larger scanning angle is required, the waveguide can be bent to lengthen the path. However, this will change the feeding structure and waveguide configuration, and results in larger volumn and higher cost.

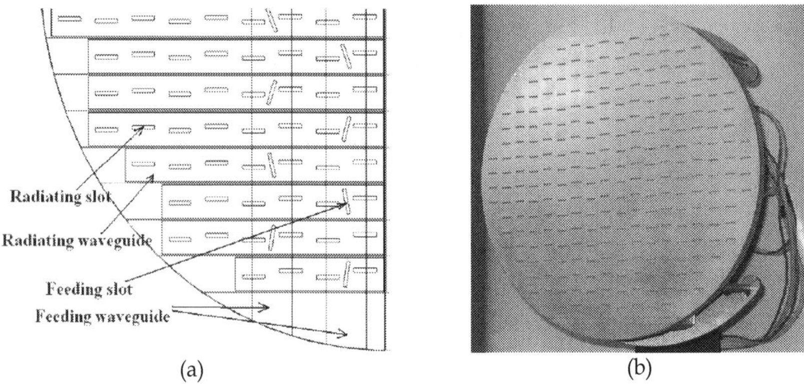

Fig. 10. A circular slotted waveguide array antenna: (a) antenna configuration and (b) antenna photo.

The frequency scanning array antenna can support 3-D scanning while it uses mechanical scan in the azimuth plane and electrical scan in the elevation plane. The advantages include small volume, light weight, and good mobility. Several engineering concepts should be considered to evaluate the phased array antenna performance:

1. To satisdy the requirements of broadband and low sidelobe, the traveling wave arrays using the rectangular waveguide are generally adopted.
2. To suppress the cross-polarization of the planar array, the inclined angles between waveguides should be in the opposite direction. The phase of the excitation source for each waveguide needs be carefully designed.
3. To lower the weight of the waveguide, the materials with low specific gravity can be used.
4. While electrically scanning in the elevation plane, the feeding waveguide forms an array in the azimuth plane so the mainbeam in the azimuth plane will be affected. This angle deviation has to be corrected to avoid detection error.
5. The temperature variaion due to the envioronment or the materials heated by the radiation power will result in the mainbeam direction shift and thus has to be considered for high power applicaitons.

Besides, some of the second-order effects such as the slot alternating displacement, inclined angle alternation, edge effects, non-uniform aperture illumination, slot length error, and mutual coupling through the high-order modes would degrade the array performance than what expected. Manufacturing tolerance also affects the accuracy of the array. These parameters should not be ignored in the phased array antenna design. It is normally to overdesign the array antenna for analysis and fabrication deficiencies that would rely on experience. Figure 10 shows a Ku-band slotted waveguide array antenna for airborne radar applications (Sekretarov & Vavriv, 2008).

3.2 Antennas for the ground penetrating radar

Ground-penetrating radar (GPR) is a geophysical method that uses radar to image the subsurface. This non-destructive method uses electromagnetic radiation in the microwave band (3-30 GHz) of the radio spectrum, and detects the reflected signals from subsurface structures. GPRs can be used in a variety of media, including rock, soil, ice, fresh water, pavements and structures. It can detect objects, changes in material, and voids and cracks. Landmine detection is also an emergent application for GPRs. Unfortunately, there is little international effort to detect and clear the landmines due to shortage of funding. Before they can be removed, they must be located and the ultra-wideband (UWB) GPR technology is ideal for finding the location of these deadly devices. Here we present the antennas popularly used in the GPR applications, which can provide desired bandwidth.

A) TEM horn antenna

For air-launched GPR systems, TEM horn antenna is a good choice because it meets the demand in the aspects of a wide bandwidth, small distortion, directional radiation pattern, and small reflection. It is usually made from two tapered metal plates, including exponentially tapered or linearly tapered. There are a narrow feed point and a wide open end at both sides of the TEM horn antenna (Milligan, 2004). Figure 11 shows a TEM horn antenna which is most widely-used as the GPR antenna (Chung & Lee, 2008).

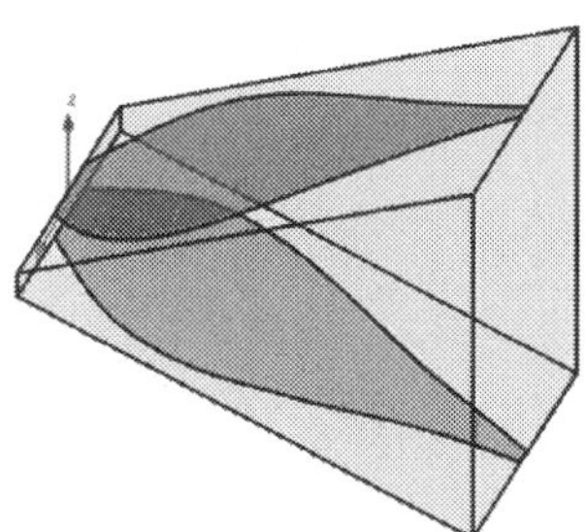
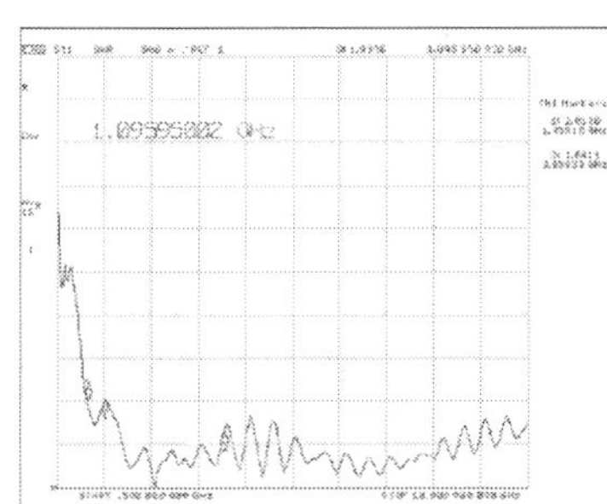

Fig. 11. A TEM horn antenna which can cover 1-10 GHz (VSWR < 2).

B) Monopole antennas

Plenty modifications of monopole antennas are used for broadband applications. The cone antenna is a typical representative, which is usually made as three dimensional structures with a direct coaxial feeding, as shown in Figure 12. The basic electric property of elementary monopole antennas is linearly polarized omni-directional radiation, which changes to a common one with increasing frequency. From this reason, these antennas are used especially in mobile terminals. Typical fractional impedance bandwidth of the monopole antenna is about 70-80%. Further improvements can be made by means of forming the ground plane shape.

The development of high-frequency substrates has enabled the planar design of the monopole antennas. Planar monopole antennas are made single-sided, fed by a coplanar waveguide (CPW), or dual-sided with a microstrip feed line. An example is shown in Figure 13. The advantages are in a manufacture and in an occupied space. However, it is noted that antennas on the planar substrate may have lower radiation efficiency/gain.

Fig. 12. Cone antenna.

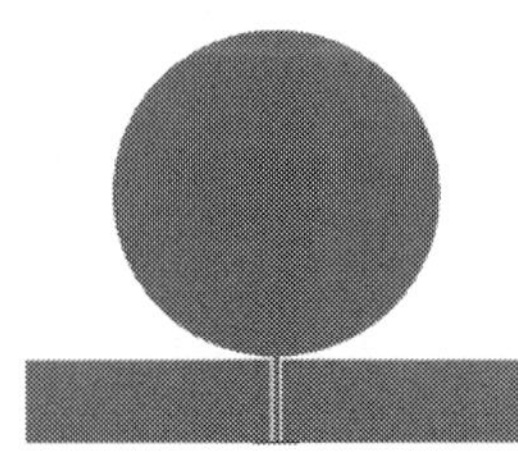

Fig. 13. Planar CPW-fed monopole antenna.

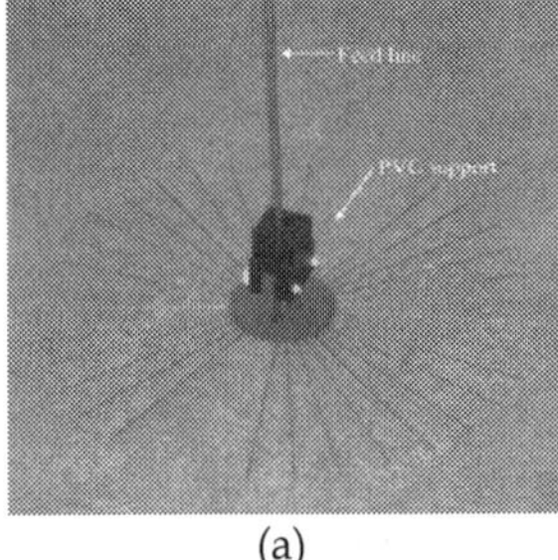

(a)

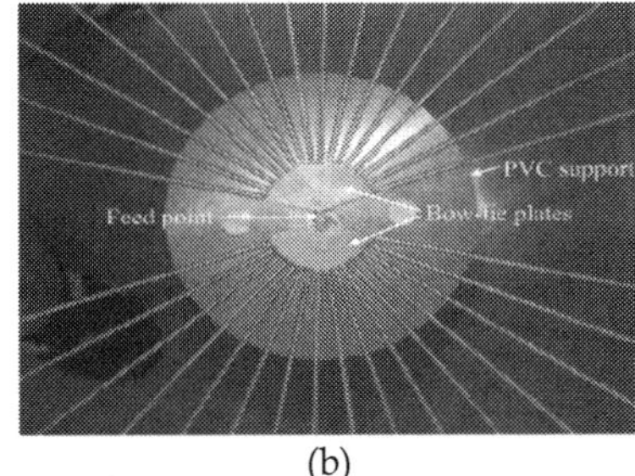

(b)

Fig. 14. (a) Experimental adaptive wire bow-tie antenna and (b) its feed region. The wires are hold together by a PVC support.

It has been known that when an antenna is situated in a proximity to the ground, its input impedance varies significantly with the antenna elevation and the type of the ground. From the system point of view, such input impedance variation is disadvantageous since it leads to difficulty in maintaining a matched condition at the antenna's terminal. When the maximum power transferring from the transmitter to the antenna, via a transmission line, is considered critical, one should provide a variable matching device which is capable of coping with such impedance variation to achieve a matched termination of the antenna for different antenna elevations and ground types. The new design of the GPR antenna allows such a fast and convenient way for controlling the bow-tie flare angle required in a real GPR survey, while the antenna could promptly adapt to any changes in antenna elevation and soil type, as shown in Figure 14 (Lestari *et al.*, 2005).

3.3 Antennas for the noise radar

Wideband random noise radar technology works by transmitting a random noise waveform and cross-correlating the reflected echoes with a time-delayed (to obtain range information) and frequency-shifted (to ensure phase coherence) replica of the transmit signal. Random noise signals are inherently difficult to detect and jam. Since signal reflected from the target is compared with reference from noise generator. Any distortions added to the reflected signal from the target to receiver will have an influence on radar properties, for example, accuracy of target position measurement or resolution of the target. One of the important elements of distortions is antenna. Noise radars usually have separated antennas for transmit and receive due to the need of good cross-coupling isolations. Specific features of noise radars causes that requirements for its antenna systems are similar to other wideband radars.

Some types of antennas are especially compatible with requirements for wideband noise radars, including horn antennas, notch antennas, log-periodic antennas and spiral antennas. The common properties of those antennas are their wide bandwidth. However, most of the noise radar systems will need high gain and high directivity to increase the probability of detection for the short-distance target which is less than 200 meter in front of the radar. The noise radars usually require CW amplifier which has its limitation of maximum transmit power. High gain (> 7 dBi) and wideband antennas are always preferred.

Below there are presented the most popular antennas for wideband applications with described their main properties from the point of view of the noise radar systems. The distortions introduced to the system by antennas can be divided in two main parts: amplitude and phase distortions. The amplitude distortions are caused mainly by antenna gain fluctuations or VSWR properties. In case of phase distortions the main reason are dispersive properties of the antenna. Only non-dispersive antennas are free of that problem (Wisniewski, 2006).

A) Ridged horn antenna

To extend the maximum bandwidth of horn antennas, ridges are introduced in the flared part of the antenna. This is commonly implemented in waveguides to decrease the cutoff frequency of the dominant propagating mode and thus expands the fundamental mode range before high-order modes occur (Hopfer, 1955; Chen, 1957). While the bandwidth of a standard horn is typically up to 2:1, the bandwidth of ridged horns can be up to 18:1. One ridged horn antnena is shown in Figure 15. The ridged horn antenna has high gain directional pattern and shaped radiation pattern in both planes. It can be linear (double-ridge) or dual-polarized (quard-ridged) and dispersive that depend on the ridge design. The main application for noise radars lies in ground penetrating, foliage penetrating, SAR imaging, and tracking.

B) Log-periodic antenna

One good candidate for the noise radar availiable in the market is the Log-Periodic Yagi antenna, as shown in Figure 16. From our radar experiment (Lai & Narayanan, 2005), it was found that this antenna has very good performance over UHF frequency band (400–1000 MHz). It can provide an extended bandwidth of 50 MHz beyond the mentioned lower cutoff frequency of the antenna. Their maintenance free construction on rugged FR-4 material boasts of the added advantages of tuning-free, non-fragile antenna elements. The small size and wide bandwidth makes them ideal for feeding reflector antennas such as the easily constructed corner reflectors or parabolic grids. This antenna also features an SMA Female connector, and their wide bandwidth characteristic ensures total reliability within their respective ranges.

C) Spiral antennas

The bandwidth of spiral antennas is even up to 40:1 and directional in both E- and H-planes. Beamwidth has small changes along with frequency changes. It is usually circular polarized and dispersive. Possibility of application is for noise radar antenna arrays and penetrating detection. More information about the spiral antennas will be presented in next section.

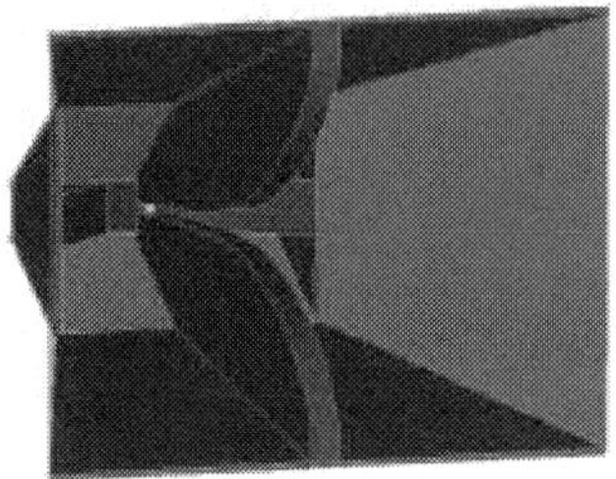

Fig. 15. Construction of the ridged horn antenna   Fig. 16. Planar log-periodic antenna

3.4 Antennas for the UWB system

The development of the ultra-wideband (UWB) sources and antennas has been significantly progressed. UWB antennas need a constant phase center and low VSWR across the whole operation bandwidth. A change in phase center may result in distortion of the transmitted signals and worse performance at the receiver. Hence, UWB antenna design is one of primary challenges in the design of UWB radar systems, especially when low-cost, geometrically small, and radio-efficient structures are required for typical radar applications. It is difficult to compare a UWB antenna with a conventional antenna because the performance considerations for the conventional antenna are based on continuous wave or narrowband theory. However, basic concepts for antenna design should be kept in mind when a conventional approach is used for UWB antennas (Allen *et al.*, 2007).

In the UWB antenna desing, perhaps the most significantly question is: what kind of structures result in ultra-wideband radiation? It has been knonw that several mechanisms can give such a wide bandwidth, and their combinations would be an alternative way to implement (Wiesbeck *et al.*, 2009). These includes: 1) frequency independent antennas (the antenna shape is in terms of angles); 2) self-complementary antennas; 3) traveling-wave antennas; 4) multiple resonance antennas. However, none of anyone structure can guarantee constant radiation patterns over the whole bandwidth. In most cases, the radiation patterns distort from the omni-directional pattern or dipole-like pattern as the operation frequency increases. Also, current commercial UWB systems with impulse signals is seeking for compact antenna solution so that the antenna can be easily packaged with the transceiver module. Therefore, it is obviuosly that the design of UWB antennas should consider the following requirements. First, the bandwidth of the antenna has to satisfy the specification and not interfere the frequency bands used by other wireless systems. Recently, many frequency-notched UWB antennas have been reported to avoiding the conflict between the ISM band. Secondly, the size of the antenna should be miniaturized and compatible to the UWB unit. Thirdly, the UWB antenna should have a omni-directional coverage, indepent of the frequency. Forthly, the antenna would require good time-domain characteristics subjected to the UWB system specification. In this seciton, we will introduce two UWB antennas with the performance qualified the requirements mentioned above.

A) Coupled sectorial loop antenna

The coupled sectorial loop antenna (CSLA) is designed based on the magnetic coupling of two adjacent sectorial loop antennas in a symmetric geometry (Behdad & Sarabandi, 2005). Figure 17(a) shows the topology of a coupled loop antenna arrangement. Two loop antennas

(Antenna 1 and Antenna 2) are placed in close proximity of one another. The antennas are located in the near field of each other and thus a strong mutual coupling between them exists. This mutual coupling can be controlled by changing the separation distance between the two loops and the shape of the loops. If the two antennas are fed with a single source in parallel at the feed location, then

$$Z_{in} = (Z_{11} + Z_{12})/2 \tag{19}$$

where Z_{in} is the input impedance of the double-element antenna, Z_{11} is the self impedance of each loop in the presence of the other one and Z_{12} is the mutual coupling between the two loop. The self impedance of each loop (Z_{11}) is mainly a function of its geometrical dimensions and does not strongly depend on the location and separation distance of the other loop. On the other hand, the mutual coupling (Z_{12}) is mainly a function of separation distance between the two loops as well as the geometrical parameters of each loop. It is possible to choose the dimensions of each loop and the separation distance between the two loops in a manner such that the variations of Z_{11} vs. frequency cancel the variations of Z_{12}. Consequently, a relatively broadband constant input impedance as a function of frequency can be obtained.

Since the antenna topology shown in Figure 17(a) needs a balanced feed, a coaxial cable is used to fed half of the antenna along the plane of zero potential ($z = 0$) over a ground plane, as shown in Figure 17(b). The CSLA is printed on a dielectric substrate of 3 cm × 1.5 cm,

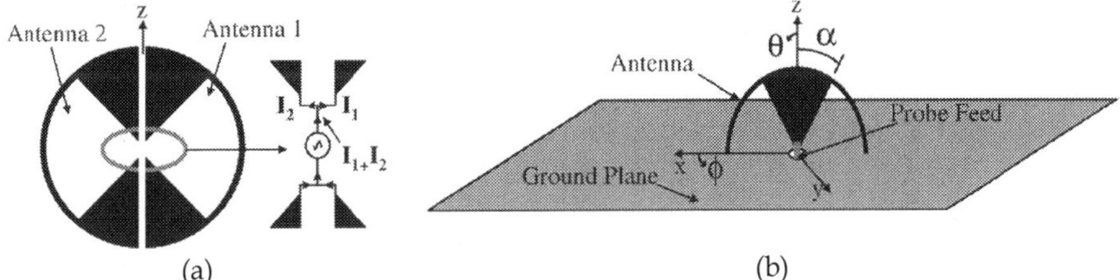

Fig. 17. (a) Topology of the CSLA; (b) Half of the CSLA is used, which is balanced fed by a coaxial probe.

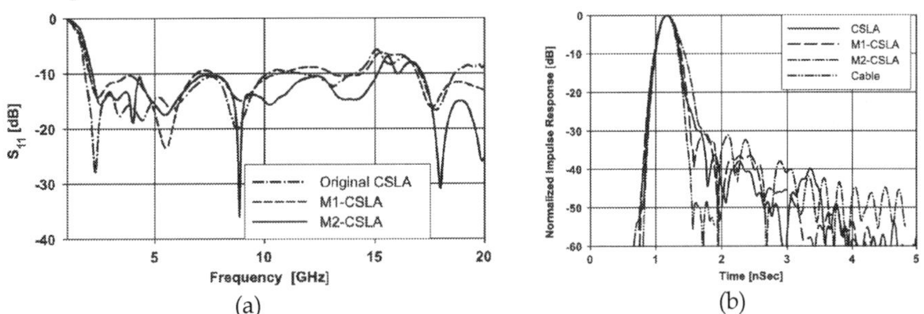

Fig. 18. (a) Measured return losses of the CSLAs. The original CLSA is shown in Fig. 17(b). M1-CSLA and M2-CSLA are the modifed CLSAs with controlled arch angle or notched metal radiator (topologies not shown); (b) time-domain impulse response using two identical CLSAs.

with a relative dielectric constant of 3.4 and thickness of 0.5 mm, and the ground plane size is 10 cm × 10 cm. The self and mutual impedance of the CSLA can be changed by changing the antenna dimensions including the arch angle, α, which gives an additional degree of flexibility in shaping the frequency response. It has been shown that this compact topology can provide an impedance bandwidth in excess of 10:1, as shown in Figure 18(a). Figure 18(b) shows the normalized time-domain impulse responses using two identical CLSAs, which are very close to that of the coaxial cable with the same electrical length. Figure 19 shows the azimuthal radiation patterns and these remains similar up to 8 GHz. They start to change as the frequency is higher than 10 GHz. The gain of this antenna is between -5 dBi to +4 dBi when the operation frequency is lower than 10 GHz.

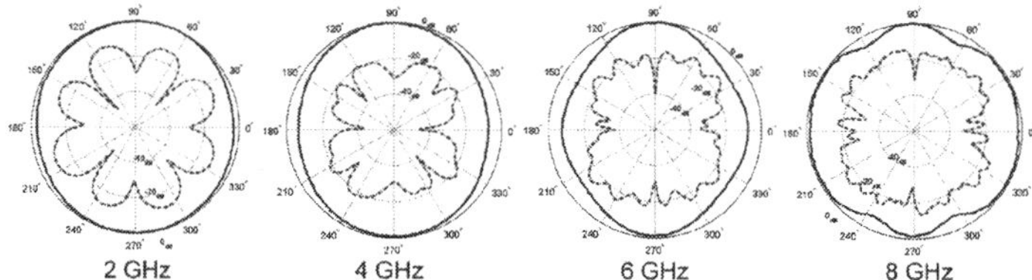

Fig. 19. Measured radiation patterns in the azimuth plane. The solid line and the dash line are co-polarized (E_θ) and cross-polarized (E_ϕ) components, respectively.

B) Frequency independent antenna

An UWB antenna that is able to cover the lower frequency bands such as HF-UHF is challenging due to the requirement and limitation of the bandwidth and physical-size. The spiral antenna is a frequency independent antenna and is a good candidate for UHF and below applications. To have a broadband operation frequency, the integration of multi-arm spiral antennas with broadband balun and feed structures can present numerous challenges. Many innovative designs have been implemented that provide the broadband differential excitation to the complimentary arms of the spiral. In these designs, the Dyson balun provides a very compact form that can be integrated into the structure of the antenna. However, since this balun's implementation typically involves coaxial line, its implementation can create difficulties within the context of IC fabrication techniques. Thus, a design that integrates the spiral antenna, balun, and feed network into a multi-layer structure can provide many possibilities, including antenna miniaturization and bandwidth maximization.

The design of spiral antennas has four primary components: (1) the geometry of antenna, (2) the finite-ground stripline structure, (3) a stripline impedance transformer, and (4) the balun. The spiral antenna is created by winding the finite-ground stripline structure into one arm of the spiral, and extending the center-conductor through the middle of the spiral (the balun) into the complimentary arm of the spiral. The stripline grounds in this configuration will act as both a part of the feed structure and a distributed arm of the spiral antenna. The impedance transformer can then be positioned along the feed section into the center of the spiral to provide the impedance match.

The spiral antenna properties and performance of the finite-ground stripline will have an intimate relationship based around the arm-width constant K (or slot-to-metal ratio). This parameter impacts the input impedance and radiated fields of the spiral, but can also be treated as a degree of freedom between the spiral and stripline geometries. For the finite-ground stripline, this parameter will impact the guiding properties and effectively create a lower bound for its characteristic impedance. Therefore, the parameter K can be used to create a spiral with an input impedance that corresponds to a desired width of the stripline center-conductor for both guidance and obtaining an impedance match with the spiral antenna. By using a high impedance line that corresponds to a narrower line width, small feature sizes can be created. The use of an impedance transformer along the winding of spiral will match the input impedance of the spiral with the system impedance.

An Archimedean spiral with a slot-to-metal ratio of 0.75 has been designed on Rochelle foam for the experimental validation of the stripline-based spiral antenna design using Dyson-style balun (Huff & Roach, 2007). The antenna is designed to operate from 800 MHz to 4.5 GHz. Figure 20 shows the spiral geometry with the simulated input impedance and VSWR. Note that this spiral only has single turn. With more turns, the lowest operation frequency of the antenna can be decreased. Theoretically, the effective frequency of the spiral should not have an upper limit.

The radiation pattern of the spiral antenna is stable within the whole operation frequencies due to its constant proportionality ratio, as shown in Figure 21. The planar spiral antenna exhibits a bidirectional radiation property. By placing a cavity-backed absorbing material, or using a ground plane, behind the antenna, the antenna will have a unidirectional pattern. This constant pattern characteristic avoids the use of beamforming for certain incident angles and reduces the attenuation due to the multi-paths. Considering the role of the mobile platform where the antenna installed, it can behave as a special reflector to

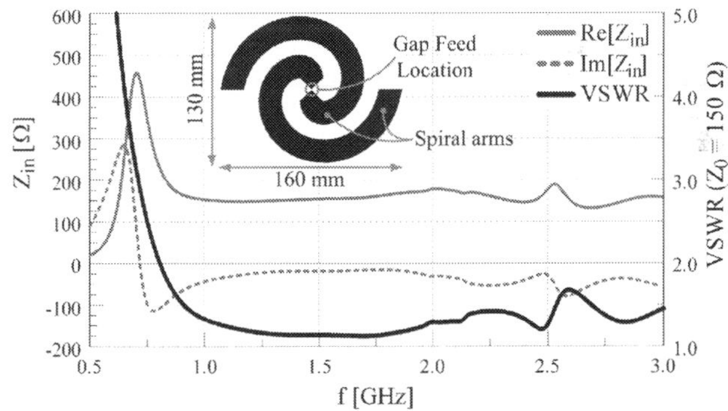

Fig. 20. Geometry of the spiral antenna, with its simulated input impedance and VSWR.

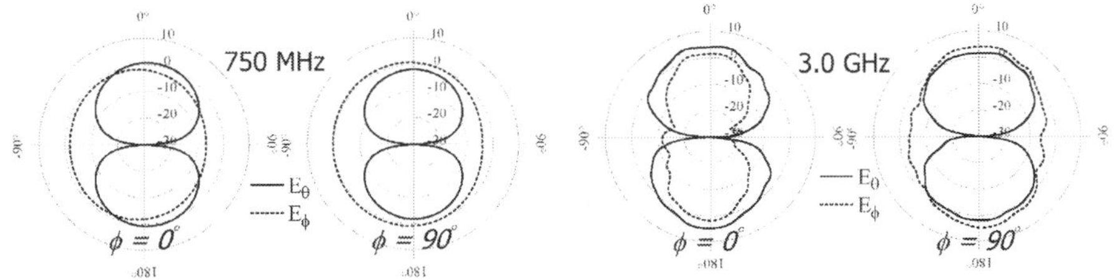

Fig. 21. Measured radiation patterns of the fabricated spiral antenna.

constructively add signals in phase and hence strengthen the radiation gain, which can increase the efficiency of the radio link and decrease power consumption to amplify the transmitted or received signals. Also, the spiral antenna can be used to from a high gain array by using a well-designed impedance matching network.

C) Polarization and antenna miniaturization

In some wireless applications, circular polarization is preferred to linear polarization due to the following reasons. First, while using a linear polarized antenna, the strength of the reflected signal from an object will depend on the azimuthal position of the antenna relative to the object. Secondly, if the direction of the transmitting antenna and the receiving antenna is orthogonal, the receiving antenna cannot detect the reflected signal from the object. Broadband dipole antennas, log-periodic antennas, and Yagi-Uda antennas are linear polarized broadband antennas and broadband circular polarization is relatively difficult to implement, compared to realize the circular polarization of spiral antennas.

Since the physical size of the low frequency antenna would be very large, the antenna miniaturization should be considered. The resonant frequency of the antenna can be reduced by using pieces of dielectric material to sandwich the antenna. In other words, it is also possible to reduce the size of the antenna at a given resonant frequency. On the other hand, both the thickness and permittivity of the substrate influence the antenna performance. A thick substrate with a low dielectric constant provides better efficiency and wider bandwidth at the expense of larger antenna element size. The same result can be obtained using a thin substrate with a high dielectric constant. Therefore, there is a trade-off between the efficiency and the physical size of a designed structure.

4. Antenna protection and filtering

4.1 Radome

For radar applications, a radome (a contraction of radar and dome) capable of protecting the antenna from the effect due to harsh environmental and other physical conditions must be developed. The radome should be capable of shielding the antenna from various hazardous conditions and not adversely affect the performance of the antenna. Therefore, investigation of the antenna protecting issue is an important task that must be carried out concurrently with development of the radar systems.

The antenna radome has many important parameters to be considered. Firstly, it has to withstand specified vibration, shock, and acceleration levels, including explosion and gunfire. In general, the radome may be used in the following scenarios:

1. For the ground use, it works form -50°C to 60°C, with relative humidity of 0-95%. It can not be damaged in wind speed of 67m/s and can afford ice/hail/snow load by 300 kg/m² at least.
2. For the airborne use, it works from -50°C to 70°C, with relative humidity of 0-95%. The altitude ranges from 0 to 12,000 meter.
3. For air fighter, the requirement is harder. The temperature is -50°C to 180°C and the altitude is 0-20,000 meter.

Besides, it has to be resistant for rain adhesion, rain erosion, and hail impact, should be fire retardant and lightening-protected, and is able to withstand solar radiation and even damage from nuclear explosions, if applicable.

The parameters list below can be used to evaluate the performance of the radome after integrated with the antenna:

1. Transmission and reflection coefficients: determines how much energy may loss or be reflected from surfaces, both internal and external diffraction and refraction effects.
2. Half-power beamwidth of the antenna: affects the coverage range and resolution.
3. Error in broadside: the mainbeam of the antenna may deviate from the broadside direction.
4. Pattern distortion and sidelobe change: the mainbeam shape of the antenna may change and the sidelobe level of the antenna may becomes worse. Also, the reflection due to the radome may result in new sidelobes.
5. Cross-polarized isolation: the cross-polarized pattern of the antenna may change, and the cross-polarized isolation may be enhanced/degraded.
6. Scanning error: the scanning range may be changed.

Several kinds of wall contruction have been used in radomes such as homogeneous single-layer (monolithic), multilayer and metallic-layer:

1. Monolithic wall includes thin wall and half-wave wall, which consists of a single slab whose thickness is less than $0.1\lambda_0/(\varepsilon_r)^{0.5}$ (thin wall), or is an multiple of half-wavelength $(n\lambda_0/2)$, where n is the order of the radome wall. More strictly, the thickness of the monolithic wall is given by

$$t = \frac{n\lambda}{2} \cdot \frac{1}{\sqrt{\varepsilon_r - (\sin\theta)^2}} \tag{20}$$

The thin wall can provide good electrical property, adequate strength, and regidity. However, it can be structrally weak. The half-wave wall should have no reflection at its incidence angle θ and thus the energy can be transmitted with minimal loss within a range of incidence angle. The ohmic loss is the primary loss of such material.

2. Multilayer wall includes A-sandwich, B-sandwish, C-sandwich, and the sandwich with even more layers. The A-sandwich have two thin, high-dielectric skins seperated by a low-dielectric core material of foam or honeycomb. The B-sandwich is a inverse version of the A-sandwich whose core is a high-dielectric skin. The C-sandwich consists of two back-to-back A-sandwichs for greater strength and rigidity.
3. Metallic radome can provide the necessary frequency filtering, which can be combined with the dielectric radomes. Conducting metal material with features of periodic patterns can be designed to have the characteristics of the lumped element shunted across a transmission line (thin layer), or of the waveguide with cuf-off frequency. Such elements can results in a planar or conformal frequency selective surface (FSS), which will be introduced in next section. Metallic radomes have better potential for higher power handling, suppressed signal interference, reduced static charge buildup, and strong resistance to rigid weather such as heat, rain, and sand.

It is expected that the use of radomes should have the following goals and advantages for the antenna operation: (a) the antenna can work in all-weather operation, (b) a cheaper antenna is produced because it requires less maintenance, (c) the overall system performance may be more accurate and reliable, and (d) the structural load of the antenna is reduced due to the removal of environmental forces. In terms of the fabrication process and tolerance, the radome has to be designed and optimized concurrently with the platform geometry and critical design parameters for the antenna systems. Furthermore, through the radome to reduce the antenna RCS is an efficient way to protect the antenna and extend it lifetime, because the antenna may be detected and attacked by the enemy. More information of the performance analysis and materials for radomes can be found in (Rudge *et al.*, 1983; Kozakoff, 1997; Skolnik, 2001).

4.2 Frequency selective surface

A frequency selective surface (FSS) is a periodic surface that exhibits different transmission and reflection properties as a function of frequency. Two basic types are the array of wires (dipoles) and the array of slots, which are followed by a dielectric slab to form the FSS. As shown in Figure 22, an array of resonant dipoles acts as a spatial bandstop filter; an array of slots acts a spatial bandpass filter. However, unlike microwave filters, the frequency response of an FSS is not only a function of frequency but also a function of the polarization and incidence angle. In general, a FSS can be designed based on the transmission line theory to predict the desired transmission characteristics.

Figure 23 shows a newly developed low profile muti-layer FSS (Behdad *et al.*, 2009), which applied the third-order bandpass frequency response, with an overall thickness of only $\lambda/24$. Each layer is a two-dimensional periodic structure with sub-wavelength unit cell dimensions and periodicity, where both resonant and non-resonant elements have been combined and miniaturized. It also shows that the multi-layer FSS can be analyzed using the transmission line model to calculate required parameters for the periodic structures. This new design demonstrates a rather stable frequency response as a function of the incidence angle without the aid of any dielectric superstrates that are commonly used to stabilize the frequency response of FSS's for oblique incidence. The polarization has to be considered in the FSS deisng. Various periodic structures and unit elements have been proposed for single- and dual-polarized FSS's (Wu, 1995; Munk, 2000).

Besides, to minimize the size and weight of the FSS structures, high-dielectric ceramic materials can be used. Dielectric breakdown measurement could be first performed at macro-level to give the guideline on the FSS element designs. When properly designed, these ceramic materials are capable of reducing significantly the size of the FSS element and can handle high-power energy. To understand the microscopic structures of materials, mixing-law can be used to obtain microscopic-level field distribution within the materials that help further understand the dielectric breakdown mechanism. In addition, by mixing

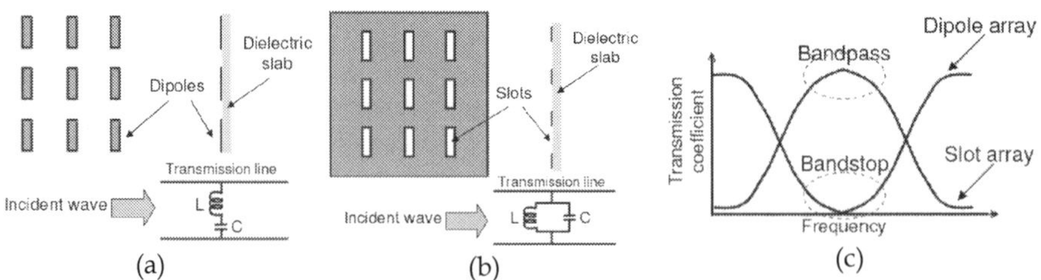

Fig. 22. Basic frequency selective surfaces: (a) dipole array, (b) slot array, and (c) frequency responses.

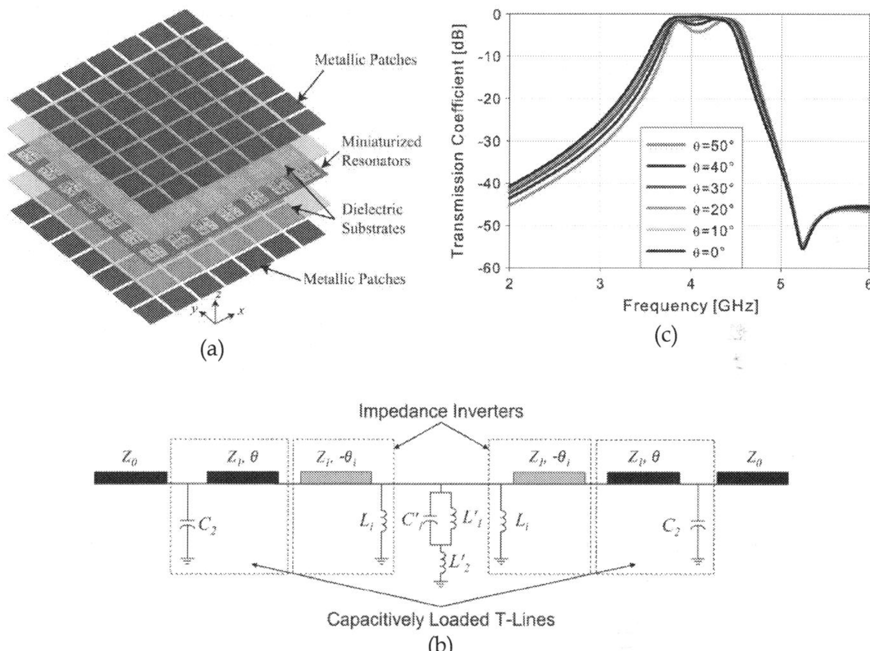

Fig. 23. A multi-layer third-order FSS: (a) topology, (b) equivalent circuit, (c) measured transmission coefficients for various incidence angles.

materials of two or more different dielectrics, synthetic dielectrics with different textured substrates can be developed for impedance matching. Combining all of these characteristics with the FSS configuration, it is possible to improve upon past benchmarks of bandwidth and power handling while maintaining significant miniaturization.

FSS's can be used to limit electromagnetic interference (EMI) and radar cross section (RCS) associated with antennas. When designing high power microwave (HPM) systems, problems associated with air and material breakdown must be addressed. However, at the present time, many advanced materials, including FSS's, are not employed in HPM systems due to breakdown concerns. The major reason for this is that both the macro-level electric field distribution within FSS unit element and the micro-level electric field distribution within dielectric materials have to be modeled simultaneously. For example, even two composite dielectric materials may have the same dielectric constant, due to the internal microscopic level construction, the breakdown voltages of these materials can be significantly different. Innovative designs are needed to allow FSS's to be employed in HPM systems. In addition to the power handling capability, other performance parameters, such as resonant frequencies and bandwidth, as well as compactness of FSS structure are also important. Good methodology to predict, test, and validate the power handling capability of FSS designs are hence highly desired for future FSS applications.

5. Conclusion

In this chapter, the basics of the antenna and phased array are reviewed and different wideband antennas for modern radar systems are presented. The concepts of the radome and frequency selective surface are also reviewed. The main contents include important parameters of the antenna, and theory and design consideration of the array antenna. Various wideband antennas are introduced and their performances are demonstrated, including: (1) for the phased array radar, the slotted waveguide array antenna has been widely used in airborne and marine radar systems; (2) for the ground penetrating radar, TEM horn antenna, broadband monopole antennas, and adaptive bow-tie antenna are presented; (3) for the noise radar, ridged horns and log-periodic Yagi antenna are introduced; (4) for the UWB systems, radiation characteristics of two newly developed UWB antennas (a coupled sectorial loop antenna and a spiral antenna with Dyson-style balun) are demonstrated. Furthermore, the radome and frequency selective surface used to protect the antenna/radar systems are brought in, whose functions, requirements, structures, and performance evaluation are presented, including state-of-the-art designs. The design considerations and future trend of the radar antennas and associated devices have been discussed and suggeted. More detailed can be referred to given references.

The development of the wideband antenna technology triggers the advancement of the antenna in the radar system. The progress of the antenna design also reveals applications of many innovative materials and structures. Designing low-profile, ultra-wideband, and electrically small antennas have become one of the primary goals in the antenna society. Future radar systems will require ultra-wide/multi-band antennas with capabilities of high-speed, high resolution, and high reliability. Wideband antennas covering ELF-VLF and HF-UHF will be challenging but eagerly desired. Smart impedance matching network should integrate with the miniaturized wideband antennas to form high efficiency antenna system. Also, fast and low-cost conductive printing technology can assist antennas conformal to various mobile platforms and hence become invisible. These innovative technologies will benefit future radar systems that may initiate a new research field for the wideband antennas.

6. References

Allen, B. *et al. (2007). Ultra-wideband Antennas and Propagation for Communications, Radar and Imaging,* John Wiley & Sons, ISBN 0470032553, England.

Balanis, C. (1982). *Antenna Theory: Analysis and Design,* John Wiley & Sons, ISBN 047160352X, Hoboken, NJ.

Behdad, D. & Sarabandi, K. (2005). A compact antenna for ultrawide-band applications. *IEEE Trans. Antennas and Propagation,* Vol. 53, No. 7, July 2005, pp. 2185-2192.

Behdad, D. *et al.* (2009). A low-profile third-order bandpass frequency selective surface. *IEEE Trans. Antennas and Propagation,* Vol. 57, No. 2, February 2009, pp. 460-466.

Chen, T. (1957). Calculation of the parameters of ridge waveguides. *IRE Trans. Microwave Theory and Techniques,* Vol. 5, No. 1, January 1957, pp. 12-17.

Chung, B. & Lee, T. (2008). UWB antenna assists ground-penetrating radar. *Microwave & RF Magazine,* December 2008, pp. 59-65.

Hopfer, S. (1955). The design of ridged waveguides. *IRE Trans. Microwave Theory and Techniques,* Vol. 3, No. 5, October 1955, pp. 20-29.

Huff, G. & Roach, T. (2007). Stripline-based spiral antennas with integrated feed structures, impedance transformer, and Dyson-style balun, *IEEE Antennas and Propagation Society International Symposium,* pp. 2698-2701, Honolulu, HI, USA, June 2007.

Kozakoff, D. (1997). *Analysis of Radome-Enclosed Antennas,* Artech House, ISBN 0890067163, Norwood, MA.

Lai, C. & Narayanan, R. (2005). Through-wall imaging and characterization of human activity using ultrawideband (UWB) random noise radar, *Proc. SPIE Conference on Sensors, and Command, Control, Communications, and Intelligence (C3I) Technologies for Homeland Security and Homeland Defense IV,* vol. 5778, pp. 186–195, Orlando, FL, USA, March-April 2005.

Lestari, A. *et al.* (2005). Adaptive wire bow-tie antenna for GPR applications. *IEEE Trans. Antennas and Propagation,* Vol. 53, No. 5, May 2009, pp. 1745-1754.

Rudge, A. *et al.* (1983). *The Handbook of Antenna Design,* Vol. 2, IET, ISBN 0906048877, UK.

Mailloux, R. (2005). *Phased Array Antenna Handbook,* Artech House, ISBN 1580536905, Norwood, MA.

Milligan, T. (2004). A design study for the basic TEM horn antenna. *IEEE Antennas and Propagation Magazine,* Vol. 46, No. 1, February 2004, pp. 86-92.

Milligan, T. (2005). *Modern Antenna Design,* John Wiley & Sons, ISBN 0471457760, New York, NY.

Munk, A. (2000). *Frequency Selective Surface : Theory and Design,* John Wiley & Sons, ISBN 0471370479, New York, NY.

Sekretarov, S. & Vavriv, D. (2008), Circular slotted antenna array with inclined beam for airborne radar application, *Proceedings of the German Microwave Conference,* pp. 475-478, Hamburg, Germany, March 2008.

Skolnik, M. (1990). *Radar Handbook,* McGraw-Hill, ISBN 007057913X, New York, NY.

Skolnik, M. (2001). *Introduction to Radar Systems,* McGraw-Hill, ISBN 0072909803, New York, NY.

Stevenson, R. (1948). Theory of slots in rectangular waveguides. *J. App. Physics,* Vol. 19, 1948, pp. 24-38.

Volakis, J. *et al.* (2007). *Antennas Engineering Handbook,* McGraw-Hill, ISBN 0071475745, New York, NY.

Watson, W. (1949). *The Physical Principles of Waveguides Transmission and Antenna Systems.* pp. 122-154, Clarendon Press, Oxford.

Wisniewski, J. (2006). Requirements for antenna systems in noise radars, *International Radar Symposium 2006,* pp.1-4, Krakow, Poland, May 2006.

Wiesbeck, W. *et al.* (2009). Basic properties and design principles of UWB antennas. *IEEE Proceedings,* Vol. 97, No. 2, February 2009, pp. 372-385.

Wu, T. (1995). *Frequency Selective Surface and Grid Array*, John Wiley & Sons, ISBN 0471311898, New York, NY.

18

Reconfigurable Virtual Instrumentation Design for Radar using Object-Oriented Techniques and Open-Source Tools

Ryan Seal and Julio Urbina
The Pennsylvania State University
USA

1. Introduction

Over the years, instrumentation design has benefited from technological advancements, and, simultaneously suffered adverse effects from increasing design complexity. Among these advancements has been the shift from traditional instrumentation, defined by a physical front-panel interface controlled and monitored by the user; to modern instrumentation, which utilize specialized hardware and provide interactive, software-based interfaces, from which the user interacts using a general-purpose computer (GPC). The term virtual instrumentation (VI) (Baican & Necsulescu, 2000) is used to describe such systems, which generally use a modular hardware-based stage to process or generate signals, and a GPC to provide the instrument with a layer of reprogrammable software capabilities. A variety of instruments can be emulated in software, limited only by capabilities of the VI's output stage hardware. In more recent years, the popularity of reconfigurable hardware (Hsiung et al., 2009), more specifically field programmable gate arrays (FPGAs) (Maxfield, 2004), have revolutionized capabilities of electronic instrumentation. FPGAs are complex integrated circuits (ICs) that provide an interface of digital inputs and outputs that can be assembled, using specialized development tools, to create customized hardware configurations. By incorporating FPGA technology into traditional VI, the instrument's capabilities are expanded to include an additional layer of hardware-based functionality.

In this chapter we explore the use of such enhanced instruments, which are further augmented by the use of object-oriented programming techniques and the use of open-source development tools. Using low-cost FPGAs and open-source software we describe a design methodology enabling one to develop state-of-the-art instrumentation via a generalized instrumentation core that can be customized using specialized output stage hardware. Additionally, using object-oriented programming (OOP) techniques and open-source tools, we illustrate a technique to provide a cost-effective, generalized software framework to uniquely define an instrument's functionality through a customizable interface, implemented by the designer. The intention of the proposed design methodology is to permit less restrictive collaboration among researchers through elimination of proprietary licensing; making hardware and software development freely available while building upon an existing open-source structure. Beginning with section 2, the concept of a virtual instrument, its inherent properties, and improvements through reconfigurable hardware will be discussed. Section 3 will cover object-oriented software design principles and its application to instrumentation. Finally, section 4 will provide an application of these techniques applied to radar instrumentation, where we present the design of a radar pulse generator used to synchronize transmit and receive stages for the 430 MHz radar at the Arecibo Observatory (AO).

2. Virtual instrumentation

In this section, an overview of VI and related concepts will be provided in conjunction with an introduction to reconfigurable hardware and tools necessary for design.

2.1 Virtual instruments

VI, like traditional instrumentation, is used for systematic measurement and control of physical phenomena; but, unlike its counterpart, VI contains additional software layer which defines the VI's functionality. This additional functionality is used to emulate traditional instrumentation and is bounded solely by physical limitations imposed by the instrument's output stage hardware. A block diagram illustrating the virtual instrument is shown in Figure 1. Popularity of VI has flourished in scientific and research-oriented environments, largely driven by the necessity and ubiquitousness of general-purpose computing. Many commercially available (Baican & Necsulescu, 2000) systems exist today and all provide some form of modular, specialized hardware configurations to augment the number of emulable instruments available to the user. These specialized hardware units commonly utilize some form of analog-to-digital (A/D) or digital-to-analog (D/A) circuitry combined with a standard interface to communicate information to and from the GPC.

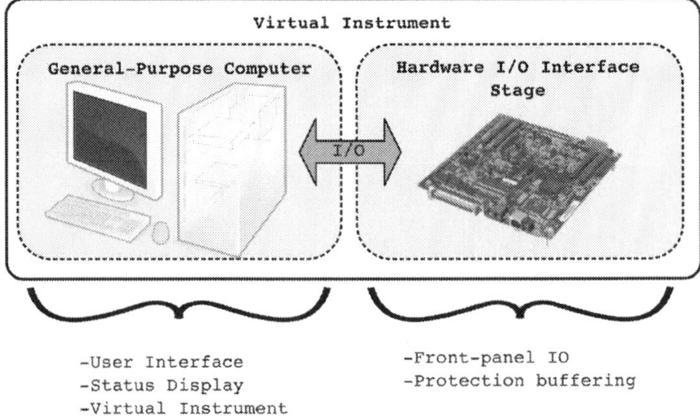

Fig. 1. A block diagram illustrating the concept of the virtual instrument

2.2 Reconfigurable logic

Reconfigurable, or programmable, logic devices come in many forms including: simple programmable logic devices (SPLDs), complex programmable logic devices (CPLDs), field programmable logic devices (FPGAs), application-specific integrated circuits (ASICs), and application-specific standard parts (ASSPs). The general term, programmable logic device (PLD), is used to describe both SPLDs and CPLDs, which contain coarsely defined groups of logic, determined by the manufacturer, that can be customized by the end-user. These devices are more beneficial for special-purpose hardware applications, and are typically chosen when the design engineer has a well-defined set of specifications to work from, but the potential for variability in specifications or the instrument's functionality exists. At the opposite end of the programmable logic spectrum lies both ASICs and ASSPs. These devices are also used in special-purpose hardware applications, but these applications require stringent design specifications, allowing the design engineer to finalize and send the design to a special-purpose ASIC/ASSP manufacturer. Any engineering changes made to an ASIC/ASSP require retooling by the manufacturer, thus making this technology more costly in design environments with variability. FPGAs fall somewhere between the more complex CPLD design and the rigidness of the ASIC design, providing smaller groups of internal logic that can be combined to create CPLD-like designs and even larger, complex designs like that of the typical ASIC application. This amount of flexibility, combined with costs much lower than ASIC design, and comparable to CPLDs, make the FPGA a strong candidate for reconfigurable VI design.

In addition to these advantages, many FPGA vendors provide free software design tools that can be used with many of the vendor's low-cost devices, which are suitable for most VI applications. FPGA designs follow a well-defined design structure, known as the *design flow* (Navabi, 2006), which consists of several stages including:

1. Design Entry

In this stage the design engineer translates the design requirements into a language recognized by FPGA vendor tools. There are several languages to choose from and the field of FPGA language design is growing at a rapid rate. The five most common of these are: 1) Verilog, 2) SystemVerilog, 3) VHDL, 4) SystemC, and 5) pure C/C++ (Maxfield, 2004). Future trends appear to be moving towards higher-level (i.e. object-oriented) languages to better accommodate the ever-growing size and complexity of FPGAs.

2. Design Validation

After entering the design, the design engineer must create a simulation test-bench to ensure that the design entry functions as expected. A number of tools are available for test-bench simulation, and, depending on the language used for design entry, many of these tools are available in open-source, including: iverilog and gplcver for Verilog; ghdl for VHDL; and SystemC.

3. Synthesis

The Synthesis stage compiles the given design entry into a bitstream file format that can be used to program the device. Several sub-stages in the synthesis process optimize the design to meet requirements using vendor-specific tools.

4. Post Synthesis Verification

During the synthesis stage, timing information and loading effects for each electrical path in the design are collected. This information is contained in the bitstream file generated at the final stage of synthesis and is used to perform secondary simulations to ensure that timing delays and loading effects, not apparent in the design entry phase, are understood.

2.3 Reconfigurable virtual instruments

Combining the concepts presented in sections 2.1 and 2.2, an even more powerful form of VI can be constructed, providing the designer with the ability to utilize a reconfigurable hardware module to assist in time-critical applications or application-specific algorithms (e.g. FFTs, image processing, filtering, etc...) more suitable for hardware-based platforms. An illustration of reconfigurable virtual instrumentation (RVI) is depicted in Figure 2 along with each component's responsibilities. RVIs distribute the virtual instrument component among the GPC and reconfigurable hardware module, with the partitioning ratio dependent upon the application and the intentions of the design engineer. RVIs provide opportunities for areas of instrumentation development unachievable using standard VI, due to the GPC's computational inefficiencies and scheduling delays inherent in GPC-based systems. High-performance computing is now commonplace using GPC platforms assisted by reconfigurable hardware (Bishof et al., 2008), with results often tens to hundreds of times faster than equivalent multiprocessor systems. The benefits of such systems can also be applied to radar-based problems, including computational problems and instrumentation design; an example of the latter will be detailed in section 4.

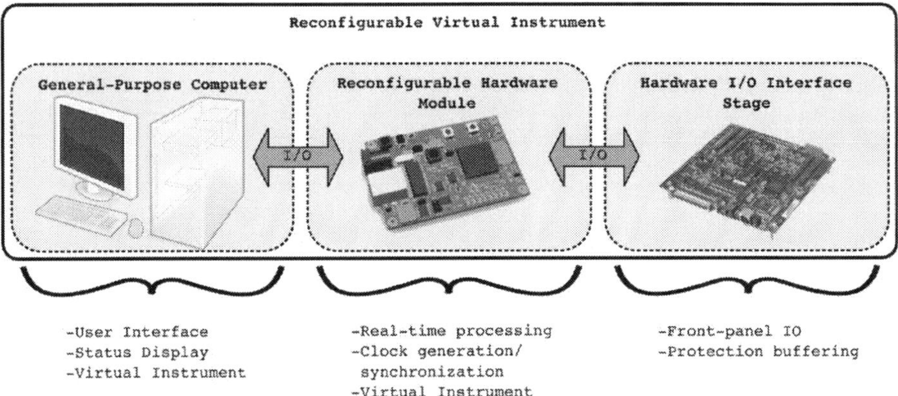

Fig. 2. A block diagram illustrating the introduction of a reconfigurable hardware module to enhance the basic VI concept.

3. RVI design methodology using a generic framework

In this section, additional specifications are given to the RVI definition provided in section 2.3 to further generalize the design and alleviate difficulties typically encountered in the implementation phase. Furthermore, relevant details concerning principles of software re-use and object-oriented principles will be discussed.

RVIs provide a basic instrumentation platform for a user to interface with the instrument through a software interface, typically supplied by a Personal Computer (PC), which also controls the reconfigurable hardware module and output stage hardware through a common communication peripheral. Although useful, in this section we present a more robust solution providing a stand-alone instrumentation core that can be customized to a variety of applications, including radar instrumentation.

3.1 Small form factor general-purpose computers

In recent years, GPC technology has become an imposing competitor to more traditional standards (e.g. PC104 and Single Board Computers) in the embedded market, primarily due to price/performance ratios, but also from the surge in small, custom hardware modules supporting standard communication interfaces (e.g. PCI, Ethernet, USB 1.0/2.0). Small form factor (SFF) GPCs provide four form factors, shown in Figure 3, suitable for compact, stand-alone instrumentation. All SFF GPCS are commercial-off-the-shelf (COTS) products and

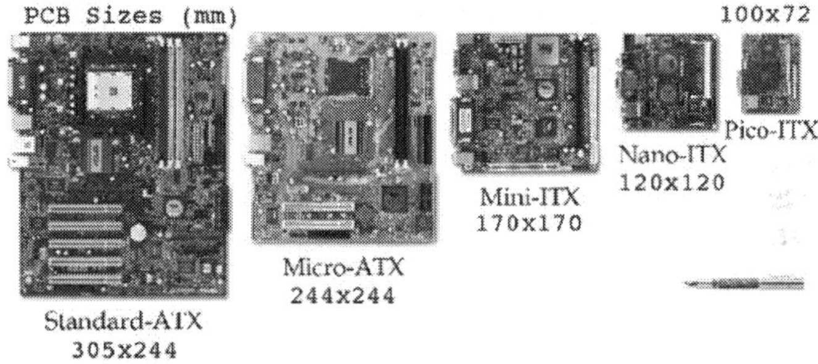

Fig. 3. Comparison of Small Form Factor Motherboards available today

incorporate standard peripherals, such as integrated video cards, keyboard/mouse connections, USB 1.0/2.0, IDE connectors, RS-232, and 100Mbps Ethernet ports. These systems are competively priced, provide impressive processing capabilities, and can be readily replaced in the event of failure. Capitalizing on these factors, we propose a generic RVI core that can be integrated into a standard rack-mount chassis; providing a packaged instrument accessible directly by keyboard, video, and mouse; or remotely accessible using an Ethernet-based communication protocol. The instrument provides a generalized framework from which customizations can be made to meet specifications determined by the designer. The overall design is intended to provide a packaged, customizable system, that can operate in remote locations accessible to Internet. This concept is illustrated in Figure 4

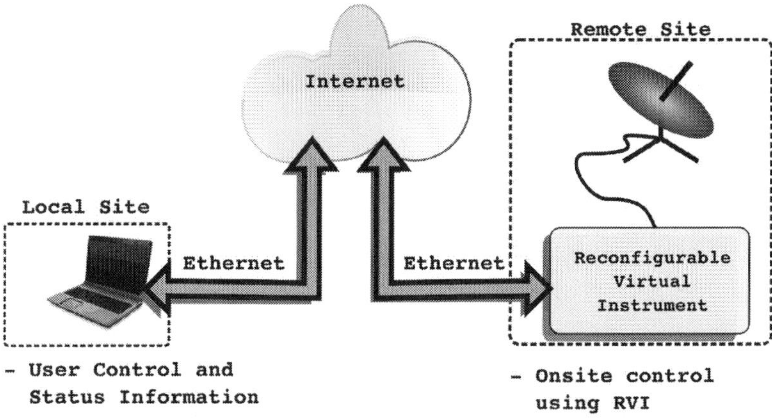

Fig. 4. An illustration of a reconfigurable virtual instrument with a remote interface.

3.2 Software description and structure overview

Designing a generalized software framework conducive to customization is a crucial component in VI designs. In this section, we will provide a framework for generic, object-oriented interfaces that can be easily extended to meet requirements of custom applications.

3.3 Object-oriented programming and open-source software

Object-oriented programming, popularized in the mid 1980s, has long been considered a method for managing ever-increasing software complexity through a structuring process known as *object-oriented decomposition* (Booch, 1994). *Objects*, with respect to software design, can be considered tangible entities providing an interface through which interactions take place. The description, or definition, of the *object* is usually referred to as a *class*. In a program, a *class* is *instantiated* to create an *object*. Details of these actions are hidden behind the interface using a technique referred to as *encapsulation*, which separates the behavior of the *object* from its interface. Concerning instrumentation development, we can use these techniques to define software utilizing well-defined interfaces (Pugh, 2007), and then subsequently define its behavior to emulate that of the instrument being designed.

An instrument is defined by its inputs, outputs, control parameters, and its function. Mathematically, an instrument can be represented by

$$\mathbf{y} = T\left(\mathbf{x}, \mathbf{c}\right), \qquad \mathbf{y} \in \mathbb{R}^{l}, \mathbf{x} \in \mathbb{R}^{m}, \mathbf{c} \in \mathbb{R}^{n}, \tag{1}$$

where $\mathbf{y}$ is the output vector, $\mathbf{x}$ represents the input vector, and $\mathbf{c}$ is the control vector. Design of the instrument requires definition of the instrument's function $T(\cdot, \cdot)$ and $\mathbf{c}$. Additionally, careful consideration must be given to users of the instrument by ensuring that the instrument's control parameters use familiar terminology and provide expected behavior. These criteria are typically provided in the instrument's planning stages, known as design requirements. Modern VI designs commonly use graphical user interfaces (GUIs) to simplify the instrument's input interface. Although useful, designs utilizing command-line interfaces provide important advantages over GUIs, primarily relating to system debugging, upgrades, and, in some cases, performance. In addition to these benefits, the command-line system can be easily augmented to accommodate a GUI after completing the initial design, which also removes the tendency for the designer to couple code describing system functionality with the GUI framework, leading to difficulties when making necessary revisions. For this reason, focus will be given to a command-line driven system.

To provide a generic framework for customization, the instrument's design is divided into two categories:

- Instrument configuration
  Describes the instrument's static configuration, which includes the instrument's type and its initial control parameter values.
- Instrument run-time operation
  Defines the dynamic portion of the instrument, namely its dynamic command-set (e.g. start, stop, reset, etc...) as well as control parameter values that can be changed dynamically, while the system is in operation.

3.4 Instrument configuration design

The instrument's configuration is defined by the user through a plain-text input file, either located directly on the instrument, or, in the case of a remote-interface, the user's local machine. Reasons (Hunt & Thomas, 1999) for using plain-text input files are numerous, including:

- Universally understood by all Operating Systems (OS) and programming languages.
- Consists of printable characters interpretable by people with the use of special tools.
- Can be made more verbose, more representative of human language, thus making it self-describing (i.e. no documentation required) and obsolescent-proof.
- Expedites debugging and development.

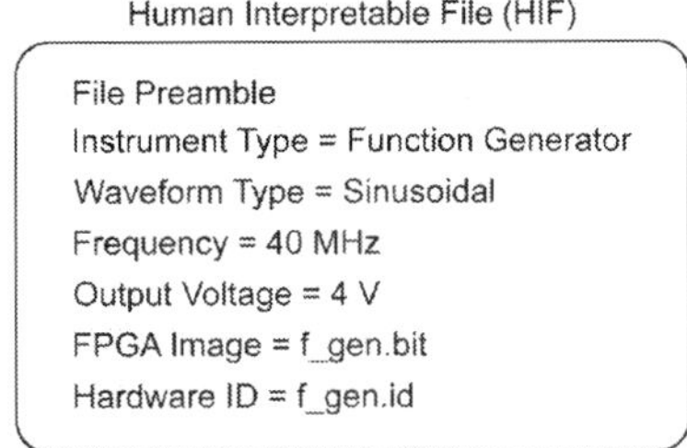

Fig. 5. Sample Human Interpretable File illustrating simplicity and ease of use.

This plain-text file is referred to as a Human Interpretable File (HIF) and contains a number of parameters necessary to define the instrument and its configuration. These parameters include, but are not limited to: instrument type, control parameters, FPGA bitstream file, and, if necessary, the output stage hardware. An example HIF is shown in Figure 5. The HIF structure and content are chosen at the designer's discretion, but the instrument's type must be consistent, as it is required to load the proper interface for parsing and translating the remaining portion of the file, which contain control parameters particular to the specified instrument type. This interface is referred to as the Instrument Definition Interface (IDI) and its responsibilities include:

- Contains the instrument's control parameter specifications, which are used to verify settings requested in the HIF.
- Formats the HIF into an intermediate format more suitable for use in the verification and post-verification processes.
- Creates an Instrument Interpretable File (IIF) formatted for direct use by the reconfigurable hardware.

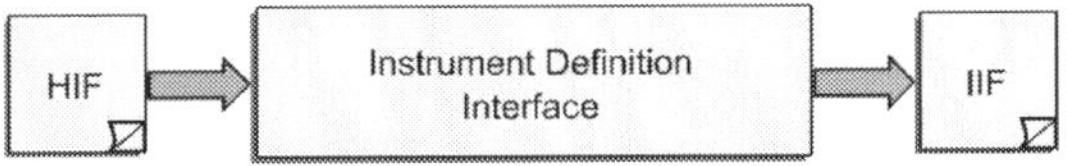

Fig. 6. Block diagram illustrating the Instrument Definition Interface

A block diagram of the IDI is shown in Figure 6. The IDI, in terms of OOP-style programming, defines an interface that accepts an input file and produces an output file. The implementation details are coded by the instrument's designer, who must provide a customized instrumentation *class* that implements the IDI. The resulting output file, or IIF, is structured to provide information concerning the required bitstream file, output stage hardware, and a custom data structure directly readable by the reconfigurable hardware. A sample IIF file is illustrated in Figure 7. As an example, a class diagram, depicted in Figure 8, is used to illustrate how a specialized instrument *class* implements the IDI. Additionally,

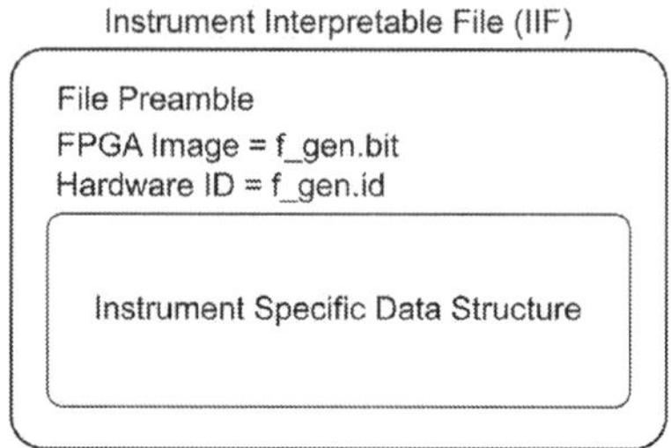

Fig. 7. Sample Instrument Interpretable File illustrating structure.

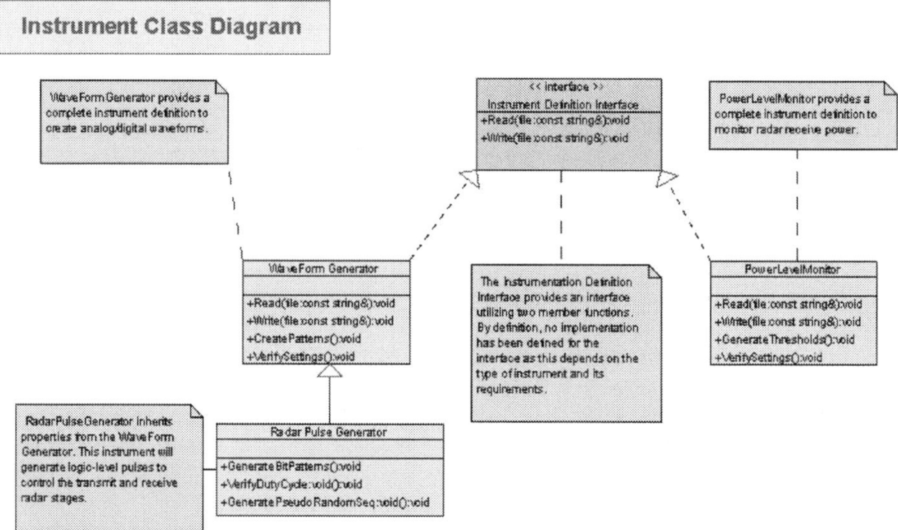

Fig. 8. Block diagram illustrating the Instrument Definition Interface

the concept of inheritance can also be seen, where the *WaveFormGenerator class* is further specialized into a *RadarPulseGenerator class*, which, in turn, is a derivative of the *WaveFormGenerator class*. Several instrumentation *classes* can coexist in the RVI's storage and a particular *class* is chosen by the instrument type, which is defined by the HIF. It should be noted that all information provided by the IDI is considered *static* information, meaning this file is used to initialize the system; including the instrument's type, the instrument's control parameters, and the FPGA bitstream file. These concepts are realized through creation of a *generator* program, which is a command-line program used to read an HIF, validate its inputs, and create an IIF readable by the *run-time* program described in section 3.5.

3.5 Instrument run-time operation design

After defining the instrument's specifications and control parameters, a mechanism is needed from which the user can select and load the virtual instrument and reconfigurable hardware definitions into the system. All information necessary to accomplish this task is contained in the IIF, which was produced by the configuration program presented in section 3.4. These operational tasks are performed by the *run-time* program, which provides a command-line interface through which the user interacts with the system. The *run-time* program begins with a minimal interface, and, after the user chooses an IIF, this interface is expanded to include functionality provided by the instrument described in the file. A specialized helper *class* is used to parse the IIF, determine the instrument's type, and load the specified Instrument Operating Interface (IOI) into the system. Next, the system programs the reconfigurable hardware with a generalized bitstream file, whose purpose is to retrieve identification information (e.g. reading an EEPROM device) from the output stage hardware and verify compatibility with the user's requested instrumentation type. After verification, the system initializes the reconfigurable hardware with the bitstream file provided in the requested IIF. Finally, the system can accept commands provided by the loaded virtual instrument. A diagram illustrating this process is shown in Figure 9. Software encompassing these concepts is referred to as the *run-time* program, which operates in a shell environment incorporating a command-line user interface. As previously mentioned, these concepts can easily be expanded to include a GUI-based display system, depending on the user's requirements.

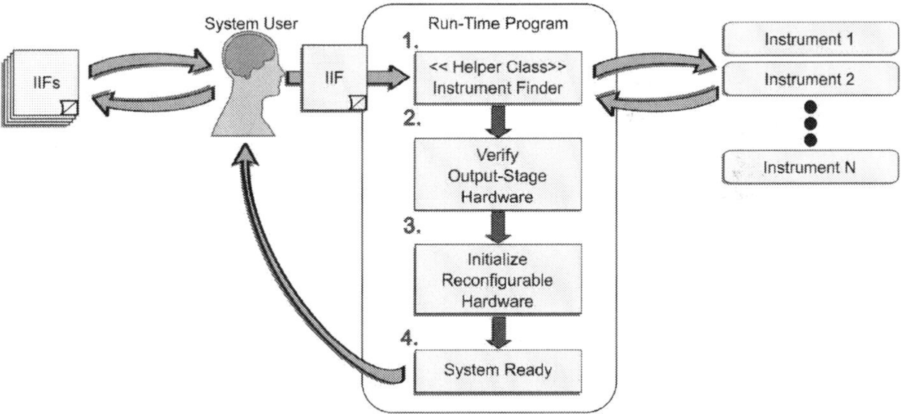

Fig. 9. Block diagram illustrating the interactive run-time interface.

4. Implementation of a radar pulse generator

In this section, discussion of a specialized radar instrument, referred to as the radar pulse generator (RPG), is developed using concepts defined in earlier sections. The sections that follow will provide details of hardware components, software, and all tools necessary for implementation. The RPG (Seal, 2008) was designed for use at AO and served as a starting point for concepts developed in sections 2 and 3.

RPGs provide logic-level signals to control and synchronize a radar's transmit and receive stages in pulsed radar applications as shown in Figure 10. These pulses ensure the

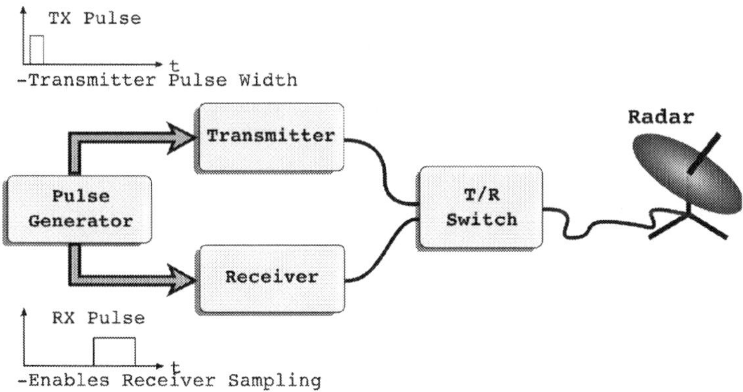

Fig. 10. Monostatic radar overview illustrating the pulse generator used to control both transmitter and receiver.

transmitter operates within a safe regime while simultaneously controlling the receiver's sampling mechanism to record user-defined ranges. The transmitted pulse is given by

$$p(t) = \Re\{m(t)e^{j\omega_0 t}\}, \tag{2}$$

where ω_0 is the transmitter's carrier frequency and $m(t)$ is a periodic waveform that modulates the carrier signal with pulse width δ_t and period T_t. For monostatic radars, a pulse is transmitted, and after a specified delay, the radar switches into receive mode. The RPG, in its most basic form, supplies the receiver's gating circuitry with a periodic pulse (gating pulse) of width δ_r and period T_t, as specified by $m(t)$. The rising edge of δ_r represents the first range sample r_0. The radar's received signal is given by

$$r(t) = \alpha p(t - \tau) + w(t), \tag{3}$$

where α is a generalized attenuation factor reflecting power loss in the returned signal, $w(t)$ represents additive noise, and τ is a specified delay determined by

$$\tau = 2r/v_p, \tag{4}$$

where r is the range in meters, and v_p is the wave's velocity in meters per second. In addition to timing control between transmit and receive modes, the RPG must also generate the transmitter's modulating waveform $m(t)$, which can potentially consist of moderately complex pulse modulation patterns. The RPG's design must accommodate a large number of these patterns, as well as an efficient, user-friendly interface to create, modify, and store them.

4.1 Hardware

Specifications for the design were taken from existing hardware in operation at AO and new capabilities were suggested from the scientific staff. Primary suggestions included: an efficient, easy-to-use interface for designing moderately complex bit patterns; the ability to store and cycle bit patterns without clocking glitches; flexibility, with minimal hardware modifications, to expand system functionality for new research as well as future needs; and the ability to monitor, configure, and control the system remotely. Given these specifications and requirements, an RVI-based system, utilizing a GPC and COTS FPGA module, was chosen, and a block diagram illustrating these components is shown in Figure 11.

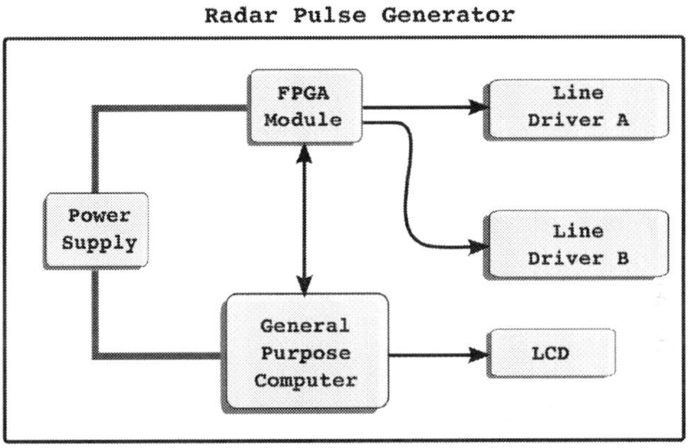

Fig. 11. Block diagram illustrating radar pulse generator components.

Small form factor general-purpose computer

Remote configuration and control were an important aspect of the system; allowing engineers to operate, diagnose, and monitor the instrument's status from an off-site location (e.g. home). To accomplish this, a small on-board computer, supplied with an Ethernet interface, was chosen. The computer, a Mini-ITX form factor mainboard, depicted in Figure 12, contains a number of standard peripherals (e.g. USB, RS-232, video, etc...) commonly found in PCs. The Linux Operating System was chosen for the design due to scalability and availability of existing open-source tools (e.g. VNC, ssh, sockets) for remote access.

Fig. 12. VIA ML6000 Mini-ITX form factor mainboard

On-site system status

To assist on-site monitoring, debugging, and maintenance; a USB-controlled liquid crystal display (LCD) was used. Software was written to display a number of parameters to indicate the system's operational state.

System power

The instrument's mainboard and FPGA carrier are powered from a standard PC power supply. Although smaller, more-efficient solutions were available, these supplies are more readily available and less costly to replace. LCD power is derived directly from the mainboard's USB interface, and output stage hardware is powered from the FPGA carrier. A momentary pushbutton switch mounted on the rear panel is used to power the system.

FPGA module

For experiments requiring sub-microsecond timing, reconfigurability, and the ability to produce lengthy bit patterns, a low-cost COTS FPGA module was chosen. The module, shown in Figure 13, uses a consumer-grade Xilinx Spartan-3A FPGA and provides an on-board clock oscillator and phase lock loop (PLL) circuitry. The module is controlled using a USB 2.0 interface with greater than 32 MBPS downstream (GPC to FPGA) transfer rates and nearly 18 MBPS upstream data rates. *Verilog HDL* was used for coding and verification was performed using *icarus* and *gtkwave*. FPGA synthesis was performed using the Xilinx Webpack development suite, which is freely available and supports a number of Xilinx FPGAs. Further details of the FPGA code implementation is described in section 4.3

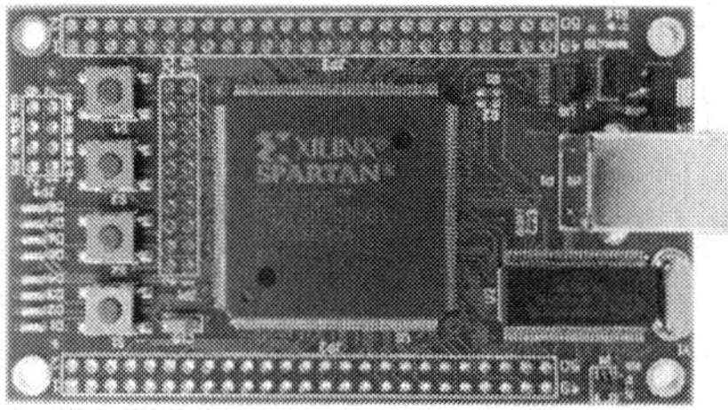

Fig. 13. XEM3001V2 Opal Kelly FPGA module

FPGA carrier

The FPGA carrier board was designed to provide both signal buffering and access using 20-pin dual-header connectors. In total, 32 signal outputs, divided into two 16-bit ports labeled *PORTA* and *PORTB*, are provided, along with external clock inputs, synchronization sources, and an external trigger. These modifications allow the FPGA module to synchronize with an external system clock, and provide precision timing through an on-site atomic clock. A 256Kx16 SRAM module is mounted on-board, allowing *PORTA* to optionally function as a dedicated SRAM controller. Additionally, *PORTB* can be dedicated to use a quadrature 10-bit AD9761 DAC. A block diagram of the board is shown in Figure 14. Schematic capture, board layout, and overall design were completed using the GPL'd Electronic and Design Automation software (gEDA), a popular open-source development suite for schematic capture and printed circuit board (PCB) design.

Output stage hardware

The output stage hardware is composed of two boards, each providing 8 signals, to drive a 50-ohm coaxial transmission line using TTL-level signals. Each board mounts to the instrument's front-panel using BNC pcb-to-panel connectors and provides socketed ICs for quick, in-field replacement. Power and signals are supplied from ribbon cables connected to the FPGA carrier board's 20-pin port connectors.

4.2 System design and operation

Software design consists of two independent programs: 1) the system's *generator* program, which is responsible for bit pattern creation and system configuration; and 2) the *run-time*

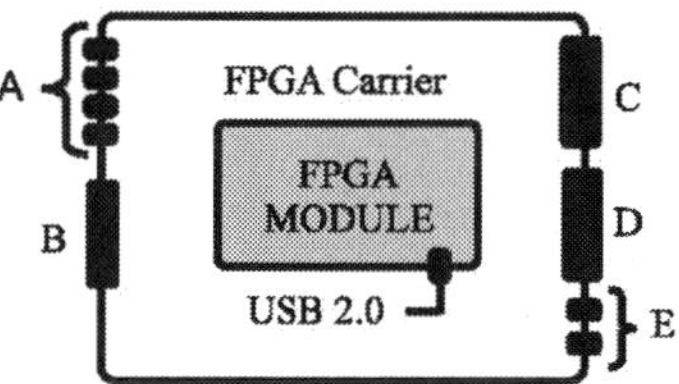

A: Ext. Clock / Sync Inputs
B: ATX Power Conn.
C: 16 Channel PORT A Digital Outputs
D: 16 Channel PORT B Digital Outputs
E: 10-bit DAC Quadrature Outputs

Fig. 14. FPGA carrier board block diagram

shell, which allows the user to control the system via command-line. These programs follow design approaches presented in sections 3.4 and 3.5 and all software was written using the C++ program language (Stroustrup, 2000).

Generator program

The RPG is capable of producing any number of arbitrary, indefinitely repeatable, bit patterns. Generation of such patterns must account for operating limits of the radar system, and an effective, efficent entry method who use the system. To lessen complexity, the approach discussed in section 3.4 was utilized by designing an HIF to store common system parameters specific to the radar's transmitter. These parameters are passed to the IDI, where they are analyzed using a language parser. The parser translates the data into a bit-vector format which passes through rule-checking and verification stages that contain custom-defined transmitter specifications. If verification succeeds, an IIF is written to the system's hard drive; otherwise the system exits and reports the error to the user. This particular IIF contains an ASCII-based structure of 1's and 0's, representing digital logic levels. The FPGA's bitstream file is configured to parse this structure and instruct the hardware of the requested bit pattern sequence, clock source, and synchronization method.

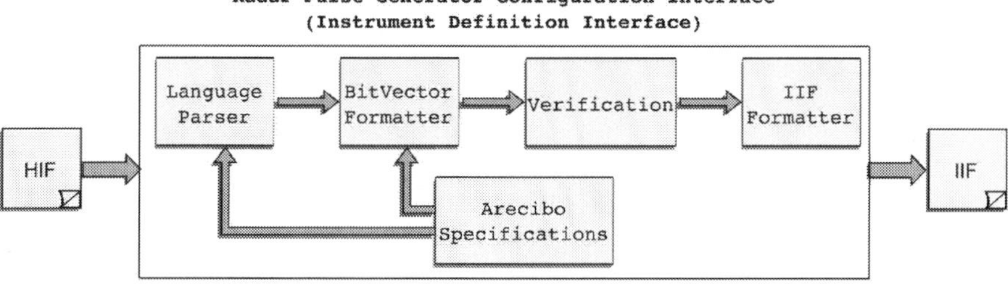

Fig. 15. The Configuration Interface implements the Instrument Definition Interface.

Run-time shell program

After bit pattern generation and verification, a *run-time shell* program is used to operate the instrument; communicating directly with the FPGA through a library-based interface provided by the manufacturer. This library is free and currently available in the following languages: C/C++, Python, Ruby, Matlab, and LabView. Although we present an implementation using C++, any of these languages can be substituted to independently develop a customized system that may be more suitable for a particular development group, depending on the developer's experience and skill set. The program resides in the on-board, Linux-based computer and provides the following control commands: 1) load and unload any number of bit patterns; 2) start and stop the system; 3) request hardware status; 4) switch bit patterns; and 5) select the desired clock and trigger source. Using a separate thread of execution (i.e. multi-threading), system status is polled from the FPGA at a user-defined rate (25 ms default) and an overall status is displayed on the instrument's LCD. Figure 16 illustrates an overview of program operation.

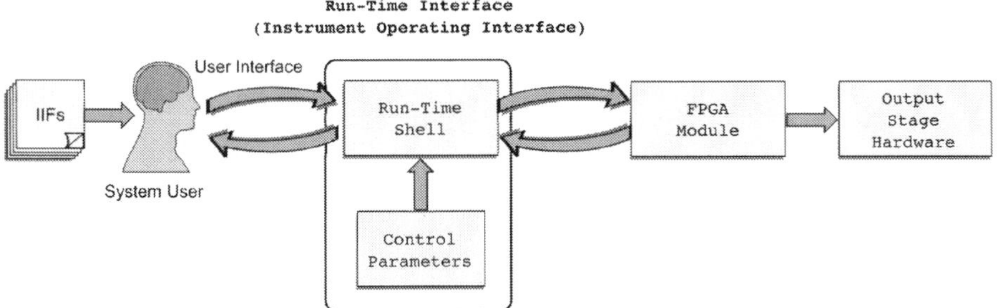

Fig. 16. Block diagram illustrating interaction between the user and the radar pulse generator's run-time shell.

4.3 HDL design and operation

Code for FPGA implementation was written in *Verilog HDL* and functionality was divided into a number of small, well-defined modules designed to improve readability and alleviate code maintenance. The Opal Kelly XEM3001v2 FPGA module, used in this design, utilizes a Xilinx xc3s400 FPGA containing 400,000 gates, 208 I/O, and 288 kbits of on-board RAM. Other features provided by the XEM3001v2 limit the potential number of I/O to 90. Considering the system's requirements, it was determined that the memory structure implemented in the design and clock routing choices would be the primary factors determining performance. Design of the FPGA module began with an analysis of communication methods provided by the Opal Kelly library's Application Programming Interface (API). The Opal Kelly API operates using firmware to establish FPGA/PC communication and a small HDL module integrates into the user's FPGA design to provide communication with the host API. For data transfers implemented in this design, two types of communication were chosen: 1) multi-byte data transfers using the Opal Kelly *PIPE* modules, and 2) simple status/control commands using the Opal Kelly *WIRE* modules. Opal Kelly *PIPE* modules are designed to efficiently transfer a known number of 16-bit wide integers between the host PC and FPGA module while the *WIRE* modules are more suitable for controlling or monitoring a single 16-bit state. An overview of the FPGA's HDL data flow design is depicted in Figure 17 and a description of relevant modules is given in the sections that follow.

InputControl module

This module was designed to act as a simple state machine and makes use of a single Opal Kelly *PIPE* input module to transfer data into the FPGA. When data is present on the *PIPE*

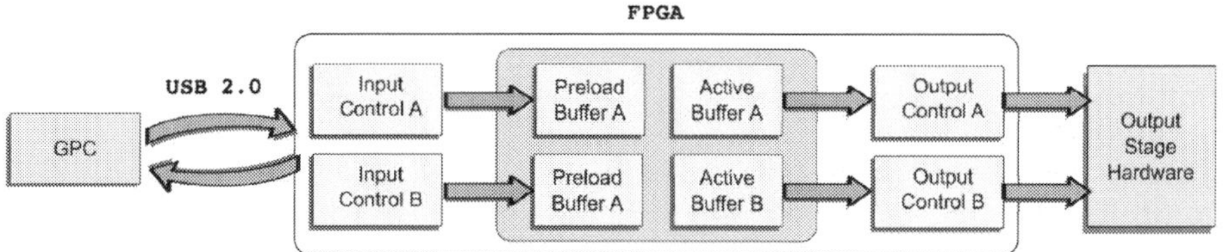

Fig. 17. Overview of FPGA HDL model and illustration of data flow from the module's input to output.

port, the *InputControl* module routes data to the proper memory buffer and maintains a count of bytes transferred. When the proper number of bytes have been transferred, the module sends a signal back to the host PC and returns to the idle state. The module also communicates with other modules in the FPGA to perform synchronization and hand-shaking internal to the system.

OutputControl module

After system initialization is complete (i.e. bit patterns are loaded), the user can send a *start* command to the FPGA. This signal is sent to the *OutputControl* module and the first state of the active buffer's bit pattern is loaded onto the output ports. An internal state counter is decremented at a rate determined by *clock divider* variable until the count reaches zero. At this point, the next state is loaded onto the output ports and the state counter is initialized with the new state count. When the process reaches the last state in the bit pattern, the system checks for a *switch* or *stop* signal. The *switch* signal tells the module to switch buffers and continue with the new bit pattern. Additionally, the host is alerted of the successful switch and prompted to load a new bit pattern into the *preload* buffer. This dual-layer buffering allows the system to cycle, glitch free, through any number of bit patterns without delay. The *stop* signal will halt the module's execution and go into an idle state. If neither signal is present, the module will reload the first state in the current buffer and repeat the bit pattern. Since this module monitors the *switch* signal, it controls the buffering system, alternating the designation of *preload* and *active* buffers. The *OutputControl* module is responsible for sending bit patterns to the output ports, incrementing through the states via the state counter, and selecting the active and preload buffers, which change when signaled from the PC using a command. Additionally, the module synchronizes the states to the user selected source and implements a clock divider module for seamless switching of the state machine's operating rate.

ControlSignals module

The *ControlSignals* module relies on a 16-bit word that is responsible for controlling the FPGA's changeable states. The word uses 14 bits, with each bit controlling states on the port's input and output modules. The 2 remaining bits provide custom functionality specific to the 430 MHz transmitter at AO. This behavior is easily customizable to match needs for any system. Among the various controls are: 1) system start, 2) system stop, 3) switch patterns, and 4) sync source selection.

StatusSignals module

Similar to the *ControlSignals* module, the *StatusSignals* module monitors the status of both *InputControl* and *OutputControl* modules. This 16-bit word is read by the calling program as often as necessary. Currently, the shell program creates a separate thread and reads the status word at 25 ms intervals. The module's primary function is to provide feedback to the

PC and synchronize data transfers; preventing any aliasing of data buffers and providing verification of the control commands sent the FPGA.

Results

System performance can be characterized in terms of resource usage, clock speed, and GPC-to-FPGA communication rates. At the start of the design, specifications were given requiring the FPGA module to accept 3 possible clock sources: an on-board oscillator and 2 external clock sources. The on-board oscillator was provided by the FPGA module and is used for isolated system debugging. The 2 external clock sources both required a maximum reference frequency of 20 MHz, which specified a minimum pulse resolution of 100 ns (i.e. one half the clock rate). The memory requirements were taken from the existing equipment and specified by 4kx16 blocks for each port. Loading new bit patterns into the machine required real-time scheduling and had to be scheduled precisely with the site's clock reference (i.e. atomic clock). After the initial design entry and functional verification, synthesis and various optimizations were performed using the provided Xilinx Webpack tools. All of the system's timing requirements were met and clock inputs up to $\approx$ 68 MHz were achieved; well above the specified 20 MHz clock rate. According to generated reports, approximately 25% of the FPGA's resources were utilized. All available block RAM was allocated to pattern storage and communication rates were limited by the 25 ms refresh rate imposed on the status and control lines. The FPGA employed a dual-stage buffer to provide real-time mode switching. Precision of the pattern switching was accomplished through

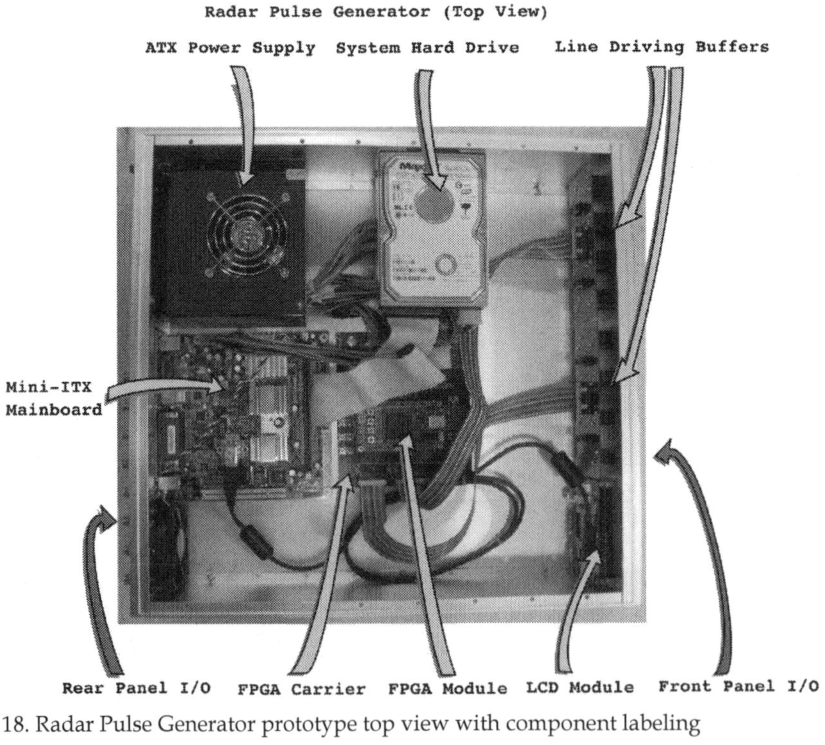

Fig. 18. Radar Pulse Generator prototype top view with component labeling

synchronization with a 1 PPS timing signal derived from the on-site atomic clock. This enabled the user to request a new pattern from the non-real-time *run-time shell* program, and still maintain real-time scheduling requirements. Machining of the prototype took place in-house at AO and the FPGA carrier and output stage hardware boards were sent to a manufacturing facility. The device is contained in a 3U size rack-mount chassis and three different views of the instrument are illustrated in Figures 18, 19 and 20.

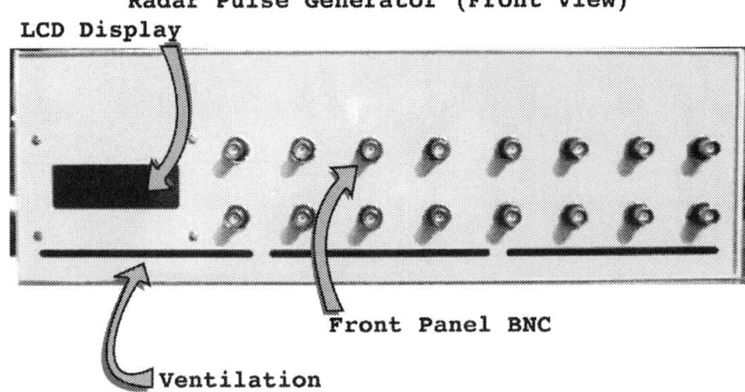

Fig. 19. Radar Pulse Generator prototype front view with component labeling

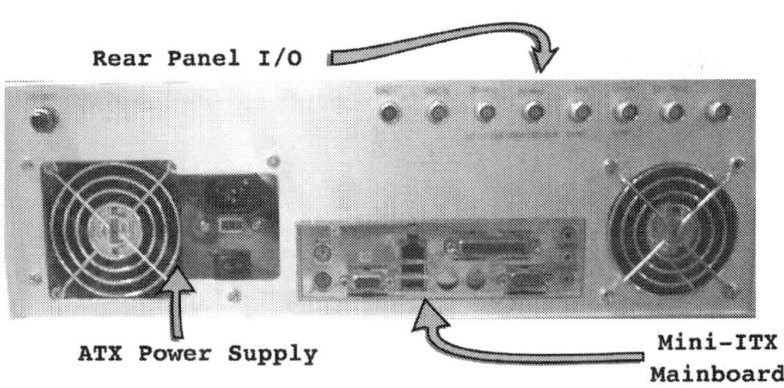

Fig. 20. Radar Pulse Generator prototype rear view with component labeling

5. Conclusion

In this chapter, we have presented principles and techniques enabling the design and implementation of high-speed, low-cost instrumentation employing COTS components and reconfigurable hardware. Several open-source tools were introduced to provide viable alternatives to expensive, proprietary software requiring maintenance fees and limiting the ability to promote code sharing due to licensing restrictions. Emphasis was also given to object-oriented programming techniques, which help reduce design complexity and promote software reuse through interface-oriented techniques. Finally, a radar-based application was presented, demonstrating a number of concepts developed throughout the chapter.

6. References

Baican, R. & Necsulescu, D. (2000). *Applied Virtual Instrumentation*,WIT Press, UK.

Bishof, C., Bucker,M., Gibbon, P., Joubert, G., Lippert, T.,Mohr, B. & Peters, F. (2008). *Advances in Parallel Computing*, IOS Press.

Booch, G. (1994). Object-Oriented Analysis and Design, The Benjamin/Cummings Publishing Company, Inc. Hsiung, P., Santambrogio, D. & Huang, C. (2009). Reconfigurable System Design and Verification, CRC Press, USA.

Hunt, A. & Thomas, D. (1999). *The Pragmatic Programmer*, Addison-Wesley Professional, USA.

Maxfield, C. (2004). *The Design Warrior's Guide to FPGAs*, Newnes, USA.

Navabi, Z. (2006). *Verilog Digital System Design*, McGraw-Hill Publishing Companies, Inc., USA. Pugh, K. (2007). *Interface-Oriented Design*, The Pragmatic Programmers LLC., USA.

Seal, R. (2008). Design and implementation of a multi-purpose radar controller using opensource tool, *Proceedings of the IEEE Radar Conference 2008*, Rome, May 2008, pp. pp. 1–4.

Stroustrup, B. (2000). *The C++ Programming Language*, Addison-Wesley, USA.

19

Superconducting Receiver Front-End and Its Application In Meteorological Radar

Yusheng He and Chunguang Li
Institute of Physics, Chinese Academy of Sciences, Beijing 1000190,
China

1. Introduction

The discovery of superconductivity is one of the greatest achievements in fundamental sciences of 20th century. Nearly one hundred year ago, the Dutch physicist H. Kamerlingh Onnes found that the resistance of pure mercury abruptly dropped to zero at liquid helium temperature 4.2K (Kammerlingh Onnes, 1911). Since then scientists tried very hard to find more superconducting materials and to understand this microscopic quantum phenomenon, which has greatly pushed forward the development of modern physics. Enormous progress has been made since 1986, when the so called high temperature superconductors (HTS) were discovered (Bednorz & Mueller, 1986). Superconductivity becomes not only a branch of fundamental sciences but also a practical technology, ready for applications.

It is commonly recognized that the first application of HTS thin films is in high performance microwave passive devices, especially filters. As a successful example the application of HTS filter in meteorological radar will be introduced. In this chapter firstly the fundamental concepts of superconductivity will be summarized and then the basic principles of superconducting technology at microwave frequency will be briefly outlined. Then the design and construction of superconducting filters and microwave receiver front-end subsystems will be discussed. Finally results of field trial of the world first superconducting meteorological radar will be introduced and conclusions will then be drawn.

2. Superconductivity and superconducting technology at microwave frequency

2.1 Basic concepts of superconductivity

Superconductivity is a microscopic quantum mechanical phenomenon occurring in certain materials generally at very low temperatures, characterized by exactly zero DC electrical resistance and the exclusion of the interior magnetic field, the so called Meissner effect (Meissner & Ochsenfeld, 1933). Superconductivity is a thermodynamic phase of condensed matters, and thus possesses certain distinguishing properties which are largely independent of microscopic details. For example, the DC resistance of a superconductor drops abruptly to zero when the material is cooled below its "critical temperature T_c ". Conventional superconductors are in a wide variety, including simple elements like Al, Pb and Nb, various metallic alloys (e.g., Nb-Ti alloy) and compounds (e.g., Nb_3Sn) as well as some strong correlated systems (Heavy Fermions). As of 2001, the highest critical temperature found for a conventional superconductor is 39 K for MgB$_2$, although this material displays enough exotic properties that there is doubt about classifying it as a "conventional" superconductor.

A revolution in the field of superconductivity occurred in 1986 (Bednorz & Mueller, 1986) and one year later a family of cuprate-perovskite ceramic materials was discovered with superconducting transition temperature greater than the boiling temperature of liquid nitrogen, 77.8 K, which are known as high temperature superconductors (hereinafter referred to as HTS). Examples of HTS include YBa$_2$Cu$_3$O$_7$, one of the first discovered and most popular cuprate superconductors, with transition temperature T_c at about 90 K and Tl$_2$Ba$_2$CaCu$_2$O$_8$, HgBa$_2$Ca$_2$Cu$_3$O$_8$, of which the mercury-based cuprates have been found with T_c in excess of 130 K. These materials are extremely attractive because not only they represent a new phenomenon not explained by the current theory but also their superconducting states persist up to more manageable temperatures, beyond the economically-important boiling point of liquid nitrogen, more commercial applications are feasible, especially if materials with even higher critical temperatures could be discovered.

Superconductivity has been well understood by modern theories of physics. In a normal conductor, an electrical current may be visualized as a fluid of electrons moving across a heavy ionic lattice. The electrons are constantly colliding with the ions in the lattice, and during each collision some of the energy carried by the current is absorbed by the lattice and converted into heat, which is essentially the vibrational kinetic energy of the lattice ions. As a result, the energy carried by the current is constantly being dissipated. This is the phenomenon of electrical resistance. The situation is different in a superconductor, where part of the electrons is bounded into pairs, known as Cooper pairs. For conventional superconductors this pairing is caused by an attractive force between electrons from the exchange of phonons whereas for HTS, the pairing mechanism still being an open question. Due to quantum mechanics, the energy spectrum of this Cooper pair fluid possesses an energy gap, meaning there is a minimum amount of energy ΔE that must be supplied in order to break the pair into single electrons. In the superconducting state, all the cooper pairs act as a group with a single momentum and do not collide with the lattice. This is known as the macroscopic quantum state and explains the zero resistance of superconductivity. More over, when a superconductor is placed in a weak external magnetic field H, the field penetrates the superconductor only a small distance λ, called the London penetration depth, decaying exponentially to zero within the bulk of the material. This is called the Meissner effect, and is a defining characteristic of superconductivity. For most superconductors, the London penetration depth λ, is on the order of 100 nm.

2.2 Superconductivity at microwave frequency

As mentioned above, in a superconductor at finite temperatures there will be normal electrons as well as Cooper pairs, which has been successfully depicted by an empirical two-fluid model, proposed by Gorter and Casimir (Gorter & Casimir, 1934). In this model it simply assumes that the charge carriers can be considered as a mixture of normal and superconducting electrons, which can be expressed as $n=n_n+n_s$, where n, n_n, n_s are the numbers of total, normal and superconducting electrons, respectively, and $n_s=fn$, $n_n=(1-f)n$. The superconducting fraction f increases from zero at T_c to unity at zero temperature. At microwave frequencies, the effect of an alterative electromagnetic field is to accelerate both parts of carriers or fluids. The normal component of the current will dissipate the gained energy by making collisions with lattice. In the local limit the conductance of a superconductor can be represented by a complex quantity

$$\sigma = \sigma_1 - j\sigma_2 \tag{1}$$

where the real part σ_1 is associated with the lossy response of the normal electrons and the imaginary part is associated with responses of both normal and superconducting electrons. In his book (Lancaster, 1997) M. J. Lancaster proposed a simple equivalent circuit, as shown in Fig. 1, to describe the physical picture of the complex conductivity. The total current J in the superconductor is split between the reactive inductance and the dissipative resistance. As the frequency decreases, the reactance becomes lower and more and more of the current flows through the inductance. When the current is constant this inductance completely shorts the resistance, allowing resistance –free current flow.

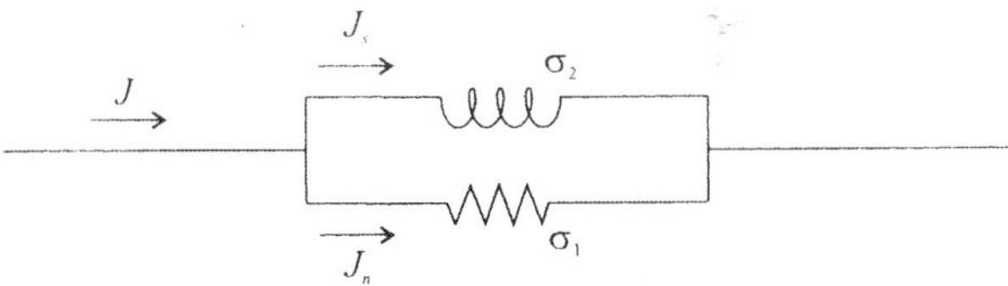

Fig. 1. Equivalent circuit depicting complex conductivity (Lancaster, 1997)

The microwave properties are determined by currents confined to the skin-depth, in the normal metal state and to the penetration depth, λ, in the superconducting state. The surface impedance of a superconductor should be considered, which is defined as

$$Zs = \left(\frac{Ex}{Hy}\right)_{z=0} = Rs + jXs \tag{2}$$

where Ex and Hy are the tangential components of the electric and magnetic fields at the surface (z=0). For a unit surface current density, the surface resistance Rs is a measure of the rate of energy dissipation in the surface and the surface reactance Xs is a measure of the peak inductive energy stored in the surface. Relating to the complex conductivity, the surface impedance can be then written as

$$Zs = \left(\frac{j\omega\mu_0}{\sigma_1 - j\sigma_2}\right)^{\frac{1}{2}} \tag{3}$$

In the case of the temperature is not too close to T_c, where more superconducting carriers are present, and σ_1 is finite but much smaller than σ_2, approximation of $\sigma_1 << \sigma_2$ can be made. It then turns out that

$$Zs = \frac{1}{2}\omega^2\mu_0^2\lambda^3\sigma_1 + j\omega\mu_0\lambda \tag{4}$$

with the frequency independent magnetic penetration depth $\lambda = \left(m / \mu_0 n_s e^2\right)^{1/2}$, where m is the mass of an electron and e is the charge on an electron.

From Equation (2), important relations for surface resistance and reactance can be easily derived,

$$Rs = \frac{1}{2}\omega^2 \mu_0^2 \lambda^3 \sigma_1 \qquad (5)$$

and

$$Xs = \omega\mu_0\lambda \qquad (6)$$

It is well known that for a normal metal, the surface resistance increases with $\omega^{1/2}$, whereas for a superconductor, that increases with ω^2 (Equation 5). Knowledge of the frequency dependence of the surface resistance is especially important to evaluate potential microwave applications of the superconductor. It can be seen from Fig. 2 that $YBa_2Cu_3O_7$ displays a lower surface resistance compared to Cu up to about 100 GHz.

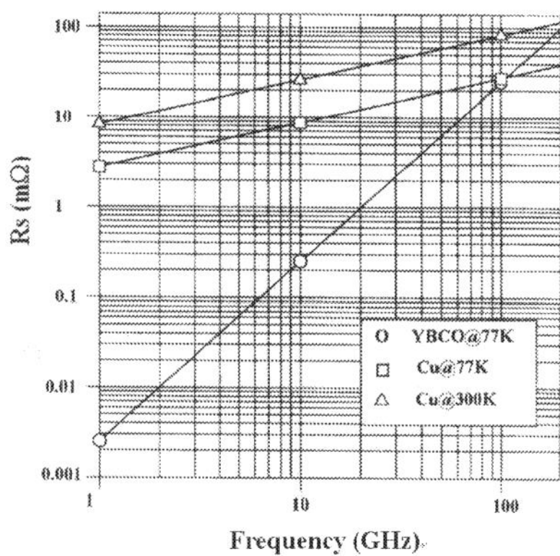

Fig. 2. Frequency dependence of surface resistance of $YBa_2Cu_3O_7$ compared with Cu .

2.3 Superconducting thin films for microwave device applications

Superconducting thin films are the main constituent of many of the device applications. It is important for these applications that a good understanding of these films is obtained. Superconducting thin films have to be grown on some kind of substrates which must meet certain requirements of growth compatibility, dielectric property and mechanical peculiarity in order to obtain high quality superconducting films with appropriate properties for microwave applications. First of all the substrates should be compatible with film growth. In order to achieve good epitaxial growth the dimensions of the crystalline lattice at the surface of the substrate should match the dimensions of the lattice of the superconductor, otherwise strains can be set up in the films, resulting in dislocations and defects. Chemical compatibility between superconductors and the substrates are also important since impurity levels may raise if the substrate reacts chemically with the superconductor or atomic migrations take place out of the substrate. Appropriate match between the substrate and the film in thermal expansion is also vital so as to avoid cracks during repeated cooling and warming processes. For dielectric properties, the most important requirement is low enough dielectric tangent loss, without which the advantage of using superconducting film will be counteracted. The high value of dielectric constant is also good for the miniature of the microwave devices. It should be convenient and easier for the design if the dielectric constant is isotropic in the plane of the film. From the machining and processing point of view, the surface of the substrate should be smooth and free from defects and twinning if possible. It should also be mechanically strong, environmentally stable, and capable of being thinned to a certain extent. As a matter of fact, no existing substrate can meet all of the above requirements. Some of the problems can be overcome by certain technologies, e.g., introduction of buffer layer between the films and the substrates. There are indeed a number of good substrates being used and excellent superconducting films with satisfactory microwave properties were produced on them. Table 1 shows the common used substrates for HTS thin films in microwave device applications.

Substrate	ε_r (typical)	$tg\delta$ (typical)
LaAlO$_3$	24.2@77K	7.6×10^{-6}@77K and 10GHz
MgO	9.6@77K	5.5×10^{-6}@77K and 10GHz
Sapphire	11.6 $\parallel$ c-axis @77K 9.4 $\perp$ c-axis @77K	1.5×10^{-8}@77K and 10GHz

Table 1. Substrates of HTS thin films for microwave device applications

The properties of superconducting thin films and their suitability for microwave devices are critically dependent on processing. So far almost all HTS thin film microwave devices have been fabricated from two materials, i.e., YBa$_2$Cu$_3$O$_7$ (T_c~85-93 K in thin film form) of which element Y may be substitute by some other rare-earth elements, e. g., Dy and Tl$_2$Ba$_2$CaCu$_2$O$_8$ (T_c ~100-110 K in thin film form). There is a variety of methods for preparing superconducting thin films, including pulsed laser ablation, magnetron sputtering, and evaporation etc. In principle, a film in some kind precursor phase is grown first and then followed by in-situ or post-deposition anneal treatments. For example, oxygen deficient YBa$_2$Cu$_3$O$_6$ phase can be grown at about 850°C, which is later converted into the YBa$_2$Cu$_3$O$_7$ phase by a low temperature (~500°C) anneal in oxygen atmosphere. Using this process YBa$_2$Cu$_3$O$_7$ (or more accurately YBa$_2$Cu$_3$O$_{7-\delta}$, with δ<1) films can be made with surface resistance at the order of a few hundred micron ohms.

Handling of HTS films must be with great care. Once formed the properties of the films can degrade under some environmental conditions. Aqueous environments are very deleterious. Therefore the superconducting films should be always kept dry and away from water vapour. Care must be also exercised during all fabrication processes. A protection layer on the top of the superconducting patterns of the devices will be helpful. It should be advised, however, to keep the devices always away from humidity.

2.4 Superconducting transmission lines

In microwave circuits, a transmission line is the material medium or structure that forms all or part of a path from one place to another for directing the transmission of electromagnetic waves. It not only serves as interconnections between components in the circuits, but also often forms the basic element of components and devices. It is indeed the foundation of the microwave circuits and most microwave theories are originated directly or indirectly from transmission theory, which is also true for superconducting devices.

The most important effect of superconducting transmission lines is their very low loss. A superconducting transmission line is dispersionless, provided the wave propagated is in a TEM mode. This is due to that the penetration depth is not varying with frequency, contrasting with normal conductors where the skin depth is a function of frequency.

The realistic transmission line model represents the transmission line as an infinite series of two-port elementary components, each representing an infinitesimally short segment of the transmission line with distributed resistance R, inductance L, and capacitance C. Commonly used transmission lines include wires, coaxial cables, dielectric slabs, optical fibers, electric power lines, waveguides, and planar transmission lines etc. Considering most superconducting filters are based on superconducting films, in this book we will concentrate on the planar transmission lines, i.e., microstrip, coplanar, and stripline.

Microstrip is a widely used type of microwave transmission line which consists of a conducting strip with a width w and a thickness t separated from a ground plane by a dielectric layer known as the substrate with a dielectric constant ε and a thickness d. The general structure of a microstrip is shown schematically in Fig. 3 (a). The fields in the microstrip extend within two media: air above and dielectric below and in general, the dielectric constant of the substrate will be greater than that of the air, so that the wave is traveling in an inhomogeneous medium. In consequence, the propagation velocity is somewhere between the speed of microwaves in the substrate and the speed of microwaves in air. This behavior is commonly described by stating the effective dielectric constant (or effective relative permittivity) of the microstrip; this being the dielectric constant of an equivalent homogeneous medium. Due to this inhomogeneous nature, the microstrip line will not support a pure TEM wave; both the E and H fields will have longitudinal components. The longitudinal components are small however, and so the dominant mode is referred to as quasi-TEM. The field propagation velocities will depend not only on the material properties but also on the physical dimensions of the microstrip. Comparing to other transmission lines, e.g., the traditional waveguide technology, the advantage of microstrip is much less expensive as well as being far lighter and more compact. Whereas the disadvantages of microstrip compared with waveguide are the generally lower power handling capacity, and higher losses, of which, the later can be greatly improved by employing superconducting materials.

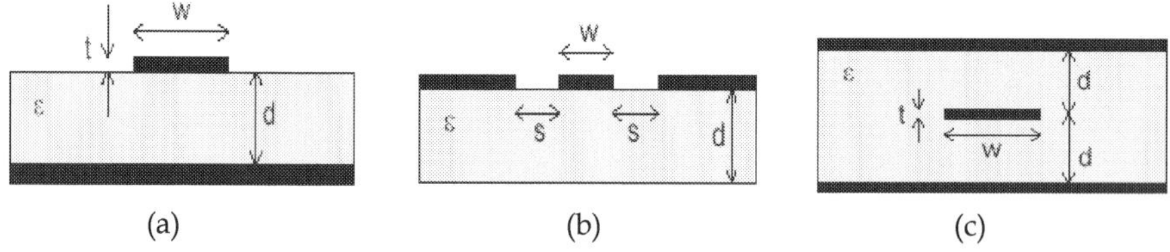

(a)   (b)   (c)

Fig. 3. Cross-section geometries of microstrip (a), coplanar line (b), and stripline (c).

A coplanar line is formed from a conductor separated from a pair of ground planes, all on the same plane, at the top of a dielectric medium. In the ideal case, the thickness of the dielectric is infinite; in practice, it should be thick enough so that EM fields die out before they get out of the substrate. The advantages of coplanar line are that active devices can be mounted on top of the circuit, like on microstrip. More importantly, it can provide extremely high frequency response (100 GHz or more) since connecting to coplanar line does not entail any parasitic discontinuities in the ground plane. For superconducting

coplanar line it means only single side superconducting film is required, which will greatly reduce the difficulties in film making since double-sided films need special technology, especially for those newly discovered superconductors. One of the disadvantages of the coplanar line is the imbalance in the ground planes, although they are connected together at some points in the circuit. The potential of the ground planes can become different on each side, which may produce unwanted mode and interference. The problem can be solved in some extent by connecting the two ground planes some where in the circuit with "flying wire" (air bridge) over the conducting plane. It, however, will bring complexity and other problems.
A standard stripline uses a flat strip of metal which is sandwiched between two parallel ground planes. The width of the strip, the thickness of the substrate (the insulating material between the two ground plans) and the relative permittivity of the substrate determine the characteristic impedance of the transmission line. The advantage of the stripline is that the radiation losses are eliminated and a pure TEM mode can be propagated in it. The disadvantage of stripline is the difficulty in construction, especially by superconducting films. It requires two pieces of superconducting films of both double-sided or one single-plus one double-sided film. The air gap, though extremely small, will cause perturbations in impedance and to form contacts to external circuits is also very difficult. Besides, if two pieces of double-sided films employed, the precise aiming between the very fine patterns of two films brings another problem.

2.5 Application of transmission lines: Superconducting resonators

A microwave resonator is a simple structure which can be made of cavity or transmission line and is able to contain at least one oscillating electromagnetic field. If an oscillating field is set up within a resonator it will gradually decay because of losses. By using superconductors the losses can be greatly reduced and high Q-values can be than achieved. Resonators are the most fundamental building blocks in the majority of microwave circuits. By coupling a number of resonators together, a microwave filter can be constructed. A high Q-resonator forms the main feed back element in a microwave oscillator.
The quality factor for a resonator is defined by

$$Q_0 = 2\pi \frac{w}{p} \qquad (7)$$

where w is the energy stored in the resonator, and p is the dissipated energy per period in the resonator.
The losses in a resonator arise due to a number of mechanisms. The most important are usually the losses associated with the conduction currents in the conducting layer of the cavity or transmission line, the finite loss tangent of dielectric substrate and radiation loss from the open aperture. The total quality factor can be found by adding these losses together, resulting in

$$\frac{1}{Q_0} = \frac{1}{Q_c} + \frac{1}{Q_d} + \frac{1}{Q_r} \qquad (8)$$

where Q_c, Q_d, and Q_r are the conductor, dielectric and radiation quality factors, respectively. For conventional cavity or transmission line resonators at microwave frequency, usually Q_c dominates. This is why superconductor can greatly increase the Q-value of the resonators.

For superconducting resonators, the resonant frequency and Q value are temperature dependent and the variation is remarkable when close to T_c, as shown in Fig.4 (Li, H. et al., 2002). It is therefore preferable to operator at a temperature below 80% of T_c, where more stable resonant frequency and high enough Q value being achieved.

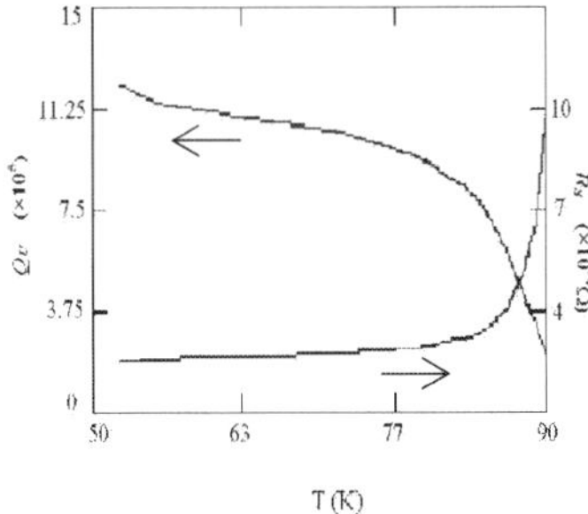

Fig. 4. Temperature dependence of resonate frequency and Q value of superconducting resonator (Li, H. et al., 2002)

Beside the Q-value, size is also important to HTS resonator not only because the commercial consideration for reducing the cost but also because that the HTS filters need to work at low temperature and large size will increase burden of the cryo-cooler. However, the high quality factor and the minimized size are often a pair of contradiction, because the reduction of the size always causes the reduction of the stored energy and hence the Q-vales. The quality factors and volumes of various kinds of HTS resonators are shown in Fig.5 in dual-log coordinate at frequency around 5 GHz. All the data come from literatures published over the last decades. Here the volume of the planar resonators is calculated by assuming it has a height of 8 mm.

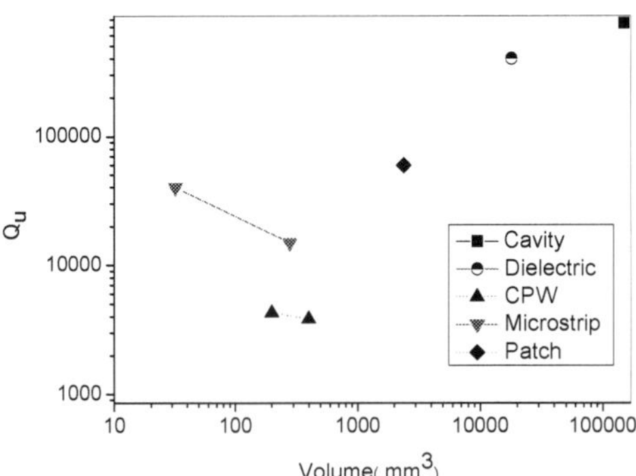

Fig. 5. The quality factor *vs.* volume of different kind of HTS resonators

HTS Cavity resonators are usually designed to resonate in the TE_{011} mode. In this mode, the current flow is circumferential, so there is no current flow across the joint of the side wall and the end plates. Although such structures have very high Q-values, their volume becomes prohibitively large at low frequencies. HTS dielectric resonators usually consist of a dielectric rod of single crystal material sandwiched between two planar HTS films. They can have very high Q-values (if very large area high quality superconducting films applied), but are smaller than unloaded resonant cavities. Due to their large volume and difficulty in realizing cross couplings, HTS Cavity resonators and HTS dielectric resonators are often used to construct low phase-noise oscillators and, for the dielectric ones, to measure the surface resistance of HTS thin films. They are seldom used in filters. Patch resonators, which are two dimensional resonators, are of interest for the design of high power handling HTS filters. An associated advantage of them is their lower conductor losses as compared with one dimensional line resonators. However, patch resonators tend to have stronger radiation, so they are not suitable for designing narrow band filters. The frequently used one dimensional HTS transmission line resonators in applications are Coplanar Waveguide (CPW) resonators and Microstrip resonators. Because the circuit structure and the ground planes are fabricated on a single plane, CPW resonators offer the possibility of integration with active components. But the electromagnetic energy in the CPW resonators are more confined to a small area between the HTS line and the grounds so they tend to have lower quality factors compared to other type of HTS resonators. Microstrip resonators are the most popular type of resonators used in HTS microwave filters. Besides their relatively high quality factor and small size, the ease with which they can be used in a variety of coupling structures is a big advantage. For most applications the basic shape of a HTS microstrip resonator is a half-wavelength (Fig. 6 (a)) line instead of the grounded quarter-wavelength $(\lambda/4)$ line, due to the difficulty in realizing ground via HTS substrate. The conventional $\lambda/2$ resonators can be miniaturized by folding its straight rectangular ribbon. The height of a folded resonator, called a hairpin resonator (Fig. 6 (b)), is usually less than $\lambda/4$. Meander-line resonators (Fig. 6 (c)), meander-loop resonators (Fig. 6 (d)), spiral resonators (Fig. 6 (e)) and spiral-in-spiral-out resonators (Fig. 6 (f)) can be considered as modifications of conventional straight ribbon $(\lambda/2)$ resonators by folding the ribbons several times and in different ways. In principle, the $\lambda/2$ resonators can be folded or transformed to any other shapes according to the demands of miniaturization or other purpose.

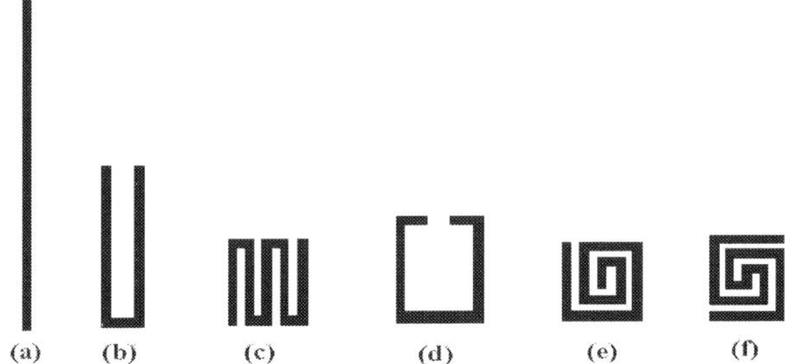

Fig. 6. Typical one dimensional microstrip resonators: (a) half-wave length line resonator; (b) hairpin resonator; and (c) to (f) other open loop resonators.

2.6 A brief survey of superconducting microwave passive devices

Passive microwave device is an ideal vehicle for demonstrating the advantage of HTS and in deed the first commercial application of the HTS thin film was in passive microwave device. The main advantage gained from HTS is the low surface resistance, which is tens to hundreds times lower than those of copper under the same condition at frequencies from 100 MHz up to, say close to 40 GHz. Such a low resistance can be directly converted into small insertion loss or high Q-values for the microwave devices. Another unique property of superconductors is the dispersionless due to the penetration depth not varying with frequency, which is extremely useful in many devices, e.g., wide band delay lines. Further more the substrates with high dielectric constants, used for superconducting films, make device miniaturization possible. In this brief survey, only some of the passive devices, i.e., superconducting delay line, antenna, and filter will be discussed.

A delay line consists of a long transmission line, which is used to delay the incoming signals for a given amount of time while minimizing the distortion caused by crosstalk, dispersion and loss. The conventional delay lines, e.g., coaxial cables, printed circuit delay lines etc., have the problem of large volume and unacceptable insertion loss. HTS delay lines with extremely small losses and compact sizes provide a promising solution for solving these problems. The advantages of using HTS coplanar lines to make delay lines are distinctive. Firstly, the ground planes in between the signal line provide a shield to reduce the crosstalk. Secondly, the line width of the signal line is quite flexible since the impedance of a coplanar line is mainly determined by the ratio between the signal line width and the gap width. Whereas the disadvantage is that the unwanted odd mode can be excited by turnings, curvatures, or any asymmetry with respect to the signal line, which can be resolved by bonding "flying wires" (air bridges) between the ground planes over the top of the signal line. Stripline is the only planar transmission line without modal dispersion. It has longer delay per unit length since the signal propagation velocity is slower than that in the other two lines. It also has less cross talk than the microstrip line for the existence of the top plane. An excellent example of the application of stripline can be found from the work reported by Lyons et al. (Lyons et al., 1996), where HTS delay line was used for chirp-response tapped filter and installed in a 3 GHz band width spectral analysis receiver.

Antennas can be benefit in a number of ways when superconductors are used. The obvious application is in the improvement of the radiation efficiency of small antennas. For antennas with the size comparable with wavelength their efficiency are fairly high. However, for those with dimensions being small compared with wavelength, which is defined as electrically small antenna, the efficiency reduces due to the increasing dominance of the ohmic losses. By using extremely low loss superconductors, reasonable efficiency can be achieved. Superconductors are also useful in feeding and matching networks for super-directive antennas, which are very inefficient. High Q value matching networks provided by superconducting circuit helps considerably in performance improvement. Another application of superconductors to antenna is in the feed networks of millimeter-wave antennas. The losses associated with long narrow microsrip feed lines ca be improved considerably if superconductors are used. This is especially true for arrays with a large number of elements.

The most attractive applications of HTS in passive devices are those of filers. Due to the very small surface resistance of HTS films, filters can be constructed with remarkably high performance, e.g., negligible insertion loss, very large out-of-band rejection, and extremely steep skirt slope. It can reduce the band width and makes ultra-narrow band filter possible. With the above excellent performance HTS filters can eliminate unwanted interference while keeping the system with minimized noise figure. More over, by using HTS films the filter can be miniaturized due to not only special substrates being employed but also new geometry designs being invented. In the past twenty years, various kinds of HTS filters have been constructed and successful applications in many fields have been realized, including those in direct data distribution between earth to satellite and satellite to satellite (Romanofsky el al., 2000), detection of deep space of radio astronomy (Wallage et al., 1997; Li, Y. et al., 2003), base stations of mobile communications (STI Inc., 1996; Hong et al., 1999) and meteorological radar for weather forecasting (Zhang et al., 2007).

3. Superconducting filter and receiver front-end subsystem

3.1 Principles and theories on filter design

Electrical filters have the property of frequency-selective transmission, which enables them to transmit energy in one or more passbands and to attenuate energy in one or more stopbands. Filters are essential elements in many areas of RF/microwave engineering.

For linear, time-invariant two port filter networks, the transmission and reflection function may be defined as rational functions,

$$S_{21}(p) = \frac{N(p)}{\varepsilon D(p)}; \quad S_{11}(p) = \frac{E(p)}{D(p)} \tag{9}$$

where ε is a ripple constant, $N(p)$, $E(P)$ and $D(p)$ are characteristic polynomials in a complex frequency variable $p=\sigma+j\Omega$, Ω is the normalized frequency variable. For a lossless passive network, $\sigma=0$ and $p=j\Omega$. The horizontal axis of the p plane is called the real axis, while the vertical axis is called the imaginary axis. The values of p at which $S_{21}(p)$ becomes zero are the transmission zeros of the filter, the values at which $S_{11}(p)$ becomes zero are the reflection zeros of the filter, and the values at which $S_{21}(p)$ becomes infinite are the poles of the filter.

Filters may be classified into categories in several ways. The main categories are defined in terms of the general response type of low-pass, bandpass, high-pass, and bandstop. Filters can also be classified by the transmission zeros and poles. Butterworth filter is called maximum flat response filter because its amplitude-squared transfer function has maximum number of zero derivatives at $\Omega=0$. Chebyshev filter exhibits equal-ripple passband response and maximally flat stopband response. All the transmission zeros of $S_{21}(p)$ of Chebyshev and Butterworth filters are located at infinite frequencies. When one or more transmission zeros are introduced into the stopband of Chebyshev filter at finite frequencies the filter is known as a generalized Chebyshev filter or as a quasi-elliptic filter. The special case where the maximum number of transmission zeros are located at finite frequencies such that the stopbands have equal rejection level is the well-known elliptic function filter. This is now rarely used since it has problems in practical realization. Gaussion filter, unlike those mentioned above, has poor selectivity but a quite flat group delay in the passband.

Chebyshev filters have, for many years, found frequent application within microwave space and terrestrial communication systems. The generic features of equal ripple passband characteristics, together with the sharp cutoffs at the edges of the passband and hence high selectivity, give an acceptable compromise between lowest signal degradation and highest noise/interference rejection.

As the frequency spectrum becomes more crowded, specifications for channel filters have tended to become very much more severe. Very high close-to-band rejections are required to prevent interference to or from closely neighboring channels; at the same time, the incompatible requirements of in-band group-delay and amplitude flatness and symmetry are demanded to minimize signal degradation. With generalized Chebyshev filter it is easy to build in prescribed transmission zeros for improving the close-to-band rejection slopes and/or linearizing the in-band group delay. Therefore generalized Chebyshev filters have been extensively investigated by many researchers over the last few decades. Coupled resonator circuits are of importance for design of generalized Chebyshev filters. This design method is based on coupling coefficients of inter-coupled resonators and the external quality factors of the input and output resonators. One of the fundamental methods of synthesis of generalized Chebyshev filters, based on cross-coupled resonators and proposed by Atia & Williams (Atia & Williams, 1972; Atia et al., 1974), is still commonly used. Alternative synthesis techniques were also advanced by many scientists (Cameron & Rhodes, 1981; Chambers & Rhodes, 1983; Levy, 1995; and Cameron, 1999).

The first step of design a generalized Chebyshev filter is to determine all the positions of the transmission zeros and reflection zeros of the filter that satisfy the specifications.

The amplitude-squared transfer function for a lossless low-pass prototype is defined as

$$S_{21}^2(j\Omega) = \frac{1}{1 + \varepsilon^2 C_N^2(\Omega)} \tag{10}$$

where ε is related to the passband return loss (RL) by $\varepsilon = \left[10^{RL/10} - 1\right]^{-\frac{1}{2}}$, and $C_N(\Omega)$ represents a filtering or characteristic function with N being order of the filter.

For a generalized Chebyshev filter $C_N(\Omega)$ may be expressed as

$$C_N(\Omega) = \cosh\left[\sum_{n=1}^{N} \cosh^{-1}(x_n)\right] \tag{11}$$

where $x_n = \Omega - \dfrac{1}{\Omega}\bigg/ 1 - \dfrac{\Omega}{\Omega_n}$, and Ω_n is the position of the nth prescribed transmission zero.

It can be proved (Cameron & Rhodes, 1981) that $C_N(\Omega)$ can be expressed as a ration of $N(p)$, $E(P)$. Given all the transmission zeros and RL various methods exist for evaluating the characteristic polynomials $N(p)$, $E(P)$ and $D(p)$ (Cameron, 1999; Amari, 2000; and Macchiarella, 2002). Then the reflection zreos and poles of the filter may be found by rooting $E_N(p)$ and $D(p)$, respectively. Then the complete transmission response L_A, reflection response L_R and phase response Φ_{21}, including the group delay τ_d, can be computed by

$$\begin{cases} L_A(\Omega) = 10\log\dfrac{1}{\left|S_{21}(j\Omega)\right|^2} \quad dB \\[2mm] L_R(\Omega) = 10\log\left[1 - \left|S_{21}(j\Omega)\right|^2\right] \quad dB \\[2mm] \varphi_{21}(\Omega) = ArgS_{21}(j\Omega) \quad radians \\[2mm] \tau_d(\Omega) = \dfrac{d\varphi_{21}(\Omega)}{-d\Omega} \quad sec\,onds \end{cases} \tag{12}$$

and so does the lowpass to bandpass frequency transformation $\Omega = \dfrac{1}{FBW}\cdot\left(\dfrac{\omega}{\omega_0} - \dfrac{\omega_0}{\omega}\right)$, where ω_0 is the center frequency of the bandpass filter, FBW is the fractional bandwidth. Knowing all the positions of the zeros and poles, the next step of filter synthesis is to choose an appropriate topology of the coupled resonators and to synthesis the coupling matrix.

Fig. 7 shows the schematic illustration of a filtering circuit of N coupled resonators, where each circle with a number represents a resonator which can be of any type despite its physical structure. $m_{i,j}$ with two subscript numbers is the normalized coupling coefficient between the two (i^{th} and j^{th}) resonators And q_{e1} and q_{e2} are the normalized external quality factors of the input and output resonators. Here we assume $\omega_0=1$ and $FBW=1$. For a practical filter, the coupling coefficient $M_{i,j}$ and external quality factors Q_{e1}, Q_{e2} can be computed by

$$\begin{cases} M_{i,j} = FBW \cdot m_{i,j} \\ Q_{ei} = q_{e1} \Big/ FBW \end{cases}$$

(13)

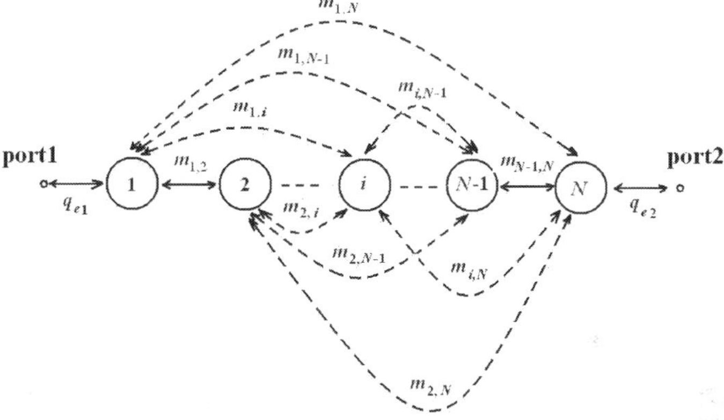

Fig. 7. The general coupling structure of a filtering circuit of N coupled resonators

The transfer and reflection functions of a filter with coupling topology in Fig. 7 are

$$\begin{cases} S_{21} = 2 \dfrac{1}{\sqrt{q_{e1} \cdot q_{e2}}} [A]^{-1}_{N1} \\ S_{11} = \pm \left(1 - \dfrac{2}{q_{e1}} [A]^{-1}_{11} \right) \\ [A] = [q] + p[U] - j[m] \end{cases}$$

(14)

where p is complex frequency variable as defined before, $[U]$ is the $N{\times}N$ identity matrix, $[q]$ is an $N{\times}N$ matrix with all entries zero except for $q_{11}=1/q_{e1}$ and $q_{NN}=1/q_{e2}$, and $[m]$ is an $N{\times}N$ reciprocal matrix being named as the general coupling matrix. The nonzero values may occur in the diagonal entries of $[m]$ for electrically asymmetric networks representing the offsets from center frequency of each resonance (asynchronously tuned). However, for a synchronously tuned filter all the diagonal entries of $[m]$ will be zero and $[m]$ will have the form

$$
\begin{bmatrix}
0 & m_{1,2} & \cdots & m_{1,i} & \cdots & m_{1,N-1} & m_{1,N} \\
m_{1,2} & 0 & \cdots & m_{2,i} & \cdots & m_{2,N-1} & m_{2,N} \\
\cdots & \cdots & \cdots & \cdots & \cdots & \cdots & \cdots \\
m_{1,i} & m_{2,i} & \cdots & 0 & \cdots & m_{i,N-1} & m_{i,N} \\
\cdots & \cdots & \cdots & \cdots & \cdots & \cdots & \cdots \\
m_{1,N-1} & m_{2,N-1} & \cdots & m_{i,N-1} & \cdots & 0 & m_{N-1,N} \\
m_{1,N} & m_{2,N} & \cdots & m_{i,N} & \cdots & m_{N-1,N-1} & 0
\end{bmatrix}
$$

The minor diagonal entries of $[m]$ all have nonzero values. They represent couplings between adjacent resonators, and in the remainder of the paper, they are called direct couplings. The other entries represent couplings between nonadjacent resonators. It is this type of couplings who create transmission zeros of filter response. Those entries may all have nonzero values in theory, which means that in the filter network that $[m]$ represents, couplings exist between every resonator and every other nonadjacent resonator. This is clearly impractical in the case of cavity type resonators. In the case of planar resonators such as HTS microstrip resonators, couplings may exist between every resonator and every other nonadjacent resonator. However, it is still impractical to handle all those couplings as expected. So it is important to choose a realizable coupling topology to design a practical filter, which means some particular nonadjacent couplings are chosen to exist and the others are not. Generally, the desired nonadjacent couplings are called cross-couplings while the others called parasitical couplings.

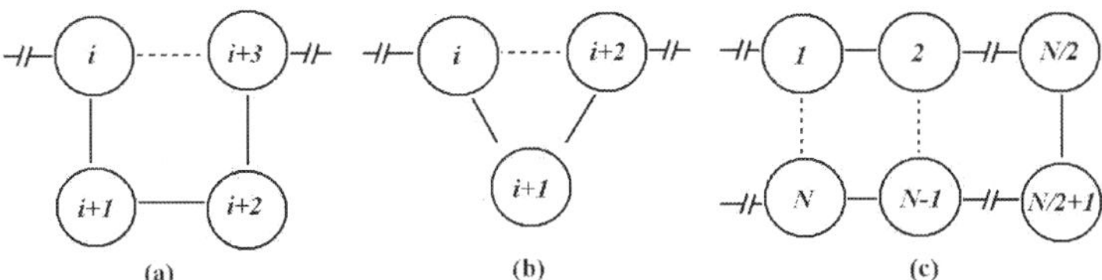

Fig. 8. Three frequently used coupling topology in generalized Chebyshev filters

There are three frequently used coupling topology in design of generalized Chebyshev filters as shown in Fig. 8, where each circle with a number represents a resonator, full lines indicate the direct couplings, and the broken lines denote cross-couplings. In the cascaded quadruplet (CQ) topology, four resonators construct a quadruplet section as illustrated in Fig. 8 (a), and then it can be cascaded to other resonators or quadruplets. Each quadruplet section will create a pair of transmission zeros which are symmetric with respect to the center frequency. Similar to the CQ filter, three resonators can construct a trisection structure as shown in Fig. 8 (b), and it can be then cascaded to other resonators or trisections to form a cascaded trisection (CT) filter. Each trisection structure, however, will only create one transmission zero. So the frequency response in a CT filter could be asymmetrical. The cross-couplings between different CQ or CT sections are independent to each other, which make the tuning of filter relatively easy. The coupling topology shown in Fig. 8 (c) is the so-called canonical fold form. The canonical fold coupling section always contains even numbers of resonators. And for $2n$ resonators, $2n$-2 transmission zeros can be realized. So for a given out-of-band rejection specification the canonical fold filter can meet it by fewer resonators than the CQ or CT structure. However, the effect of each cross-coupling in a canonical fold filter is not independent, witch makes the filter tuning more difficult.

Once an appropriate coupling topology is determined, the synthesis from characteristic polynomials to coupling matrix may follow the steps of Atia and Williams (Atia & Williams, 1972; Atia et al. 1974). They used the orthonormalization technique to obtain the general coupling matrix with all possible cross-couplings present and repeated similarity transformations are then used to cancel the unwanted couplings to obtain a suitable form of the prototype. Unfortunately, the process does not always converge. In that case optimization technique is a good alternative to derive the sequence of transformations allowing annihilation of the unwanted elements and provide the matrix with the required topology (Chambers & Rhodes 1983; Atia et al., 1998; and Levy, 1995). Generally, the desired couplings in the optimization are evaluated by minimizing a cost function involving the values S_{11} and S_{21} at specially selected frequency points such as those transmission and reflection zeros. The entries of the coupling matrix are used as independent variables in the optimization process.

The final step in the design of a filter is to find an appropriate resonator and convert the coupling matrix into practical physical structures. Readers can find detailed instructions on how to establish the relationship between the values of every required coupling coefficient and the physical structure of coupled resonators in the book by J. Hong & M. J. Lancaster (Hong & Lancaster, 2001).

It is essential in this step to employ helps of computer-aided design (CAD), particularly full-wave electromagnetic (EM) simulation software. EM simulation software solves the Maxwell equations with the boundary conditions imposed upon the RF/microwave structure to be modeled. They can accurately model a wide range of such structures.

3.2 Resonators for superconducting filter design

Resonators are key elements in a filter which decide the performance of the filter in a large part. The HTS resonators suitable for developing narrow-band generalized Chebyshev filters should have high-quality factors, minimized size, and conveniency to realize cross couplings. For a n-pole bandpass filter, the quality factors of resonators directly related to the insertion loss of a filter as

$$\Delta L_A = 4.343 \sum_{i=1}^{n} \frac{\Omega_C}{FBW \cdot Q_{Ui}} g_i \quad dB \tag{15}$$

where ΔL_A is the dB increase in insertion loss at the center frequency of the filter, Q_{ui} is the unloaded quality factor of the ith resonator evaluated at the center frequency, and g_i is either the inductance or the capacitance of the normalized lowpass prototype.

In design of narrow band microstrip filters, one important consideration in selecting resonators is to avoid parasitical couplings between nonadjacent resonators. Those parasitical couplings may decrease the out-of-band rejection level or/and deteriorate the in-band return loss of the filter. For demonstration, consider a 10-pole CQ filter which has a pair of transmission zeros (the design of this filter will be detailed in next section). The transmission zeros are created by two identical cross-couplings between resonators 2, 5 and between resonators 6, 9. All the other nonadjacent couplings should be zero according to the coupling matrix. However, because all the resonators are located in one piece of dielectric

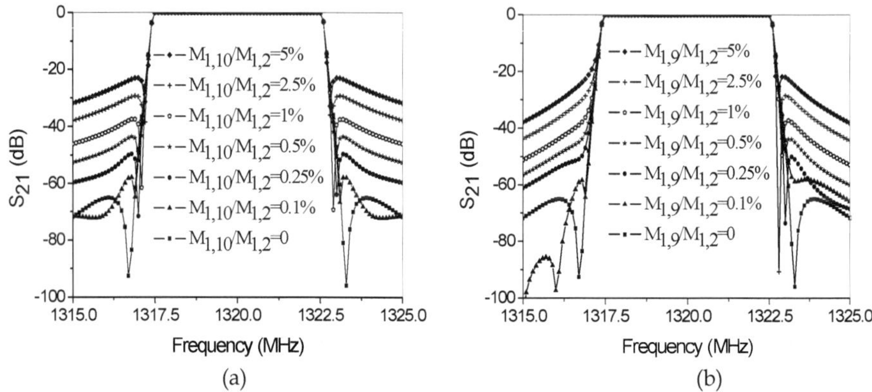

Fig. 9. Illustration of how the parasitical couplings affect the response of a narrow band bandpass filter.

substrate and in one metal housing, parasitical couplings between nonadjacent resonators are inevitable. We have performed a series of analyses by computer simulation on how the parasitical couplings affect the transmission response of this 10-pole CQ filter and the results are plotted in Fig. 9. In Fig.9 (a) different levels of parasitical coupling between resonator 1 and 10 ($M_{1,10}$) were assumed. It's magnitude is scaled to the main coupling $M_{1,2}$. Significant degeneration of the out-of-band rejection level can be seen when $M_{1,10}$ is higher than 0.1 percent of $M_{1,2}$. In Fig. 9 (b) the effects of parasitical coupling between resonator 1 and 9 ($M_{1,9}$) with different strengths were also simulated. It can be seen that in addition to the degeneration of the out-of-band rejection level, the presence of $M_{1,9}$ also causes the asymmetric transmission response of the filter. To solve these parasitical coupling problems, it is necessary to employ some unique resonators. For example, the hairpin and spiral-in-spiral-out resonators, as depicted in Fig. 6, behave as they have two parallel-coupled microstrips excited in the odd mode when resonating. In this case the currents in these two microstrips are equal and opposite resulting in cancellation of their radiation fields. This characteristic should be of great help in reducing parasitical couplings when design narrow band filters.

3.3 Superconducting filter and receiver front-end subsystem

So far the theory and principles for HTS filter design have been discussed generally. In this section the design, construction and integration of a real practical HTS filter and HTS receiver front-end subsystem will be introduced as an application example with one special type of meteorological radar, the wind profiler. The frequencies assigned to the wind profiler are in UHF and L band, which are very crowded and noisy with radio, TV, and mobile communication signals and therefore the radar is often paralyzed by the interference. To solve this problem, it is necessary to employ pre-selective filters. Unfortunately, due to the extremely narrow bandwidth ($\leq 0.5\%$) no suitable conventional device is available. The HTS filter can be designed to have very narrow band and very high rejection with very small loss, therefore it is expected that HTS filter can help to improve the anti-interference ability of the wind profiler without even tiny reduction of its sensitivity. In fact, because the low noise amplifier (LNA) in the front end of the receiver is also working at a very low temperature in the HTS subsystem, the sensitivity of the radar will actually be increased.

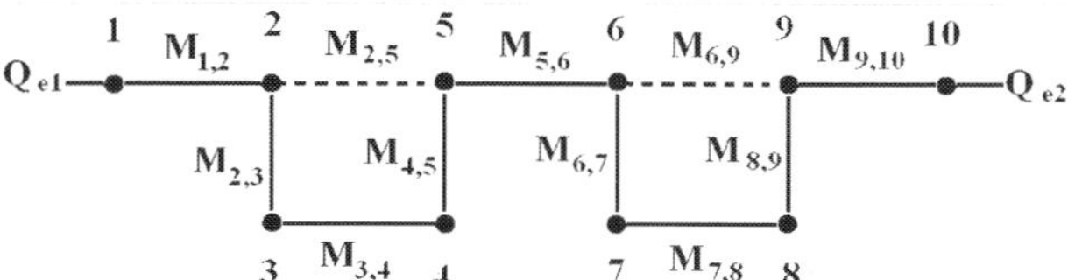

Fig. 10. Coupling topology of the 10-pole CQ filter designed for the wind profiler.

The center frequency of the wind profiler introduced here is 1320 MHz. In order to reject the near band interference efficiently the filter was expected to have a bandwidth of 5 MHz and skirt slope as sharp as possible. It has been decided in the real design to employ a 10-pole generalized Chebyshev function filter with a pair of transmission zeros placed at $\Omega=\pm1.3$ so as to produce a rejection lobe better than 60 dB on both side of the passband. For the implementation of this filter, the CQ coupling topology shown in Fig.10 was employed. The cross couplings $M_{2,5}$ and $M_{6,9}$ in Fig.10 are introduced to create the desired transmission zeros. In the present design they are set to be equal to each other to create the same pair of transmission zeros. Introducing two identical cross couplings can make the physical structure of the filter symmetric. With this strictly symmetric physical structure, only half part (e.g., the left half) of the whole filter is needed to be simulated in the EM simulation process, which will simplify the EM simulation and save the computing time remarkably.

The transfer and reflection functions and the coupling matrix can then be synthesized following the instructions in section 3.1. For this filter with topologic structure shown in Fig. 10, the finally coupling parameters are: $Q_{e1}= Q_{e2}=237.3812$, and

$$
M = 0.01 \times
\begin{bmatrix}
0 & 0.330 & 0 & 0 & 0 & 0 & 0 & 0 & 0 & 0 \\
0.330 & 0 & 0.218 & 0 & 0.0615 & 0 & 0 & 0 & 0 & 0 \\
0 & 0.218 & 0 & 0.262 & 0 & 0 & 0 & 0 & 0 & 0 \\
0 & 0 & 0.262 & 0 & 0.188 & 0 & 0 & 0 & 0 & 0 \\
0 & 0.0615 & 0 & 0.188 & 0 & 0.198 & 0 & 0 & 0 & 0 \\
0 & 0 & 0 & 0 & 0.198 & 0 & 0.188 & 0 & 0.0615 & 0 \\
0 & 0 & 0 & 0 & 0 & 0.188 & 0 & 0.262 & 0 & 0 \\
0 & 0 & 0 & 0 & 0 & 0 & 0.262 & 0 & 0.218 & 0 \\
0 & 0 & 0 & 0 & 0 & 0.0615 & 0 & 0.218 & 0 & 0.330 \\
0 & 0 & 0 & 0 & 0 & 0 & 0 & 0 & 0.330 & 0
\end{bmatrix}
$$

The synthesized response of the filter is depicted in Fig.11. The designed filter shows a symmetric response which gives a rejection lobe of more than 60 dB on both side of the passband as expected. The passband return loss is 22 dB and the band width is 5 MHz centered at 1320 MHz. Two transmission zeros locate at 1316.75 MHz and 1320.25 MHz respectively.

The resonator used in this filter is spiral-in-spiral-out type resonator as shown in Fig.12 (a), which is slightly different from that shown in Fig.6 (f). The main change is that both end of the microstrip line are embedded into the resonator to form capacitance loading, making the electric-magnetic field further constrained. More over the middle part of the microstrip line where carries the highest currents at resonance was widened to increase the quality factor of

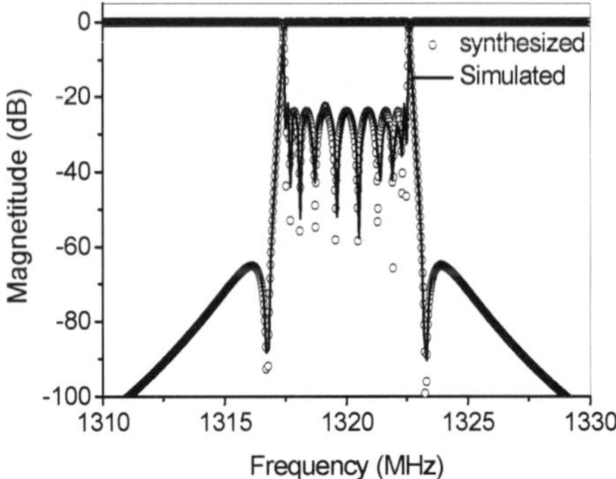

Fig. 11. Synthesized (circles) and simulated (solid line) responses of the 10-pole CQ filter.

the resonator. The resonator with center frequency of 1320 MHz is 10.64 mm long and 2.86 mm wide. The cross coupling needed for the transmission zeros can be introduced by a microstrip line as shown in Fig. 12 (b). Fig. 12 (c) shows the simulated coupling coefficient κ between two resonators using a full-wave EM simulation software Sonnet as a function of the space s. For the simulation, the substrate is MgO with thickness of 0.50 mm and permittivity of 9.65. It can be seen that κ decreases rapidly with s. When s changes from 0.2 mm to 3 mm (about a resonator's width), the coupling coefficient k becomes more than 3 orders on magnitude less than its original value, making this resonator very suitable for ultra-narrow bandpass filter design.

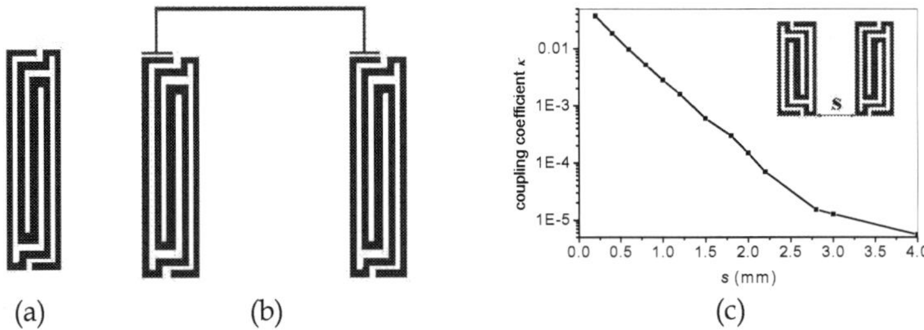

(a)   (b)   (c)

Fig. 12. The resonator used in the present work (a), the way of cross coupling being introduced (b), and the simulated coupling coefficient k between two adjacent resonators as a function of the separation space s

The filter layout was simulated and optimized using Sonnet and the final layout of the filter is shown in Fig.13. The final full-wave EM simulated responses of the filter is shown in Fig. 11 as solid lines. Comparing the full-wave EM simulated responses with the synthesized theoretical responses, the out-of-band response are very similar. The passband response of the EM simulated return loss is 21dB, only slightly worse than the theoretical return loss of 22dB.

The filter was then fabricated on a 0. 5 mm thick MgO wafer with double-sided YBCO films. The YBCO thin films have a thickness of 600 nm and a characteristic temperature of 87K.

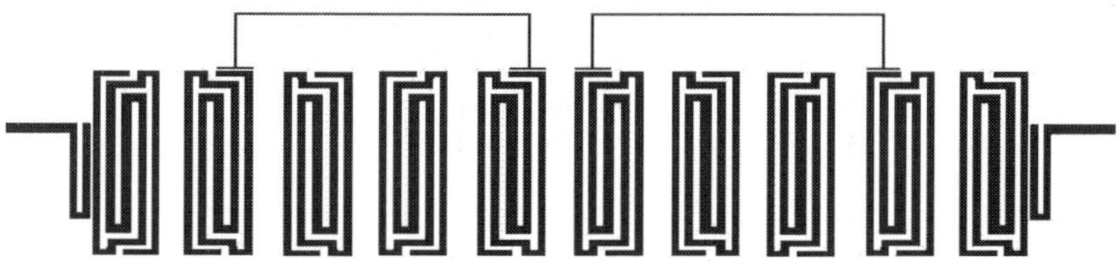

Fig. 13. The final layout of the 10-pole quasi-elliptic filter (not to scale).

Both sides of the wafer are gold-plated with 200 nm thick gold (Au) for the RF contacts. The whole dimension of the filter is 60 mm×30 mm×20 mm including the brass housing. The RF measurement was made using a HP 8510C network analyzer and in a cryogenic cooler. Fig.14 shows the measured results at 70K and after tuning the filter. The measured center frequency is 1.319 GHz, and the return loss in the passband is better than 15 dB. The insertion loss at the passband center is 0.26 dB, which corresponds to a filter Q of about 45,000. The transmission response is very similar to the theoretically synthesized and the full-wave EM simulated responses as shown in Fig. 11. The fly-back values in the S21 curve are 60.7 dB and 62 dB at the lower and upper frequency sides, respectively. Steep rejection slopes at the band edges are obtained and rejections reach more than 60 dB in about 500 kHz from the lower and upper passband edges.

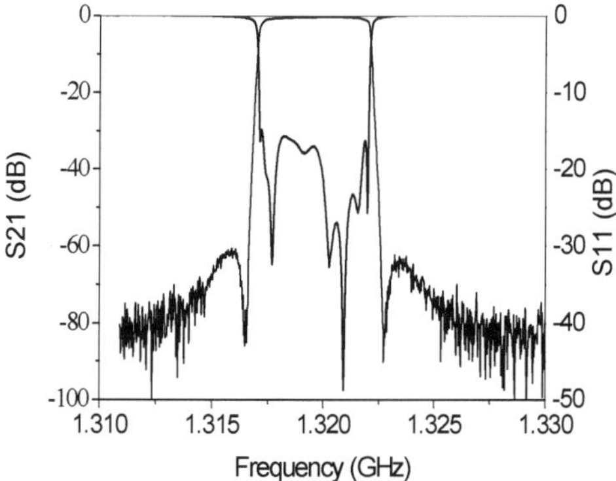

Fig. 14. The measured response of the 10-pole filter at 70K

Combining this filter together with a low noise amplifier (LNA) as well as a Sterling cryo-cooler, a HTS subsystem was then constructed as shown schematically in Fig. 15 (a). There are many types of cryo-coolers in the market which are specially built for a long life under outdoor conditions in order to provide high cooling power at a cryogenic temperature. What we chose for the HTS wind profiler subsystem is Model K535 made by Ricor Cryogenic & Vacuum Systems, Israel, for its considerably compact volume and longer life time (Fig. 15 (b)). The advantage of integrating the LNA together with HTS filter inside the cryo-cooler is obvious, as the noise figure of LNA is temperature dependent (Fig. 16 (a)).

Since the cost of a cryo-cooler is inevitable for HTS filters. Extra benefit of significant reduction of noise figure can be obtained by also putting the LNA in low temperature without any new burden of cooling devices. The noise figure of the HTS receiver front-end subsystem measured using an Agilent N8973A Noise Figure Analyzer at 65 K is about 0.7 dB in major part (80%) of the whole passband, as shown in Fig. 16 (b).

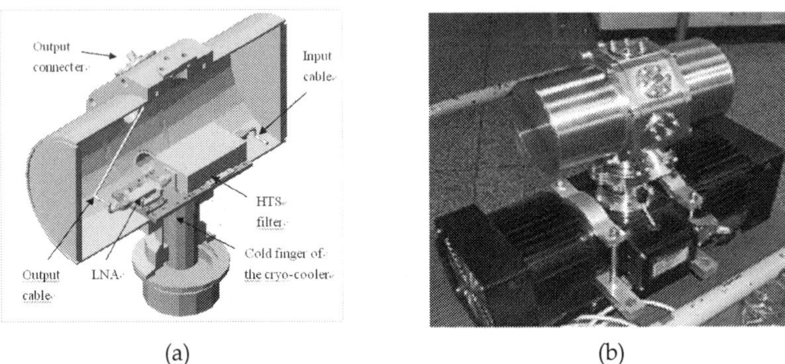

(a)   (b)

Fig. 15. Sketch and photograph of the HTS receiver front-end subsystem

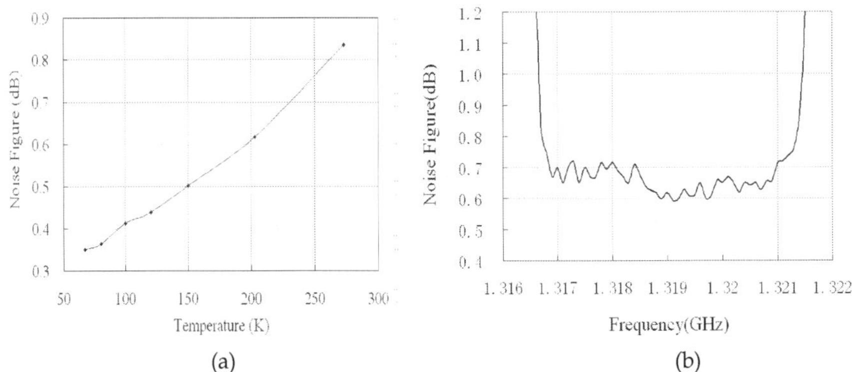

(a)   (b)

Fig.16. Temperature dependence of the noise figure of the LNA used in the HTS subsystem (a) and the noise figure of the HTS front-end subsystem measured at 70K (b)

4. Superconducting meteorological radar

4.1 Basic principle and configuration of wind profiler

A wind profiler is a type of meteorological radar that uses radar to measure vertical profiles of the wind, i.e., detecting the wind speed and direction at various elevations above the ground. The profile data is very useful to meteorological forecasting and air quality monitoring for flight planning. Pulse-Doppler radar is often used in wind profiler. In a typical profiler, the radar can sample along each of three beams: one is aimed vertically to measure vertical velocity, and two are tilted off vertical and oriented orthogonal to one another to measure the horizontal components of the air's motion. The radar transmits an electromagnetic pulse along each of the antenna's pointing directions. Small amounts of the transmitted energy are scattered back and received by the radar. Delays of fixed intervals are built into the data processing system so that the radar receives scattered energy from discrete altitudes, referred to as range gates. The Doppler frequency shift of the backscattered energy is determined, and then used to calculate the velocity of the air toward or away from the radar along each beam as a function of altitude. The photograph of a typical wind profiler is shown in Fig. 17. Two antennas can be clearly seen in the middle of the photograph, one is vertical and the other is tilted. The third antenna is hidden by one of the four big columns, which are used for temperature profile detection by sound waves (SODAR). The circuit diagram of a typical wind profiler is shown in Fig 18. The "transmitting system" transmits pulse signal through the antenna, then the echoes come back through the LNA to the "radar receiving system", that analyzes the signal and produces wind profiles.

Fig. 17. Photograph of a typical wind profiler

The wind profiler measures the wind of the sky above the radar site in three directions, i.e., the roof direction, east/west direction and south/north direction, and produces the wind charts correspondingly. The collected wind chart data are then averaged and analyzed at every 6 minutes so as to produce the wind profiles. Typical wind profiles are shown in Fig. 19 (a). In the profile the horizontal axis denotes the time (starting from 5:42 AM to 8:00 AM with intervals of every 6 minutes); the vertical axis denotes the height of the sky. The arrow-like symbols denote the direction and velocity of the wind at the corresponding height and in corresponding time interval. The arrowhead denotes the wind direction (according to the provision: up-north, down-south, left-west, right-east), and the number of the arrow feather denotes the wind velocity (please refer to the legend).

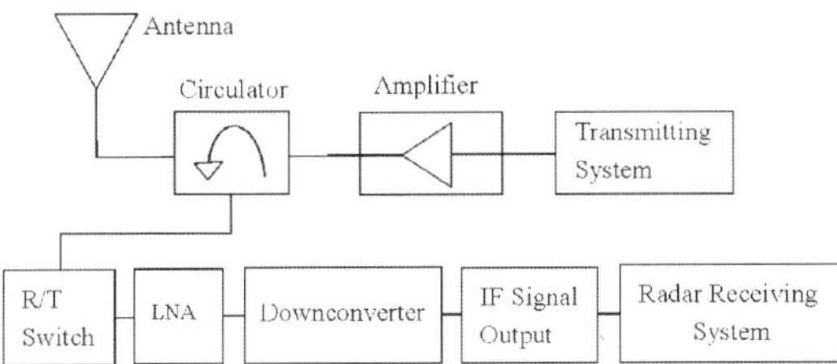

Fig. 18. Circuit diagram of a typical wind profiler.

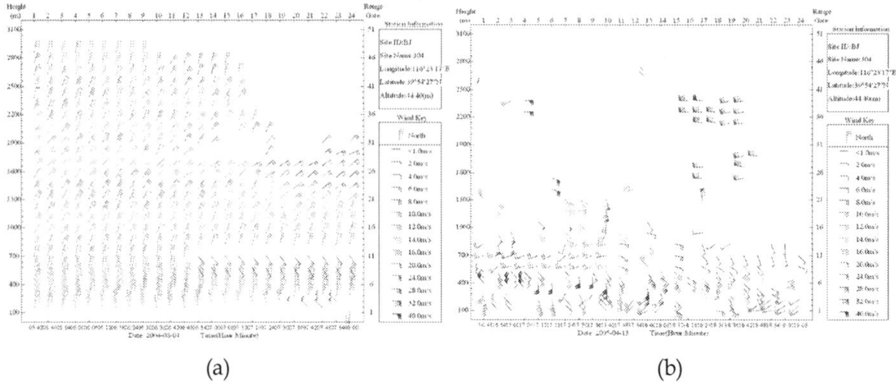

(a)   (b)

Fig. 19. Wind profiles produced by a wind profiler of a weather station in the suburb of Beijing. Data in (a) was collected in the morning of August 4, 2004 from 5:42 to 8:00 in a interval of every 6 minute and no reliable data in (b) can be seen which was collected in the afternoon of April 13, 2005, demonstrating clearly that this radar was paralyzed by interference.

The frequencies assigned to the wind profiler are in UHF and L band, which are very crowded and noisy with radio, TV, and mobile communication signals and therefore the radar is often paralyzed by the interference, which did happened, especially in or near the cities. For example, the wind profiles presented in Fig. 19 are actually real observation data recorded by a weather station in the suburb of Beijing. It is interesting to point out that the detecting range (or the height above the radar) is gradually getting shorter (from 3000 m down to 1700 m or so) after 6:30 AM in the morning as people were getting up and more and more mobile phones switched on, indicating the sensitivity of the radar was affected by increasing interference (Fig 19 (a)). Eight months later, with rapid expanding of the number of mobile phones in Beijing, the electromagnetic environment became much worse and this radar was blocked by the massive interference noise and totally lost the ability of collecting reliable data at all, as shown in Fig 19 (b).

4.2 Laboratory tests of superconducting meteorological radar

To solve the problem above, it is necessary to employ pre-selective filters. Unfortunately, due to the extremely narrow bandwidth ($\leq$0.5%) no conventional device is available. The HTS filter can be designed to have very narrow band and very high rejection with very small loss, so it is expected that it can help to improve the anti-interference ability of the wind profiler without even tiny reduction of its sensitivity. In fact, because the LNA was also working at a very low temperature in the HTS subsystem, the sensitivity of the whole system will actually be increased. To prove these, two stages of experiments have been conducted and the performance of the conventional wind profiler was compared with the so-called HTS wind profiler, i.e., the corresponding part (the front-end, i.e., the LNA) of a conventional radar being substituted by the HTS subsystem. The first stage experiments are sensitivity comparison tests and anti-interference ability comparison tests, by measuring the sensitivity and the anti-interference ability with quantitative instruments such as signal generator and frequency spectrometer, etc. The second stage experiments are the field trail of superconducting meteorological radar with the conventional counterpart, which will be introduced in next section.

4.2.1 Sensitivity comparison experiments

The circuit diagram for the experiment is shown in Fig. 20. In this experiment, an IFR2023B signal generator was used as a signal source whose output frequency was set to the wind profiler operating frequency. The signal is emitted from a small antenna and received by the antenna of the radar system. The received signal reaches the radar receiver front-end (in Fig. 20 between B and C) via the R/T switch, then passes the down-converter and finally is converted as the intermediate frequency (IF) signal. The IF signal is then sent to a frequency spectrometer (HP E44118) and finally being measured. The wind profiler sensitivity is defined as the signal source output power (in dBm) when the measured signal-to-noise ratio of the intermediate frequency signal is equal to 1.

During the experiment the sensitivity of the conventional system was measured first. Then the conventional front-end (i.e., the LNA, in Fig. 20 between B and C) of the wind profiler was replaced by the HTS subsystem (in Fig. 20 between B' and C'). Here a "HTS Filter + LNA" configuration was used instead of a "LNA + HTS Filter" configuration in order to avoid saturation of the LNA. Due to the very small insertion loss of the HTS filter, this configuration should not have any noticeable effect to the dynamic range of the LNA. The measured data show that the sensitivity of the profiler employing HTS subsystem is –43.6 dBm, the sensitivity of the system employing the conventional front-end is –39.9 dBm. Thus we get that the sensitivity of the HTS subsystem is 3.7 dB higher than that of the conventional subsystem.

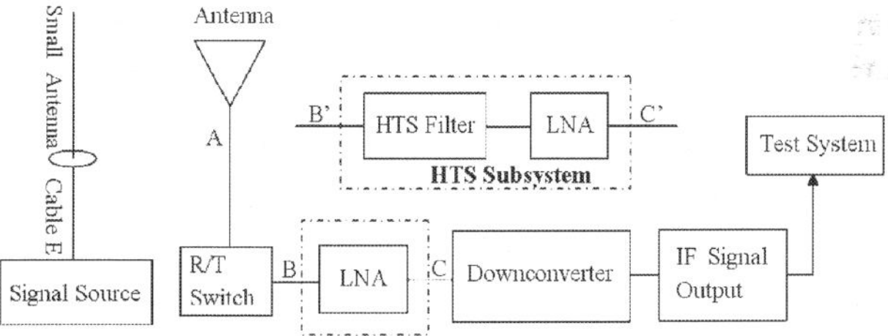

Fig. 20. The circuit diagram for the sensitivity comparison experiment

4.2.2 Anti-interference ability comparison experiments

The circuit diagram for the experiment is similar to Fig. 20. The difference is that in this experiment the signal source was linked to point A, bypassing the receiving antenna and introducing an interference signal with a frequency of 1323 MHz. Similar to Experiment 4.2.1, the IF signal is monitored by a spectrum analyzer while the interference signal is gradually increased. The anti-interference ability is defined as the power of the interference signal (in dBm) when the IF output is about to increase. The measured data show that for the conventional subsystem, an interference signal as low as –92.4 dBm brings influence to the wind profiler, whereas for the HTS subsystem the corresponding value is –44 dBm. It can thus be concluded that the anti-interference ability of the HTS subsystem is 48.4 dB higher than that of the conventional subsystem.

4.3 Field trail of superconducting meteorological radar

As already mentioned the wind profiler measures the wind of the sky above the radar site in three directions and produces the wind charts correspondingly. A typical wind chart of the east/west direction is presented in Fig. 21, where the horizontal axis is the wind velocity (m/s, corresponding to the frequency shift due to the Doppler effect) and the vertical axis is the height of the sky above the radar site (from 580 m to 3460 m with intervals of every 120 m) corresponding to the time when different echo being received. The negative values of the wind velocity indicate the change in wind direction. Each curve in Fig 21 represents the spectrum of corresponding echo. The radar system measures the wind charts of the sky above every 40 seconds. The collected wind chart data are then averaged and analyzed at every 6 minutes so as to produce the wind profiles.

The procedure of the second stage experiments (field trail) is as follows: firstly, the wind charts and the wind profiles were measured using the conventional wind profiler without any interference signal. Then an interference signal with a frequency of 1322.5 MHz and power of -4.5 dBm was applied and a new set of wind charts and wind profiles being obtained. Finally, the wind charts and the wind profiles were measured using the HTS wind profiler (the LNA at the front end of the conventional wind profiler being replaced by the HTS subsystem) with the same frequency but much stronger (+10 dBm) interference signal.

It can be seen that a series of interference peaks appeared in the wind chart measured using the conventional wind profiler while the interference signal was introduced (Fig. 21 (b)). However, no influence of the interference can be seen at all in the wind charts produced by the HTS wind profiler (Fig. 21 (c)), even when much stronger (up to +10 dBm) interference signal being applied. Moreover, the ultimate height of the wind in the sky being able to be detected reaches to 4000 meters when HTS subsystem was used (Fig. 21 (c)), in contrast with those of 3400 meters of conventional ones (Fig. 21 (a) and (b)), which is consistent with the picture that sensitivity of the HTS radar system is higher than that of the conventional system.

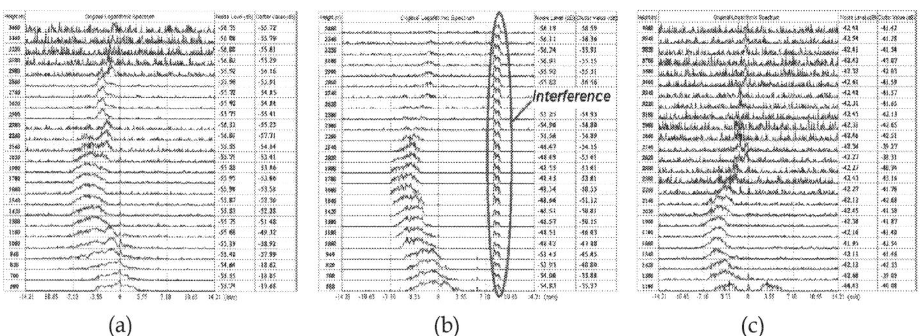

(a)   (b)   (c)

Fig. 21. Wind charts produced by (a) using the conventional wind profiler without interference, (b) using the conventional wind profiler with interference (1322.5 MHz, -4.5 dBm), (c) using the HTS wind profiler with interference (1322.5 MHz, +10 dBm)

Fig. 22 shows the wind profiles produced from the wind charts measured in the same day by the conventional wind profiler and the HTS wind profiler, respectively. It can be clearly seen that the conventional wind profiler cannot attain wind profiles above 2000 m due to the influence of the interference signal, which reproduced the phenomena observed in Fig 19.

On the contrary the HTS wind profiler functioned well even with much serious interference signal.

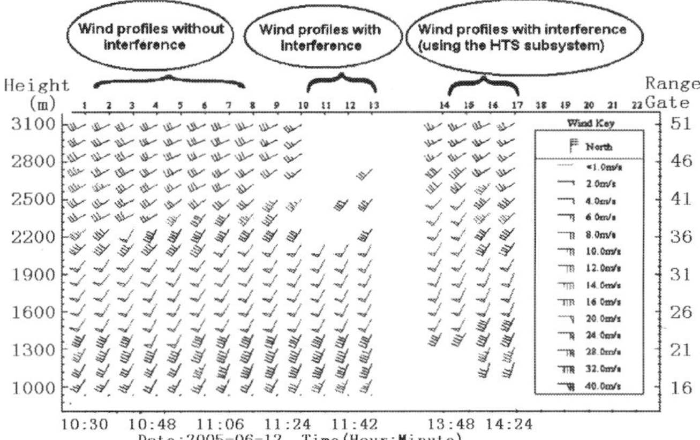

Fig. 22. Wind profiles obtained under three different conditions by the conventional wind profiler and the HTS wind profiler, respectively

5. Summary

Due to the atrocious electromagnetic environment in UHF and L band, high performance pre-selective filters are requested by the meteorological radar systems, e.g. the wind profilers. Unfortunately, due to the extremely narrow bandwidth ($\leq$0.5%) no conventional device is available. To solve this problem, an ultra selective narrow bandpass 10-pole HTS filter has been successfully designed and constructed. Combining this filter together with a low noise amplifier (LNA) as well as a cryocooler, a HTS receiver front-end subsystem was then constructed and mounted in the front end of the receiver of the wind profiler. Quantitative comparison experiments demonstrated that with this HTS subsystem an increase of 3.7 dB in the sensitivity and an improvement of 48.4 dB in the ability of interference rejection of the radar were achieved. Field tests of this HTS wind profiler showed clearly that when conventional wind profiler failed to detect the velocity and direction of the wind above 2000 meters in the sky of the radar site due to interference, the HTS wind profiler can produce perfect and accurate wind charts and profiles. A demonstration HTS wind profiler is being built and will be installed in a weather station in the suburb of Beijing. Results of this HTS radar will be reported in due course.

6. References

Amari, S. (2000). Synthesis of Cross-Coupled Resonator Filters Using an Analytical Gradient-Based Optimization Technique, *IEEE Trans. Microwave Theory and Techniques*, Vol. 48, No.9, pp.1559-1564

Atia, A. E. & Williams, A. E. (1972). Narrow-bandpass waveguide filters, *IEEE Trans. Microwave Theory and Techniques*, Vol. 20, No.4, pp.258-264

Atia, A. E.; Williams, A. E. & Newcomb, R. W. (1974). Narrow-band multiple-coupled cavity synthesis, *IEEE Trans. Microwave Theory and Techniques*, Vol. 21, No.5, pp.649-656

Atia, A. W., Zaki, K. A. & Atia, A. E. (1998). Synthesis of general topology multiple coupled resonator filters by optimization, *IEEE MTT-S Digest*, Vol. 2, pp. 821-824

Bednorz, J. G. & Mueller, K. A. (1986). Possible high T_c superconductivity in the Ba-La-Cu-O system, *Z. fur Phys.* Vol. 64, pp. 189-192

Cameron, R. J. (1999). General coupling matrix synthesis methods for Chebyshev filtering functions, *IEEE Trans .Microwave Theory and Techniques*, Vol. 47, No.4, pp.433-442

Cameron, R. J. & Rhodes, J. D. (1981). Asymmetric realizations for dual-mode bandpass filters, *IEEE Trans. Microwave Theory and Techniques*, Vol. 29, No.1, pp.51-59

Chambers, D. S. G. & Rhodes, J. D. (1983). A low-pass prototype network allowing the placing of integrated poles at real frequencies, *IEEE Trans. Microwave Theory and Techniques*, Vol. 33, No.1, pp.40-45

Gorter, C. J. & Casimir, H. B. (1934). On superconductivity I, *Physica*, Vol. 1, 306

Hong, J.-S. & Lancaster, M. J. (2001). *Microstrip filters for RF/microwave applications*, John Wiley & Sons, Inc., ISBN 0-471-38877-7 New York

Hong, J. S.; Lancaster, M. J.; Jedamzik, D. & Greed, R. B. (1999). On the development of superconducting microstrip filters for mobile communications applications, *IEEE Trans on Microwave Theory and Techniques* , Vol. 47, pp. 1656-1663

Kammerlingh Onnes, H. (1911). The resistance of pure mercury at helium temperature, *Commun. Phys. Lab. Uni. Leiden*, 120b. 3

Lancaster, M. J. (1997). *Passive Microwave devices applications of high temperature superconductors*, Cambridge University Press, ISBN 0-521-48032-9, Cambridge

Levy, R. (1995). Direct synthesis of cascaded quadruplet (CQ) filters, *IEEE Trans. Microwave Theory and Techniques*, Vol. 43, No.12, pp.2940-2945

Li, H.; He, A. S.& He, Y. S. et al. (2002). HTS cavity and low phase noise oscillator for radar application , *IEICE transactions on Electronics* , Vol. E85-C , No.3 , pp. 700-703

Li, Y. Lancaster, M. J. & Huang, F. et al (2003). Superconducting microstrip wide band filter for radio astronomy, *IEEE MTT-S Digest*, pp. 551-553

Lyons W. G.; Arsenault, D. A. & Anderson, A. C. et al. (1996). High temperature superconductive wideband compressive receivers, *IEEE Trans. on Microwave Theory and Techniques*, Vol. 44, pp. 1258-1278

Macchiarella, G. (2002). Accurate synthesis of inline prototype filters using cascaded triplet and quadruplet sections, *IEEE Trans. Microwave Theory and Techniques*, Vol. 50, No.7, pp.1779-1783

Meissner, W. & Ochsenfeld, R. (1933). *Naturwissenschaften,* Vol. 21,pp.787-788

Romanofsky, R. R.; Warner, J. D. & Alterovitz, S. A. et al. (2000). A Cryogenic K-Band Ground Terminal for NASA'S Direct-Data-Distribution Space Experiment, *IEEE Trans. on Microwave Theory and Techniques*, Vol.. 48, No. 7, pp. 1216-1220

STI Inc. (1996). A receiver front end for wireless base stations, *Microwave Journal*, April 1966, 116.

Wallage, S.; Tauritz, J. L. & Tan, G. H. et al (1997). High-T_c superconducting CPW band stop filters for radio astronomy front end. *IEEE Trans. on Applied Superconductivity*, Vol. 7 No. 2, pp. 3489-3491.

Zhang, Q.; Li, C. G. & He, Y. S. et al (2007). A HTS Bandpass Filter for a Meteorological Radar System and Its Field Tests, *IEEE Transactions on Applied Superconductivity*, Vol. 17, No2, pp. 922-925